Algorithms for Approximation II

Algorithms for Approximation II

Based on the proceedings of the Second International Conference on Algorithms for Approximation, held at Royal Military College of Science, Shrivenham, July 1988

Edited by

J. C. Mason
Professor of Computational Mathematics,
Royal Military College of Science, Shrivenham, UK
and

M. G. Cox
Senior Principal Scientific Officer,
National Physical Laboratory, Teddington, UK

London New York
CHAPMAN AND HALL

First published in 1990 by Chapman and Hall Ltd
11 New Fetter Lane, London EC4P 4EE

Published in the USA by Chapman and Hall
29 West 35th Street, New York NY 10001

Printed in Great Britain by
T. J. Press (Padstow) Ltd, Padstow, Cornwall

ISBN 0 412 34580 3

British Library Cataloguing in Publication Data

International Conference on Algorithms for Approximation (2nd: 1988: Royal Military College of Science)
Algorithms for approximation II.
1. Mathematics. Approximation. Algorithms
I. Title II. Mason, J. C. III. Cox, M. G.
511'.4

ISBN 0-412-34580-3

Library of Congress Cataloging in Publication Data

International Conference on Algorithms for Approximation (2nd: 1988: Royal Military College of Science, Shrivenham, England)
Algorithms for approximation II: based on the proceedings of the Second International Conference on Algorithms for Approximation, held at Royal Military College of Science, Shrivenham, July 1988 / edited by J. C. Mason and M. E. Cox.
p. cm.
Conference held July 12–15, 1988, and sponsored by the Cranfield Institute of Technology.
Includes bibliographical references.
ISBN 0-412-34580-3
1. Approximation theory–Data processing–Congresses.
I. Mason, J. C. II. Cox, M. E. (Malcolm E.)
III. Cranfield Institute of Technology. IV. Title.
V. Title: Algorithms for approximation, 2.
VI. Title: Algorithms for approximation, two.
QA221.I54 1988
511'.4–dc20

89-23868
CIP

We dedicate this book to the memory of Professor Jerry L. Fields of the University of Alberta, who died recently. Jerry made significant contributions to rational approximation and special functions, and was a very generous friend.

Contents

**Invited Speaker*
†*Speaker*

Contributors

R. W. Allen
Allen Clarke Research Centre, Plessey Research Ltd, Caswell, Towcester, Northamptonshire, UK.
G. T. Anthony
Division of Information Technology and Computing, National Physical Laboratory, Teddington, Middlesex, UK.
E. Arge
Institut fur Informatikk, University of Oslo, Blindern, Oslo, Norway.
I. Barrodale
Barrodale Computing Services Ltd, Victoria, British Columbia, Canada.
T. B. Boffey
Department of Statistics and Computational Mathematics, University of Liverpool, Liverpool, UK.
K. W. Bosworth
Department of Mathematics and Statistics, Utah State University, Logan, Utah, USA.
M. Bozzini
Dipartimento di Matematica, Università di Lecce, Lecce, Italy.
L. Brutman
Department of Mathematics and Computer Science, University of Haifa, Haifa, Israel.
M. D. Buhmann
Department of Applied Mathematics and Theoretical Physics, University of Cambridge, Cambridge, UK.
M. G. Cox
Division of Information Technology and Computing, National Physical Laboratory, Teddington, Middlesex, UK.
M. Dæhlen
Institut fur Informatikk, University of Oslo, Blindern, Oslo, Norway.
W. Dahmen
Fachbereich Mathematik (WE3), Freie Universität Berlin, Berlin, FRG.
L. M. Delves
Department of Statistics and Computational Mathematics, University of Liverpool, Liverpool, UK.

N. Dyn
School of Mathematical Science, Tel Aviv University, Tel Aviv, Israel.
T. A. Foley
Computer Science Department, Arizona State University, Tempe, Arizona, USA.
A. B. Forbes
Division of Information Technology and Computing, National Physical Laboratory, Teddington, Middlesex, UK.
W. Freeden
Institut für Reine und Angewandte Mathematik, R.W.T.H. Aachen, Aachen, FRG.
F. N. Fritsch
Computing and Mathematics Research Division, Lawrence Livermore National Laboratory, University of California, Livermore, California, USA.
M. Frontini
Dipartimento di Matematica, Politecnico di Milano, Milano, Italy.
R. H. J. Gmelig Meyling
Department of Applied Mathematics, University of Twente, Enschede, The Netherlands.
M. von Golitschek
Institut für Angewandte Mathematik, Universität Würzburg, FRG.
J. A. Gregory
Department of Mathematics and Statistics, Brunel University, Uxbridge, Middlesex, UK.
E. Grosse
A. T. & T. Bell Laboratories, Murray Hill, New Jersey, USA.
P. M. Harris
Division of Information Technology and Computing, National Physical Laboratory, Teddington, Middlesex, UK.
M. Heilmann
Lehrstuhl Mathematik VIII, Universität Dortmund, Dortmund, FRG.
A. Iserles
Department of Applied Mathematics and Theoretical Physics, University of Cambridge, Cambridge, UK.
H. M. Jones
Division of Information Technology and Computing, National Physical Laboratory, Teddington, Middlesex, UK.
P. E. Koch
Division of Mathematical Sciences, Norwegian Institute of Technology, Trondheim, Norway.
M. A. Lachance
Department of Mathematics and Statistics, University of Michigan-Dearborn, Dearborn, Michigan, USA.

L. Lenarduzzi
Istituto Applicazioni Matematica ed Informatica, C.N.R., Milano, Italy.
D. Levin
School of Mathematical Sciences, Tel Aviv University, Tel Aviv, Israel.
T. Lyche
Institut für Informatikk, University of Oslo, Blindern, Oslo, Norway.
J. C. Mason
Applied and Computational Mathematics Group, Royal Military College of Science, Shrivenham, Swindon, Wiltshire, UK.
J. G. Metcalfe
Allen Clarke Research Centre, Plessey Research Ltd, Caswell, Towcester, Northamptonshire, UK.
K. Mørken
Institut fur Informatikk, University of Oslo, Blindern, Oslo, Norway.
M. W. Müller
Lehrstuhl Mathematik VIII, Universität Dortmund, Dortmund, FRG.
S. P. Nørsett
Division of Mathematical Sciences, Norwegian Institute of Technology, Trondheim, Norway.
G. D. Olaofe
Mathematics Department, University of Ibadan, Ibadan, Oyo State, Nigeria.
G. Opfer
Institut für Angewandte Mathematik, Universität Hamburg, Hamburg, FRG.
C. Potier
Department Informatique, E.N.S.T., Paris-Cedex, France.
M. J. D. Powell
Department of Applied Mathematics and Theoretical Physics, University of Cambridge, Cambridge, UK.
C. J. Pursglove
Department of Statistics and Computational Mathematics, University of Liverpool, Liverpool, UK.
E. Quak
Department of Mathematics, Vanderbilt University, Nashville, Tennessee, USA.
C. Rabut
INSA, Service de Mathematiques, Toulouse-Cedex, France.
C. Rademacher
Institut für Angewandte Mathematik, Universität Bonn, Bonn, FRG.
S. Rippa
School of Mathematical Sciences, Tel-Aviv University, Tel-Aviv, Israel.
G. Rodriguez
Dipartimento di Matematica, Università di Cagliari, Cagliari, Italy.

A. Ron
Department of Computer Science, University of Wisconsin, Madison, Wisconsin, USA.
J. M. Sanz-Serna
Departamento de Matemática Aplicada y Computación, Facultad de Ciencias, Universidad de Valladolid, Valladolid, Spain.
K. Scherer
Institut für Angewandte Mathematik, Universität Bonn, Bonn, FRG.
L. L. Schumaker
Department of Mathematics, Vanderbilt University, Nashville, Tennessee, USA.
S. Seatzu
Dipartimento di Matematica, Università di Cagliari, Cagliari, Italy.
P. W. Smith
IMSL, Houston, Texas, USA.
P. T. P. Tang
Division of Mathematics and Computer Science, Argonne National Laboratory, Argonne, Illinois, USA.
F. de Tisi
Dipartimento di Matematica, Università di Milano, Milano, Italy.
L. N. Trefethen
Department of Mathematics, Massachusetts Institute of Technology, Cambridge, Massachusetts, USA.
A. E. Trefethen
Thinking Machine Corporation, Cambridge, Massachusetts, USA.
C. Vercken
Department Informatique, E.N.S.T., Paris-Cedex, France.
G. Walz
Fakultät für Mathematik und Informatik, Universität Mannheim, Mannheim, FRG.
G. A. Watson
Department of Mathematics and Computer Science, University of Dundee, Dundee, UK.
S. J. Wilde
Applied and Computational Mathematics Group, Royal Military College of Science, Shrivenham, Swindon, Wiltshire, UK.
J. Williams
Department of Mathematics, University of Manchester, Manchester, UK.
C. Zala
Barrodale Computing Services Ltd, Victoria, British Columbia, Canada.

Preface

This book tells the story of 4 summer days in Shrivenham in 1988, when 88 people came to the Second International Conference on Algorithms for Approximation at the Royal Military College of Science (RMCS) from July 12 to 15. Of course the book only tells part of the story. It cannot easily convey the good humour and very happy atmosphere that prevailed, and it does not catalogue the many friendships that were made, the excellent meals and refreshments that were enjoyed, and the plethora of social activities that took place.

The conference was organized for two main reasons. First, there were many requests for a Second Shrivenham Conference from those who attended the First in 1985. Secondly, it had clearly been established at the First Conference that there was a demand and need for a regular meeting which emphasized the algorithmic aspects and applications of approximation. There was also a continuing need for more conferences in the UK in the general area of approximation.

The meeting itself was run under the auspices of Cranfield Institute of Technology, of which RMCS is a faculty, and financial support was gratefully received from the US Air Force European Office of Aerospace Research and Development. The organizing committee was Maurice Cox (NPL) and Anne Daman and John Mason (RMCS), and the RMCS local organizers were very ably assisted by Elizabeth Smith and Pamela Moore. It was a loss to RMCS and to the conference organization when Anne Daman left early in 1988 to seek her fortune in the USA, but there were celebrations when she returned in July as Mrs Anne Trefethen to receive her Ph.D. degree and attend the conference with her husband.

The ten invited speakers, chosen by the organizing committee, covered a broad spectrum of topics and came from a wide range of countries. All of their papers appear in these proceedings. The European speakers, including the UK contingent of Dr John Gregory (Brunel), Professor John Mason (RMCS), Professor Michael Powell (Cambridge) and Professor Alastair Watson (Dundee), were Professor Wolfgang Dahmen (Berlin) and Professor Tom Lyche (Oslo). The speakers from the USA were Dr Eric Grosse (Bell Labs), Professor Larry Schumaker (then Texas A and M, now Vanderbilt) and Professor Lloyd Trefethen (MIT). Finally from

Canada came the opening speaker, Professor Ian Barrodale (Barrodale Associates, Victoria), who set the tone of all the invited talks with a paper of excellent content and very entertaining delivery. It is also appropriate at this point to mention Dr James Lyness (Argonne), who, as guest speaker at the conference dinner, added further academic distinction and good humour to the occasion.

The remainder of the conference programme was made up of submitted papers, of which 40 were selected for two parallel sessions. Within this book you will find the 33 of these papers which were finally offered and accepted for publication after a refereeing procedure. There was general agreement on the high standard of the conference talks, and we believe that this is reflected in the overall quality of the published papers.

It will be noted that the last paper in the book, by Eric Grosse, is a substantial catalogue of algorithms for approximation, and we believe that this will prove to be a very useful and popular reference. Moreover we should not be surprised to see this becoming the catalyst for a series of such offerings. Unlike the proceedings of the first conference, the present volume does not include a 'software panel'. However, this is more than compensated for by the inclusion of the above-mentioned catalogue and the separate publication (by Chapman and Hall) of a volume of proceedings of a Symposium on 'Scientific Software Systems', held at RMCS on the day before the conference (July 11). The latter volume of 17 papers covers the broad area of numerical software, as well as a wide range of software and hardware requirements. It includes, amongst many contributions which might interest current readers, two papers with immediate application to approximation, namely 'Linear algebra support modules for approximation and other software' by M. G. Cox and 'An advisory expert system for curve and surface fitting' by Anne E. Trefethen.

The 41 papers in the present volume have been arranged into 3 primary Parts: One, Development of Algorithms; Two, Applications; Three, Catalogue of Algorithms. The first two sections have been subdivided into eight groups: 1, Spline approximation; 2. Polynomial and piecewise polynomial approximation; 3. Interpolation; 4. Smoothing and constraint methods; 5. Complex approximation; 6. Computer-aided design and geometric modelling; 7. Applications in numerical analysis; 8. Applications in other disciplines.

Such a division into sections, while giving the book a useful structure, is somewhat arbitrary, and we apologise to any authors who may feel that their work has been incorrectly categorized. Several papers could have been placed in up to three groups (especially spline approximation, piecewise polynomial approximation and computer-aided design). Moreover the CAD group, which we have placed in the Applications section could perfectly well have been placed in Part One. Although there is no group headed 'nonlinear approximation', there are several nonlinear algorithms (in Part Two in particular), and of course the complex algorithms (in group 5 and elsewhere) could have come under this heading.

We must conclude with some essential but broad expressions of our gratitude. First, we thank the multitude of staff at Royal Military College of Science, National Physical Laboratory, US Air Force European Office of Aerospace Research and Chapman and Hall (publishers) who contributed in so many different ways to make the conference a great success and to print the abstracts and proceedings. Secondly, we thank all the speakers and authors of invited and contributed papers, without whose industry and patience this volume would not have existed.

As soon as conference number II was over, we were tempted to start thinking in terms of a conference series. We certainly intend that there should be further conferences on 'Algorithms for Approximation', since the first two generated so much good research and goodwill. All that we need are energy, time and resources and, of course, your support.

John Mason
Shrivenham

Maurice Cox
Teddington

May 1989

PART ONE
Development of Algorithms

1. Spline Approximation

Constrained spline approximation of functions and data based on constrained knot removal

E. Arge, M. Dæhlen, T. Lyche and K. Mørken

Institut fur Informatikk, University of Oslo, Norway

Abstract Two of the authors (Lyche and Mørken) have recently developed a knot removal strategy for splines in B-spline format, which has been applied successfully to the approximation of functions and data. In this paper we show how general constraints can be incorporated in this strategy and also in approximation methods based on knot removal. In our implementation of the knot removal strategy, two fundamental properties of B-splines were central—knot insertion or subdivision and the fact that the usual L^p-norms for splines can be approximated well by some simple discrete norms. Together with the fact that a B-spline expansion is a convex combination of the coefficients, these properties are also the key features in our treatment of constraints.

Key words: Splines, B-splines, Constrained approximation, Knot insertion, Knot removal, Discrete norms, Quadratic optimization.

1. Introduction

In Lyche and Mørken (1988), a strategy for removing knots from a B-spline function without perturbing the spline more than a prescribed tolerance was developed. It was also shown how this strategy could be used successfully to compute approximations to large sets of discrete data, by applying knot removal to some simple initial approximation like a cubic spline representation of the piecewise linear interpolant to the data. This also provides a method for approximating functions as long as they can be sampled sufficiently often. The knot removal technique was extended to parametric curves and surfaces in Lyche and Mørken (1987b).

Our general approach to constrained spline approximation will follow the same pattern as for unconstrained approximation. First we compute an initial approximation satisfying the constraints, but in general requiring a large number of parameters for its representation. Then we remove knots from this spline, but all the time making sure that we do not violate the constraints or the prescribed tolerance.

The literature on constrained approximation has grown considerably in the last few years, see Fontanella (1987) and Utreras (1987) for two recent surveys, and also the bibliography compiled by Franke and Schumaker (1987). We will consider a general class of constraints which requires one or more derivatives or the integral of the approximation to be bounded by general functions.

In the remaining part of this section we introduce our notation and give some fundamental results concerning splines. In Section 2 we discuss how various constraints can be written in terms of linear inequality and equality constraints involving the B-spline coefficients on a suffiently large knot vector. In order to carry out constrained knot removal, it is desirable to have an initial approximation which satisfies the constraints. The initial approximation scheme should also cope with the situation where the data do not satisfy the constraints. We discuss these questions in Section 3. Knot removal is the topic of Section 4. We review the unconstrained knot removal strategy in Lyche and Mørken (1988) and explain how it can be extended to handle constraints. The paper ends with a section discussing examples of constrained approximation using the proposed technique.

1.1 Notation

All splines in this paper are represented in terms of B-splines. We denote the i'th B-spline of order k on the knot vector $\boldsymbol{t}$ by $B_{i,k,t}$ and assume the B-splines to be normalized to sum to one. If $\boldsymbol{t}$ contains $m+k$ elements with none occurring more than k times, then we can form m linearly independent B-splines on $\boldsymbol{t}$. These B-splines span a linear space of splines defined by

$$\mathbf{S}_{k,t} = \Big\{ \sum_{i=1}^{m} d_i B_{i,k,t} \;\Big|\; d_i \in \mathbf{R} \text{ for } i = 1, 2, \ldots, m \Big\}.$$

In this paper we will assume that $m \geq k$, and that $t_k < t_{k+1}$ and $t_m < t_{m+1}$, and we will only be interested in the spline functions on the interval $[t_k, t_{m+1}]$.

1.2 Knot insertion

The central ideas of this paper are consequences of the fact that the B-spline coefficients can model the spline they represent with arbitrary precision. This is based on so called subdivision or knot insertion techniques, so let us consider the basis for this. A spline on a knot vector $\boldsymbol{\tau}$ can also be represented as a spline on any knot vector $\boldsymbol{t}$ that contains all the knots of $\boldsymbol{\tau}$. The reason for this is that if $\boldsymbol{\tau}$ is a subsequence of $\boldsymbol{t}$, then $\mathbf{S}_{k,\tau}$ is a subspace of $\mathbf{S}_{k,t}$. Let g be a spline in $\mathbf{S}_{k,\tau}$ with coefficient vector c relative to $\boldsymbol{\tau}$ and coefficient vector b relative to $\boldsymbol{t}$. Then c and b are related by the equation $b = Ac$, where A is the knot insertion matrix of order k from $\boldsymbol{\tau}$ to $\boldsymbol{t}$. This matrix has properties very similar to the B-spline collocation matrix, see Jia (1983), Lyche and Mørken (1987a, 1988), and Lyche (1988). One consequence of these properties is that as more and more knots are inserted, the B-spline coefficients converge to the given spline g, see Cohen and Schumaker (1985) and Dahmen (1986), where it is shown that the convergence is quadratic in the knot spacing. This means that if we work on a sufficiently fine knot vector, we do not lose much by using the coefficients as a representation of the spline.

1.3 Discrete norms

The ability of the B-spline coefficients to mimic the spline they represent is also exemplified by the fact that certain simple combinations of the B-spline coefficients provide good approximations to the L^p-norms of the spline. Specifically, if $f = \sum_{i=1}^m d_i B_{i,k,t}$ is a spline on a knot vector of length $m+k$ as above, we can define a family of discrete norms called the $(\ell^p, \boldsymbol{t})$-norms on $\mathbf{S}_{k,t}$ by

$$\|f\|_{\ell^p,t} = \begin{cases} \left\{\sum_{i=1}^m |d_i|^p (t_{i+k} - t_i)/k\right\}^{1/p}, & \text{for } 1 \le p < \infty; \\ \max_i |d_i|, & \text{for } p = \infty. \end{cases}$$

Equivalently, if we define the diagonal $m \times m$ matrix $E_t^{1/p}$ with the i'th diagonal element given by

$$(E_t^{1/p})_{i,i} = \begin{cases} \left\{(t_{i+k} - t_i)/k\right\}^{1/p}, & \text{for } 1 \le p < \infty; \\ 1, & \text{for } p = \infty; \end{cases}$$

then we also have

$$\|f\|_{\ell^p,t} = \|E_t^{1/p} d\|_p.$$

Here $\|\cdot\|_p$ denotes either the usual L^p norm of a function or the ℓ^p norm of a vector. The significance of the $(\ell^p, \boldsymbol{t})$-norms is due to the fundamental inequalities

$$D_k^{-1} \|f\|_{\ell^p,\tau} \le \|f\|_p \le \|f\|_{\ell^p,t} \le \|f\|_{\ell^p,\tau},$$

where $\boldsymbol{\tau} \subseteq \boldsymbol{t}$ and $f \in \mathbf{S}_{k,\tau}$. The leftmost inequality is due to de Boor (1976a and 1976b). The number D_k depends only on k, and numerical experiments have shown that $D_k \sim 2^k$. As an example we have $D_4 \approx 10$.

The $(\ell^p, \boldsymbol{t})$-norm is of course also a norm on any subspace of $\mathbf{S}_{k,t}$. If a spline in a subspace of $\mathbf{S}_{k,t}$ is given, then to compute its $(\ell^p, \boldsymbol{t})$-norm it must be represented as a spline in $\mathbf{S}_{k,t}$ by knot insertion or degree elevation.

We will be using the $(\ell^\infty, \boldsymbol{t})$-norm to estimate the relative importance of the interior knots during knot removal, and the error of an approximation will be measured in this norm since it gives an upper bound on the L^∞-norm. The method we will employ to compute spline approximations will be best approximation in the $(\ell^2, \boldsymbol{t})$-norm. This norm converges to the L^2-norm for continuous functions as the knot spacing goes to zero, see Lyche and Mørken (1988), and it can even be shown that the best approximation in the $(\ell^2, \boldsymbol{t})$-norm converges to the best approximation in the L^2-norm. The main advantage of working with this discrete norm instead of the L^2-norm is computational efficiency.

Compared with other $(\ell^p, \boldsymbol{t})$-norms, the $(\ell^2, \boldsymbol{t})$-norm has the advantage that best approximation leads to a simple linear system of equations. When it comes to constrained approximation this may not be so crucial, and approximation in the $(\ell^1, \boldsymbol{t})$-norm and the $(\ell^\infty, \boldsymbol{t})$-norm are interesting alternatives. To develop efficient algorithms for constrained spline approximation in these norms will be the subject of future work by the authors.

2. Constraints

Our approach to constrained approximation is also based on the fact that the B-spline coefficients model the spline they represent. Let $f = \sum_{i=1}^{m} d_i B_{i,k,t}$ be a spline in $\mathbf{S}_{k,t}$ to be approximated by a spline $g = \sum_{i=1}^{n} c_i B_{i,k,\tau}$ in a subspace $\mathbf{S}_{k,\tau}$, and denote by b the coefficient vector of g relative to $\boldsymbol{t}$, so that $b = Ac$. We will consider constraints of the form

$$\begin{aligned} E_1 b &\geq v_1, \\ E_2 b &= v_2, \end{aligned} \tag{1}$$

where E_1 and E_2 are rectangular matrices and v_1 and v_2 are vectors. (The notation $u \geq w$ for vectors u and w denotes the component-wise inequalities $u_i \geq w_i$ for $i = 1, 2, \ldots$.) In other words, the constraints can be expressed directly as restrictions on linear combinations of the coefficients relative to the $\boldsymbol{t}$ knot vector. The constraints characterize the set of permissible spline functions,

$$\mathsf{F}_{k,t} = \left\{ \psi = \sum_{i=1}^{m} b_i B_{i,k,t} \;\middle|\; E_1 b \geq v_1 \;\&\; E_2 b = v_2 \right\}.$$

As mentioned above, we will use best approximation in the $(\ell^2, \boldsymbol{t})$-norm to compute spline approximations. The typical optimization problem to be solved is therefore

$$\min_{\psi \in \mathsf{F}_{k,t} \cap \mathbf{S}_{k,\tau}} \|f - \psi\|_{\ell^2, t}, \tag{2}$$

and a matrix formulation of this problem is given in Section 4.2. This problem is a quadratic minimization problem with linear constraints. Such problems are studied extensively in the optimization literature, see e.g. Fletcher (1987). We emphasize that even though the approximation g is a spline on the knot vector $\boldsymbol{\tau}$, we minimize a weighted ℓ^2-norm of the B-spline coefficients of the error on the $\boldsymbol{t}$ knot vector which for good results should contain 'many' knots.

In practice, any constraints that can be handled by the quadratic minimization routine that is used to solve (2) are admissible, but in this paper we restrict our attention to linear constraints of the type (1). It should be noted here that constrained problems for parametric curves and surfaces often lead to nonlinear constraints, cf. Ferguson, Frank and Jones (1988).

It has been observed that bounds on a spline g or its derivatives can be replaced by a finite number of linear inequality constraints on the B-spline coefficients of g, see e.g. Cox (1987). In general though, the replacing constraints will be stronger than the original ones so that the set of feasible solutions is overly restricted. Recall that during knot removal we also have a requirement that the error should be less than the tolerance. This additional constraint together with the constraint on the B–spline coefficients may result in a problem having no feasible solutions. To weaken the constraints, we propose to use knot insertion and constrain the B-spline coefficients on a knot vector which has many knots compared to the number of oscillations in the spline to be approximated. In this way the discrepancy between the original and replacing constraints is reduced.

As an example, consider the polynomial $T(x) = 4x^3 - 3x + 1$ on the interval $[-1, 1]$ as a given function to be approximated on the Bezier knot vector $\boldsymbol{\tau}$ with four knots at -1 and 1, with L^∞-error less than ϵ and with nonnegativity as the constraint. The function T is the cubic Chebyshev polynomial with the constant 1 added to make it nonnegative. If we choose the approximation g to be T, then we solve the problem exactly. However, a general purpose algorithm for solving this type of problem would probably not discover that T is nonnegative and a feasible solution. Our approach implements the ideas above. A sufficient condition for nonnegativity is that the B-spline coefficients on some knot vector are nonnegative, and instead of requiring the L^∞-error to be less than ϵ we require the $(\ell^\infty, \boldsymbol{t})$-error to be less than ϵ for a suitable $\boldsymbol{t}$. Note that on the Bezier knot vector $\boldsymbol{\tau}$, one of the B-spline coefficients equals -4 so this knot vector is not suitable as $\boldsymbol{t}$, since the nonnegativity constraint would lead to a $(\ell^\infty, \boldsymbol{t})$-error of at least 4. If we insert the $2q-1$ interior knots $(-1+1/q, -1+2/q, \ldots, -1/q, 0, 1/q, 2/q, \ldots, 1-1/q)$ into $\boldsymbol{\tau}$, where q is a positive integer, then it is easily seen that on this refined knot vector $\boldsymbol{t}^q$, all the B-spline coefficients of T are greater than $-2/q^2$. Therefore, for the set of feasible solutions

$$\mathsf{F}_{4,\boldsymbol{t}^q} = \left\{ \psi = \sum_i b_i B_{i,4,\boldsymbol{t}^q} \;\middle|\; \|T - \psi\|_{\ell^\infty, \boldsymbol{t}^q} \le \epsilon \;\&\; c_i \ge 0 \text{ for all } i \right\}$$

to be nonempty, we must have $2/q^2 \le \epsilon$ or $q \ge \sqrt{2}/\sqrt{\epsilon}$.

In the rest of this section we assume that we have a knot vector $\boldsymbol{t}$ on which the original constraints have been replaced by constraints on the B-spline coefficients in a satisfactory way. We now discuss some specific constraints, and we start by considering nonnegativity constraints in general. As was indicated in the example, the obvious way to implement straightforward nonnegativity is to require the B-spline coefficients on $\boldsymbol{t}$, of an approximation $g = \sum_{i=1}^m b_i B_{i,k,\boldsymbol{t}}$, to be nonnegative,

$$b_i \ge 0, \qquad \text{for } i = 1, 2, \ldots, m.$$

In many cases one only wants nonnegativity locally, say on an interval $[a, b]$ which is contained in $[t_k, t_{m+1}]$. This is accomplished by requiring nonnegativity of only the coefficients that multiply a B-spline with part of its support in $[a, b]$. To be able to strictly enforce such conditions, it may be necessary to refine $\boldsymbol{t}$ by including a and b as knots of multiplicity $k-1$.

A simple generalization of the nonnegativity constraint is to require

$$b_i \geq e, \qquad \text{for } i = 1, 2, \ldots, m,$$

for some real number e. The next step is then to require $g \geq h$ for a spline $h = \sum_{i=1}^{m_1} \hat{e}_i B_{i,k,t_1}$ on a knot vector $\boldsymbol{t}_1$. To be able to compare h, f and an approximation g we then replace $\boldsymbol{t}$ by $\boldsymbol{t} \cup \boldsymbol{t}_1$ and assume that on the new $\boldsymbol{t}$ knot vector the constraining spline is given by $h = \sum_{i=1}^{m} e_i B_{i,k,t}$. Then we would simply require

$$b_i \geq e_i, \qquad \text{for } i = 1, 2, \ldots, m.$$

These more general constraints are easily restricted to some interval just as above.

All of these constraints have restricted the approximation from below. If a restriction from above is required, this is most conveniently expressed as $-b_i \geq -e_i$.

There are many other types of constraints that can be written as linear equality and inequality constraints.

- Monotonicity and convexity.
  A monotone approximation g is characterized by $g' \leq 0$ or $g' \geq 0$ and a convex approximation by $g'' \geq 0$. By the well known differentiation formula for B-splines, these constraints can be implemented similarly to the nonnegativity constraints and also generalized in the same way. In general, any derivative of order less than k can be constrained in this manner.

- Interpolation.
  The approximation can be required to interpolate a given value y at a point x leading to an equality constraint of the form,

$$\sum_{i=1}^{m} b_i B_{i,k,t}(x) = y.$$

  It is well known that at most k B-splines are nonzero at x so that this sum contains at most k terms. Any derivative can of course be interpolated in a similar way.

- Integral constraints.
  The integral of the approximation can be constrained by requiring

$$\int g = \sum_{i=1}^{m} b_i (t_{i+k} - t_i)/k \geq e.$$

The equality follows because the integral of $B_{i,k,t}$ is known to be $(t_{i+k} - t_i)/k$. If only the integral over a small interval $[a,b]$ is to be constrained, one could insert $k-1$-tuple knots at a and b.

- Smoothness.
An approximation g with a required smoothness is obtained by restricting the multiplicity of the knot vector on which g is defined.

- Discontinuities.
Jumps in derivatives can be enforced by using some of the above techniques. If we want a sharp edge at the point a for instance, we must first ensure that a occurs with multiplicity $k-1$ in the knot vector. We can then for example enforce $g'(x) \geq 1$ for $x < a$ and $g'(x) \leq 0$ for $x > a$. In fact, it is easily seen that it is sufficient to restrict only three coefficients.

Any linear combination of the constraints above is also a valid constraint. As an example one could force the approximation to take on the value of its integral at a given point by combining interpolation and integration constraints.

It should be noted that the requirement that the $(\ell^\infty, \boldsymbol{t})$-error should be less than the tolerance can be implemented as a constraint of the above type. However, it seems more efficient to implement this constraint in the same manner as in unconstrained knot removal, cf. Section 4.

3. The initial approximation

In general, constrained approximation of data by constrained knot removal, will require two steps as described in the introduction. First, an initial spline approximation to the data satisfying the constraints is computed, and then constrained knot removal is applied to this initial approximation. The purpose of the second step is to remove those knots in the initial approximation that are redundant relative to a given tolerance. This step will be described in Section 4. Here we will discuss the problem of computing an initial approximation.

In general we also determine an initial approximation in two stages. First, the data are converted to a spline ϕ which does not necessarily satisfy the constraints. Then, on the knot vector $\boldsymbol{t}$ of ϕ, we compute a spline approximation f to ϕ satisfying the constraints.

Let us discuss each stage in more detail. For the first stage we are given some data to be approximated. These data can be in different forms. It may be a set of discrete data, a general function or a spline function. The purpose of Stage 1 is to convert the data into a spline ϕ of order k on a knot vector $\boldsymbol{t}$. If the data is a set of discrete points (x_i, y_i) with the abscissae increasing, we can let ϕ be either the linear interpolant to the data, the Schoenberg variation diminishing spline, or we can determine ϕ by using unconstrained knot removal or some other suitable method. Similarly, if the data is a general function or a spline of order greater than k, we can approximate it by a spline of order k by using the same method. In either case we end up with a spline function ϕ of order k on a knot vector $\boldsymbol{t}$ of length $m+k$ with B-spline coefficients $(a_i)_{i=1}^m$.

In constrained approximation there will in general be a conflict between satisfying the constraints and making the error smaller than the tolerance, since the original data may not satisfy the constraints. As an example consider the case where the original data take on the value -1 at a point, but the constraints require the approximation to be nonnegative everywhere and the tolerance is 0.01. To cope with situations of this kind, we need Stage 2 to adjust the spline ϕ above to a spline f on the same knot vector that satisfies the constraints. This is done by simply minimizing the $\ell^2, \boldsymbol{t}$-norm of the error and enforcing the constraints. In other words, the spline f solves the quadratic optimization problem

$$\min_{\psi \in \mathsf{F}_{k,t}} \|\phi - \psi\|_{\ell^2,t} \tag{3}$$

which is just a special case of (2). This is equivalent to the problem

$$\min_{z \in \mathbf{R}^m} \|E_t^{1/2}(z-a)\|_2$$

subject to

$$E_1 z \geq v_1,$$
$$E_2 z = v_2.$$

The attitude taken here is that it is more important to satisfy the constraints than to keep the error smaller than the tolerance. This spline will then be given as input to the knot removal process.

It should be noted that this step is not always necessary. Consider once again a discrete set of nonnegative data. If we take the piecewise linear interpolant to the data as ϕ, then this spline will automatically satisfy the constraints and we can set $f = \phi$. The same applies with global monotonicity and convexity constraints if we approximate the discrete data by piecewise linear interpolation or by the Schoenberg variation diminishing spline.

In some situations the minimization problem (2) will succeed in finding a g in $\mathsf{F}_{k,t}$ even if $f \notin \mathsf{F}_{k,t}$. This would for example be the case when $\phi \geq -10^{-6}$, the tolerance is 0.01 and the constraint is nonnegativity.

4. Knot removal

With the initial approximation f, the tolerance, and the constraints given, we can start to remove knots. More precisely we have the following problem.

Constrained knot removal problem. Given a polynomial order k, a knot vector $\boldsymbol{t}$, a set of linear constraints characterizing a subset $\mathsf{F}_{k,t}$ of $\mathbf{S}_{k,t}$, a spline f in $\mathbf{S}_{k,t}$ and a tolerance ϵ; find a knot vector $\boldsymbol{\tau} \subseteq \boldsymbol{t}$ which is as short as possible and a spline $g \in \mathbf{S}_{k,\tau} \cap \mathsf{F}_{k,t}$ such that $\|f - g\|_{\ell^\infty,t} \leq \epsilon$.

To discuss this problem we first review unconstrained knot removal.

4.1 Review of unconstrained knot removal

In knot removal one is given a spline f in $\mathbf{S}_{k,t}$ and a tolerance ϵ; the goal is to determine a spline g in a subspace $\mathbf{S}_{k,\tau}$ of $\mathbf{S}_{k,t}$ of lowest possible dimension, such that $\|f-g\| \leq \epsilon$. The norm used here is in principle arbitrary, but in Lyche and Mørken (1988) we used the $(\ell^\infty, \boldsymbol{t})$-norm.

Since finding the shortest possible $\boldsymbol{\tau}$ seems very difficult, the method only attempts to find an approximate solution. The idea is to compute a ranking of the interior knots of $\boldsymbol{t}$ according to their significance in the representation of f, and then try to remove as many knots as possible according to this ranking. More specifically, for each interior knot t_i its weight w_i is computed as the absolute error in best $(\ell^\infty, \boldsymbol{t})$ approximation to f from the space $\mathbf{S}_{k,t\setminus\{t_i\}}$. Clearly, if t_i is redundant then $w_i = 0$. (This description is correct if all the interior knots are simple; multiple interior knots cause some complications.) We could then rank the knots according to their weights w_i, but this would not work very well when many knots have more or less equal weights. To compensate for this, knots with similar weights are grouped together. The first group consists of the knots with weights between 0 and $\epsilon/2$, the second group those with weights between $\epsilon/2$ and ϵ and so on. The knots in each group are listed in the order in which they occur in the knot vector. Suppose that there are 40 knots in the first group and 30 knots in the second group, and that we want to remove 50 knots. Those 50 knots would then be the 40 knots in the first group plus 10 knots from the second group. The 10 knots from the second group would be every third knot in the order that they occur in the group.

It is not possible to determine from the weights themselves how many knots can be removed without the error exceeding the tolerance. Therefore, the exact number of knots to be removed is determined by a binary search. First, half the knots are removed and an approximation is computed as the best approximation in the $(\ell^2, \boldsymbol{t})$-norm, together with the error in the $(\ell^\infty, \boldsymbol{t})$-norm. If the error is too large, the approximation is discarded and we try to remove 1/4 of the knots instead. If the error is acceptable, we save the approximation and try to remove 3/4 of the knots. This process is continued until the exact number of knots that can be removed has been determined.

By running through this process, a spline approximation g with knot vector $\boldsymbol{\tau}$ is determined. However, it turns out that it is usually possible to remove even more knots. The knots $\boldsymbol{\tau}$ of g are of course also knots of f and now we can compute a ranking of the knots of $\boldsymbol{\tau}$ as their significance in the representation of g. This ranking can then be used to remove more knots from $\boldsymbol{t}$ to obtain an approximation to f with even fewer knots than $\boldsymbol{\tau}$. The process terminates when no more knots can be removed without the error exceeding the tolerance.

The knot removal process outlined above constitutes a convenient basis for general spline approximation methods. Since we have a method for removing knots from a spline, we just have to produce a good approximation to a function or to discrete data without worrying too much about the number of knots since this can be reduced afterwards. Possible initial approximation schemes include piecewise linear approximation, cubic Hermite interpolation, Schoenberg variation diminishing approximation and many more.

4.2 Constrained knot removal

The general philosophy used in unconstrained knot removal can also be applied to the constrained case. The following changes in the strategy used in the last subsection are necessary for solving the problem stated at the beginning of this section.

- The fixed knot approximation method used during knot removal has to preserve the constraints.
- The ranking procedure also includes spline approximation, and in general the constraints must be taken into account when the weights are computed.

Constrained ranking is currently under investigation by the authors, and we will not discuss it any further in this paper. In the examples in Section 5 we will use the same ranking procedure as in the unconstrained case.

The fixed knot approximation method was introduced in Section 2. Suppose that we have decided to remove some knots and have a knot vector $\boldsymbol{\tau}$ which is a subsequence of $\boldsymbol{t}$. We want to find an approximation g on $\boldsymbol{\tau}$ to f which satisfies the constraints. We determine g by the optimization problem (2). If $\boldsymbol{\tau}$ has length $n+k$, then the B-spline coefficients of g must solve the quadratic minimization problem

$$\min_{z \in \mathbf{R}^n} \|E_t^{1/2}(Az - d)\|_2$$

subject to

$$E_1 Az \geq v_1,$$
$$E_2 Az = v_2.$$

The knot removal process will continue until it is not possible to remove any more knots without either violating the tolerance or the constraints.

5. Examples

In this section we give three examples of the use of the constrained approximation scheme outlined in the preceding sections. The quadratic optimization problems involved have been solved by general purpose library software. Since most of the constraints cause the matrices E_1 and E_2 to be very sparse, performance can be improved considerably both in terms of storage and CPU-time by tailoring the optimization routines to this type of problems. This we consider to be an important area of future research.

5.1 One-sided approximation

The method outlined in the previous sections can be used to construct one-sided approximations to functions. In this example a tolerance $\epsilon = 10^{-2}$ is given, and we want to find a cubic spline approximation g to the function

$$\Phi(x) = \max(\sin(\pi x), 0), \qquad \text{for } x \in [0, 4],$$

subject to the constraint $g(x) - \Phi(x) \geq 0$ for all $x \in [0, 4]$. To construct an initial cubic spline approximation to Φ, let

$$f(x) = \sum_{i=1}^{209} (\Phi(t_i^*) + 10^{-3}) B_{i,4,t}(x),$$

where

$$t_i^* = \frac{1}{3}(t_{i+1} + t_{i+2} + t_{i+3}), \qquad i = 1,\, 2,\, \ldots,\, 209,$$

and the knot vector $\boldsymbol{t}$ consists of the numbers $(0.02 * j)$ for $j = 0,\, 1,\, \ldots,\, 200$, with knots of multiplicity four at 0 and 4, and knots of multiplicity three at 1, 2 and 3.

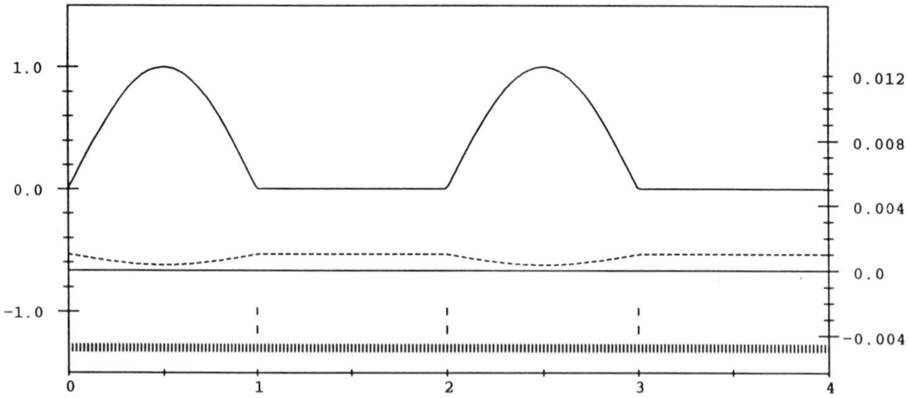

Figure 1. Initial one-sided approximation to a given function.

The spline f is the Schoenberg variation diminishing approximation to $\Phi + 10^{-3}$ which satisfies the constraints, but has far too many knots. Figure 1 shows the approximation f (solid curve) and the error $f - \Phi$ (dashed curve). The location of the knots of f are indicated at the bottom of Figure 1.

With the constrained knot removal procedure described in the previous sections, an approximation g to f is found subject to the constraint $g - f \geq 0$ and such that $g - f \leq \epsilon - 10^{-3}$. That is, the approximation g to Φ is within the given tolerance $\epsilon = 10^{-2}$ and $g - \Phi$ is positive. Figure 2 shows g (solid curve)

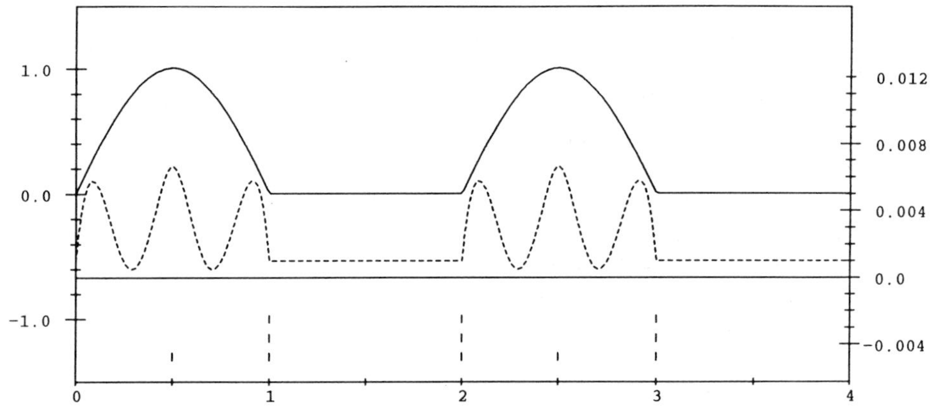

Figure 2. Final one-sided approximation after knot removal.

and the error $g - \Phi$ (dashed curve) together with the location of the interior knots left by the constrained knot removal method. The initial knot vector has been reduced to

$$\boldsymbol{\tau} = (0, 0, 0, 0, 0.5, 1, 1, 1, 2, 2, 2, 2.5, 3, 3, 3, 4, 4, 4, 4).$$

5.2 Convex approximation.

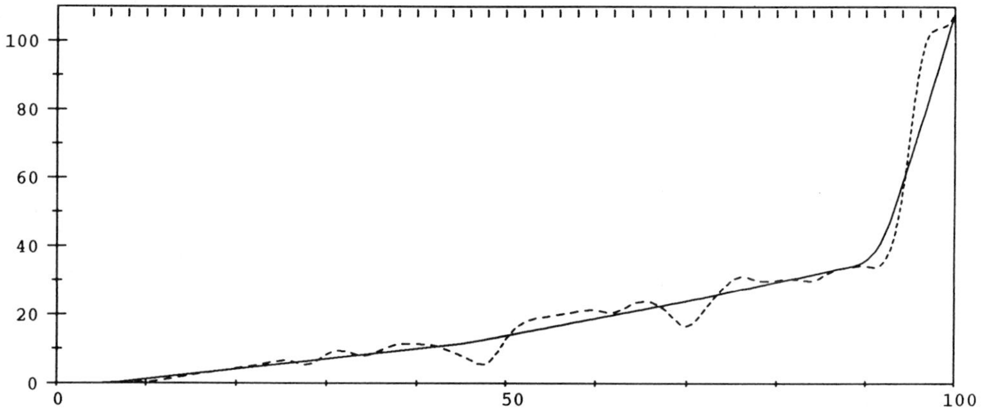

Figure 3. The initial convex approximation to the turbine data.

The dashed curve shown in Figure 3 is a cubic spline approximation to a set of data points

$$(x_i, y_i) = (2i, y_i), \qquad \text{for } i = 1,\ 2,\ \ldots,\ 50,$$

where y_i is measured electricity when $2i$ percent of maximum power is forced upon a particular water turbine. A large number of turbines have been tested and data measured, and all turbines have similar curves depending on the turbine type, the size of the turbine and various other factors. These curves make it possible to choose the right turbines given the water resources and the need for electric power. Physical considerations indicate that such curves should be convex, and with a tolerance of $\epsilon = 1.0$ there is clearly scope for knot removal, so it is natural to apply knot removal with convex constraints.

The dashed curve ϕ shown in Figure 3 is in fact the Schoenberg variation diminishing approximation to the data on the knot vector

$$\boldsymbol{t} = (t_i)_1^{54} = (2, 2, 2, 2, 6, 8, \ldots, 94, 96, 100, 100, 100, 100).$$

The interior knots are indicated at the top of Figure 3. Our problem is to construct a spline approximation g to ϕ subject to the constraints

$$g(t_1) = \phi(t_1), \qquad g(t_{54}) = \phi(t_{54}), \qquad g''(x) \geq 0 \quad \text{for all } x \in [2, 100].$$

First, we computed an initial spline approximation f on $\boldsymbol{t}$ which solves (3) with

$$\mathsf{F}_{4,t} = \Big\{ \sum_j d_j B_{j,4,t} \;\Big|\; d_1 = a_1 \;\&\; d_{50} = a_{50} \;\&\; \Delta d_j \geq 0, \text{ for } 3 \leq j \leq 50 \Big\},$$

where (a_i) and (d_i) are the B-spline coefficients of ϕ and f on $\boldsymbol{t}$, and Δd_j is defined by

$$\Delta d_j = \sigma_{j-1} d_j - (\sigma_{j-1} + \sigma_j) d_{j-1} + \sigma_j d_{j-2},$$

where $\sigma_i = (t_{i+3} - t_i)/3$. The spline f is shown in Figure 3 (solid curve).

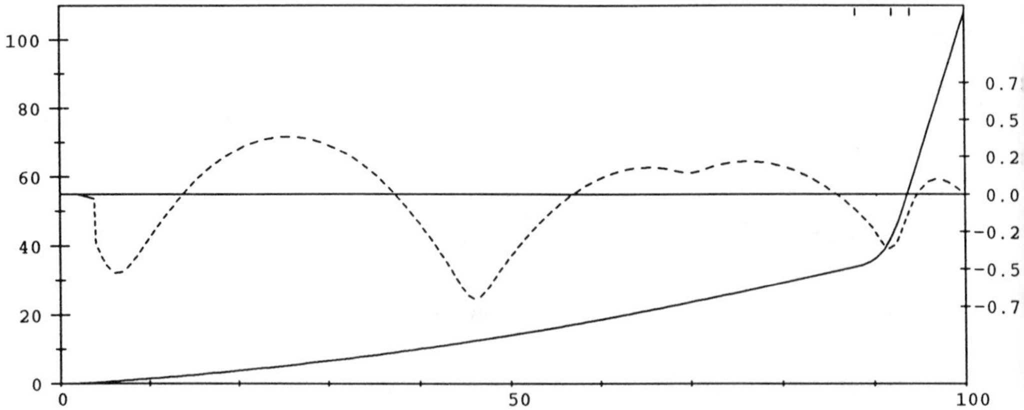

Figure 4. The final convex approximation after knot removal.

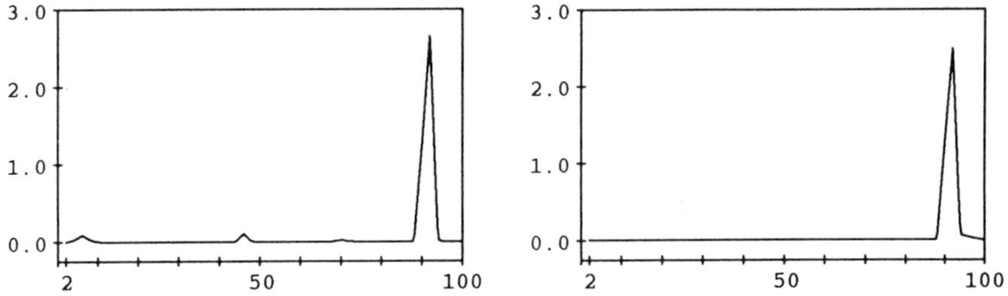

Figure 5. The second derivative of the initial (left) and final convex approximations.

Applying constrained knot removal to f with the same constraints and with tolerance $\epsilon = 1.0$, we get the final cubic spline approximation g. This spline is shown in Figure 4 (solid curve) together with the error function $g - f$ (dashed curve). The three remaining interior knots are indicated at the top of the figure.

Figure 5 shows the second derivative of f and g, and we observe that g is smoother than f. We also observe that the computed convex approximation to the given data consists almost of two straight lines.

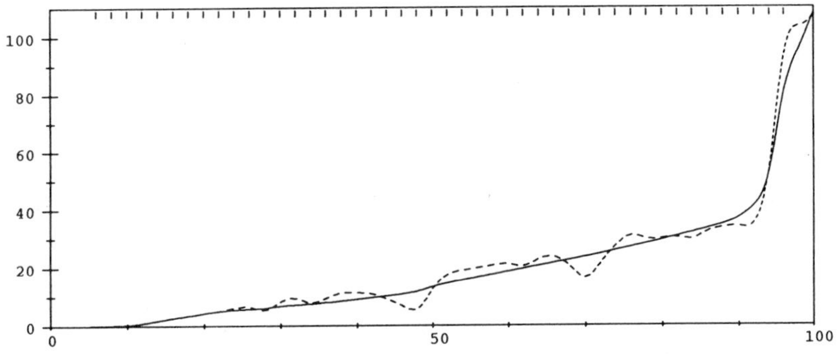

Figure 6. Initial approximation to the turbine data with the first derivative bounded from below by a positive spline.

5.3 Curved constraints

This example illustrates that it is possible to introduce curved constraints into the knot removal method. The initial data are the values of the cubic spline ϕ on the knot vector $\boldsymbol{t} = (t_i)_1^{54}$ of the previous example. We are also given a quadratic spline h on the same knot vector (the dashed curve in Figure 8). Our problem is to construct a spline approximation g to ϕ with as few knots as possible, subject to the constraints

$$g(t_1) = \phi(t_1), \qquad g(t_{54}) = \phi(t_{54}), \qquad g'(x) \geq h(x) \quad \text{for all } x \in [2, 100].$$

The tolerance is 1.0 as in the previous example.

First, we compute an initial spline approximation f on $\boldsymbol{t}$ which solves (3) with

$$\mathsf{F}_{4,t} = \Big\{ \sum_j d_j B_{j,4,t} \,\Big|\, d_1 = a_1 \;\&\; d_{50} = a_{50} \;\&\; (d_j - d_{j-1})/\sigma_j \geq e_j \text{ for } 2 \leq j \leq 50 \Big\},$$

where a_j and (e_j) are the B–spline coefficients of ϕ and h on $\boldsymbol{t}$, and $\sigma_j = (t_{j+3} - t_j)/3$. The result is shown in Figure 6, with ϕ dashed and f as a solid curve. We note that f is monotone since we have chosen a positive h.

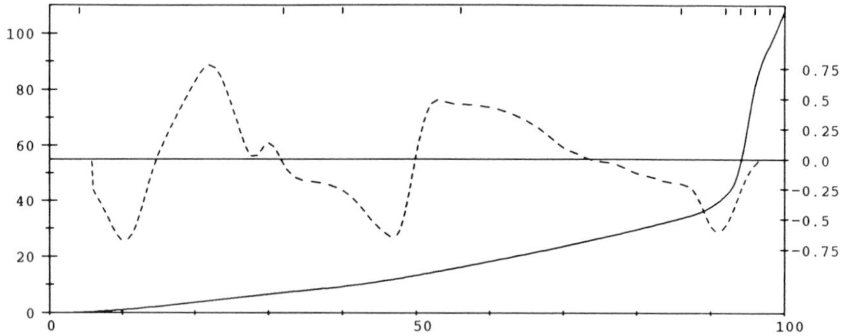

Figure 7. The final approximation with bound on the derivative after knot removal.

Figure 7 shows the result of applying constrained knot removal to f with the given tolerance $\epsilon = 1.0$. The dashed curve is the error function $g - f$, and the solid curve is the final spline approximation g to ϕ. The number of interior knots was reduced from 46 to 9, and the location of the remaining interior knots is marked at the top of the figure.

In Figure 8 the constraining function h (dashed curve) and g' (solid curve) are plotted. We note that the peak in g' is induced from the rapid increase in the data ϕ in that area.

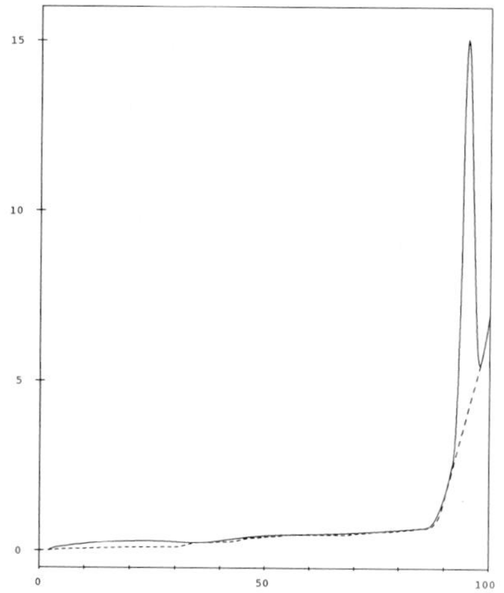

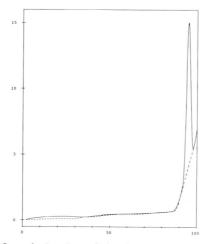

Figure 8. The first derivative of the final approximation and its lower bound.

References

de Boor, C. (1976a), Splines as linear combinations of B-splines. A survey. In *Approximation Theory II* (Lorentz, G. G., Chui, C. K., and Schumaker, L. L., Eds.), Academic Press, New York, pp. 1–47.

de Boor, C. (1976b), On local linear functionals which vanish at all B-splines but one. In *Theory of Approximation with Applications* (Law, A. G., and Sahney, B. N., Eds.), Academic Press, New York, pp. 120–145.

Cohen, E., and Schumaker, L. L. (1985), Rates of convergence of control polygons. *CAGD* **2,** 229–235.

Cox, M. G. (1987), Data approximation by splines in one and two independent variables. In *The State of the Art in Numerical Analysis,* (Iserles, A. and Powell, M. J. D., Eds.), Clarendon Press, Oxford, pp. 111–138.

Dahmen, W. (1986), Subdivision algorithms converge quadratically. *J. Comp. Appl. Math.* **16,** 145–158.

Ferguson, D. R., Frank, P. D., and Jones, A. K. (1988), Surface shape control using constrained optimization on the B-spline representation. *CAGD* **5,** 87–103.

Fletcher, R. (1987) *Practical Methods of Optimization,* Second edition, John Wiley & Sons, Chichester.

Fontanella, F. (1987), Shape preserving surface interpolation. In *Topics in Multivariate Approximation,* (Chui, C. K., Schumaker, L. L. and Utreras, F. I., Eds.), Academic Press, Orlando Florida, pp. 63–78.

Franke, R., and Schumaker L. L. (1987), A bibliography of multivariate approximation. In *Topics in Multivariate Approximation,* (Chui, C. K., Schumaker, L. L. and Utreras, F. I., Eds.), Academic Press, Orlando Florida, pp. 275–335.

Jia, R. Q. (1983), Total positivity of the discrete spline collocation matrix. *J. Approx. Theory* **39,** 11–23.

Lyche, T. (1988), Note on the Oslo Algorithm. *Computer Aided Design* **20,** 353–355.

Lyche, T., and Mørken, K. (1987a), A discrete approach to knot removal and degree reduction algorithms for splines. In *Algorithms for Approximation* (Mason, J. C. and Cox, M. G., Eds.), Clarendon Press, Oxford, pp. 67–82.

Lyche, T., and Mørken, K. (1987b), Knot removal for parametric B-spline curves and surfaces. *CAGD* **4,** 217–230.

Lyche, T., and Mørken, K. (1988), A data-reduction strategy for splines with applications to the approximation of functions and data. *IMA J. Numer. Anal.* **8,** 185–208.

Utreras, F. I. (1987), Constrained surface construction. In *Topics in Multivariate Approximation,* (Chui, C. K., Schumaker, L. L. and Utreras, F. I., Eds.), Academic Press, Orlando Florida, pp. 233–254.

Near real-time spline fitting of long sequences of uniformly-spaced data

G. T. Anthony and M. G. Cox

National Physical Laboratory, Teddington, Middlesex, UK

Abstract Data is frequently captured in very long sequences at equal intervals in the independent variable. This situation occurs, for instance, in medical diagnosis where various quantities (eg intestinal pressure and acidity) are regularly recorded at short time intervals (seconds) over periods of many hours. There are consequent difficulties in storing and analysing the resulting very large amounts of data. Another difficulty in analysing the data (to discover peaks, zeros, etc) is that the data is often noisy. These difficulties are reduced by first fitting a smooth mathematical function to the data and then working instead with that function. We consider here an approximate least-squares fitting method using polynomial splines with uniformly spaced knots at some of the data points. Advantage is taken of the structure of the normal matrix to develop a technique that is near real-time in that it produces a "running fit" over successive blocks of data. The degree to which the method is approximate can be controlled. The algorithm is fast in operation and the fits have local properties.
Key words: B-splines, data approximation, least squares, medical data analysis

1. Introduction

Tijskens, Janssens, Vantrappen, De Bondt and Vandewalle [11], in considering the automatic analysis of medical data gathered from (digitised) multi-channel chart recorders, have presented a method for fitting extremely long sequences of uniformly spaced data. They seek a fit in the form of a cubic spline with prescribed uniform knots, because this function is sufficiently versatile to represent such data adequately. Moreover, this form is convenient for subsequent (medical) analysis of the tracings. Data reduction is important because it is necessary to avoid storing the very large amounts of data involved. This implies that the fitting should be performed essentially in "real time", that is as the data is gathered. To this end they use a "window" technique. Data in a window spanning an odd number of spline intervals is fitted by a spline in its B-spline form in the least-squares sense, using the singular value decomposition (SVD) [5, page 293 *et seq*]. The central coefficient of the fit is retained, the window advanced one interval and the process repeated. The window is chosen large enough so that, to the required accuracy, the coefficients obtained are identical to those that would have been calculated in a single fit to all the data.

The SVD is a powerful tool that, in this case, takes no advantage of the *structure* of the spline-fitting problem. Consequently, the computation time is unnecessarily long. Since the knots of the spline are prescribed, each B-spline coefficient is a linear combination of the data values in the window. The multipliers or weights in the linear combination can be pre-computed using an approach described below. Thus the method is reduced to the evaluation of a single inner product for each coefficient, with the same set of multipliers in each case. This paper describes this alternative approach.

2. Solution method

We shall let x denote the independent variable and y the dependent variable (in this case time and the medical response indicated on the chart recorder, respectively). Data values $\{y_r\}$ are assumed to be provided sequentially at corresponding points $\{x_r\}$ having a uniform spacing h. This data is to be approximated in near real-time by a polynomial spline $s(x)$ of order m (degree $m-1$) with uniformly spaced knots. A knot will be placed at every pth data point, and thus the knot spacing will be ph. It is algebraically convenient to scale the independent variable so that $ph = 1$. If this is done the scaled data spacing is p^{-1} and the knots are at the integers. p is then the data density, that is the number of data abscissae per unit of the independent variable x. Henceforth we shall work only in the scaled variables.

The spline s will be represented as a linear combination of B-splines,

$$s(x) = \sum_j c_j N_m(x-j),$$

where $N_m(x)$ is the normalized B-spline of order m with the knots $0, 1, \ldots, m$ [2].

A (linear) data fitting problem, in which the fitting function is expressed as a linear combination of basis functions (here the B-splines), can be formulated by writing the model equation for each data point, giving an overdetermined system of linear equations:

$$Ac = y. \tag{1}$$

In (1), each row of the observation matrix A contains values of the basis functions at a data point, y is the vector of corresponding data values and c is the vector of coefficients to be determined. The normal-equations approach [6, page 121 *et seq*] to obtaining the least-squares solution to (1) consists in forming and solving the equations

$$Hc = z, \tag{2}$$

where $H = A^TA$ and $z = A^Ty$.

The system of equations (2) has a unique solution and, equivalently, H is invertible if and only if A has full column rank.

The computation falls broadly into two parts: forming H and inverting it. In the present application it is particularly simple to form H because, with uniform knots, all B-splines (of a given order) are translations of one another. So the element in row r and column j of A is $N_m(x_r - j)$, and it follows from the compact support property of B-splines [2] and the uniform spacing of the data that H is a $(2m-1)$-diagonal Toeplitz matrix (that is the elements on each diagonal are constant). A row of H takes the form

$$\{\ldots,\ 0,\ h_{m-1},\ h_{m-2},\ \ldots,\ h_1,\ h_0,\ h_1,\ \ldots,\ h_{m-1},\ 0,\ \ldots\},$$

with

$$h_j = \sum_r N_m(x_r)N_m(x_r - j).$$

Because of the uniform knots and data, H is well conditioned and invertible for moderate values of m provided p is greater than unity. So the usual objection that the formation and solution of the normal equations can introduce errors that are greater in magnitude than those we expect with a stable algorithm is of no significance here.

Consider, as an example, the piecewise linear case, $m = 2$. Here

$$N_2(x) = \begin{cases} x, & x \in [0,\ 1), \\ 2 - x, & x \in [1,\ 2], \\ 0, & \text{otherwise,} \end{cases}$$

so that

$$\begin{aligned} h_1 &= \sum_r N_2(x_r)N_2(x_r - 1) \\ &= \sum_{r=p}^{2p} \left(2 - \frac{r}{p}\right)\left(\frac{r}{p} - 1\right) \\ &= \frac{1}{3!p}(p^2 - 1) \end{aligned}$$

and

$$\begin{aligned} h_0 &= \sum_r N_2(x_r)N_2(x_r) \\ &= \sum_{r=0}^{p} \left(\frac{r}{p}\right)^2 + \sum_{r=p}^{2p} \left(2 - \frac{r}{p}\right)^2 - 1 \\ &= \frac{1}{3!p}(4p^2 + 2). \end{aligned}$$

The typical row of H for B-splines of other orders can be obtained similarly, using the basic recurrence relation for B-splines [1] and formulae for the sums of powers of

Linear						
$3!ph_0$	$=$	$4p^2$	$+$	2		
$3!ph_1$	$=$	p^2	$-$	1		
Quadratic						
$5!p^3h_0$	$=$	$66p^4$	$-$	6		
$5!p^3h_1$	$=$	$26p^4$	$+$	4		
$5!p^3h_2$	$=$	p^4	$-$	1		
Cubic						
$7!p^5h_0$	$=$	$2416p^6$	$+$	$\frac{112}{3}p^2$	$+$	$\frac{200}{3}$
$7!p^5h_1$	$=$	$1191p^6$	$-$	$21p^2$	$-$	$\frac{150}{3}$
$7!p^5h_2$	$=$	$120p^6$	$+$		$+$	$\frac{60}{3}$
$7!p^5h_3$	$=$	p^6	$+$	$\frac{7}{3}p^2$	$-$	$\frac{10}{3}$

TABLE 1

Expressions for the nonzero elements in the typical row of the normal matrix H.

the natural numbers, or by more sophisticated methods. The linear, quadratic and cubic cases are given in Table 1, where, for order m, each element is multiplied by $(2m-1)!p^{2m-3}$ giving even polynomials of degree $2m-2$.

Various properties of these polynomials are readily demonstrated. In particular, we note that the leading coefficients are the values of N_{2m} at the knots, the trailing coefficients are proportional to the binomial coefficients of order $2m-2$, and the coefficient of p^{2m-4} is zero (if $m > 2$). The first observation is analogous to the integral result [10]

$$\int_{-\infty}^{\infty} M_r(x-j)M_t(x-k)dx = M_{r+t}(j-k), \tag{3}$$

with $r = t$. In (3), $M_r = rN_r$ is the B-spline with integral normalization $\int_{-\infty}^{\infty} M_r(x)dx = 1$ [8].

The third observation means that (for $m > 2$) the leading term dominates by a factor $O(p^4)$. This factor gives an indication of the convergence of the solution of the *discrete* least-squares spline approximation problem, as considered here, to that of the *continuous* uniform-knot spline approximation problem of minimizing $\int_{-\infty}^{\infty}(y(x)-s(x))^2dx$.

The second part of the calculation, obtaining the inverse G of H, can also be variously accomplished. Yamamoto and Ikebe [12] give a direct formulation for the inversion of band matrices which is particularly simple for tridiagonal matrices. In our linear case, $m = 2$, this gives for the i,j-th element of G (symmetric)

$$g_{i,j} = -\frac{(\alpha^i - \beta^i)(\alpha^{n+1}\beta^j - \alpha^j\beta^{n+1})}{(\alpha-\beta)(\alpha^{n+1}-\beta^{n+1})}, \quad i \le j,$$

where α and β are the roots of

$$h_1t^2 + h_0t + h_1 = 0$$

and n is the dimension of H.

This is equivalent, for sufficiently large n, to a row

$$\{\ldots,\ g_2,\ g_1,\ g_0,\ g_1,\ g_2,\ \ldots\}$$

of G being given by

$$g_0 = \frac{1}{\sqrt{(p^2+2)/3}},$$

$$g_{j-1} = -kg_j - g_{j+1}, \quad j \geq 2, \quad k = h_0/h_1 = (4p^2+2)/(p^2+1).$$

The sequence $\{g_j\}$ can be calculated by setting $g_{N+1} = 0,\ g_N = 1$ for some (notional) N, and applying the recurrence above for decreasing values of the suffix until a g_{N-M} is calculated with a sufficient number of significant figures. The sequence is then renumbered and normalized so that g_{N-M} is $g_0 = 1/\sqrt{(p^2+2)/3}$.

The typical row of G for this order of spline is shown to 6 decimal places for p = 2, 4, 10 in the left-hand part of Table 2. For $p = 10$, the decay factor k is close to its lower limit of $2 + \sqrt{3} \approx 3.732$.

The Yamamoto and Ikebe algorithm is less straightforward for $m > 2$. An alternative approach is to use software from any of the standard linear algebra algorithms libraries such as NAG [4] or LINPACK [3]. For demonstration purposes we used the INV function from PC-MATLAB [7]. This employs an LU decomposition of H and then inverts the factors. The typical (symmetric half-) rows of G for $m = 4$ (cubic) with $p = 2, 4, 10$ are shown in the right-hand part of Table 2. The decay factor is close to 2 in all three cases. In fact we can see from Table 1 that the decay tends to $2416/1191 \approx 2.0285$ as p increases.

Finally, the coefficients of the fit are obtained by multiplying an appropriate (symmetric) section of the typical row of G into the corresponding section of z. The choice of this section is equivalent to choosing the number of spline intervals in the window of the procedure of Tijskens *et al* [11]. It should be emphasised that once the order m and data density p have been chosen, the calculation of the typical row of G is only required once and can be pre-computed. This gives an *a priori* estimate of the number of data points in a window needed to achieve a given accuracy of fit.

A refinement of the method is to compute $X = GA^T$ (the pseudoinverse [6, page 36 *seq*] of A), either directly or as the product of G and A^T. The coefficients of the fit are then obtained as the scalar product of an appropriate part of a row of X and the corresponding part of y. Table 3 shows part of the (symmetric half-) rows of X for $m = 4$ and $p = 2, 4, 10$ produced using PC-MATLAB. The elements of X exhibit a "piecewise" decay, and it is again straightforward to determine the length of the section of a row of X (equal to the number of data points in a window) required to achieve a given accuracy.

In the above we have implicitly assumed that the range over which the response is measured (and fitted) is of infinite extent. It follows that the matrices A, H, etc are of infinite order. Obviously, the data is gathered on a finite range. Equally

$p=2$	$p=4$	$p=10$
⋮	⋮	⋮
-0.000000	-0.000001	-0.000001
0.000000	0.000004	0.000003
-0.000003	-0.000018	-0.000015
0.000018	0.000078	0.000057
-0.000105	-0.000327	-0.000217
0.000612	0.001363	0.000824
-0.003571	-0.005672	-0.003132
0.020815	0.023595	0.011893
-0.121320	-0.098146	-0.045163
0.707106	0.408248	0.171498
-0.121320	-0.098146	-0.045163
0.020815	0.023595	0.011893
-0.003571	-0.005672	-0.003132
0.000612	0.001363	0.000824
-0.000105	-0.000327	-0.000217
0.000018	0.000078	0.000057
-0.000003	-0.000018	-0.000015
0.000000	0.000004	0.000003
-0.000000	-0.000001	-0.000001
⋮	⋮	⋮

$p=2$	$p=4$	$p=10$
⋮	⋮	⋮
0.000001	0.000000	0.000000
-0.000001	-0.000001	0.000000
0.000002	0.000002	0.000001
-0.000005	-0.000003	-0.000001
0.000009	0.000006	0.000002
-0.000017	-0.000010	-0.000004
0.000032	0.000019	0.000008
-0.000060	-0.000036	-0.000015
0.000114	0.000068	0.000027
-0.000215	-0.000127	-0.000051
0.000407	0.000237	0.000095
-0.000768	-0.000443	-0.000178
0.001450	0.000828	0.000333
-0.002737	-0.001547	-0.000622
0.005168	0.002891	0.001162
-0.009758	-0.005403	-0.002170
0.018425	0.010098	0.004055
-0.034791	-0.018872	-0.007575
0.065690	0.035270	0.014151
-0.124025	-0.065913	-0.026434
0.234093	0.123144	0.049365
-0.441263	-0.229778	-0.092074
0.827044	0.426393	0.170791
-1.511359	-0.771993	-0.309097
2.445129	1.240342	0.496466
-1.511359	-0.771993	-0.309097
0.827044	0.426393	0.170791
⋮	⋮	⋮

TABLE 2

The left-hand part of the table gives the values of the elements in the typical row of G, the inverse of the normal matrix H, in the case $m = 2$ (spline of degree 1) for three values of p, the data density. The right-hand part is the counterpart of this in the case $m = 4$ (cubic spline), showing (symmetric) half-rows.

$p=4$	$p=10$
⋮	⋮
0.000005	-0.002426
0.000000	-0.002362
-0.000007	-0.002016
-0.000011	-0.001360
-0.000010	-0.000466
0.000000	0.000569
0.000013	0.001652
0.000021	0.002685
0.000018	0.003573
0.000000	0.004221
-0.000024	0.004533
-0.000040	0.004413
-0.000034	0.003766
0.000001	0.002541
0.000045	0.000872
0.000074	-0.001063
0.000063	-0.003085
-0.000001	-0.005015
-0.000084	-0.006675
-0.000138	-0.007886
-0.000118	-0.008469
0.000002	-0.008245
0.000157	-0.007037
0.000259	-0.004750
0.000220	-0.001632
-0.000005	0.001982
-0.000294	0.005758
-0.000484	0.009365
-0.000411	0.012466
0.000008	0.014730
0.000549	0.015823
0.000904	0.015410
0.000769	0.013159
-0.000016	0.008894
-0.001026	0.003079
-0.001690	-0.003666
continued ...	

$p=2$	$p=4$	$p=10$
	... continued	
⋮	⋮	⋮
0.000030	0.002686	-0.029575
-0.000041	-0.000055	-0.028844
-0.000057	-0.003582	-0.024690
0.000078	-0.005902	-0.016789
0.000107	-0.005020	-0.005991
-0.000147	0.000103	0.006555
-0.000202	0.006695	0.019704
0.000278	0.011030	0.032307
0.000382	0.009383	0.043218
-0.000525	-0.000191	0.051290
-0.000722	-0.012511	0.055375
0.000992	-0.020614	0.054327
0.001363	-0.017540	0.046998
-0.001873	0.000347	0.032772
-0.002573	0.023371	0.013145
0.003536	0.038528	-0.009852
0.004859	0.032814	-0.034192
-0.006676	-0.000567	-0.057847
-0.009174	-0.043585	-0.078789
0.012606	-0.072007	-0.094988
0.017324	-0.061596	-0.104418
-0.023800	0.000387	-0.105050
-0.032720	0.080694	-0.094856
0.044916	0.134580	-0.072627
0.061847	0.117300	-0.040441
-0.084623	0.004759	-0.001194
-0.117319	-0.144547	0.042219
0.158243	-0.251472	0.086899
0.225926	-0.236873	0.129951
-0.286154	-0.053637	0.168479
-0.462210	0.217217	0.199584
0.433175	0.462635	0.220372
1.126300	0.569564	0.227945
0.433175	0.462635	0.220372
-0.462210	0.217217	0.199584
⋮	⋮	⋮

TABLE 3

The values of the elements in the typical (symmetric) half-row of X, the pseudoinverse of the observation matrix A, in the case $m = 4$ (cubic spline) for three values of p, the data density. Note the "piecewise decay" in the magnitude of the elements.

obviously, the observation matrix A is finite and not all its rows have the same non-zero elements, because the first few and last few rows are curtailed. These end effects lead to a matrix G whose first few and last few rows are not shifted copies of a typical row. However, for sufficiently large orders of matrix, the rows converge to a central section of the matrix whose rows do have this property to working precision. Furthermore, to that precision the typical row is the typical row of the infinite case. The required order is quite modest; the examples given above were computed with matrices of order 40.

3. Conclusion

We have described a near real-time algorithm for fitting arbitrarily long sequences of uniformly-spaced data. The method uses splines with uniform knots and achieves an approximation to the least-squares fit to all the data. If sufficient data values "local to" any particular B-spline coefficient in the approximant s are employed then, to any specified precision, s can be regarded as *the* least-squares fit. In particular, for practical purposes, the accuracy of the data indicates the number of data values to be employed. For instance, if the data density p is 10, the data is accurate to approximately two significant figures, and a cubic spline is used, each B-spline coefficient can be formed as a linear combination of about 150 of the data values. However, there is one "replacement" B-spline coefficient in this case for every 10 data values. Hence each data value is used in each of 15 B-spline coefficient evaluations. The total number of floating-point operations is therefore only 15 times the number of data points. Similar statements can be made for other values of m and p. The multipliers in the linear combination are formed by a straightforward pre-computation that depends on p and the data accuracy.

The case p equals 1 corresponds to placing a knot at every data point and hence to *interpolation.* Of course, this is a much-studied problem dating from the seminal work of Schoenberg [8, 9]. In one sense, the ideas here are a straightforward generalization of Schoenberg's work. Considerable work needs to be done, particularly on its theoretical aspects, to develop this generalization fully.

Acknowledgments

We thank our colleagues Alistair Forbes and Peter Harris for useful discussions and for reading a draft of this paper.

References

1. M. G. Cox. The numerical evaluation of B-splines. *J. Inst. Math. Appl.*, 10:134–149, 1972.
2. M. G. Cox. Practical spline approximation. In P. R. Turner, editor, *Notes in Mathematics 965: Topics in Numerical Anlaysis*, pages 79–112, Berlin, 1982. Springer-Verlag.
3. J. J. Dongarra, C. B. Moler, J. R. Bunch, and G. W. Stewart. *LINPACK User's Guide.* Society for Industrial and Applied Mathematics, Philadelphia, USA, 1979.
4. B. Ford, J. Bently, J. J. du Croz, and S. J. Hague. The NAG Library 'machine'. *Software - Practice and Experience*, 9:56–72, 1979.
5. G. H. Golub and C. F. Van Loan. *Matrix Computations.* North Oxford Academic, Oxford, 1983.
6. C. L. Lawson and R. J. Hanson. *Solving Least Squares Problems.* Prentice-Hall, Inc., Englewood Cliffs, New Jersey, 1974.
7. C. Moler, J. Little, and S. Bangert. *PC-MATLAB.* The MathWorks, Inc., Sherborn, Ma, USA, 1987.
8. I. J. Schoenberg. Contributions to the problem of approximation of equidistant data by analytic functions. Part A–On the problem of smoothing or graduation. A first class of analytic approximation formulae. *Qu. Appl. Math.*, 4:45–99, 1946.
9. I. J. Schoenberg. Contributions to the problem of approximation of equidistant data by analytic functions. Part B–On the problem of osculatory interpolation. A second class of analytic approximation formulae. *Qu. Appl. Math.*, 4:112–141, 1946.
10. I. J. Schoenberg. Cardinal interpolation and spline functions. *J. Approx. Theory*, 2:167–206, 1969.
11. G. Tijskens, J. Janssens, G. Vantrappen, F. De Bondt, and J. Vandewalle. Spline functions used for on-line data reduction and analysis of motility tracings. Presented at the Contact Group on Numerical Analysis, Splines and their Applications in Modelling, Namur, Belgium, 4 December, 1987.
12. T. Yamamoto and Y. Ikebe. Inversion of band matrices. *Linear Alg. Appl.*, 24:105–111, 1979.

An algorithm for knot location in bivariate least squares spline approximation

M. Bozzini

Dipartimento di Matematica, Università di Lecce, Italy

F. de Tisi

Dipartimento di Matematica, Università di Milano, Italy

Abstract An automatic algorithm for the determination of a lattice suitable for the construction of bivariate least squares splines is studied. Some numerical examples are quoted.

Key words: Bivariate splines, Knot location, Least squares approximation.

1. Introduction

The problem of constructing a surface $F(x, y)$, $(x, y) \in D \subset \mathcal{R}^2$, from a discrete number of points (x_i, y_i, f_i) $(i = 1, \ldots, N)$ has been studied by many authors. In the literature there appears to be greater concentration on the development of algorithms for data interpolation than for approximation of data subject to errors, where we are faced with additional difficulties.

When spline functions are employed, it is well-known that cubic splines afford a good balance between efficiency and goodness of fit. These functions are generally written in terms of the tensor product of B-spline basis functions $\{B_\ell(x)\}$ with $\{B_m(y)\}$ defined on the Cartesian axes.

In the case of cubic spline interpolation, an interpolant to data on any rectangular grid can be constructed. If the grid is defined by the lines $x = x_i$ $(i = 1, \ldots, N_x)$, $y = y_j$ $(j = 1, \ldots, N_y)$, then the points x_i and y_j determine the knots for the interpolating spline.

When the data values contain random errors, we may construct an approximation by the method of least squares. The approximating function is written in terms of basis functions, fewer in number than the data points, defined on a knot set which is different from the original data point set. The choice of knots can greatly affect how well the surface fits the data.

In this paper an algorithm is presented for the automatic placement of the knots in the case of bivariate least squares spline approximation. Before describing this algorithm, we briefly review in the next section how to construct such approximations.

2. Bivariate least squares spline approximation

Consider a rectangle $R = \{(x, y) \in \mathcal{R}^2 : x \in [a, b], y \in [c, d]\}$ and the lattice defined by the knots $\{\lambda_i\}_{i=0}^{h+1}$ and $\{\mu_j\}_{j=0}^{g+1}$, where

$$a = \lambda_0 < \lambda_1 < \ldots < \lambda_{h+1} = b$$

and

$$c = \mu_0 < \mu_1 < \ldots < \mu_{g+1} = d.$$

A bivariate spline of order k on R with interior knots $\{\lambda_i\}_{i=1}^{h}$ and $\{\mu_j\}_{j=1}^{g}$ is a function $s(x, y)$ with the properties that:

i) On the subrectangle $R_{ij} = [\lambda_{i-1}, \lambda_i] \times [\mu_{j-1}, \mu_j]$ $(i = 1, \ldots, h+1;\ j = 1, \ldots, g+1)$, $s(x, y)$ is a polynomial of total order $2k$ in x and y.

ii) The functions

$$\frac{\partial^{p+q}}{\partial x^p \partial y^q} s(x, y) \qquad (0 \leq p, q \leq k-2)$$

are continuous on R.

A generally well-conditioned basis for $s(x, y)$ is given by the tensor product of kth order univariate B-splines $M_i(x)$ with knots λ_u $(u = 1, \ldots, h)$ and $N_j(y)$ with knots μ_v $(v = 1, \ldots, g)$:

$$s(x, y) = \sum_{i=1}^{h+k} \sum_{j=1}^{g+k} \nu_{ij} M_i(x) N_j(y).$$

The least squares spline fit to the given data is the solution of the problem

$$\min_{\nu_{ij}} \sum_{r=1}^{N} \left(f_r - \sum_{i,j} \nu_{ij} M_i(x_r) N_j(y_r) \right)^2 .$$

Equivalently, we may solve, in the least squares sense, the overdetermined system of equations

$$A\boldsymbol{\nu} = \boldsymbol{f}, \tag{1}$$

where $\boldsymbol{\nu} = \{\nu_{ij}\}_1^{h+k,g+k}$, expressed as a $(h+k)(g+k)$-vector, $\boldsymbol{f} = \{f_r\}_1^N$ and A is the $N \times (h+k)(g+k)$ matrix whose r^{th} row contains the basis function values $M_i(x_r)N_j(y_r)$, with

$$N > (h+k)(g+k). \tag{2}$$

A solution can be found using a numerically stable method which employs Householder transformations. These reduce the matrix A to triangular form by means of orthogonal transformations. Another method is based on Givens rotations. These are particularly useful in the case of structured matrices, are numerically stable and reduce the storage requirements.

The major difficulty that occurs in determining the least-squares solution of (1) is the possible rank deficiency of the matrix A which depends on the lattice used. In the univariate case, A has full rank if the Schoenberg-Whitney conditions are satisfied (Cox, 1986; de Boor, 1978), while in the bivariate case conditions exist only for particular configurations of the points and knots. Rank deficient matrices occur very frequently and lead to non-unique solutions. We can determine a particular solution selecting either the minimum norm solution (which, however, is not invariant under translation) or a geometrically invariant solution (Cox, 1986).

3. Knot placement

From what we have said above, a need clearly arises for subdividing the rectangle R into a lattice that affords a good balance between the closeness of the fit to the data and the smoothness of the fit, whilst reducing the number of empty panels.

In the literature we find results for data on a mesh and scattered data. In the former case an algorithm due to Dierckx (1981) can be used. This algorithm constructs a lattice in such a way as to control the above-mentioned balance.

When the data are scattered over the domain, the literature suggests methods which are based on sequences of "trials". More precisely, one forms an initial rectangular lattice and constructs the relevant spline; if the fit is not satisfactory in some regions, the lattice is refined appropriately (see Cox (1986)).

4. The proposed algorithm

Our objective is to construct an algorithm which determines a suitable lattice for any point configuration. In order that the behaviour of the fitted surface is sensible and in particular sufficiently smooth, our algorithm is required

i) to provide a subdivision which achieves near-uniformity in the point distribution between the mesh lines,

ii) to impose a limitation on the knot spacing.

The limitation in ii) is controlled by an index L, as described below.

We first determine a functional relationship between L and the number of data points N. To this end, we consider the real variable n, which coincides with N at the points of integer abscissa, and require that the following conditions are satisfied:

$$\begin{aligned} \lim_{n\to 0} L &= \infty, \\ \lim_{n\to\infty} L &= \text{constant}, \\ L &> 0 \quad \forall\, n \in \mathcal{R}^+, \\ L' &< 0 \quad \forall\, n \in \mathcal{R}^+. \end{aligned}$$

As a first approximation, we write

$$L = \beta \frac{1}{n^\alpha} + \gamma, \tag{3}$$

where β and γ are positive constants which depend on the measure $\mu(R)$ of the domain, and $\alpha \in \mathcal{R}^+$. The relationship (3) must also satisfy the following two conditions:

1) The decrement in L must not in magnitude be much larger than the relevant increment in N.
2) When the given points are dense in the domain, L must be nearly insensitive to variations in N.

If we consider a value n, an increment Δn in n and the corresponding values $L(n)$ and $L(n + \Delta n)$, we have

$$\Delta L = L(n + \Delta n) - L(n) = L'(n)\Delta n + R_2(\Delta n),$$

where $R_2(\Delta n)$ is the remainder from the Taylor formula. From this, for Δn sufficiently small,

$$\Delta L = L'(n)\Delta n = -\frac{\alpha\beta}{n^{\alpha+1}}\Delta n.$$

In order to satisfy the first of the above conditions, it is necessary that $0 < \alpha \leq 1$; whereas, for the second, it is convenient to take $\alpha \geq 1$. As a consequence, a suitable value is $\alpha = 1$, giving

$$L = \beta \frac{1}{n} + \gamma. \tag{4}$$

This conjecture is corroborated by practice. Therefore, when forming the lattice, we shall consider (4) and the first assumption i). In order to satisfy the latter, we proceed in the following way.

Let us think of the points as point masses and determine their barycentre. If the points are regularly distributed, the barycentre is the centre of symmetry of the system and its coordinates subdivide the domain into four equal and homogeneous parts. Otherwise, the barycentre lies in the zone where the density is the greatest;

in this case, the subdivision makes the four zones obtained more homogeneous than any other subdivision.

The algorithm works in the following way. For a given set I_N of points $P_i = P(x_i, y_i)$ $(i = 1, \ldots, N)$, we determine the point $P^{(0)} \in I_N$ which has the minimum Euclidean distance from the barycentre B. We then find a circular neighbourhood $U(P^{(0)})$ about $P^{(0)}$ with radius r depending on the mean density of the data in the domain R:

$$r = k\frac{\mu(R)}{N}.$$

The factor k depends on the point configuration. If we have some information on the lattice, the user can assign an initial value k^*. Otherwise, the following rule can be used.

Suppose the N data points are placed on a rectangular grid as follows:

$$a = x_1 < x_2 < \ldots < x_{N_x} = b,$$

$$c = y_1 < y_2 < \ldots < y_{N_y} = d,$$

and define

$$h_1 = x_{i+1} - x_i = (b - a)/(N_x - 1),$$

$$h_2 = y_{i+1} - y_i = (d - c)/(N_y - 1).$$

In order that the neighbourhoods cover the domain, it is sufficient to choose the radius such that

$$r \geq \sqrt{h_1^2 + h_2^2}.$$

Hence,

$$k \geq \frac{N}{\mu(R)}\sqrt{\frac{(b-a)^2}{(N_x - 1)^2} + \frac{(d-c)^2}{(N_y - 1)^2}}. \tag{5}$$

Thus, we shall thus choose an initial value k^* for k satisfying (5).

Once we have determined $U(P^{(0)})$, we calculate the barycentre B_0 of the points $P_i \in I_N \cap U(P^{(0)})$ and call (x_{B_0}, y_{B_0}) the coordinates of B_0. We proceed by considering the set of points $P_j \in (I_N - U(P^{(0)})) = I_{N_1}$; the algorithm determines the neighbourhood $U(P^{(1)})$, with radius r, of the point $P^{(1)} \in I_{N_1}$ which has the minimum distance from the barycentre B. Then we calculate, as above, the barycentre B_1 of the points $P_i \in I_N \cap U(P^{(1)})$. This procedure is repeated until there are no more points P_i remaining. If, for some $P^{(s)}$, we have $U(P^{(s)}) = \emptyset$, this point is not taken into consideration, and we proceed to determine the next point closest to B.

When the procedure ends, we see if (2) is satisified; if not, we repeat the procedure iteratively by incrementing the value of k until (2) is satisfied.

The set A_{B_j} of the coordinates (x_{B_j}, y_{B_j}) of the barycentres B_j $(j = 0, \ldots, m)$ can be considered as a potential set of knots on the relevant coordinate axes, defining a first lattice. At this point, the algorithm controls the knot spacing by considering the sequences $\{x_{B_0}, \ldots, x_{B_m}\}$ and $\{y_{B_0}, \ldots, y_{B_m}\}$ of coordinates. For any couple B_j, B_{j+1} consider the inequalities:

$$|x_{B_{j+1}} - x_{B_j}| > L, \tag{6}$$

$$|y_{B_{j+1}} - y_{B_j}| > L. \tag{7}$$

If (6) and (7) are satisfied, the knots are accepted. Otherwise, if one of the above relationships is not verified for at least one of the couples B_j and B_{j+1}, we proceed as follows.

Suppose (6) is not verified. We consider the knots

$$x_{B_{j-1}},\ x_{B_j},\ x_{B_{j+1}} \text{ and } x_{B_{j+2}},$$

and define $\overline{x}_j$, $\overline{x}_{j+1}$ and $\overline{x}_{j+2}$ to be the midpoints of the intervals $[x_{B_l}, x_{B_{l+1}}]$ $(l = j-1, \ldots, j+2)$. If $\overline{N}_{x_j}$ and $\overline{N}_{x_{j+1}}$ are, respectively, the number of points (each with its multiplicity) belonging to the intervals $[\overline{x}_j, \overline{x}_{j+1}]$ and $[\overline{x}_{j+1}, \overline{x}_{j+2}]$, we determine in the interval $[x_{B_j}, x_{B_{j+1}}]$ the point x^* for which the equality

$$x^* \overline{N}_{x_j} = (x_{B_{j+1}} - x_{B_j} - x^*) \overline{N}_{x_{j+1}}$$

holds. The sequence of knots on the x-axis is then modified to

$$x_{B_0}, \ldots, x_{B_{j-1}},\ x^*,\ x_{B_{j+2}}, \ldots, x_{B_m}.$$

5. Numerical results

In this section we use an example found in the literature, namely the test function $F(x, y) = \exp(-x^2 - y^2)$. The function is assumed specified by a sample of values

$$f_i = F_i + \epsilon_i, \qquad \epsilon_i \in U[-0.5 \times 10^{-3}, 0.5 \times 10^3],$$

where U is the uniform distribution and the approximating spline is that of minimum norm.

The error indices we have considered are

$$\text{L.S.E.} = \sqrt{\textstyle\sum e_i^2/N}, \qquad \text{A.E.} = \textstyle\sum |e_i|/N \text{ and } \text{MAX} = \max_{1 \le i \le N} |e_i|,$$

where $e_i = F(x_i, y_i) - s(x_i, y_i)$.

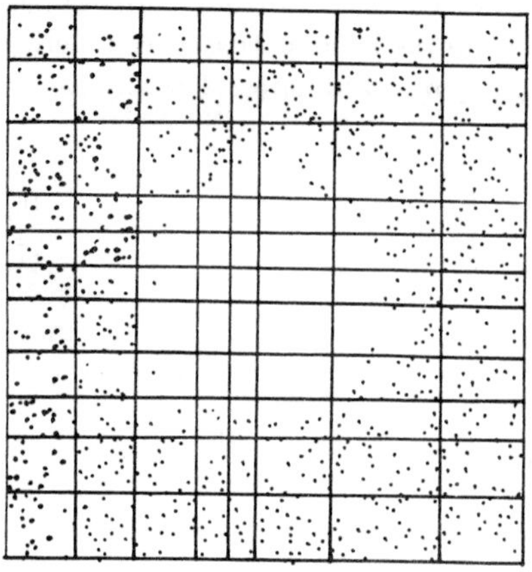

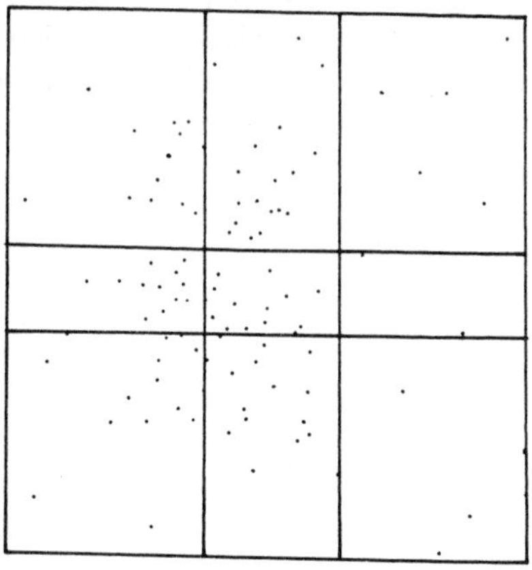

Example 1

$R = [0, 1] \times [0, 1]$

$N = 1000$

$\text{L.S.E.} = .1179E - 03$

$\text{A.E} = .9284E - 04$

$\text{MAX} = .4642E - 03$

Example 2

$R = [-1, 1.5] \times [-1, 1]$

$N = 100$

$\text{L.S.E.} = .1091E - 02$

$\text{A.E.} = .8577E - 03$

$\text{MAX} = .3078E - 02$

6. Concluding remarks

For small data sets, the algorithm presented provides good results both for data on a mesh and for scattered data. For large data sets, our results compare favourably with those provided by methods discussed in the literature. The comparison holds also for computation time. Moreover, memory requirements are not excessive, so the algorithm can be used on a personal computer and can be modified for interactive use.

References

1. M.G. Cox. Data approximation by splines in one and two independent variables. Technical Report DITC 77/86, National Physical Laboratory, 1986.
2. C. de Boor. *A Practical Guide to Splines.* Springer-Verlag, New York, 1978.
3. P. Dierckx. An algorithm for surface fitting with spline functions. *I.M.A. J. Num. Anal.*, 1:267–283, 1981.

A knot placement strategy for least squares spline fitting based on the use of local polynomial approximations

M. G. Cox, P. M. Harris and H. M. Jones

National Physical Laboratory, Teddington, Middlesex, UK

Abstract We are concerned with the use of univariate spline functions in fitting noisy data in the least-squares sense. When defining the linear space of approximating functions from which our fit is to be taken, there is freedom both in the number and locations of the interior knots. The distribution of these knots can have a profound influence on how well the spline fits the data.

We describe an algorithm for determining an initial estimate of the number of knots, and their distribution, from which an initial fit to the data is obtained. The algorithm exploits in a natural way the piecewise polynomial nature of the approximating function by isolating subsets of the data which are adequately described by simple polynomials. Once an initial set of knots is known, adaptive knot placement strategies may be used to improve this set.

Key words: Knot placement, Least squares, Local polynomial, Polynomial spline.

1. Introduction

Polynomial spline functions are widely used for fitting "noisy" data; that is, data in which there are random errors in the dependent variable values. In practice, such data can arise when these values are measurements of the response of a physical system to known discrete inputs (the independent variable values). Using the method of *least squares*, a polynomial spline approximation to the underlying function represented by the data can be constructed.

When defining the linear space of splines from which the fit is to be taken, there is considerable freedom of choice in both the order n of the spline and in the number and locations of the interior knots λ_j, $j = 1, \ldots, N$. We are interested in constructing strategies and algorithms for automatically choosing $\boldsymbol{\lambda} = (\lambda_1, \ldots, \lambda_N)$ so that the spline fit of given order with these knots is, in some sense, an "acceptable" fit to the data.

In Cox, Harris and Jones (1987) we discuss various approaches to solving the problem of automatic knot placement. Algorithms for deriving an *initial* trial set of knots and *adaptive* knot placement strategies are both considered. In particular, we present there a new version of an algorithm given in de Boor (1978) which we summarise below.

Let the data consist of the points (x_i, f_i), $i = 1, \ldots, m$, where $x_i \in [x_{min}, x_{max}]$ for all i. Suppose that we require a spline fit of order n (degree at most $n-1$) to the data and we wish to position N interior knots in (x_{min}, x_{max}). We first define *exterior* knots $\lambda_j = x_{min}$ for $j < 1$, $\lambda_j = x_{max}$ for $j > N$ and then, according to this strategy, we try to choose λ_j, $j = 1, \ldots, N$, such that

$$\int_{\lambda_k}^{\lambda_{k+1}} |f^{(n)}(x)|^{1/n}\, dx = \frac{1}{N+1} \int_{x_{min}}^{x_{max}} |f^{(n)}(x)|^{1/n}\, dx, \tag{1}$$

for $k = 0, \ldots, N$. We assume that f, the (unknown) underlying function, belongs to the continuity class $C^n[x_{min}, x_{max}]$. The criterion (1) gives an interior knot distribution which is asymptotically equivalent, as $N \to \infty$ with $\max_{k=0,\ldots,N} |\lambda_{k+1} - \lambda_k| \to 0$, to that obtained from solving

$$\min_{\boldsymbol{\lambda}} \left\{ \max_{k=0,\ldots,N} \operatorname{dist}_{[\lambda_k,\lambda_{k+1}]} (f, S^n_{\boldsymbol{\lambda}}) \right\}. \tag{2}$$

In (2), "$\operatorname{dist}_{[\lambda_k,\lambda_{k+1}]}(f, S^n_{\boldsymbol{\lambda}})$" represents the "distance", measured in the infinity norm over $[\lambda_k, \lambda_{k+1}]$, between f and $S^n_{\boldsymbol{\lambda}}$, the linear space of splines of order n with interior knot vector $\boldsymbol{\lambda}$. Criterion (1) is derived from (2) by replacing this distance function by a suitable bound.

Our version of the algorithm differs from de Boor's in that we approximate $f^{(n)}$ by a piecewise linear function rather than by a piecewise constant function. As a result we hope to achieve a better, and certainly smoother, approximation. The strategy is adaptive since the approximation to $f^{(n)}$ is obtained from a spline fit to the data based on a *current* set of interior knots. Thus, one application of the criterion results in this current set being updated, and the new knot vector forms the current knots at the next application.

In order to use (1) adaptively, we need

(a) an initial interior knot vector $\boldsymbol{\lambda}$,
(b) a means of approximating $f^{(n)}$ using the current set of knots, and
(c) a means of terminating the algorithm.

Details of how the approximation to $f^{(n)}$ is constructed is given in Cox et al (1987). Observe that (a) will need to be found *independently* of a current knot set since, to begin with, no such set will exist. Moreover, the effectiveness of this algorithm will depend on the quality of the initial approximation.

In this paper we present an algorithm for determining initial values for N and λ_j, $j = 1, \ldots, N$. The idea is to try to deduce information about the underlying function f from simple polynomial fits to subsets of the complete data set. These "local" fits give us "local" information about f independently of a set of interior knots. We then use this information to construct $\boldsymbol{\lambda}$. We assume throughout that there is enough data to ensure that proper subsets which adequately describe f exist. Once an initial set of knots is known, the adaptive knot placement strategy described above may be used to improve this set.

In Section 2 we present an algorithm which associates a local data set and a corresponding local polynomial fit with each point "sufficiently interior" to the complete data set. We show in Section 3 how this local information may be used in a strategy for knot placement. Results of using the strategy are also presented in this section. Finally, in Section 4 we summarise the work.

2. Deriving Local Information About f

Given m data points $(x_i, f_i), i = 1, \ldots, m$, with $x_1 < \ldots < x_m$, representing an unknown underlying function f, we show how we may associate with each data point *sufficiently interior* (see later) a *local data set* and a *local polynomial fit* of order n.

Consider the ith point (x_i, f_i). We first construct a subset of contiguous points taken from the complete data set, with indices centred on i, and consisting of $n+1$ points for n even or n points for n odd. We regard this subset as *local* to the ith point. It is assumed that these points can be "adequately" fitted by a simple polynomial of order n. This will certainly be true in the case when n is odd since we can *interpolate* the local data by such a polynomial. It is now also evident what we mean by a point being sufficiently interior – it is a point for which this initial local data set can be constructed.

Our next step is to increase the size of the subset by adding the neighbouring point at each end if this can be done. A polynomial fit of order n to the new local data is computed and Powell's trend test (Powell, 1970) is used to assess the "acceptability" of the fit.

The largest subset so formed for which the polynomial fit to this data satisfies the test is taken to be *the local data set* associated with the ith point. The procedure is then repeated for all sufficiently interior points.

In the use of the algorithm as described, there can be a problem, particularly for large data sets, with regard to the time taken for its execution. This is because we may need to compute a large number of polynomial fits (and evaluate these at every point of the corresponding data set for use in the trend test) for each of a large number of points.

To reduce the amount of work we use the size of the final local data set associated with the $(i-1)^{th}$ data point as a starting value for the size of the local data set for the ith point. This set is then enlarged or reduced depending on whether it can adequately be fitted by a polynomial in the sense described. It is not clear whether we can conclude that if the fit to this set of data satisfies the trend test then the fits to all smaller subsets must also satisfy the test. But equally, if we always start with the smallest subset and look for the first local data set to fail the trend test, it is not clear that all larger sets must also fail the test. Thus, we believe that reducing the work in this way is not detrimental to the results produced by the algorithm.

The execution time is also affected by the manner in which the polynomial fits are computed. Traditionally, the methods used to generate a least-squares polynomial fit to data are those of Forsythe (Forsythe, 1957) and Forsythe-Clenshaw (Clenshaw, 1960). We use Forsythe's method in preference to that of Forsythe-Clenshaw because for our purposes it is appreciably faster.

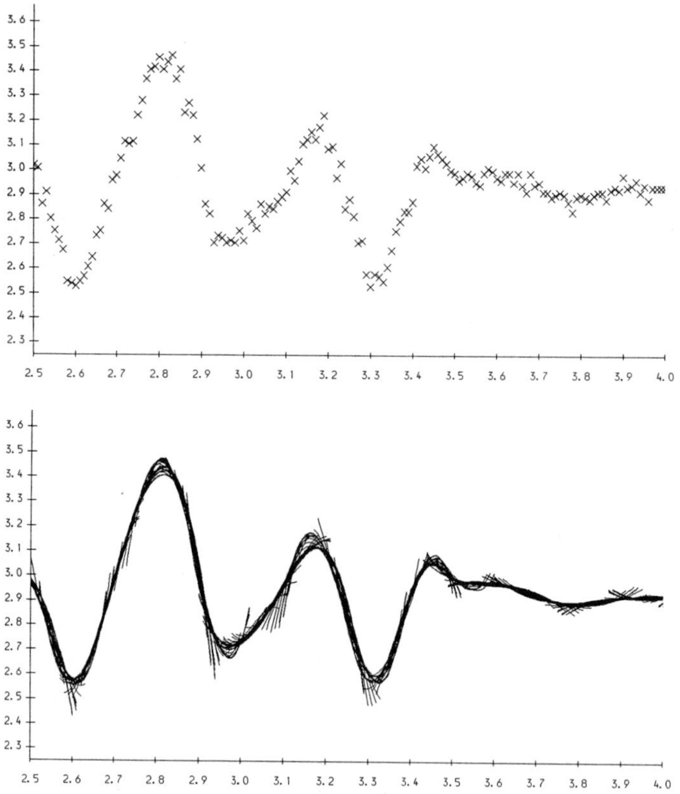

FIG. 1. *A subset of the data obtained from measurements of the response of an oscilloscope to a step input function and local polynomial fits obtained using a trend test.*

Figure 1 shows the local polynomial fits as produced by the algorithm for a set of data generated by measuring the response of an oscilloscope to a step input function. The complete data set consists of 512 points with equispaced abscissa values but we show here only a subset of the data in order to illustrate more clearly the local polynomials. The first diagram shows this subset and the second the local polynomial fits of order 4 associated with these and neighbouring points. We notice that the aggregate of the simple polynomial fits appears to be adequately describing the underlying function. In the following section we use the complete data set in our knot placement strategy.

3. An Algorithm for Knot Placement

In this section we show how we may use the information generated by the algorithm presented in Section 2 to determine an initial trial set of knots. These knots are used in a least squares spline fit to the data, and may be subsequently updated according to (1) to give improved fits.

Suppose we compute the local data sets associated with all sufficiently interior data points using polynomial fits of order n. Let the first and last data point in each local data set mark the *lower extent* and *upper extent*, respectively, of that set. Then, we position a knot at any point which is the lower (or upper) extent of more than two local data sets associated with consecutive data points. In this way we derive not only an estimate for N but also for the initial knot positions λ_j, $j = 1, \ldots, N$.

If the local data set for the ith point is determined independently of those for the other points, then this strategy clearly gives a vector $\boldsymbol{\lambda}$ whose elements do not depend on the order in which the points are processed. However, we observed in Section 2 that for reasons of efficiency, we use the size of the final local data set for the $(i-1)^{th}$ point as a starting value for the size of the local data set for the ith point. Thus, the points are necessarily processed from left to right and there is a dependence between the local data sets for consecutive data points. A better criterion is to locate a knot at an extent of a local data set which is *greater than or equal to* the extents of more than two subsequent and consecutive local data sets. Similarly, if the points were processed from right to left, we would use a "less than or equal to" test here.

Figure 2(a) shows the result of applying the knot placement strategy to the oscilloscope data, part of which is illustrated in Figure 1. The data points (small crosses) are shown together with the knot positions (vertical bars) and the fourth order spline fit to the data with these knots. There are 70 knots and the root mean square residual in the fit is 0.0543. We notice that in some regions there are *clusters* of knots and these can cause *overfitting* of the data. We say that overfitting occurs when the spline fit passes so close to the data that it begins to model the noise.

To remove the clusters we use a simple strategy which requires that each knot interval contains at least n ($= 4$, in this example) data points. If there are fewer than n points in an interval, the right-hand knot is removed (excepting the last knot interval where it is the left-hand knot which is removed). As a result, we are left with 44 knots and a spline fit as shown in Figure 2(b). The root mean square residual in this fit is 0.0605. The fit looks much more acceptable than that of Figure 2(a) because the oscillations in the spline that were previously present have been eliminated or reduced.

Finally, if we use these 44 knots to initialize the adaptive knot placement algorithm (1), and continue to remove clusters of knots as they are generated, we produce after 4 iterations the knots and associated spline fit shown in Figure 2(c). There are 42 knots here and the root mean square residual in the fit is 0.0618. Indeed, the use of (1) has produced little improvement over the set of 44 knots. We note, however, that the distribution of the reduced set of knots is markedly different.

Figure 3 shows the results at the same stages for another practical set of data.

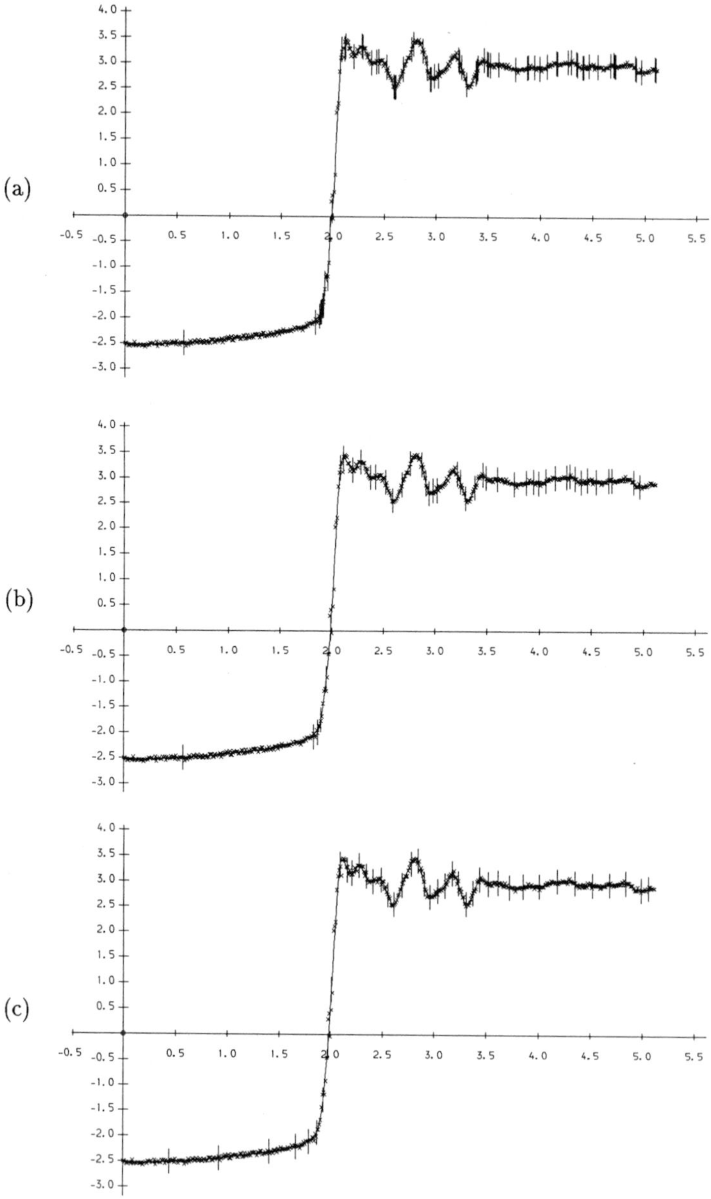

FIG. 2. *Automatic knot placement applied to oscilloscope data:*
(a) Initial knots (70 knots, r.m.s. residual = 0.0543*).*
(b) After clusters are removed (44 knots, r.m.s. residual = 0.0605*).*
(c) After adaptive knot placement (42 knots, r.m.s. residual = 0.0618*).*

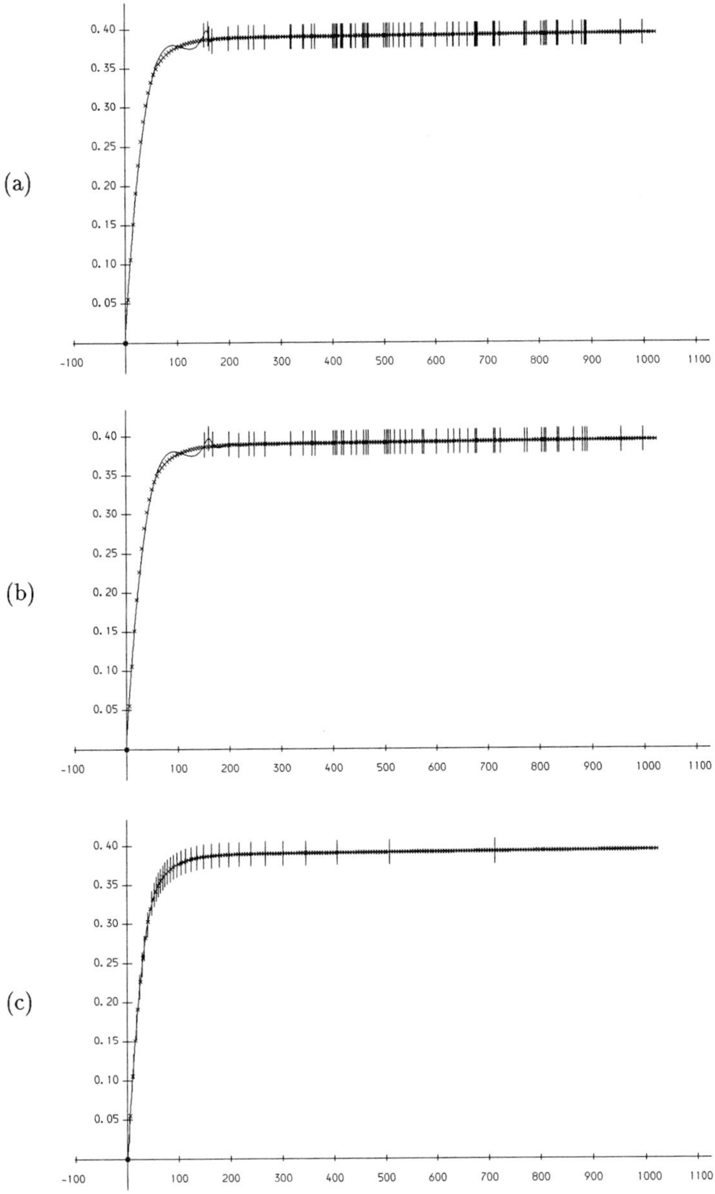

FIG. 3. *Automatic knot placement applied to photodiode data:*
(a) Initial knots (79 knots, r.m.s. residual = 0.00280*).*
(b) After clusters are removed (55 knots, r.m.s. residual = 0.00295*).*
(c) After adaptive knot placement (34 knots, r.m.s. residual = 2.73×10^{-7}*).*

This arises from very accurate measurements of the response of a photodiode. The data set consists of 1024 points with equispaced x_i-values. For clarity, we have plotted only every fifth point in the diagrams of Figure 3. A set of 79 knots is initially generated and this is reduced to 55 when clusters are removed as described above. The corresponding fits are illustrated in Figures 3(a) and (b) and the root mean square residuals are 0.00280 and 0.00295, respectively. In this example, the use of (1) gives a new set of knots which results in a very substantial improvement in the root mean square residual. Figure 3(c) shows the knots and corresponding spline fit after 4 iterations of adaptive knot placement. There are 34 knots here, and the root mean square residual in the fit is 2.73×10^{-7}. Adaptive knot placement has reduced the root mean square residual by a factor of 10,000 without overfitting. We observe that much of this improvement results from the strategy repositioning the knots in such a way as to remove the severe oscillation that previously existed in the "shoulder" of the curve.

4. Summary

In this paper we have presented a strategy for automatically deriving a knot vector $\boldsymbol{\lambda}$ to be used in constructing a least squares spline fit of given order to noisy data. We determine subsets of the complete data set which are adequately fitted by simple polynomials. These polynomials provide local information about the underlying function which the data represents. By considering how these subsets, or local data sets, interact we generate initial estimates for N and λ_j, $j = 1, \ldots, N$. With these initial values we may then use an adaptive knot placement algorithm to produce improved knot distributions. The particular adaptive strategy we use here relocates knots according to the criterion (1) and removes knots to prevent clustering. The examples we have given illustrate some of the features of our approach. For further details and examples, see Cox, Harris and Jones (1988).

Acknowledgement

We would like to thank Mr. G. T. Anthony for his comments on an earlier draft of this paper.

References

1. C.W. Clenshaw. Curve fitting with a digital computer. *Comp. J.*, 2:170–173, 1960.
2. M.G. Cox, P.M. Harris, and H.M. Jones. Strategies for knot placement in least squares data fitting by splines. Technical Report DITC 101/87, National Physical Laboratory, 1987.
3. M.G. Cox, P.M. Harris, and H.M. Jones. A knot placement strategy for least squares spline fitting based on the use of local polynomial approximations. National Physical Laboratory Technical Report, in preparation, 1988.

4. C. de Boor. *A Practical Guide to Splines.* Springer-Verlag, New York, 1978.
5. G.E. Forsythe. Generation and use of orthogonal polynomials for data fitting with a digital computer. *J. Soc. Indust. Appl. Math.*, 5:74–88, 1957.
6. M.J.D. Powell. Curve fitting by splines in one variable. In J.G. Hayes, editor, *Numerical Approximation to Functions and Data*, pages 65–83. The Athlone Press, London, 1970.

An algorithm for nonlinear splines with non-negativity constraints

G. Opfer

Institut für Angwandte Mathematik,
Universität Hamburg, FRG

Abstract The problem of minimizing the exact form of the strain energy of a "spline" subject to non-negativity constraints is treated. Several necessary optimality conditions in terms of systems of differential equations are derived. Due to the constraints, additional boundary conditions have to be introduced at points which are not known in advance. An algorithms is outlined, which we call a *local* algorithm, which is based on only two grid points.

Key words: Curvature functional, Elastica with constraints, Global spline algorithm, Local spline problem, Nonlinear splines, Non-negativity constraints, Obstacle problem, Splines with constraints.

1. Introduction

In many applications, splines should not only pass through given data points (t_j, x_j), $j = 1, 2, \ldots, n$ but should also satisfy some additional conditions which are prescribed for the given problem. For example, if one wants to model a density function [cf. Feller (1971, p.3)] the model is useless unless it can produce non-negative values. A more general type of restriction is one which requires a fit to be "visually pleasing" [cf. Carlson (1987)] and the meaning of this may vary from problem to problem.

One approach to such a problem is motivated by physical arguments. The curve one is interested in is regarded as a flexible ruler or beam (draughtsman's spline) (cf. Malcolm (1977), de Boor (1978, p.67)) and thus, the strain energy has to be minimized, of course subject to the given restrictions.

The second approach is motivated more by practical arguments. In this approach one constructs a spline directly, such that it fits all the given requirements. This is ordinarily done by choosing additionally introduced parameters suitably.

For splines with no restrictions apart from the interpolatory conditions, the first approach was followed for example by Golomb and Jerome (1982). For non-negativity constraints in connection with the simplified strain energy $\int_a^b \ddot{x}(t)^2 dt$ see Opfer and Oberle (1988). For results in the case of minimizing $\int_a^b \{x^{(k)}(t)\}^2 dt$ restricted to $x^{(m)} \geq 0$ see Dontchev and Kalchev (1988). Recent contributions to the second approach are given by Rentrop and Wever (1988) and Schaback (1989). The case where the functional to be minimized and the constraints are all of general form is treated by Opfer (1989).

In this paper we shall follow the concept of minimizing the strain energy, which is expressed by the integral over the square of the curvature of the corresponding curve, subject to non-negativity constraints. Thus the curves to be investigated cannot penetrate the x-axis, and this type of problem may be regarded as an *obstacle* problem.

For brevity and simplicity our problem will be posed only for functions rather than curves.

2. The Problem

Given *data* (t_j, x_j), $j = 1, 2, \ldots, n \geq 2$, with $a = t_1 < t_2 < \cdots < t_n = b$ and n fixed, we require a function $x \in W_2^2$ with $x(t_j) = x_j$, $j = 1, 2, \ldots, n$, which, if parametrized with respect to arc length s, minimizes

$$f(x) = \int_0^\ell \kappa^2(s)ds \tag{1}$$

subject to

$$x(s) \geq 0, \tag{2}$$

where κ is the curvature of x and ℓ the length of the curve described by x. It should be pointed out that in this setting neither the length ℓ of the resulting curve x nor the interpolation conditions (which have the form $x(s_j) = x_j$) are known in advance, since the arc lengths s_j, which are the lengths from the first point to the point numbered j measured on x, are not known beforehand.

It is mentioned in various papers [e.g. Golomb and Jerome (1982)] that the corresponding Euler equation for the unrestricted case can be put (by suitable transformation) in the form of the pendulum equation, with the consequence that the minimum of the above mentioned functional (if it exists at all) is a piecewise elliptic integral. However, very few authors use this result. One exception is K.D. Reinsch (1981) who takes advantage of the existing procedures for elliptic functions presented by Bulirsch (1965).

Most of the numerical work in the unrestricted case was discussed by Malcolm (1977). In particular, results by Glass (1966), Larkin (1966), Lee and

Forsythe (1973), Mehlum (1969), Woodford (1969) are mentioned. We are not aware of any more recent published results.

The above setting in terms of the arc length s has the advantage of formal simplicity; the disadvantage already mentioned is that the upper bound for the integral in (1) and the interpolation conditions contain a quantity unknown a priori. In this paper we have decided to use rectangular coordinates and restrict our attention to curves which are graphs of functions.

Given $x \in C^2[a,b]$, the curvature of x at a point $t \in [a,b]$ is defined by

$$\kappa(x(t)) = \frac{\ddot{x}(t)}{(1+\dot{x}(t)^2)^{3/2}}. \tag{3}$$

If we use the transformation

$$ds = \sqrt{1+\dot{x}(t)^2}\,dt, \tag{4}$$

the above functional (1) reads

$$f(x) = \int_a^b \frac{\ddot{x}(t)^2\,dt}{(1+\dot{x}(t)^2)^{5/2}}. \tag{5}$$

By the introduction of the "spaces"

$$M = \{x \in W_2^2 : \quad x(t_j) = x_j,\ j = 1,2,\dots,n\}, \tag{6}$$

$$H = \{x \in W_2^2 : \quad x(t_j) = 0,\ j = 1,2,\dots,n\}, \tag{7}$$

$$M^+ = \{x \in M : \quad x \geq 0\}, \tag{8}$$

we can give the problem the brief form:

$$\text{minimize } f(x) \text{ given by (5) subject to } x \in M^+. \tag{9}$$

For later reference we abbreviate the integrand of (5) as

$$F(x) = F(t,x,\dot{x},\ddot{x}) = \frac{\ddot{x}(t)^2}{(1+\dot{x}(t)^2)^{5/2}}. \tag{10}$$

There is one important subproblem which we term the *local* problem.

Local Problem. *Specify problem (9), but with $n = 2$ and with prescribed additional boundary conditions*

$$\dot{x}(t_1) = \dot{x}_1, \quad \dot{x}(t_2) = \dot{x}_2, \tag{11}$$

where $\dot{x}_1$, $\dot{x}_2$ are any given values.

If we could solve the local problem satisfactorily, we could already set up an algorithm for finding the solution of the general problem, which in this context we term the *global* problem. The algorithm would have the form:

Step 1: Solve the unrestricted problem, and call the solution x.

Step 2: Replace the solution x in those intervals $I_j = [t_j, t_{j+1}]$ in which x has negative values by the corresponding solution of the local problem, by using the computed derivatives at the endpoints as prescribed boundary conditions.

Step 3: Repeat Steps 1 and 2 by taking all derivatives at the given knots as unknowns, with the aim of producing a C^2-spline.

It should be noted that the first two steps alone already produce a C^1-spline. For the simplified strain energy $\int_a^b \ddot{x}(t)^2 dt$, this algorithm has been proposed by C. Reinsch (1988) and by Fischer, Opfer and Puri (1987). These authors show that the corresponding local solution can be computed explicitly in a very simple and efficient manner. For this simplified case, the algorithm has undergone extensive numerical testing by Dauner (1987) and Kröger (1989).

The local problem is a special case of the so-called *Hermite*-problem, in which, in addition to the already stated conditions, we require that $\dot{x}(t_j) = \dot{x}_j$, where $\dot{x}_j$, $j = 1, 2, \ldots, n$ are arbitrarily specified numbers.

3. Necessary Optimality Conditions

If one of the data points should fall below the x-axis, then our problem certainly has no solution. The following assumption on the data specification is therefore immediate.

Assumption 1. *Assume* $M^+ \neq \emptyset$, *i.e.,* $x_j \geq 0$ for all $j = 1, 2, \ldots, n$.

For our first result we need two more "spaces", the definition of which depends only on $x \in M^+$, namely

$$H_x^+ = \{h \in H: \quad x + h \geq 0\}, \tag{12}$$

$$H_x^{++} = \{h \in H_x^+: \quad x - h \geq 0\}. \tag{13}$$

The latter space H_x^{++} is usually referred to as the *envelope* or *hull* of x, since it contains all those h for which $|h| \leq x$. It has the properties

$$h \in H_x^{++} \iff -h \in H_x^{++}, \tag{14a}$$

$$h \in H_x^{++}, x(\tau) = 0,\ a < \tau < b \implies h(\tau) = \dot{h}(\tau) = 0 \leq \ddot{h}(\tau). \tag{14b}$$

Theorem 1. *Suppose that* x_0 *solves the problem stated in (9). Then*

$$f'(x_0, h) := \lim_{\substack{\alpha>0 \\ \alpha\to 0}} \frac{1}{\alpha}\{f(x_0 + \alpha h) - f(x_0)\} = \int_a^b \{F_{\dot{x}}(x_0)\dot{h}(t) + F_{\ddot{x}}(x_0)\ddot{h}(t)\}dt$$

$$\geq 0 \ \text{for all } h \in H_{x_0}^+, \tag{a}$$

and

$$= 0 \ \text{for all } h \in H_{x_0}^{++}. \tag{b}$$

Proof: (a) By the definition of $H_{x_0}^+$ we have $x_0 + \alpha h \in M^+$ for all $0 \le \alpha \le 1$. Since in that case, by definition of x_0, we have $f(x_0) \le f(x_0 + \alpha h)$, the assertion follows from the definition of the derivative of f.
(b) This follows from (a), by using (14a) in connection with the linearity of $f'(x_0, \cdot)$ with respect to h. ■

In order to proceed from Theorem 1 we choose a fixed subinterval $I_j = [t_j, t_{j+1}]$, $j = 1, 2, \dots, n-1$ and an $h \in H_{x_0}^{++}$ which vanishes outside I_j. If $j > 1$ and $j + 1 < n$ then $h \in W_2^2$ implies $h^{(\ell)}(t_j) = h^{(\ell)}(t_{j+1}) = 0$ for $\ell = 0, 1$. By applying partial integration, Theorem 1 implies that for all $j = 1, 2, \dots, n-1$

$$\int_{t_j}^{t_{j+1}} \tilde{F}(x_0)\ddot{h}(t)dt = 0, \text{ for } h \in H_{x_0}^{++}, \text{ and } h(t) = 0 \text{ for } t \notin I_j, \tag{15}$$

where in each interval

$$\tilde{F}(x_0) = F_{\ddot{x}}(x_0) - v, \tag{16a}$$

$$\dot{v} = F_{\dot{x}}(x_0), \tag{16b}$$

and

$$v(a) = v(b) = 0, \tag{16c}$$

where F is defined in (10). In order to obtain (15) we use the relation

$$\begin{aligned} \int_{t_j}^{t_{j+1}} F_{\dot{x}}(x_0)\dot{h}(t) &= \dot{h}(t)v(t)\Big|_{t_j}^{t_{j+1}} - \int_{t_j}^{t_{j+1}} v(t)\ddot{h}(t)dt \\ &= -\int_{t_j}^{t_{j+1}} v(t)\ddot{h}(t)dt. \end{aligned}$$

If $1 < j < n-1$ then $\dot{h}(t_j) = 0$, and thus the above constant parts vanish. If, however, $j = 1$ or $j = n$ then $\dot{h}(t_j)$ may take any value. In this case we note the fact that v defined in (16b) is determined only up to a constant, and this allows us to set $v(a) = v(b) = 0$, at least provided that we have $n > 2$.

For later use we compute the partial derivatives of F:

$$F_{\dot{x}} = \frac{-5\dot{x}\ddot{x}^2}{(1+\dot{x}^2)^{7/2}}, \quad F_{\ddot{x}} = \frac{2\ddot{x}}{(1+\dot{x}^2)^{5/2}}. \tag{17}$$

Theorem 2. *(a) For $n = 2$ the solution x_0 of (9) is the linear interpolant of the two data (t_1, x_1) and (t_2, x_2). (b) For $n > 2$, if x_0 solves the problem (9) and if v is defined as in (16b) and (16c), then*

(i) $F_{\ddot{x}}(a) = F_{\ddot{x}}(b) = 0$,
(ii) $F_{\ddot{x}} - v$ *is linear in* $I_j = [t_j, t_{j+1}]$, *provided* $x_0 > 0$ *in* I_j,
(iii) $F_{\ddot{x}} - v$ *is linear between interior zeros of* x_0, *and linear between interior zeros and knots, where an interior zero of* x_0 *is a zero in the open interval* $]t_j, t_{j+1}[$,
(iv) $\frac{d}{dt}(F_{\ddot{x}} - v)$ *always has non-negative jumps at the interior zeros of* x_0.

Proof: (a) In this case the proof follows because $x_0 \in M^+$ and $f(x_0) = 0$. (b) An application of a theorem of du Bois-Reymond tells us that in each subinterval I_j the expression $\tilde{F}$ is a polynomial of degree 1 in t. In addition, if τ is a zero of x_0 then τ has to be introduced as a *new knot*, since we have $h \in H_{x_0}^{++}$ together with the derivative $\dot{h}$ vanishing at τ. The remaining details of the proof follow the corresponding proofs in Opfer and Oberle (1988). ■

Condition (i) corresponds to the so-called "natural boundary conditions" of an ordinary spline. In the case of an ordinary cubic spline, condition (ii) is the condition that the second derivative should be piecewise linear. The remaining two conditions cover the constraints. Condition (iii) implies that any zero of the solution x_0 must be considered as a new knot. The jump condition (iv) means that the difference between the right and left derivative at a new knot must always be non-negative.

Condition (ii) reads explicitly

$$2\ddot{x}(t)/(1+\dot{x}(t)^2)^{5/2} - v(t) = At + B, \tag{18}$$

where

$$-5\dot{x}\ddot{x}^2/(1+\dot{x}^2)^{7/2} = \dot{v}.$$

Differentiation with respect to t yields

$$(2(1+\dot{x}^2)x^{(iii)} - 5\dot{x}\ddot{x}^2)/(1+\dot{x}^2)^{7/2} = A. \tag{19}$$

If we differentiate once more we obtain Euler's equation

$$(1+\dot{x}^2)\{2(1+\dot{x}^2)x^{(iv)} - 20\dot{x}\ddot{x}x^{(iii)} - 5\ddot{x}^3\} + 35\dot{x}^2\ddot{x}^3 = 0. \tag{20}$$

If $\ddot{x} \neq 0$ we may multiply equation (19) by $\ddot{x}$ and obtain

$$\frac{d}{dt}\frac{\ddot{x}^2}{(1+\dot{x}^2)^{5/2}} = \frac{2\ddot{x}x^{(iii)}}{(1+\dot{x}^2)^{5/2}} - \frac{5\dot{x}\ddot{x}^3}{(1+\dot{x}^2)^{7/2}} = A\ddot{x}.$$

Thus, by integrating we obtain

$$(A\dot{x} + C)(1+\dot{x}^2)^{5/2} = \ddot{x}^2, \tag{21}$$

which is Woodford's (1969, eq. (5)) equation.

The first three differential equations, valid in each subinterval and connected by suitable boundary values can be written as first order systems which are more convenient for computing. Equation (18) reads

$$\dot{x} = u, \tag{18a}$$

$$\dot{u} = 0.5(1+u^2)^{5/2}(v + At + B), \tag{18b}$$

$$\dot{v} = -1.25u(1+u^2)^{3/2}(v + At + B)^2. \tag{18c}$$

Equation (19) takes the form

$$\dot{x} = u, \tag{19a}$$

$$\dot{u} = v, \tag{19b}$$

$$\dot{v} = 0.5A(1+u^2)^{5/2} + 2.5uv^2/(1+u^2). \tag{19c}$$

Finally, Euler's equation (20) is equivalent to

$$\dot{x} = u, \tag{20a}$$
$$\dot{u} = v, \tag{20b}$$
$$\dot{v} = w, \tag{20c}$$
$$\dot{w} = (10uvw + 2.5v^3)/(1+u^2) - 17.5u^2v^3/(1+u^2)^2. \tag{20d}$$

In addition to these equations we have the boundary conditions (16c), part *(i)* of Theorem 2, the interpolation conditions and the smoothness conditions at the interior knots. Moreover we have to introduce new knots τ, $t_j < \tau < t_{j+1}$ (for some j), whenever τ is an interior zero of x_0. For such points τ, we have the additional conditions $\dot{x}_0(\tau) = 0$, $\ddot{x}_0(\tau) \geq 0$. The latter conditions reflect the fact that x_0 has not only a zero at τ but also a minimum.

The implied algorithm requires the repeated solution of one of the above sets of differential equations with additional known and unknown boundary conditions. Results in this direction will be published elsewhere.

Acknowledgments

The author would like to thank the editors for valuable help in preparing this manuscript.

References

1. de Boor, C., *A Practical Guide to Splines*, Springer, New York, Heidelberg, Berlin, 1978.
2. Bulirsch, R., Numerical Calculation of Elliptic Integrals and Elliptic Functions, Numer. Math. **7**, 1965, 78–90.
3. Carlson, R.E., Shape Preserving Interpolation, in: Mason, J.C. and Cox, M.G. (eds.): *Algorithms for Approximation*, Proceedings of IMA Conference on Algorithms for the Approximation of Functions and Data, Shrivenham, England (1985), Clarendon Press, Oxford, 1987, 97–114.
4. Dauner, H., Formerhaltende Interpolation mit kubischen Splinefunktionen, Diploma Thesis, Technische Universität München, 1987.
5. Dontchev, A.L. and Kalchev, Bl., On Convex and Nonnegative Best Interpolation, C. R. Acad. Bulgare Sci. **41** (1988), 21–24.
6. Feller, W., *An Introduction to Probability Theory and its Applications*, Vol. 2, 2nd ed., Wiley, New York, London, Sydney, Toronto, 1971.
7. Fischer, B., Opfer, G. and Puri, M.L., A Local Algorithm for Constructing Non-negative Cubic Splines, Hamburger Beiträge zur Angewandten Mathematik, Reihe A, Preprint 7, 1987, 18p, to appear in J. Approx. Theory.
8. Glass, J.M., Smooth-Curve Interpolation: A Generalized Spline-Fit Procedure, BIT **6** (1966), 277–293.

9. Golomb, M. and Jerome, J., Equilibria of the Curvature Functional and Manifolds of Nonlinear Interpolating Spline Curves, SIAM J. Math. Anal. **13** (1982), 421–458.
10. Kröger, N., Nicht-negative Splines und deren numerische Berechnung, Diploma Thesis, Universität Hamburg, 1989.
11. Larkin, F.M., An Interpolation Procedure Based on Fitting Elastica to Given Data Points, COS Note 5/66, Theory Division, Culham Laboratory, Abingdon, England, 1966, 15 p.
12. Lee, E.H. and Forsythe, G.E., Variational Study of Nonlinear Spline Curves, SIAM Review **15** (1973), 120–133.
13. Malcolm, M.A., On the Computation of Nonlinear Spline Functions, SIAM J. Numer. Anal. **14** (1977), 254–282.
14. Mehlum, E., Curve and Surface Fitting Based on Variational Criteriae for Smoothness, Central Institute for Industrial Research, Oslo, 1969, 83 p.
15. Opfer, G., Necessary Optimality Conditions for Splines under Constraints, in: Chui, C.K., Schumaker, L.L., Ward, J.D. (eds.): *Approximation Theory VI*, Academic Press, New York, 1989, to appear.
16. Opfer, G. and Oberle, H.J., The Derivation of Cubic Splines with Obstacles by Methods of Optimization and Optimal Control, Numer. Math. **52** (1988), 17–31.
17. Reinsch, C., Software for Shape Preserving Spline Interpolation, Mscr. 1988, 12p.
18. Reinsch, K.D., Numerische Berechnung von Biegelinien in der Ebene, Dissertation, Technische Universität München, 1981, 246 p.
19. Rentrop, P. and Wever, U., Computational Strategies for the Tension Parameters of the Exponential Spline, in: Bulirsch, R., Miele, A., Stoer, J., Well, K. (eds.):*Optimal Contol*, Lecture Notes in Control and Information Sciences, **95** (1987), Springer, Berlin, Heidelberg, New York, London, Paris, Tokyo, 122–134.
20. Schaback, R., Convergence of Planar Curve Interpolation Schemes, in: Chui, C.K., Schumaker, L.L., Ward, J.D. (eds.): *Approximation Theory VI*, Academic Press, New York, 1989, to appear.
21. Woodford, C.H., Smooth Curve Interpolation, BIT **9**, (1969), 69–77.

Spline curve fitting of digitized contours

C. Potier and C. Vercken

Départment Informatique,
Ecole Nationale Supérieure des Télécommunications,
Paris, France

Abstract This paper presents an algorithm for finding a mathematical curve approximation to a digitized contour. The whole application consists in processing optically scanned graphic documents to store them in a data base for consultation and manipulation. For each contour, an approximating parametric cubic spline curve, with as few control points as possible, is determined by minimizing a smoothing criterion. The algorithm is particularly efficient for large data sets, the number of control points always being very small. To handle the curve, the B-spline representation is associated to a hierarchical data structure, obtained by generating the Bézier points and subdividing the corresponding Bézier segments.

Key words : Cubic splines, Curve fitting, Data compression.

1. Introduction

The algorithm presented is a part of a software project at our school concerned with the consultation and manipulation of documents archived in digital form. Documents which were not created in this form are to be converted to a compatible form. To digitize them, we use a scanner whose resolution is 300 dots per inch. The output, a binary image, is an array of pixels (approximately 2400*3500 for a A4 document) without any information about the structure of the document. For archiving purposes, data compression is necessary, and even run length encoded forms such as CCITT group III or IV, used for facsimile coding and transmitting, represent a very large amount of data and give no information about the structure of the document. For black and white graphics an analytical representation of contours is useful for data compression, shape description and geometrical transformation. To deal with complex documents it would be necessary to have preliminary processing, which is outside the scope of this paper, such as segmentation into text, black and white graphics and half tone zones as described by Ito and Saakatani (1982) and determination of the connected components of the graphic zones as described by Ronse and Devijver (1984). Since we presently have studied only very simple graphics including a single component, a minimal basic preprocessing is necessary to determine an ordered list of black pixels representing a contour or a curve and to convert it into a continuous boundary by vectorization. The next step consists in finding a good spline approximation by first determining a small accurate set of knots and then adjusting the B-spline coefficients.

2. Contour extraction and polygonal approximation

Let us recall some basics in binary image processing which are given, for instance, by Pavlidis (1982) :

- Background pixels (white) have value 0 and object pixels (black) have value 1.
- An "object" connected component is a set P of black pixels adjacent to an edge (D-neighbours) or to a corner (I-neighbours) .

A connected component is entirely determined by its external contour and possibly by internal ones if there are "holes".

2.1 Contour extraction

Definition : Let P be a connected component. The contours C of P are the sets of adjacent pixels in P which have at least one D-neighbor not in P.

A contour C can be represented by a closed path that can be determined by the following algorithm proposed by Pavlidis (1982) : "An observer walks along pixels belonging to the set and selects the rightmost available until the current pixel is the initial pixel". This initial pixel is usually found by a top-to-bottom, left-to-right scan. The algorithm must be applied for each contour of the set P.

2.2 Line thinning

Given the scanner resolution and the graphics attributes (pen-size) a line is scarcely ever one pixel wide and often has interior pixels. Since the contour is adequate to describe a filled shape but not a line, it is necessary to distinguish shapes from lines that are described by the skeleton of the component and its width. A maximum thickness T is set and the contour is peeled off by removing all contours points that are not "skeletal", following Pavlidis (1982). This connectivity-preserving peeling is repeated at most T times and, if during this process no points can be removed, the skeleton is obtained and the component is considered as a line of thickness $T' \leq T$; otherwise it is a filled shape. Henceforth, line skeletons and shape contours will be processed as closed or open lines.

2.3 Vectorization

The next step consists in converting the discrete pixel-chain representation of lines, possibly closed, into a continuous very fine polygonal approximation (vectorization) which vertices are the end points of "straight line" segments of the grid. We determine the vertices by using the sequential method, based on the minimal area deviation criterion, proposed by Wall and Danielsson (1984), with a small tolerance value S :

Algorithm 1:

(a) Let $M_0(x_0,y_0)$ be the starting point and M_1 its contour successor.
Initiate the algebraic area f_1 of $(M_0M_0M_1)$ to 0. Let $i = 2$.

(b) Calculate the algebraic area Δf_i of $(M_0M_{i-1}M_i)$ and the cumulative area $f_i=f_{i-1}+\Delta f_i$.
Calculate the length L_i from M_0 to M_i.

(c) if $f_i \leq S\, L_i$ then increment i and repeat (b).
Otherwise put M_{i-1} in the set of vertices and take it as new starting point.

The detection of "characteristic points" is done simultaneously. At any time we keep the segment's general direction, as one of the 8-connectivity, and we mark the points at which a true direction change (not temporary for a single point) is found. These "characteristic points" are taken as new starting points rather than the last point satisfying the criterion .

3. Curve fitting

We want to determine a parametric curve, fitted to the ordered sampled points $\{P_i\}_{i=1}^{N}$ obtained by the preprocessing, which can be written using B-spline basis: $M(t) = \sum_{i=1}^{n} M_i\, B_{i,4}(t)$, for $t \in [T_0, T_1]$, with the number n of basis functions as small as possible and where $B_{i,4}$ are cubic normalized B-splines .

To fit a parametric curve to a set of points $\{P_i\}_{i=1}^{N}$, any data point P_i must be assigned a parameter value u_i $(u_1<u_2<\ldots<u_N)$ that we suppose to vary from 0 to T for sake of simplicity. The chord-length parametrization needs more arithmetic operations than

uniform parametrization but it is more natural since the points P_i are unevenly spaced and the contour of the curve is very close to the polyline joining the points P_i.

3.1 Knots determination

To determine n basis functions $\{B_{i,4}\}_{i=1}^{n}$, a subdivision $t_1 \leq t_2 \leq \ldots \leq t_{n+4}$ of $[0,T]$ is necessary and is automatically searched while keeping a trade-off between the closeness of the fit and the number of basis functions. There must be as few knots as possible but the mean square distance to the data points must be no greater than a tolerance level. Three ways to obtain the knots placement have been compared:

- In the first method, the subdivision $\{t_j\}$ is determined by using the minimal area deviation criterion of Wall and Danielsson (1984), as described above, with a sufficiently large tolerance value S.

Figure 1: Subdivision obtained by algorithm 1

Figure 2: Subdivision obtained by algorithm 1 used twice

The subdivision $\{t_j\}_{j=1}^{n}$ obtained by this method is not symmetric, the "beginning" and the "end" of the sampling do not play the same part (fig.1). To avoid this, it is possible to use the algorithm twice, forward and back, and to calculate the mean of the two subdivisions. The subdivision obtained by using this algorithm twice is more regular (fig.2).

- In the second method, knots are added until, on each subinterval $I=[T_i,T_{i+1}]$, E_I the mean square distance of points $P_j \in I$ to the least-squares straight line is less than a parameter value E_{max}.

Algorithm 2

(a) Start with one subinterval $[T_1,T_2]$; call M the collection of subintervals.

(b) On each $I \in M$, calculate the rms E_I to the least-squares straight line.

(c) If $E_I > E_{max}$ then split I in two and put both pieces into M, otherwise increment I.

The algorithm is repeated from (b) until all subintervals are processed.

The subdivision obtained with this algorithm (fig.3) is not very regular, some intervals seem to be too small. This algorithm should be improved with a "merge" algorithm (see

Pavlidis,1974) to merge two consecutive intervals if possible. Split and merge algorithms produce good results but are time-consuming since they require multiple passes through the data.

- The third method is based on least-squares fitting by cubic polynomials in X and Y.

Algorihm 3

Let S be a parameter value.

(a) Start with one subinterval $[T_1,T_2]$; call M the collection of subintervals.

(b) On each interval $I \in M$, calculate 2 least-squares cubic polynomials (P_x and P_y) for X and Y.

Calculate $D_x=\Sigma_j (P_x (u_j) - X_j)^2$ and $D_y=\Sigma_j (P_y (u_j) - Y_j)^2$.

(c) If $D_x>S$ or $D_y>S$ then split the interval in two and put both pieces into M, otherwise increment I. Return to (b).

(d) Subdivide each interval $[T_i,T_{i+1}]$ into 4 to obtain the subdivision $t_1 \le t_2 \le \ldots \le t_{n+4}$.

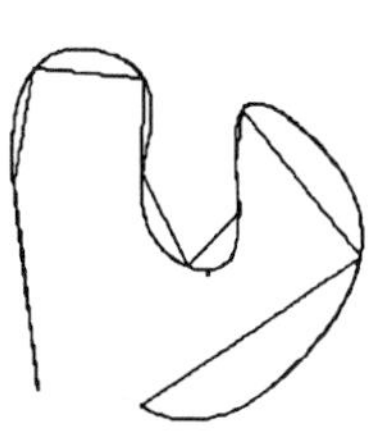

Figure 3 : Subdivision obtained by algorithm 2

Figure 4 : Subdivision obtained by algorithm 3

We use one of these algorithms to obtain a subdivision of intervals between two "characteristic" points: either endpoints or points marked during the preprocessing. After obtaining the subdivision, we add multiple (triple) knots corresponding to marked points. Multiple knots are also added at the ends of an open curve and cyclic knots are added if the curve is closed.

In all cases, the number of knots n+4 is much smaller than the number N of data points P_i and generally subdivisions are quite similar (fig.1 to 4). Nevertheless algorithm 3 should produce better results in case of lightly snaky data. With algorithm 1 the subdivision is not optimal; however this method is attractive since it is less time-consuming as shown in Tab.1.

Table 1: Time to obtain the subdivision.

Algorithm	1	2	3
Time	3 seconds	6 seconds	25 seconds

3.2 Fitting a parametric spline curve

The third step of the fitting process is to determine the parametric spline which must be close to the points P_j and "smooth". The control points $\{M_i\}_{i=1}^{n}$ are searched while minimizing the quadratic functional:

$$I_2(M) = \sum_{j=1}^{N} [M(u_j) - P_j]^2 + \mu \int_0^T \{ [X'']^2 + [Y'']^2 \} \, dt$$

The positive parameter μ controls the trade-off between the sum of the distances of points $\{P_i\}_{i=1}^{N}$ from the curve, measured by the first term of I_2, and the smoothness of the curve, measured by the integral, which is a rough approximation to the curvature. However this functional I_2 involves two independent minimization problems to determine X and Y. The solution is easily obtained (Potier and Vercken, 1985) by solving two linear systems : $Qa=z_x$ and $Qb=z_y$ where the matrix Q is a symmetric positive matrix. Moreover, in the case of open curves, Q is banded, whereas in the case of closed curves there are non-zero terms at opposite corners, as shown in figure 5 (in case of quadratic splines). The solution of both systems can be obtained by factorizing the matrix Q with a data structure adaptated to the matrix (fig.6) and takes O(n) operations.

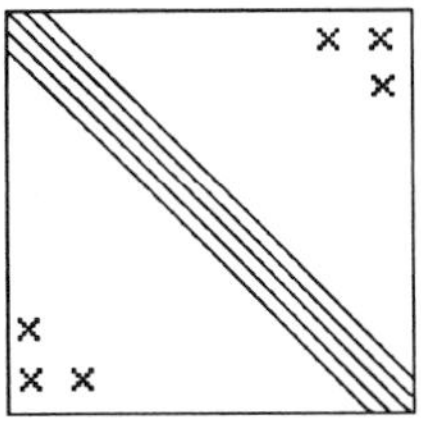

Figure 5 : Matrix Q in cyclic case

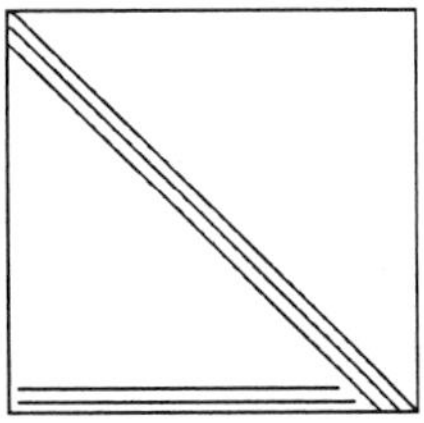

Figure 6 : Cholesky factorisation

The parameter μ is adjusted so that the mean square distance to the data points is no greater than a value T which may be a number of pixels. In figures 7 and 8 the value of T corresponds approximately to a tolerance of 2 pixels. On both figures, the initial curve, the B-spline approximation and the control polygon are drawn. The initial curve of figure 7 was designed with 9 B-splines, the digitized picture contained about 5000 pixels and the skeleton 1139 pixels, the vectorization gave 120 segments. The fitting by 11 B-splines, with knots obtained from algorithm 1, is very accurate.

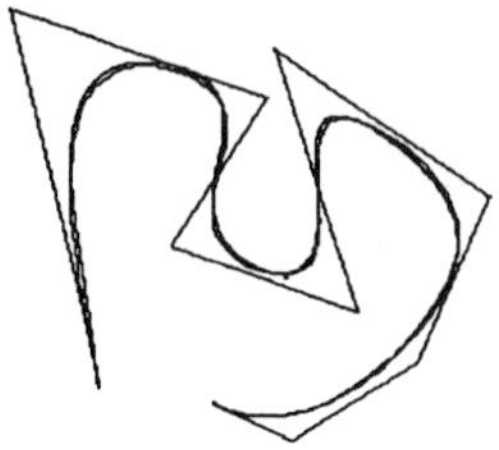

Figure 7 : 1139 pixels
and 11 B-splines

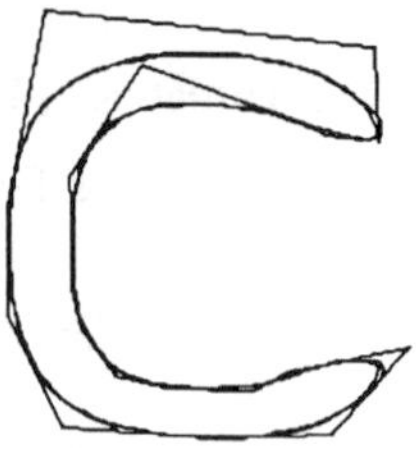

Figure 8 : 781 pixels
and 20 B-splines

However we did not succeed in finding a good approximation with fewer basis functions. However, figure 8 designed with 16 B-splines, algorithm 3 gave better subdivision than algorithm 1. The fitting with 16 B-splines kept the initial shape but smoothed the corners.

4. Interactive Curve Handling

This smoothing algorithm was integrated in an interactive graphics environment where performance of the algorithms is of fundamental importance. To make geometrical transformations on a displayed curve, the user has to identify the curve on the screen with a cursor pointed "near" the curve. The selection is done by testing the convex hull of the cubic Bézier composite representation of the curve obtained by the algorithm of Böhm (1981). By applying the De Casteljau algorithm, each cubic Bézier segment is split in two, if necessary, and this process is recursively applied. Successive splittings involve a hierarchical data representation of the curve (fig.9).

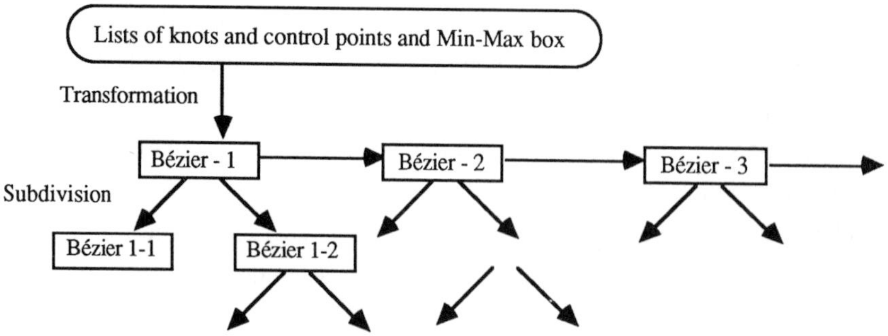

Figure 9 : Hierarchical data representation

The cubic Bézier representation of the curve can be used for printing "nicely" the curve on a PostScript-laserprinter since the PostScript operator "curveto" adds a cubic Bézier section to the current curve.

5. Conclusion

The fitting algorithm presented in this paper is very efficient and can easily be inserted in graphic documents processing software. The knots determination, which is crucial, will be developed and the knot removal algorithm of Lyche and Morken (1987) will be compared with other methods.

Moreover it would be very useful to extend this algorithm to more basis functions such as non-uniform rational B-splines to fit straight lines and conics.

References

Böhm W., Generating the Bézier points of B-spline curves and surfaces, Computer Aided Design, 13, n°6, November 1981, 365-366.

Ito S. and Saakatani S., Field Segmentation in Document Image, 6th IJCPR, Munich, 1982, 492-495.

Lyche T. and Morken K., Knot Removal for Parametric B-splines Curves and Surfaces,CAGD 4, 1987, 217-230.

Pavlidis T. and Horowitz S.T., Segmentation of Plane Curves, IEEE Transactions on Computers, 23, n°8, August 1974, 860-870.

Pavlidis T., Algorithm for Graphics and Image Processing, Washington D.C., Computer Science Press, 1982.

C. Potier and C. Vercken, Lissage de Surfaces par Eléments Finis, L'Echo des Recherches, n° 122, 1985, p. 51-58.

Ronse C. and Devijver P.A., Connected Components in Binary Images: The detection Problem, Research Studies Press,1984.

Wall K. and Danielsson P.E., A Fast Sequential Method for Polygonal Approximation of Digitized Curves, Graphics and Image Processing, 28, 1984, 220-227.

A B-spline approximation algorithm for quasi-interpolation or filtering

C. Rabut

Institut National des Sciences Appliquées, Toulouse, France

Abstract We propose some straightforward modifications to the usual B-spline algorithm for the approximation of discrete data. The modifications permit quasi-interpolation or filtering of the data. We propose a simple algorithm for the computation of "good" nodes – a linear combination of neighbourhood data, with suitable coefficients – in order to obtain desirable properties for the global approximation process. To analyse this process – and to determine the coefficients of the linear combination – we use the **transfer function** of the associated filter, which is a very good tool to obtain a global view of the process (i.e., which frequencies of the data are amplified, or attenuated, and how much, etc.).

Key words B-spline approximation, Quasi-interpolation, Filtering, Transfer function

1. Introduction

Given discrete equidistant data (x_i, y_i), it is a common requirement to determine the B-spline approximation of those data, i.e. to perform the following algorithm:

FOR wanted x,

compute $\sigma(x) = \sum_i y_i B_i(x)$

where the B_i are the usual cubic B-splines, centered at x_i; in this paper we will consider only cubic splines, but the ideas may be extended to any degree.

The important advantage of the above computation is that the curve so-obtained is very quickly determined, and follows the shape of the data (precisely, we have $\sigma(x_j) = (y_{j+1} + 4y_j + y_{j-1})/6$. But there is also an important drawback: such an algorithm allows no flexibility, in the sense that we cannot force the curve to be nearer to the data, or, conversely, we cannot use it to filter noise which may be present in the data.

To overcome this drawback, we propose splitting up the nodes, Y_i^n, of the B-spline approximation, from the data y_i; for fast computation, we will require Y_i^n to be a linear combination of 2n+1 of the y_j placed around y_i (the coefficients of the linear combination, as well as the value of n, being chosen according to the type of result to be obtained, see later). So, we propose performing the following algorithm:

For concerned i

$$\text{compute } Y_i^n = \sum_{j=-n}^{n} \alpha_j^n y_{i-j}.$$

For wanted x

$$\text{compute } \sigma^n(x) = \sum_{i\in\mathbb{Z}} Y_i^n \, B_i(x).$$

The aim of this paper is to show some way of choosing appropriate coefficients of the linear combination, and, in order to do so, to present and use the "transfer function" (which is also called "attenuation function" in some papers), which is a very convenient and comprehensive tool to obtain a global view of the approximation process.

To do so, we will suppose we have an infinite regularly spaced data set (it is easy, if necessary, to extend a finite set to an infinite one, in an appropriate way), so $\forall\, i \in \mathbb{Z}$, $x_i = ih$. Also, we will take symmetric combinations ($\forall\, j \in [1, n]$, $a_{-j}^n = a_j^n$), so we will use the equivalent form $Y_i^n = \sum_{j=0}^{n} b_j (\delta^{2j} y)_i$, where

$$(\delta^2 y)_i = y_{i+1} - 2y_i + y_{i-1}.$$

2. Tranfer function

2.1 Definition

Presentation The first idea is to study only the result of the approximation scheme at the data: we will look only at $(\sigma^n(x_i))_{i \in \mathbb{Z}}$, and not at $\sigma^n(x)$ for $x \neq x_i$. In other words, we will study the transformation $T_n : y \in \mathbb{R}^{\mathbb{Z}} \longrightarrow z \in \mathbb{R}^{\mathbb{Z}}$ defined by

$$\forall\, j \in \mathbb{Z},\; z_j = \sigma^n(x_j) = \sum_{i\in\mathbb{Z}} \left(\sum_{k=-n}^{n} b_k^n \delta^{2k} y\right)_i B(x_{j-i}).$$

Since the values $\sigma(x_i)$ provide information on $\sigma(x)$, studying this discrete transformation will provide knowledge of the whole approximation process.

Obviously T_n is a linear transformation, and the vector $y = e_\alpha = (e^{2i\pi\alpha k})_{k\in\mathbb{Z}}$ (here $i^2 = -1$) is an eigenvector of T_n.

Definition Let us call $H_n(\alpha)$ the eigenvalue of the above transformation T_n, associated with the eigenvector e_α. The application $H_n : \alpha \longrightarrow H_n(\alpha)$ is called the "transfer function" of the filter T_n.

H_n is a real-valued function which is even and periodic (of period $1/h$).

Remarks: By definition, $H_n(\alpha)$ is the coefficient of amplification of a (co)sinusoïdal signal of frequency α. $H_n(0)$ is the coefficient of amplification of constants (we will always require $H_n(0)=1$!).

As $(e_\alpha)_{\alpha\in\mathbb{R}}$ are the only eigenvectors of T_n, it is understandable that the set of eigenvalues $(H_n(\alpha))_{\alpha\in\mathbb{R}}$ - i.e. the function H_n - is indicative of the whole transformation T_n.
As a particular case, if $\forall\,\alpha \in \mathbb{R}$, $H_n(\alpha) = 1$, then T_n is the identity.

2.2. Use of transfer function

Evaluation of H_n: theorem

Let H_n be defined as above.

Let $b_{n+1}^n = 0$. Then

$$H_n(\alpha) = b_0^n + \sum_{k=1}^{n+1} (b_k^n + b_{k-1}^n/6)(-4\sin^2\pi\alpha h)^k ;$$

The proof is quite easy: since $\forall\, y \in \mathbb{R}^{\mathbb{Z}}$,

$T_n(y) = b_0^n y + \sum_{k=1}^{n+1} (b_k^n + b_{k-1}^n/6)\delta^{2k} y$; so if g is the elementary filter

$y \longrightarrow \delta^{2k} y$ (whose transfer function is $G(\alpha) = -4\sin^2\pi\alpha h$),

$T_n = b_0^n g^0 + \sum_{k=1}^{n+1} (b_k^n + b_{k-1}^n/6) g^k$, and so $H_n = b_0^n + \sum_{k=1}^{n+1} (b_k^n + b_{k-1}^n/6) G^k$.

Error majoration: theorem

Let g and h be two filters, with respective transfer functions G and H.

Let $y \xrightarrow{g} z$ and $y \xrightarrow{h} z'$.

For any periodic function f of period P, let $\|f\|_2 = \sqrt{\int_P (f(x))^2 dx}$.

If $y \in \ell^2$, $\|z-z'\|_{\ell^\infty} \leqslant \|z-z'\|_{\ell^2} \leqslant \|G-H\|_\infty . \|y\|_{\ell^2}$.

If $y \in \ell^1$, $\|z-z'\|_{\ell^\infty} \leqslant \|z-z'\|_{\ell^2} \leqslant \|G-H\|_2 . \|y\|_{\ell^1}$.

3. Choosing the coefficients

Of course, the actual values of b_j^n will determine the properties of T_n, and so of the approximation process. The aim of this section is to present some criteria in order to determine interesting values of b_j^n. To do so, having chosen a transfer function G, we want to determine b_j^n so that the associated function H_n is as close to G as possible (if we want to quasi-interpolate data, we choose $G \equiv 1$, if we want to filter data as much as possible, we will choose $G(0) = 1$, $G(\alpha) = 0$ for $\alpha \neq 0$).

3.1. First criterion: truncated development

<u>Definition</u> The b_j^n are chosen in order that the first terms of the limited development at some point α_0 (for example $\alpha_0 = 0$) of H_n coincide with those of G; so, if G is written as

$G(\alpha) = \sum_{k \in \mathbb{N}} c_k (-4 \sin^2 \pi\alpha h)^k$, we obtain ($\alpha_0 = 0$):

$$\begin{cases} b_0^n = c_0 \\ j = 1,\dots,n : b_j^n = c_j - b_{j-1}^n/6 \\ j \geqslant n+1 \quad : b_j = 0 \end{cases}$$

For quasi-interpolating data ($G \equiv 1$) , we obtain

$$\forall k \in [0,n],\ b_k^n = (-1/6)^k$$

For a high-cut filter ($G(0) = 1$, $\forall k \in \mathbb{N}$, $G^{(k)}(1/2h) = 0$), we obtain:

$$\forall k \in [0,n]\ ,\ b_k^n = \binom{n}{k}(1/4)^k \quad (\Leftrightarrow \forall k \in [0,n]\ ,\ a_k^n = \binom{2n}{n+k}(1/4)^k).$$

Error majoration: theorem

Let $n \in \mathbb{N}$

Let $\forall j \in [0,n]$, $b_j^n = (-1/6)^j$; $\forall j \in [0,n]$, $b_j^n = 0$.

Let $\sigma^n(x) = \sum_{i \in \mathbb{Z}} \left(\sum_{k=0}^{n} b_k^n \delta^{2k} y\right)_i B(x - x_i)$.

Then $\forall j \in \mathbb{Z}, \quad \sigma^n(x_j) = y_j - (-1/6)^{n+1}(\delta^{2n+2}y)_j$.

As a consequence majorations of $|\sigma^n(x_j) - y_j|$ may easily be obtained, and, in particular, σ^n interpolates any polynomial of degree at most 2n+1.

Proof: induction on n.

Figures: Every figure is provided for n = 0, 1, 2, 3, 4 and 5. Figures 1 and 2 are transfer functions of a quasi-interpolation and of a high-cut filter.
Figures 3 and 4 are σ^n for some data: (the □ are the data; the + are the computed nodes Y_i^n). Figure 3 is for step data: $i \leqslant 0 \Longrightarrow y_i = 0$, $i \geqslant 1 \Longrightarrow y_i = 1$. Figure 4 is for a parametric curve: each component is calculated separately (equidistant parameter).

3.2. Second criterion: least squares

The truncated development criterion is very interesting as the formulae obtained are very simple and lead to a many interesting theorical and practical results. But other criteria give better results in most cases. An example of such a criterion is given in this section.

3.2.1. Criterion

Looking at theorem 2.3, the idea now is to minimize $\|H_n - G\|_2$. This is quite simple as $E_n = \int_0^{1/2h}(G(\alpha) - H_n(\alpha))^2 d\alpha$ is a quadratic function of the coefficients b_j^n. Minimising E_n leads to the solution of a linear system with unknowns b_j^n.

Of course, we can add some additional constraints, such as, for example:

- reproducing constants: $H_n(0) = 1$ $\quad (\Longleftrightarrow b_0^n = 1,\ b_1^n = -1/6)$
- reproducing polynomials of degree $\leqslant 3$: $H_n(0) = 1;\ H''_n(0) = 0$ $\quad (\Longleftrightarrow b_0^n = 1,\ b_1^n = -1/6)$
- no frequency amplified: $\forall\alpha \in \mathbb{R},\ H_n(\alpha) \leqslant 1$
- no frequency inversed: $\forall\alpha \in \mathbb{R},\ H_n(\alpha) \geqslant 0$.

Here are some values obtained for quasi interpolation (i.e. $G(\alpha) \equiv 1$), with the reproducing constants constraint.

n=1		$b_0^1=1$	$b_1^1=-8/23$			
n=2	d=474	$b_0^2=1$	$b_1^2=-30/d$	$b_2^2=47/d$		
n=3	d=8838	$b_0^3=1$	$b_1^3=-1890/d$	$b_2^3=-351/d$	$b_3^3=-240/d$	
n=4	d=15443	$b_0^4=1$	$b_1^4=-22776/d$	$b_2^4=11320/d$	$b_3^4=4002/d$	$b_4^4=1137/d$

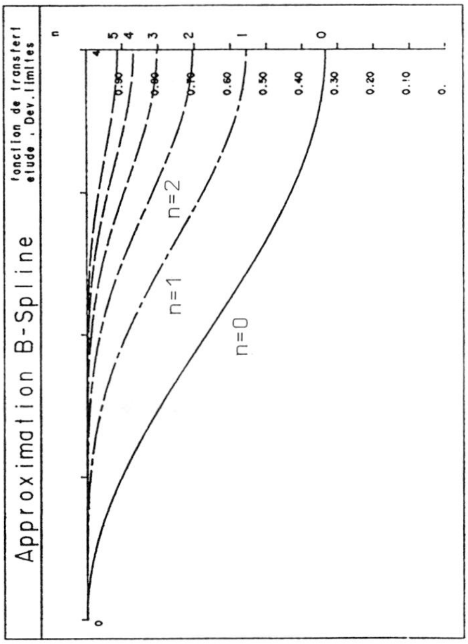

Figure 1

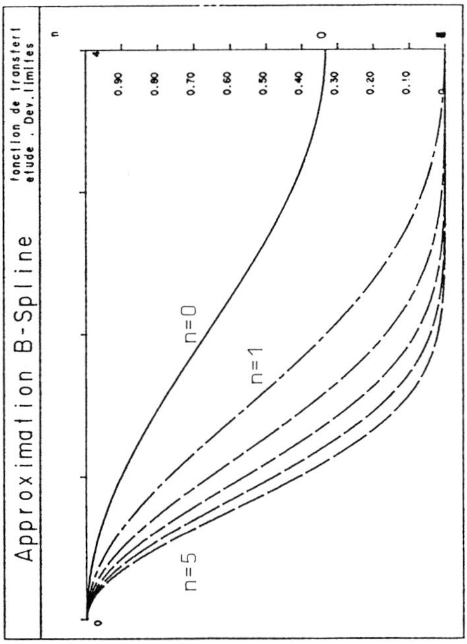

Figure 2

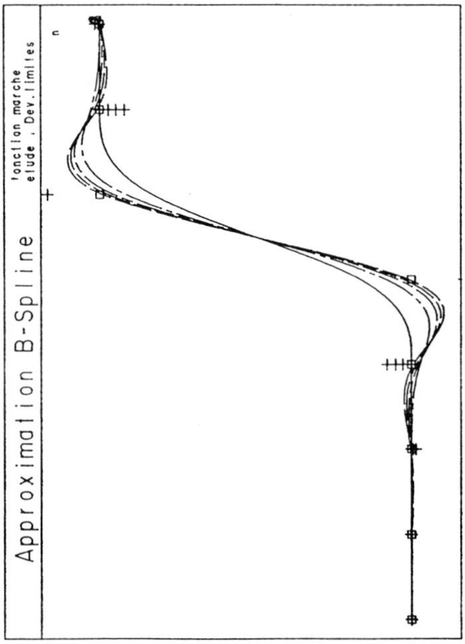

Figure 3

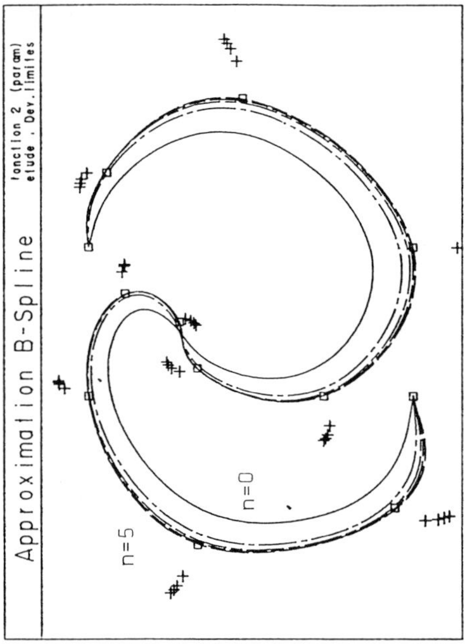

Figure 4

Figures 1 to 4 : Truncated development criterion.

We notice that, for n = 1, with $\alpha_0 = (1/\pi h)\sin^{-1}(5/8) \simeq (1/h)0.215$, we have $\underset{\alpha\in\mathbb{R}}{\mathrm{Max}}(H_1(\alpha)) = H_1(\alpha_0) \simeq 1.1415$, which may be considered as amplifying too much the frequencies around α_0. So we suggest to use, for n = 1, $b_1 = -1/4$ which lies between the truncated development case and the least squares one, and gives improved results.

3.2.2. Figures:

Figures 5 to 8 are for quasi-interpolating B-spline approximation, with the least squares criterion, each one for n = 1, 2, 3, 4 and 5 (for n = 1, $b_1 = -1/4$)

Figure 5: transfer function $H_n(\alpha)$

Figure 6: transfer function $H_n(\alpha)$ (enlarging the scale around $H_n(\alpha)$).

Figures 7 and 8: response to the same data as figures 3 and 4.

When comparing figures 5-8 with figures 1-4, we can see how much closer the least squares B-spline approximation is, for the same n, to the interpolant than the truncated development B-spline approximation is.

3.3 A pretty filter

It is worthwhile to mention the filter defined by its a_j^n coefficients (see section 1) by: $\forall j \in [-n,n]$, $a_j^n = (n + 1 - |j|)/(n + 1)^2$. For lack of space, it is not possible to detail its properties, but we can say that it is a very efficient and simple high-cut filter. Its transfer function is shown at figure 9, for n = 5, 15, 45; the curve obtained by filtering some noisy data (shares of some French company) is shown in figure 10.

4. Conclusion

4.1. **"Transfer function" is a most efficient way** to obtain a global view of an approximation process. For quasi-interpolation, it is much better than "an order of convergence h^q" (or "interpolating polynomials of degree at most q - 1").

4.2. We think that this algorithm may have many applications:
- in CAGD: thanks to this algorithm, an operator has no need to give points far from the desired curve, which is much more convenient
- for closed curves: the way to extend the points in a periodic way is obvious, and there is of course no linear system to solve.
- for filtering data: most methods need to solve large linear systems; here we get satisfactory results with little computation.
- for d-dimensional surfaces obtained by tensor product: simple

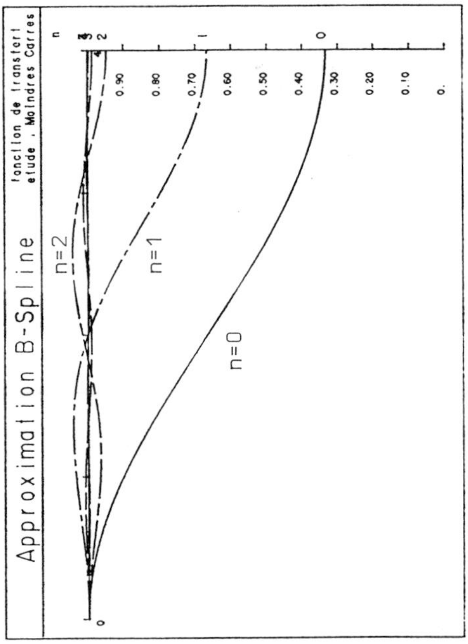

Figure 5

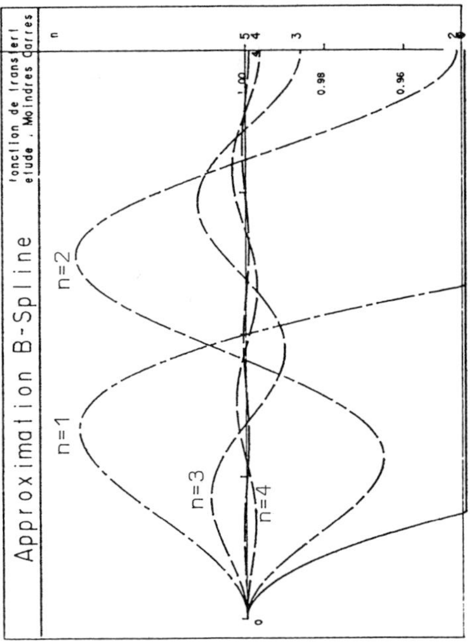

Figure 6

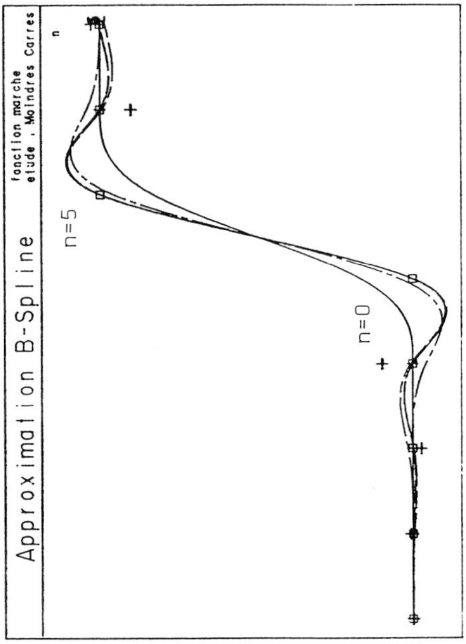

Figure 7

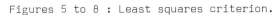

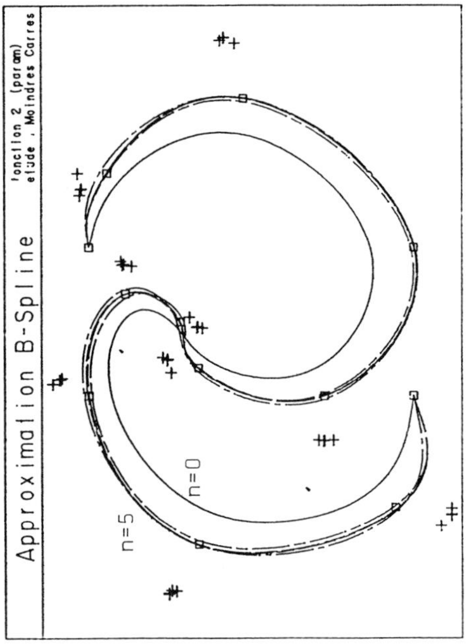

Figure 8

Figures 5 to 8 : Least squares criterion.

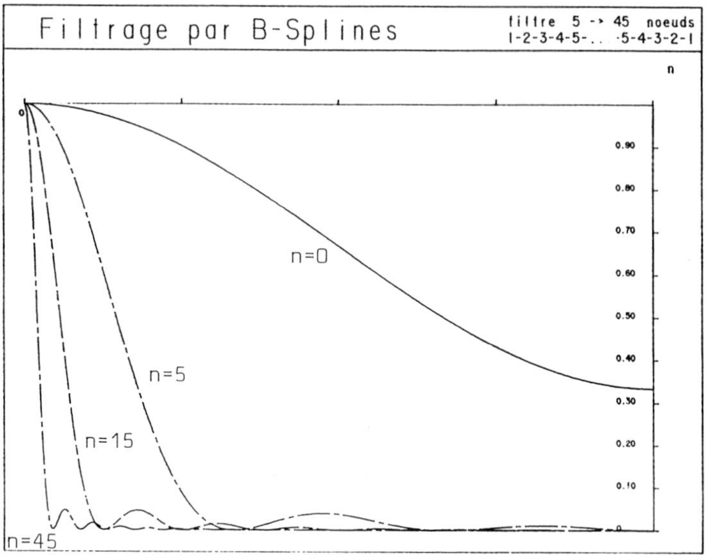

Figure 9

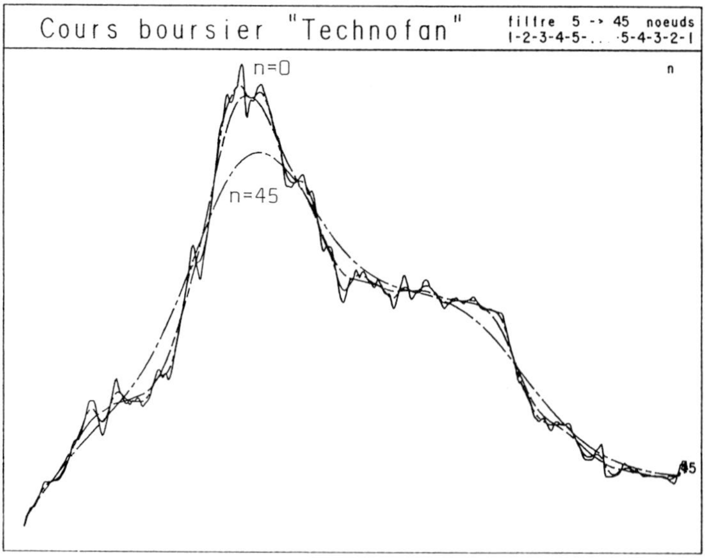

Figure 10

<u>Figures 9 and 10</u> : A very efficient filter.

quasi-interpolation is very interesting as the usual tensor product B-spline approximation smoothes the data in a quite important way, which of course is not the case of tensor product B-spline quasi-interpolation.

4.3. This method is very easy to introduce in existing algorithms: we only have to add a line to compute the nodes Y_i^n; everything else remains unchanged!

4.4. The "modified B-splines" B^n can certainly be used in place of the usual B-splines in every application (as rational splines). So, we can have the opportunity of one (or even many) degree of freedom.

4.5. At present work is carried out for determining **best high-cut filter** by cross-validation techniques. Extensions are also being made to **thin plate splines**.

References

de Boor, C "A practical guide to splines", Springer Verlag 1978

Lyche, T "Local spline approximation methods and osculatory formulae", Approximation Theory, R Schabak and K Scherer (editors), Springer lecture notes in mathematics, 556 (1976), 305-319

Schumaker, L "Spline functions: Basic Theory" 1981, John Wiley and sons

On knots and nodes for spline interpolation

P. W. Smith
IMSL, Houston, Texas, USA

Abstract This paper explores the relationships among the knots, nodes, and norm of the interpolating spline projection. In particular, we derive lower bounds for the norm of the projection in terms of the knots and nodes. We then turn our attention to the role played by equioscillating splines in deriving the best interpolating projector for a fixed spline subspace.
Keywords: equioscillating spline, B-spline, projection.

1. Introduction

The general problem of choosing knots and/or nodes for spline interpolation is addressed in this paper. Controlling the size of the interpolating projection is the criterion for 'good' knots and/or nodes. We adhere to the following notation throughout this work. The spline subspace S_t^k, $t := t_1 \le \cdots \le t_{n+k}$, is the linear span of the n normalized B-splines $B_{1,k,t}, \ldots, B_{n,k,t}$ where

$$B_{i,k,t}(x) := (t_{i+k} - t_i)[t_i, \ldots, t_{i+k}]_s (s - x)_+^{k-1}.$$

If no misunderstanding can arise, we denote $B_{i,k,t}$ by B_i. Let $\tau := \tau_1 < \cdots < \tau_n$ be n nodes of interpolation. The operator

$$P : C[t_1, t_{n+k}] \to S_t^k$$

is well defined by

$$(Pf)(\tau_i) = f(\tau_i) \quad i = 1, \ldots, n$$

if and only if $B_i(\tau_i) \ne 0$ (i.e. usually $t_i < \tau_i < t_{i+k}$; however, equality can occur if k knots are stacked together). Throughout, we assume that P is well defined.

The central issue is the relationship of the norm of the operator P, $\|P\|$, to the knots and nodes. Instead of working directly with P, we shift our attention to the finite dimensional linear operator $A : R^n \to R^n$ defined by

$$A_{ij} = (A_{n,\tau,t})_{ij} = B_{j,k,t}(\tau_i) \quad 1 \leq i, j \leq n.$$

The relationship between P and A is

$$Pf = \sum_{j=1}^{n} [A^{-1} f(\tau)]_j B_j,$$

where $f(\tau) := (f(\tau_1), \ldots, f(\tau_n))$. Furthermore, one observes [1] that

$$D_k^{-1} \|A^{-1}\| \leq \|P\| \leq \|A^{-1}\|. \tag{1}$$

where all norms in this paper are supremum norms on the appropriate space. The important feature of the above inequality is that it is independent of n or t.

Recall that A is totally positive. This means that A^{-1} is checkerboard, that is

$$(-1)^{i+j} A_{ij}^{-1} \geq 0,$$

from which it follows that

$$\|A^{-1}\| = \|A^{-1} e\|$$

where $e_i = (-1)^{i-1}$, for $i = 1, \ldots, n$. From the inequality (1), we see that upper and lower bounds on $\|A^{-1}\|$ in terms of t and τ yield similar bounds on $\|P\|$.

We focus on two problems. The first problem is fundamental but extremely difficult: Given k and τ, choose the 'best' knots for interpolation. We interpret this to mean, find knots t^* which satisfy

$$\|A_{t^*}^{-1}\| \leq \|A_t^{-1}\|$$

among all competing knot sequences t. Perhaps it would be even more practical to find 'acceptable' t^{**} which depend on local information and satisfy

$$M_k \|A_{t^{**}}^{-1}\| \leq \|A_{t^*}^{-1}\|$$

for some fixed positive constant M_k.

The second problem is, in some sense, the dual problem: Given k and t, choose the best τ for interpolation. We interpret this to mean, find nodes τ^* so that

$$\|A_{\tau^*}^{-1}\| \leq \|A_\tau^{-1}\|$$

among all competing nodes τ. S. Demko [3] has shown that there exists τ^{**} so that

$$\|A_{\tau^{**}}^{-1}\| \leq D_k.$$

In fact, it has been noted by Morken [4] that Demko's choice of nodes yields the best set of nodes for the given knot sequence.

2. Estimates for $\|A^{-1}\|$

In this section, we elaborate on a technique for obtaining lower estimates for $\|A^{-1}\|$ that first appeared in [1]. Recall [2], that if $s \in S_t^k$ and

$$s = \sum_j b_j B_{j,k}$$

then (ignoring boundary effects)

$$s' = (k-1) \sum_j \frac{b_j - b_{j-1}}{t_{j+k-1} - t_j} B_{j,k-1}.$$

We define the weighted difference operator $\nabla_{t,k}^{(1)}$ by

$$\nabla_{t,k}^{(1)} b_j := (k-1) \frac{b_j - b_{j-1}}{t_{j+k-1} - t_j}$$

which allows us to rewrite s' as

$$s' = \sum_j \nabla_{t,k}^{(1)} b_j B_{j,k-1}.$$

More generally, if we take the rth derivative then

$$s^{(r)} = \sum_j \nabla_{t,k}^{(r)} b_j B_{j,k-r}.$$

where we define $\nabla_{t,k}^{(r)}$ recursively by

$$\nabla_{t,k}^{(r)} = \nabla_{t,k-r+1}^{(1)} \nabla_{t,k}^{(r-1)}.$$

The goal in this section is to estimate the coefficients of an oscillating spline and to relate this information to bounds on $\|A^{-1}\|$. If the spline s satisfies $s(\tau_i) = (-1)^{i-1}$ for $i = 1, \ldots, n$, then the ℓ_∞ norm of the coefficients is $\|A^{-1}\|$.

Theorem 1. Let k, n, t, and τ be given. We assume that

$$A = A_{n,t,\tau}$$

is invertible and that $S_t^k \in L_\infty^r[\tau_1, \tau_n]$. Set $e(i) := (-1)^{i-1}$ for $i = 1, \ldots, n$. Then

$$\|A^{-1}\| \geq \frac{|[\tau_i, \ldots, \tau_{i+r}]e|}{\Delta_{i,r}} r!$$

where $i = 1, \ldots, n-r$ and

$$\Delta_{i,r} := \max\{|\nabla_{t,k}^{(r)} e_j| : (t_j, t_{j+k-r}) \cap (\tau_i, \tau_{i+r}) \neq \emptyset\}.$$

Proof: Let $s \in S_t^k$ satisfy

$$s(\tau_i) = e_i = (-1)^{i-1} \quad i = 1, \ldots, n.$$

Then $s = \sum_j a_j B_{j,k}$ and $Aa = e$. It follows that $\|A^{-1}\| = \|a\|$. From the Peano kernel theorem we have

$$[\tau_i, \ldots, \tau_{i+r}]s = [\tau_i, \ldots, \tau_{i+r}]e \le \frac{\|s^{(r)}\|_{[\tau_i,\tau_{i+r}]}}{r!}.$$

Now

$$s^{(r)} = \sum_j \nabla_{t,k}^{(r)} a_j B_{j,k-r}$$

and hence

$$\|s^{(r)}\|_{[\tau_i,\tau_{i+r}]} \le \max\{|\nabla_{t,k}^{(r)} a_j| : \text{supp } B_{j,k-r} \cap (\tau_i, \tau_{i+r}) \ne \emptyset\}.$$

Since the matrix A is variation diminishing, we know that $a_j(-1)^{j-1} \ge 0$ for all j. It follows that there is no cancellation in the differences $\nabla_{t,k}^{(r)} a_j$, and hence

$$|\nabla_{t,k}^{(r)} a_j| \le |\nabla_{t,k}^{(r)} e_j| \|a\|.$$

Assembling the pieces yields

$$|[\tau_i, \ldots, \tau_{i+r}]e| \le \|a\| \frac{\Delta_{i,r}}{r!},$$

completing the proof.

We mention two corollaries of this result.

Corollary 2. (de Boor [1]) Let $r = 1$, then

$$\|A^{-1}\| \ge \frac{d_i}{(\tau_{i+1} - \tau_i)(k-1)}$$

and

$$\|P\| \ge \frac{D_k d_i}{(\tau_{i+1} - \tau_i)(k-1)}.$$

where

$$d_i := \min\{t_{j+k-1} - t_j : (t_j, t_{j+k-1}) \cap (\tau_i, \tau_{i+1}) \ne \emptyset\}$$

The second inequality was derived by Ching-Ching Rojas, a student of ours.

Corollary 3. Let $r = 2$ and $k = 3$ with $S_t^k \subset L_\infty^2$. Then

$$\|A^{-1}\| \ge \frac{\delta_i}{(\tau_{i+2} - \tau_{i+1})(\tau_{i+1} - \tau_i)}$$

where

$$\delta_i = \max\{\frac{(t_{j+1} - t_{j-1})(t_{j+2} - t_j)(t_{j+1} - t_j)}{(t_{j+1} - t_{j-1}) + (t_{j+2} - t_j)} : (t_j, t_{j+1}) \cap (\tau_i - \tau_{i+2})\}.$$

3. Equioscillating Splines

In [3] it was shown that there exists a spline $T = \sum a_j^* B_{j,k} \in S_t^k$ which equioscillates maximally. That is, on $[t_1, t_{n+k}]$ there exist n points $\tau_1 < \tau_2 < \cdots < \tau_n$ so that

$$T(\tau_i) = (-1)^{i-1}$$
$$\|T\| = 1.$$

Combining this observation with the inequality

$$D_k^{-1}\|a^*\| \leq \|T\| = 1$$

yields the following result

Theorem 4. (Demko [3]) Given n, t, and k, there exist nodes τ so that the matrix $A := A_{n,t,\tau}$ satisfies

$$\|A^{-1}\| \leq D_k.$$

It is important to note that this result yields a universal upper bound (depending only on k) for these particular nodes. Indeed, this result illustrates a dichotomy between the dual problems of obtaining

$$\text{(A)} \quad \sup_t \inf_\tau \|A_{t,\tau}^{-1}\|$$
$$\text{(B)} \quad \sup_\tau \inf_t \|A_{t,\tau}^{-1}\|$$

Note that by Theorem 3.1, (A) is bounded above by D_k, while (B) is always infinite by Theorem 2.1 or Corollary 2.2, if $k > 2$ and n is sufficiently large.

As Morken [4] pointed out, the abscissae of the extreme points of the equioscillating spline yield the best points for interpolation. This can be seen by setting T to be the equioscillating spline with equioscillation points τ, as above, and letting $\zeta_1 < \cdots < \zeta_n$ be another set of nodes. Let $s \in S_t^k$ be defined by $s(\zeta_i) = (-1)^{i-1}$, then

$$T = \sum a_j^* B_{j,k}$$
$$s = \sum b_j B_{j,k}$$
$$(s - T)(\zeta_i)(-1)^{i-1} \geq 0.$$

This means that the coefficients of $s - T$, which are $b_j - a_j^*$, weakly oscillate (i.e. $(b_j - a_j^*)(-1)^{j-1} \geq 0$). But since b_j and a_j^* oscillate in the same orientation, we conclude that

$$|a_j^*| \leq |b_j| \quad j = 1, \ldots, n$$
$$\text{hence } \|a^*\| \leq \|b\|$$
$$\text{or } \|A_\tau^{-1}\| \leq \|A_\zeta^{-1}\|$$

We now discuss the computing of the nodes τ for the equioscillating spline. First, observe that T (and hence τ) is unique, since T can be viewed as an error in the best approximation of $a_1^* B_{1,k}$ from the span of $\{B_{2,k}, \ldots, B_{n,k}\}$. Furthermore, if we set $t_1 = \cdots = t_k$ and $t_{n+1} = \cdots = t_{n+k}$, then $\tau_1 = t_k$ and $\tau_n = t_{n+1}$. This follows by noting that all B-splines vanish at t_k except for $B_{1,k}$, and this spline takes on its maximum at t_k. Similarly, all B-splines vanish at t_{n+1} except $B_{n,k}$, which takes on its maximum at t_{n+1}.

The computational algorithm is a Remez exchange algorithm, where τ^1 is chosen so that A_{t,τ^1} is invertible and $\tau_1^1 = t_k$, $\tau_n^1 = t_{n+1}$. Currently, we choose τ to be the interior knot averages for the B-splines, that is

$$\tau_i := \frac{\sum_{j=i+1}^{i+k-1} t_j}{k-1}.$$

We then compute s_1 which satisfies

$$s_1(\tau_i^1) = e_i = (-1)^{i-1}.$$

We then choose τ^2 to be the unique strictly increasing vector satisfying

$$\begin{aligned} \tau_1^2 &= t_k \\ \tau_n^2 &= t_{n+1} \\ s_1'(\tau_j^2) &= 0 \quad j = 2, \ldots, n-1. \end{aligned}$$

Then τ^2 is substituted for τ^1, and the process is repeated until convergence.

References

1. Carl de Boor. On bounding spline interpolation. *J. of Approximation Theory*, 14:191–203, 1975.
2. Carl de Boor. *A Practical Guide to Splines.* Springer-Verlag, New York, 1978.
3. S. Demko. On the existence of interpolating projections onto spline spaces. *J. of Approximation Theory*, 43:151–156, 1985.
4. Knute Morken. *On Two Topics in Spline Theory: Discrete Splines and the Equioscillating Spline.* Master's thesis, Institute for Informatics, University of Oslo, 1984.

2. Polynomial and Piecewise Polynomial Approximation

A basis for certain spaces of multivariate polynomials and exponentials

W. Dahmen

Fachbereich Mathematik,
Freie Universität Berlin, FRG

Abstract A basis for certain spaces of multivariate polynomials and exponentials is constructed from the polynomial pieces of so-called multivariate truncated powers introduced by Dahmen (1980). The need for such a basis arises in connection with certain Hermite interpolation problems which were recently investigated by Dyn and Ron (1988a, b). It is indicated how to compute the elements of the basis explicitly by means of various representations of truncated powers.

Key words: Truncated powers, Exponential splines, Interpolation, Construction of bases, Common null space of families of differential operators.

1. Introduction

The literature of the past few years reflects a rapidly growing interest in multivariate interpolation problems. The numerous methods which have been developed so far are based on a diversity of function systems where an important role is played, of course, by spaces of splines and polynomials.

This paper is concerned with certain spaces of multivariate polynomials and exponentials which are obtained as common null spaces of certain differential operators (see e.g. Dahmen and Micchelli (1983a), (1985), (1987)). These spaces play a central role in the theory of multivariate splines on regular grids as well as in various related combinatorial and algebraic aspects (cf. Dahmen and Micchelli (1988)). Moreover, they arise in connection with certain Hermite interpolation problems that were recently investigated by Dyn and Ron (1988a, 1988b).

However, the practical solution of such interpolation problems would require the explicit knowledge of bases for these spaces which it seems have not previously been available.

The objective of this paper is to construct such bases by making use of the intimate relationship to the corresponding multivariate splines. More precisely, it will be shown how to obtain a basis from the pieces of certain splines called multivariate truncated powers introduced by Dahmen (1980). The computation of these pieces is facilitated by various known representations for truncated powers.

The paper is organized as follows. In Section 2 we state the relevant definitions and collect some background material. Section 3 is devoted to the general construction of bases for the above mentioned spaces. Finally, an example is discussed in Section 4.

2. Some definitions and background material

Let us start by recalling the definitions and some properties of certain multivariate (polynomial and exponential) splines which will play a crucial role in subsequent discussions. For more technical details and proofs the reader is referred to Dahmen (1980), Dahmen and Micchelli (1983b), (1987). To this end, let $X = \{x^1,\dots,x^n\}$ be a set of (not necessarily distinct) vectors in $\mathbb{R}^s \setminus \{0\}$ and let $\mu \in \mathbb{C}^n$ be fixed. For convenience the $(s \times n)$-matrix whose columns are the elements of X is also denoted by X. Under suitable assumptions on μ and X one can then define a function $T_\mu(\cdot|X)$ by requiring that

$$\int_{\mathbb{R}^s} f(x) T_\mu(x|X)\,dx = \int_{\mathbb{R}^n_+} e^{-\mu\cdot v} f(Xv)\,dv \qquad (1)$$

holds for all $f \in C_0(\mathbb{R}^s)$ (where $v \cdot z = v_1\bar{z}_1 + \ldots + v_n\bar{z}_n$ denotes the standard inner product on $\mathbb{C}^n$). A case of particular interest is $\mu = 0$. For $T(\cdot|X) := T_0(\cdot|X)$ to be well defined one has to assume that

$$0 \notin [X]\,, \qquad (2)$$

where $[X]$ denotes the convex hull of the set X. If $\langle X\rangle$, the span of X, is all of $\mathbb{R}^s$, $T(\cdot|X)$ is easily seen to be indeed a function with support

$$\operatorname{supp} T(\cdot|X) = \langle X\rangle_+\,, \qquad (3)$$

where $\langle X\rangle_+ = \{Xv : v \in \mathbb{R}^n_+\}$ is the cone spanned by X. $T(\cdot|X)$ can be shown to be a piecewise polynomial of degree at most $n - s$, where the polynomial pieces are separated by the hyperplanes in

$$\mathcal{C}(X) = \{\langle V\rangle : V \in \mathcal{B}_{s-1}(X)\}\,. \qquad (4)$$

Here

$$\mathcal{B}_\ell(X) = \{V \subset X : \dim\langle V\rangle = |V| = \ell\}$$

and $|V|$ denotes the cardinality of V. In general, when μ is some complex vector one has to assume

$$\operatorname{Re} \mu_j > 0 \qquad (5)$$

whenever x^j belongs to X', the largest subset of X such that there exists $\beta \in \mathbb{Z}_+^{|X'|}$, $\beta_i > 0$, $i = 1,\ldots,|X'|$ with $X'\beta = 0$. Again $T_\mu(\cdot|X)$ turns out to be piecewise analytic with cut regions given by (4) and support $\langle X\rangle_+$. In general its pieces are composed of polynomials and exponentials to be explained in more detail below. Specifically, one readily concludes from (1) that for $n = |X| = s$,

$\langle X\rangle = \mathbb{R}^s$,

$$T_\mu(x|X) = \chi_{\langle X\rangle_+}(x)\,|\det X|^{-1} e^{-\mu\cdot(X^{-1}x)} . \tag{6}$$

In general $T_\mu(\cdot|X)$ can be evaluated recursively. In addition, explicit analytic expressions are available for special choices of X. The following case will be needed in the next section. To this end, let

$$X = \{y^1,\dots,y^1,\dots,y^s,\dots,y^s\},$$

where each y^i occurs m_i times and $Y = \{y^1,\dots,y^s\}$ spans $\mathbb{R}^s$. Denoting by $u = Y^{-1}x$ the coordinates of x with respect to the basis Y one has (cf. Dahmen and Micchelli (1983b))

$$T(x|X) = \chi_{\langle Y\rangle_+}(x)\,u_1^{m_1-1}\cdots u_s^{m_s-1} . \tag{7}$$

When $\mu \neq 0$ note first that for any nonsingular $(s\times s)$-matrix A:

$$|\det A|\,T_\mu(Ax|AX) = T_\mu(x|X).$$

Choosing $A = Y^{-1}$ one then easily verifies that

$$T_\mu(x|X) = |\det Y|\,T_\mu(u|Y^{-1}X) = \prod_{j=1}^{s} T_{\mu^j}(u_j|1(j)), \tag{8}$$

where $\mu^j = (\mu_{m_1+\dots+m_{j-1}+1},\dots,\mu_{m_1+\dots+m_j})$ and $1(j)$ denotes the $(1\times m_j)$-matrix with all entries equal to 1. Moreover, noting that the Laplace transform of each factor $T_{\mu^j}(\cdot|1(j))$ is given by $\Pi_{i=1}^{m_j}(\mu_i^j+\lambda)^{-1}$, $\lambda > -\mu_i^j$, $i = 1,\dots,m_j$, each univariate factor in (8) is a convolution of the functions $t_+^0\, e^{-\mu_i^j t}$. Defining $D_y f = \Sigma_{j=1}^s y_j \frac{\partial f}{\partial x_j}$ and denoting by $\mu\setminus j$ the element in $\mathbb{C}^{n-1}$ obtained from μ by discarding μ_j it is also not hard to show that

$$(\mu_j + D_{x^j})T_\mu(\cdot|X) = T_{\mu\setminus j}(\cdot|X\setminus\{x^j\}). \tag{9}$$

Many of the properties stated above indeed follow from (9). In particular, setting

$$\mathcal{Y}(X) = \{V \subset X : \langle X \setminus V\rangle \neq \mathbb{R}^s\},$$

a repeated application of (9) yields, in view of (6), that

$$(\Pi_{v \in V}(\mu_v + D_v))T_\mu(x|X) = 0 \tag{10}$$

for all $V \in \mathcal{Y}(X)$, provided that x does not belong to any of the cut regions in (4). Hence $T_\mu(\cdot|X)$ belongs locally to the space

$$D_\mu(X) = \{f \in \mathcal{D}'(\mathbb{R}^s) : (\Pi_{v \in V}(\mu_v + D_v))f = 0, \forall V \in \mathcal{Y}(X)\}.$$

where $\mathcal{D}'(\mathbb{R}^s)$ denotes the space of Schwartz distributions on $\mathbb{R}^s$. The fact that $D_\mu(X)$ is finite dimensional is a consequence of the following more general result.

<u>Theorem 1</u> Suppose that for X as above $\{L_v\}_{v \in X}$ is a family of commuting endomorphisms on some linear space S. Setting

$$K(X) = \{f \in S : (\Pi_{v \in V} L_v)f = 0 , V \in \mathcal{Y}(X)\},$$

one has

$$\dim K(X) \leq \Sigma_{Y \in \mathcal{B}_s(X)} \dim K(Y). \tag{11}$$

Theorem 1 is a special case of Theorem 3.1 in Dahmen and Micchelli (1987). Moreover, conditions are given there which ensure that equality holds in (11). Specializing these facts to the case at hand gives

$$\dim D_\mu(X) = |\mathcal{B}_s(X)| \tag{12}$$

independent of $\mu \in \mathbb{C}^n$.

Defining for $Y = \{x^{i_1}, \ldots, x^{i_s}\} \in \mathcal{B}_s(X)$,

$$u_Y = Y^{-T}\mu_Y , \tag{13}$$

where $\mu_Y = (\mu_{i_1}, \ldots, \mu_{i_s})$, the common zeros of the polynomials $\Pi_{v \in V}(x \cdot v + \mu_v)$, $V \in \mathcal{Y}(X)$, are the points $- u_Y$, $Y \in \mathcal{B}_s(X)$. Hence one expects that $D_\mu(X)$ is spanned by functions of the form $\exp\{- u_Y \cdot x\}p(x)$, where $p(x)$ is some polynomial. More precisely, let us call any two elements Y , $Y' \in \mathcal{B}_s(X)$ equivalent if and only if $u_Y = u_{Y'}$, and let $\mathcal{E}_j$, $j = 1,\ldots,m$, denote the corresponding equivalence classes with representers $Y_j \in \mathcal{E}_j$, $j = 1,\ldots,m$. Moreover, define for any $Y \in \mathcal{B}_s(X)$

$$X_Y = \{z \in X : z \cdot u_Y = \mu_z\},$$

i.e. $Y \subseteq X_Y$ so that $\langle X_Y \rangle = \mathbb{R}^s$. It is pointed out in Dahmen and Micchelli (1987) and Ben-Artzi and Ron (1987) that when $p_{j,i}$, $i = 1,\ldots,\ell_j$, is a basis of $D_0(X_{Y_j})$ then the collection of functions

$$\{\exp(- u_{Y_j} \cdot x)p_{i,j}(x) : i = 1,\ldots,\ell_j , j = 1,\ldots,m\} \tag{14}$$

forms a basis for $D_\mu(X)$.

The spaces $D_0(X)$, $D_\mu(X)$ play a fundamental role for the theory of box or cube splines and exponential cube splines (cf. Dahmen and Micchelli (1987), de Boor and Höllig (1982/83)) leading among other things to interesting implications concerning related combinatorial and algebraic problems (see Dahmen and Micchelli (1988), (1987)). Moreover, certain interpolation problems were recently shown by Dyn and Ron (1988a), (1988b) to have unique solutions in these spaces.
To make any practical use of these results requires the determination of appropriate bases for these spaces. In view of (14) this reduces to constructing for any X a basis for $D_0(X)$. Since we will be mainly concerned with this space we will henceforth drop the sub-

script 0. It was shown by Dahmen and Micchelli (1983a) that D(X) contains only polynomials. In fact, one has the inclusion

$$\Pi_{d(X)}(\mathbb{R}^s) \subseteq D(X) \subseteq \Pi_{n-s}(\mathbb{R}^s), \tag{15}$$

where $d(X) + 1 = \min\{|V| : V \in \mathcal{V}(X)\}$ and $\Pi_k(\mathbb{R}^s)$ denotes the space of all real polynomials of total degree at most k on $\mathbb{R}^s$. Although the dimension of D(X) is known precisely, its structure is generally rather complicated. For instance, when s = 2 , $X = \{(1,0),(1,0),(0,1),(1,1)\}$, one easily verifies that $\{1,x,y,x^2,y^2 - 2xy\}$ is a basis for D(X) in this case. However, in general, when more directions are involved or when dealing with more than two variables no general recipe for constructing a basis seems to be known. Exploiting the intimate connection between the space D(X) and certain spline spaces a general strategy for constructing a basis for D(X) will be derived in the following section.

3. A basis for D(X)

It was pointed out by Dahmen and Micchelli (1983a) and de Boor and Höllig (1982/83) that when $X \subseteq \mathbb{Z}^s \setminus \{0\}$ the space D(X) is spanned by the translates of the cube spline $C(\cdot|X)$. If, in addition, X satisfies (2) then $C(\cdot|X)$ is related to the truncated power by

$$C(\cdot|X) = \nabla_X T(\cdot|X),$$

where $\nabla_y f(\cdot) = f(\cdot) - f(\cdot - y)$ and for $v \in V$, $\nabla_V f = \nabla_v(\nabla_{V\setminus\{v\}} f)$. Hence appropriate pieces of multivariate truncated powers should be expected to form a basis of D(X). It will be shown next that this is indeed the case even for arbitrary $X \subset \mathbb{R}^s \setminus \{0\}$.

Note first that replacing x^j in X by cx^j for any $c \in \mathbb{R}$ will leave D(X) unchanged. Thus, assigning appropriate signs to the elements of X , we may assume without loss of generality that X satisfies (2) so that $T(\cdot|X)$ is well defined. Moreover, upon rescaling the x^i if necessary we may assume throughout the following that

$$\langle x^i \rangle = \langle x^j \rangle \text{ implies } x^i = x^j . \tag{16}$$

When $|X| = n = s$ there is nothing to prove since $D(X) = \Pi_0(\mathbb{R}^s)$ This suggests proceeding by induction on $|X|$. Suppose that for some X, $|X| = n \geq s$ one has already determined a basis for $D(X)$. Furthermore, suppose that

$$X' = X \cup \{y\}$$

also satisfies (2). Then, defining

$$A(X|y) = \{Y \in \mathcal{B}_{s-1}(X) : \{y\} \cup Y \in \mathcal{B}_s(X')\},$$

one has

$$|\mathcal{B}_s(X')| = |A(X|y)| + |\mathcal{B}_s(X)|. \tag{17}$$

Thus it remains to construct $N := |A(X|y)|$ additional linearly independent elements in the quotient space $D(X')/D(X)$.

The construction of these additional polynomials will be based on an appropriate ordering for $A(X_0|y)$, where $X_0 \subseteq X$ is a maximal subset of pairwise distinct elements in X.

To this end, $S = \{S_j : j = 1,\dots,L\}$ will denote a fixed maximal sequence in $A(X_0|y)$ such that the cones $C_j = \langle S_j \cup \{y\}\rangle_+$ satisfy

$$C_j \setminus \cup_{i<j} C_i \neq \emptyset . \tag{18}$$

Without loss of generality one can assume that no element of S can be written as the union of essentially disjoint cones of the form $\langle Y \cup \{y\}\rangle_+$, $Y \in A(X_0|y)$. In fact, if such a cone C belongs to S at least one of the cones C' contained in C cannot belong to S. Swapping the corresponding elements Y, Y' of $A(X_0|y)$ would preserve (18) for a sequence of at least the same length L.

The following information about the elements of $A(X_0|y)$ which are not contained in S will be useful.

<u>Lemma 1</u> For every $Y \in A(X_0|y) \setminus S$ there exists a subset H of S such that

$$\langle Y\rangle_+ \subseteq \cup\{\langle S\rangle_+ : S \in H\},$$

i.e. $\langle Y\rangle = \langle S\rangle$, $S \in H$.

<u>Proof</u> Let $j \leq L$ be the smallest index such that

$$C = \langle\{y\} \cup Y\rangle_+ \subseteq \cup_{i\leq j} C_i , \tag{19}$$

i.e.

$$C \setminus \cup_{i<j} C_i \neq \emptyset . \tag{20}$$

In particular, this means that

$$K = \langle Y\rangle_+ \setminus \cup_{i<j} C_i \neq \emptyset$$

and $K \subset C_j$. If K intersects the interior of C_j one easily concludes that

$$C_j \setminus ((\cup_{i<j} C_i) \cup C) \neq \emptyset .$$

In view of (20) this implies that inserting Y between S_{j-1} and S_j still provides a sequence satisfying (18) and thus contradicting the maximality of S. Hence $K \subseteq \langle S_j\rangle_+$ which means $\langle S_j\rangle = \langle Y\rangle$. Let $v \in S_j \setminus Y$. Clearly $\langle Y\rangle_+ \subseteq \cup\{\langle\{v\} \cup (Y \setminus \{z\})\rangle_+ : z \in Y\}$. Since every $\langle V\rangle_+$, $V \in A(X_0|y)$, is covered by $(s-1)$-cones $\langle S\rangle_+$, $S \in S$, the assertion follows.

An immediate consequence of Lemma 1 is that $S = A(X_0|y)$ if the vectors in X_0 are in general position. In particular, this is the case for any $X \subset \mathbb{R}^2$ satisfying (2).

Now let G_j denote the (possibly empty) set of all $Y \in A(X_0|y) \setminus S$ such that the highest index j of a covering H for $<Y>_+$ given by Lemma 1 is as small as possible. Fixing any ordering for G_j and inserting G_j between S_j and S_{j+1} establishes an ordering for all of $A(X_0|y)$ which will be denoted throughout the sequel by $A(X_0|y) = \{Y_j : j = 1,\dots,M\}$. The corresponding cones $<Y_j \cup \{y\}>_+$ will again be denoted by C_j.

The role of $A(X_0|y)$ with respect to the full set $A(X|y)$ becomes clear when considering the equivalence relation

$$Y \sim V \quad \text{iff} \quad <Y \cup \{y\}>_+ = <V \cup \{y\}>_+ \ , \ Y, V \in A(X|y) .$$

In fact, recalling (16), one easily verifies that $A(X_0|y)$ is a set of representers for the corresponding equivalence classes E_j of $Y_j \in A(X_0|y)$. Note that for $Y_i = \{x^{i,1},\dots,x^{i,s-1}\}$ the cardinalities $\ell_i = |E_i|$ are given by

$$\ell_i = m_{i,1} \cdots m_{i,s-1} , \tag{21}$$

where $x^{i,k}$ occurs $m_{i,k}$ times in X . For any $I = (i_1,\dots,i_{s-1}) \in \mathbb{Z}_+^{s-1}$ let

$$Y_i(I) = \{x^{i,1},\dots,x^{i,1},\dots,x^{i,s-1},\dots,x^{i,s-1}\} ,$$

where $x^{i,k}$ occurs exactly i_k times in $Y_i(I)$. Specifically, $Y_i((1,\dots,1)) = Y_i$, while $\hat{Y}_i$ will be used as a shorthand notation for $Y_i((m_{i,1},\dots,m_{i,s-1}))$. Let $\ell(i) \le i$ be the largest integer such that $Y_{\ell(i)} \in S$. Let $\Delta_i = <Y_{\ell(i)}>_+ \setminus \cup_{j<\ell(i)} <Y_j>_+$ and define for $Y_i \in S$

$$H_i = (\cup\{\hat{Y}_j : \ell(i) \le j < i \ , <Y_j> = <Y_i>, <Y_j>_+ \cap \Delta_i \ne \emptyset\} \setminus \hat{Y}_i , \tag{22}$$

while $H_i = \emptyset$ when $Y_i \in S$. Furthermore, let

$$J_i = \{I = (i_1,\dots,i_{s-1}) : i_k \in \{0,\dots,m_{i,k}-1\} , k = 1,\dots,s-1\} , \tag{23}$$

so that $|J_i| = \ell_i = |E_i|$. Finally, defining, for $I \in J_i$,

$$V(I) = Y_i(I) \cup H_i , \tag{24}$$

one is ready to construct additional linearly independent polynomials in $D(X')\setminus D(X)$. Writing briefly $D_V f = (\Pi_{v\in V} D_v) f$ the polynomials $P_{i,I}(x)$, $I \in \mathcal{J}_i$, $i = 1,\ldots,M$, are defined to be extensions of certain polynomial pieces of truncated powers, namely

$$P_{i,I}(x) = D_{V(I)} T(x|X') \; , \; x \in (C_{\ell(i)} \setminus \cup_{j<\ell(i)} C_j) \cap C_i . \tag{25}$$

The main result of the paper may be stated as follows.

Theorem 2 Let X , $X' = X \cup \{y\}$ satisfy (2) and let $P_{i,I}$ be defined by (25). Then for any basis $\mathcal{B}$ of $D(X)$ the set

$$\mathcal{B} \cup \{P_{i,I} : I \in \mathcal{J}_i \; , \; i = 1,\ldots,M\}$$

is a basis for $D(X')$.

Proof Note that in view of (21), (23) there are exactly

$$\sum_{i=1}^{M} |\mathcal{J}_i| = \sum_{i=1}^{M} |\mathcal{E}_i| = |\mathcal{A}(X|y)| = N$$

functions $P_{i,I}$. In view of (12) and (17) it remains to show that the $P_{i,I}$ represent linearly independent elements in $D(X')/D(X)$.

To this end, recall from (9) that for any $V \subset X'$,

$$D_V T(\cdot|X') = T(\cdot|X'\setminus V) . \tag{26}$$

Since $D(W) \subset D(X)$ for $W \subset X$, (10) readily confirms that

$$P_{i,I} \in D(X') \; , \; I \in \mathcal{J}_i \; , \; i = 1,\ldots,M . \tag{27}$$

Next observe that for $x \in (C_{\ell(i)} \setminus \cup_{j<\ell(i)} C_j) \cap C_i$,

$$D_{X\setminus\langle Y_i\rangle} P_{i,I}(x) = D_{X\setminus\langle Y_i\rangle} D_{V(I)} T(x|X')$$

$$= T(x|\{y\} \cup (X \cap \langle Y_i\rangle)\setminus V(I)) .$$

By definition (24) of $V(I)$, the set $(X \cap \langle Y_i\rangle)\setminus V(I)$ still contains some element of $A(X|y)$ while also (cf. (3))

$$C_i \cap (C_{\ell(i)} \setminus \cup_{j<\ell(i)} C_j) \subseteq \langle \{y\} \cup ((X \cap \langle Y_i\rangle)\setminus V(I))\rangle_+$$

$$= \operatorname{supp} T(\cdot|\{y\} \cup ((X \cap \langle Y_i\rangle)\setminus V(I))) .$$

Hence $D_{X\setminus\langle Y_i\rangle} P_{i,I}$ does not vanish identically. But since $X\setminus\langle Y_i\rangle \in \mathcal{Y}(X)$, this means that

$$P_{i,I} \notin D(X) \quad , \; I \in J_i \quad , \; i = 1,\ldots,M . \tag{28}$$

So, in view of (27), (28), it remains to confirm the linear independence of the $P_{i,I}$ in $D(X')/D(X)$. To this end, the following observation is useful.

Lemma 2 For $i = 1,\ldots,M$, $M = |A(X_0|y)|$, let

$$W_i = \{y\} \cup V(m^i) ,$$

where $m^i = (m_{i,1},\ldots,m_{i,s-1})$. Then the functions $D_{X'\setminus W_i} P_{i,I}$, $I \in J_i$, are linearly independent.

Proof Suppose

$$\Sigma_{I\in J_i} c_I D_{X'\setminus W_i} P_{i,I}(x) = 0 \; , \; x \in \mathbb{R}^s .$$

By definition (25), (26) and (9) this implies

$$\Sigma_{I\in J_i} c_I T(x|W_i\setminus V(I)) = 0 \; , \; x \in (C_{\ell(i)} \setminus \cup_{j<\ell(i)} C_j) \cap C_i .$$

By (24) one has

$$\{W_i \setminus V(I) : I \in \mathcal{J}_i\} = \{\{y\} \cup Y_i \cup Y_i(I') : I' \in \mathcal{J}_i\}.$$

Hence the summands $T(\cdot|W_i \setminus V(I))$, $I \in \mathcal{J}_i$, range over all different products of the form (7). Since these functions are obviously linearly independent one concludes $c_I = 0$, $I \in \mathcal{J}_i$. This completes the proof of Lemma 2.

Next suppose that for a given basis $\{Q_j : j = 1,\ldots,|\mathcal{B}_s(X)|\}$ of $D(X)$

$$\Sigma_{i=1}^{M} \Sigma_{I \in \mathcal{J}_i} c_{i,I} P_{i,I}(x) + \Sigma_{j=1}^{|\mathcal{B}_s(X)|} a_j Q_j(x) = 0 , \ x \in \mathbb{R}^s .$$

Since $X' \setminus W_k \in \mathcal{Y}(X)$ one also has for any $k \leq M$,

$$\Sigma_{i=1}^{M} \Sigma_{I \in \mathcal{J}_i} c_{i,I} D_{X' \setminus W_k} P_{i,I}(x) = 0 , \ x \in \mathbb{R}^s . \tag{29}$$

<u>Lemma 3</u> For any $I \in \mathcal{J}_i$, $i > k$, one has

$$D_{X' \setminus W_k} P_{i,I}(x) = 0 , \ x \in \mathbb{R}^s . \tag{30}$$

<u>Proof</u> Note that for fixed i and $x \in (C_{\ell(i)} \setminus \cup_{j<\ell(i)} C_j) \cap C_i$,

$$D_{X' \setminus W_k} P_{i,I}(x) = D_{V(I)} T(x|X' \setminus (X' \setminus W_k)) \tag{31}$$

$$= D_{V(I)} T(x|W_k) .$$

Let $k < i$ and suppose first $Y_i \in \mathcal{S}$, i.e. $\ell(i) = i$. Note that in this case

$$C_i \setminus \langle W_k \rangle_+ \neq \emptyset . \tag{32}$$

In fact, (32) readily follows from (18) and (22) when $\langle Y_i \rangle \neq \langle Y_k \rangle$ or when $Y_k \in \mathcal{S}$. So suppose $\langle Y_i \rangle = \langle Y_k \rangle$ and $Y_k \notin \mathcal{S}$. Let R denote an (s-2)-dimensional supporting hyperplane of Δ_i that separates Δ_i and Δ_k. By

(22) those Y_m that contribute to H_k must be on the same side of R as Δ_k, again confirming (32). Thus, when $Y_i \in \mathcal{S}$ (30) follows from (3), (31) and (32). The same reasoning applies if $Y_i \notin \mathcal{S}$ but $k < \ell(i)$. So assume next that $\ell(i) \leq k < i$. Hence $\langle Y_i\rangle = \langle Y_k\rangle$. From (22), (24) one concludes that $\dim\langle W_k \setminus V(I)\rangle < s$ so that (30) follows again from (3), (31) and (26). This completes the proof of Lemma 3.

Choosing now successively $k = 1,2,\ldots,M$ in (29) and invoking Lemma 2 at each step shows that $c_{i,I} = 0$, $I \in \mathcal{J}_i$, $i = 1,\ldots,M$. This proves the linear independence of the polynomials $P_{i,I}$ in $D(X')/D(X)$ which, in view of (12), (17) finishes the proof of Theorem 2.

As pointed out in Section 2 (cf. (14)) being able to construct a basis for spaces of the type $D(X)$ immediately allows to exhibit a basis of $D_\mu(X)$ for any $\mu \in \mathbb{C}^n$. Alternatively, using the operators $\mu_j + D_{x^j}$ instead of D_{x^j} , and (8), a basis for $D_\mu(X)$ could be constructed directly from the pieces of $T_\mu(\cdot|X)$ in the same way as shown above.

4. An example

In this section possibilities of evaluating and representing the above basis functions will briefly be discussed and illustrated by an example.

One should note first that under additional assumptions on X the above construction may simplify significantly. For instance, when X has the form $X = \{y^1,\ldots,y^1,\ldots,y^s,\ldots,y^s\}$ where $Y = \{y^1,\ldots,y^s\}$ spans $\mathbb{R}^s$, one has $|\mathcal{A}(X_0|y)| = 1$. In this case (25) produces in view of the representation (7) the expected tensor product basis. When the elements of X are in general position one has $\mathcal{A}(X|y) = \mathcal{A}(X_0|y) = \mathcal{S}$ and the different pieces of $T(\cdot|V)$, $V \subset X$, form a basis of $D(X)$. In this case the following explicit representation for $T(\cdot|X)$ was established by Dahmen and Micchelli (1986):

$$T(x|X) = \frac{1}{(n-s)!}\sum_{Y\in\mathcal{B}_s(X)} a_Y|\det Y|^{-1}(x^Y \cdot x)^{n-s}\chi_{\langle Y\rangle_+}(x), \tag{33}$$

where, for $Y \in \mathcal{B}_s(X)$,

$$x^Y \cdot y = 1 \, , \; y \in Y$$

and

$$a_Y = \prod_{z \notin Y} (1 + x^Y \cdot z)^{-1} \, .$$

A similar representation can be given for $T_\mu(\cdot|X)$ when $\mu_j \neq 0$, $j = 1,\dots,n$ (cf. Dahmen and Micchelli (1981)).

When X is in general position another explicit basis was constructed by Dahmen and Micchelli (1985) without using truncated powers.

In general, one has to use the recurrence relation (cf. Dahmen (1980))

$$T(x|X) = \frac{1}{n-s} \Sigma_{j=1}^n \lambda_j T(x|X \setminus \{x^j\}) \tag{34}$$

which holds whenever

$$x = \Sigma_{j=1}^n \lambda_j x^j \, . \tag{35}$$

A general strategy would be to select first a possibly large subset X^1 of X for which a basis is easily available either via (7) or (33). The representations for the extensions of type (25) could then be obtained by applying the recursion (34). The freedom in choosing the representations (35) should then be exploited in such a way that possibly many of the previously calculated representations occur during the recursion. This is in fact strongly favoured by the fact that (25) always involves truncated powers for subsets of X .

The following bivariate example illustrates this strategy. As pointed out before, after assigning appropriate signs to the elements of X , so that $0 \notin [X]$, X_0 will be always in general position. Hence $\mathcal{S} = \mathcal{A}(X_0|y)$ and $H_i = \emptyset$ (cf. (22)) for all i , which simplifies, of course, the construction.

Let $X = \{x^1,\dots,x^6\} \subset \mathbb{R}^2$ where $x^1 = x^2 = (1,0)$, $x^3 = x^4 = (0,1)$, $x^5 = (1,1)$, $x^6 = (-1,1)$. In order to construct a basis for D(X) one can follow the lines of the previous section extending step by step bases of appropriate subspaces of D(X) . For instance, consider

$X^1 = \{x^1, \ldots, x^4\}$. As pointed out before the representation (7) already indicates the tensor product structure of $D(X^1)$ and one easily verifies that

$$D(X^1) = \mathrm{span}\{1, x_1, x_2, x_1x_2\} \quad . \tag{36}$$

Setting $X^2 = X^1 \cup \{x^5\}$, one has

$$A(X^1|x^5) = \{\{x^1\},\{x^2\},\{x^3\},\{x^4\}\} .$$

One can take $Y_1 = \{x^1\}$, $Y_2 = \{x^3\}$ as representers of the corresponding equivalence classes in $A(X^1|x^5)$, with $J_1 = J_2 = \{(0),(1)\}$. According to (25) one has

$$P_{1,(0)}(x) = T(x|X^2) \ , \ P_{1,(1)}(x) = T(x|X^2\setminus\{x^1\})$$

for $x \in C_1 = <\{x^1,x^5\}>_+$ and

$$P_{2,(0)}(x) = T(x|X^2) \ , \ P_{2,(1)}(x) = T(x|X^2\setminus\{x^3\})$$

for $x \in C_2 = <\{x^3,x^5\}>_+$. By (34) one obtains, for $x \in C_1$,

$$T(x|X^2) = \frac{1}{3}((x_1 - x_2)T(x|X^2\setminus\{x^1\}) + x_2T(x|X^1)).$$

Using (34) again yields

$$T(x|X^2\setminus\{x^1\}) = \frac{1}{2}((x_1 - x_2)T(x|x^3,x^4,x^5) + x_2T(x|x^2,x^3,x^4)),$$

which for $x \in C_1$ reduces to

$$\frac{x_2}{2} T(x|x^2,x^3,x^4) = x_2^2/2 .$$

Likewise, $T(x|X^1) = x_1x_2$, so that

$$P_{1,(0)}(x) = \frac{1}{2}(x_1x_2^2 - \frac{1}{3}x_2^3) . \tag{37}$$

Similarly, for $x \in C_1$,

$$T(x|X^2\setminus\{x^1\}) = \frac{1}{2}((x_1 - x_2)T(x|x^3,x^4,x^5) + x_2T(x|x^2,x^3,x^4)) = \frac{x_2^2}{2} ,$$

so that

$$P_{1,(1)}(x) = x_2^2/2 . \tag{38}$$

By symmetry one only has to interchange x_1 and x_2 to obtain

$$\begin{aligned} P_{2,(0)}(x) &= \frac{1}{2}(x_2x_1^2 - \frac{1}{3}x_1^3) , \\ P_{2,(1)}(x) &= \frac{1}{2}x_1^2 , \end{aligned} \tag{39}$$

so that by Theorem 2 $D(X^2)$ is spanned by the polynomials

$$1 , x_1 , x_2 , x_1x_2 , x_1^2 , x_2^2 , x_1x_2^2 - \frac{1}{3}x_2^3 , x_2x_1^2 - \frac{1}{3}x_1^3 .$$

Finally let

$$X^0 = X^2 \cup \{x^6\} ,$$

so that

$$A(X^2|x^6) = \{\{x^i\} : i = 1,\dots,5\}.$$

Setting $Y_1 = \{x^3\}$, $Y_2 = \{x^5\}$, $Y_3 = \{x^1\}$, $J_1 = \{(0),(1)\} = J_3$, $J_2 = \{(0)\}$, one has $C_1 = \langle\{x^3,x^6\}\rangle_+$, $C_2 = \langle\{x^5,x^6\}\rangle_+$, $C_3 = \langle\{x^1,x^6\}\rangle_+$. C_1,C_2,C_3 clearly satisfy (18). Hence, according to (25) one obtains

$$P_{1,(0)}(x) = T(x|X) , P_{1,(1)}(x) = T(x|X\setminus\{x^3\})$$

for $x \in C_1$,

$$P_{2,(0)}(x) = T(x|X) \ , \ x \in C_2 \setminus C_1 \ ,$$

and

$$P_{3,(0)}(x) = T(x|X) \ , \ P_{3,(1)}(x) = T(x|X \setminus \{x^1\})$$

for $x \in C_3 \setminus C_2$.

Repeated application of (34) yields

$$P_{1,(0)}(x) = (x_1 + x_2)^4/48 \ , \ P_{1,(1)}(x) = (x_1 + x_2)^3/12 \ ,$$

$$P_{2,(0)}(x) = \frac{1}{48} x_2^2(x_1 + x_2)^2 + \frac{1}{24} x_1 x_2^2(x_1 + x_2) + \frac{1}{16}(x_2 - x_1)(x_2 x_1^2 - \frac{1}{3} x_1^3),$$

$$P_{3,(0)}(x) = \frac{1}{24}(x_1 + x_2)x_2^3 + \frac{1}{8}(x_1 x_2^3 - \frac{1}{3} x_2^4) \ ,$$

$$P_{3,(1)}(x) = \frac{1}{6} x_2^3 \ .$$

One easily verifies (cf. (14)) that

$$\dim D(X) = |\mathcal{B}_2(X)| = 13 \ .$$

According to the preceding calculations a basis for D(X) is then given by the following thirteen polynomials

$$1 \ , x_1 \ , x_2 \ , x_1 x_2 \ , x_1^2 \ , x_2^2 \ , x_1 x_2^2 - \frac{1}{3} x_2^3 \ ,$$

$$x_2 x_1^2 - \frac{1}{3} x_1^3 \ , x_2^3 \ , (x_1 + x_2)^3 \ , \ (x_1 + x_2)^4 \ ,$$

$$\frac{1}{48} x_2^2(x_1 + x_2)^2 + \frac{1}{24} x_1 x_2^2(x_1 + x_2) + \frac{1}{16}(x_2 - x_1)(x_2 x_1^2 - \frac{1}{3} x_1^3) \ ,$$

$$\frac{1}{24}(x_1 + x_2)x_2^3 + \frac{1}{8}(x_1 x_2^3 - \frac{1}{3} x_2^4) \ .$$

Acknowledgement The author wishes to thank A. Ron for valuable comments and discussions that were of significant help in revising the first draft of the paper.

References

Ben-Artzi, A. and Ron, A. (1987), Translates of exponential box splines and their related spaces. Manuscript, to appear in Trans. Amer. Math. Soc.

de Boor, C. and Höllig, K. (1982/83), B-splines from parallelepipeds. J. Analyse Math., 42, 99 - 115.

Dahmen, W. (1980), On multivariate B-splines. SIAM J. Numer. Anal., 17, 179 - 191.

Dahmen, W. and Micchelli, C.A. (1981), On limits of multivariate B-splines. J. Analyse Math., 39, 256 - 278.

Dahmen, W. and Micchelli, C.A. (1983a), Translates of multivariate splines. Linear Algebra and its Applications 52/53, 217 - 235.

Dahmen, W. and Micchelli, C.A. (1983b), Recent progress in multivariate splines. In: Approximation Theory IV, ed. by C.K. Chui, L.C. Schumaker, J.D. Ward. Academic Press, 27 - 120.

Dahmen, W. and Micchelli, C.A. (1985), On the local linear independence of translates of a box spline. Studia Mathematica, 82, 243 - 263.

Dahmen, W. and Micchelli, C.A. (1988), The number of solutions to linear diophantine equations and multivariate splines. Trans. Amer. Math. Soc., 308, 509 - 532.

Dahmen, W. and Michelli, C.A. (1986), Statistical encounters with B-splines. Contemporary Mathematics, 59, 17 - 48.

Dahmen, W. and Micchelli, C.A. (1987), On multivariate E-splines. Preprint No. 267, Freie Universität Berlin, to appear in Advances in Mathematics.

Dyn, N. and Ron, A. (1988a), On multivariate polynomial interpolation. To appear in these proceedings.

Dyn, N. and Ron, A. (1988b), Local approximation by certain spaces of exponential polynomials, approximation order of exponential box splines and related interpolation problems. To appear in Trans. Amer. Math. Soc.

Monotone piecewise cubic data fitting

F. N. Fritsch
Computing and Mathematics Research Division, Lawrence Livermore National Laboratory, University of California, Livermore, USA

Abstract This paper describes PCHLS, an algorithm for least squares fitting of a monotone piecewise cubic function to data. It extends the piecewise cubic Hermite interpolation package PCHIP to situations in which the data are noisy or are adequately represented by far fewer cubic pieces than data points.
Key words: Data fitting, Least squares, Data reduction, Piecewise cubic approximation, Monotonicity preserving approximation, Shape preserving approximation

1. Introduction

Necessary and sufficient conditions for a piecewise cubic function to be monotonic were published in [1]. PCHIP, a complete package for interpolation with piecewise cubic functions and evaluating the results has been in use for many years [4]. PCHLS is a new algorithm to do least squares fitting of a monotone piecewise cubic function to given univariate data (x_k, y_k), $k=1,\ldots,m$. Its development was motivated by user-expressed needs to extend PCHIP to noisy data, for which interpolation is not appropriate, or to cases in which it is clear that an adequate approximation could be achieved with far fewer cubic pieces than data points (data reduction).

In Section 2 is given the mathematical statement of the problem solved by PCHLS, and algorithmic details are described in Section 3. Several examples to illustrate the results of the algorithm appear in Section 4. Directions for further work are indicated in Section 5.

2. The Mathematical Problem

Let $[a, b]$ be an interval containing all of the data points x_k and define breakpoints t_i so that

$$a = t_1 < t_2 < \ldots < t_{nseg} < t_n = b, \tag{1}$$

where $nseg = n - 1$. A piecewise cubic Hermite (PCH) function with breakpoints t_i has the form

$$f(x) = \sum_{i=1}^{n} \left[f_i \; H_i^{\mathrm{f}}(x) + d_i \; H_i^{\mathrm{d}}(x) \right], \tag{2}$$

where $f_i{=}f(t_i)$, $d_i{=}f'(t_i)$ and $H_i^{\mathrm{f}}, H_i^{\mathrm{d}}$ are the cubic Hermite basis functions defined by

$$H_i^{\mathrm{f}}(t_j) = \delta_{ij}, \; \frac{d}{dx} H_i^{\mathrm{f}}(t_j) = 0; \tag{3}$$

$$H_i^{\mathrm{d}}(t_j) = 0, \frac{d}{dx} H_i^{\mathrm{d}}(t_j) = \delta_{ij}, \tag{4}$$

where we have used the standard Kronecker delta notation.

2.1 Least squares equations

The least squares equations for the unknown PCH parameters f_i, d_i are obtained by evaluating (2) at each of the data points:

$$\sum_{i=1}^{n} \left[f_i \; H_i^{\mathrm{f}}(x_k) + d_i \; H_i^{\mathrm{d}}(x_k) \right] = y_k, \quad k = 1, \ldots, m. \tag{5}$$

We note that if the $2n$ unknowns are arranged in pairs $f_1, d_1, f_2, d_2, \ldots, f_n, d_n$ the k-th row of the $m \times 2n$ least squares matrix will be

$$\left[H_1^{\mathrm{f}}(x_k) \; H_1^{\mathrm{d}}(x_k) \; H_2^{\mathrm{f}}(x_k) \; H_2^{\mathrm{d}}(x_k) \; \ldots \; H_n^{\mathrm{f}}(x_k) \; H_n^{\mathrm{d}}(x_k) \right]. \tag{6}$$

Since the i-th Hermite basis function is zero unless $x_k \in (t_{i-1}, t_{i+1})$, the matrix will have a block structure with (at most) four nonzero entries in any row, as indicated in Figure 1.

2.2 Monotonicity constraints

To the least squares equations (5) must be added the monotonicity constraints

$$0 \leq s_{i-1} d_i \leq 3 s_{i-1} \frac{f_i - f_{i-1}}{t_i - t_{i-1}}, \quad i = 2, \ldots, n; \tag{7}$$

$$0 \leq \; s_i d_i \; \leq 3 s_i \frac{f_{i+1} - f_i}{t_{i+1} - t_i}, \quad i = 1, \ldots, n-1. \tag{8}$$

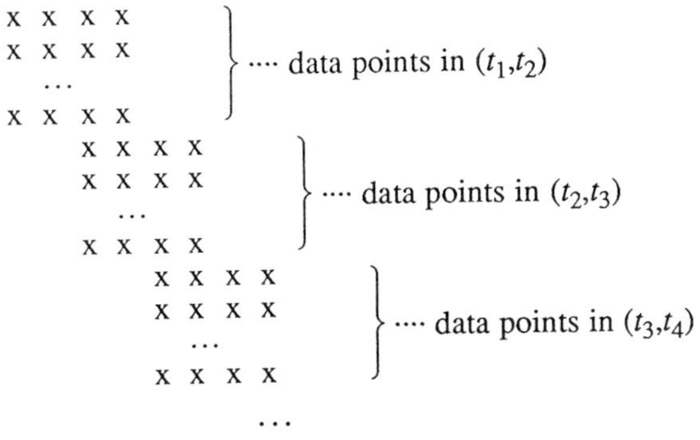

Figure 1. Structure of least squares matrix

Here s_i will either be $+1$ or -1 according as the approximation is to be monotone increasing or decreasing in (t_i, t_{i+1}). These sufficient conditions for montotonicity come from requiring that the derivatives lie in the square $[0,3]\times[0,3]$ in the monotonicity region of [1]. Conditions (7) and (8) constitute $4(n-1)$ linear inequalities[1] among the unknowns f_i, d_i. Each involves either one or three of the unknowns and has a block structure similar to that of the least squares equations.

3. The Algorithm

PCHLS is given data (x_k, y_k) and weights w_k, breakpoints t_i satisfying (1), and an array of constraint flags s_i:

$$s_i = \begin{cases} +1 & \text{for monotone increasing in } (t_i, t_{i+1}), \\ -1 & \text{for monotone decreasing in } (t_i, t_{i+1}), \\ 0 & \text{for no constraint in } (t_i, t_{i+1}). \end{cases} \tag{9}$$

PCHLS is to return PCH coefficients (f_i, d_i), $i=1,\ldots,n$ in (2) which minimize the sum of squares of weighted residuals

$$\sum_{k=1}^{m} (w_k[f(x_k) - y_k])^2 \tag{10}$$

subject to the constraints (7) and (8).

[1] These will not all be independent. In the common case in which the approximation is to be increasing throughout $[a,b]$, for example, there will be only $n + 2(n-1)$ constraints.

The algorithm operates as follows.

Step 1.
Sort the data so that $x_1 \leq x_2 \leq \ldots \leq x_m$. (This is actually done externally to PCHLS, to avoid extra work in case the data arrives ordered.)

Step 2.
Evaluate the cubic Hermite basis functions at the data points and set up the least squares matrix. Weights are included by multiplying the k-th row (6) and the right-hand side y_k by w_k.

Step 3.
Set up the monotonicity constraints (7) and (8). No constraints are generated for intervals in which $s_i{=}0$. Redundant sign constraints are eliminated when $s_{i-1}{=}s_i$.

Step 4.
Solve this linearly constrained linear least squares problem via SLATEC subroutine LSI, which is based on the algorithm described in [5].

Step 5.
Set "small" d_i-values to zero. (This postprocessing step is necessary because the current version of LSI often returns values the size of the unit roundoff, possibly with the wrong sign, when one of the sign constraints is binding.)

Since the hard part of obtaining a satisfactory fit is generally the choice of breakpoints and constraints, an interactive driver has been built on top of PCHLS. It reads data and weights, translating absent weights to $w_k{=}1$. After the user sets $nseg$, it generates either uniform or equidistributed[2] breakpoints and automatically selects the s_i by examining the data. It performs Step 1, above, calls PCHLS, and does Step 5. It plots the approximation, with data points and breakpoints superimposed, after each fit. The driver allows the user to move or delete breakpoints, change sign constraints, or change the number of segments.

4. Examples

This section contains some plots generated by the PCHLS driver, to illustrate the capabilities of the algorithm. All fits were unweighted.

The first example is the titanium data from [3]. This is uniformly spaced and appears to contain a moderate amount of noise. Figure 2(a) uses $nseg{=}7$ with all default settings and is clearly a poor fit. Adjusting the positions of breakpoints 3–5 and setting $s_i{=}{+}1$ in the first four intervals, –1 in the last three, yields 2(b). Changing s_4 to zero improves the RMS error from .0026 to .0018 (maximum relative error from 8% to 5%). (See Figure 3(a).) Deleting the second breakpoint has

[2] Same number of data points in each segment.

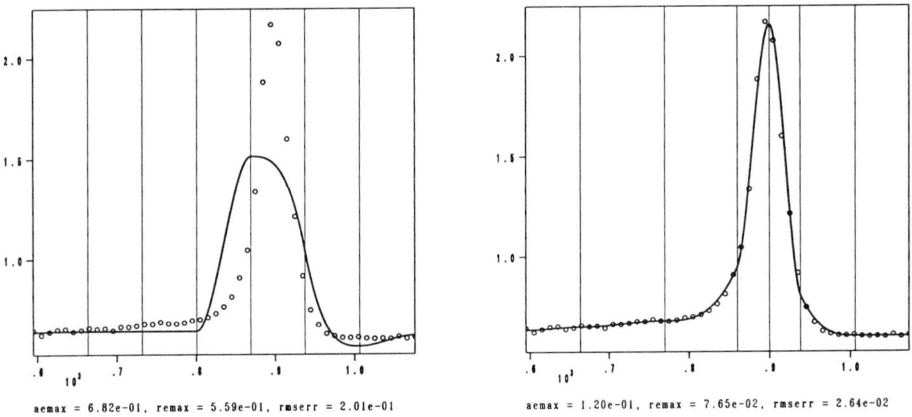

(a) (b)

Figure 2. Two fits to titanium data. (a) The result with all default settings. (b) The result after adjusting middle breakpoints and constraints.

virtually no effect on the result, as indicated in Figure 3(b). In Figure 4 this curve is compared with the PCHIP interpolant. The latter has $2m$=98 parameters, whereas the PCHLS fit has only $2n$=14, so we have smoothed out the noise and achieved a data reduction factor of 7.

The second example illustrates the use of equidistributed breakpoints. The assumption here is that if the data are not uniformly spaced, then the sample rate was dictated by the behavior of the phenomonon under study. This is certainly the case with the potentiometric titration data from [6]. In figure 5(a) is the PCHLS result with seven equidistributed breakpoints. This has $2n$=14 parameters and an RMS error of 18 (maximum relative error 1%). It looks much the same as the PCHIP interpolant in Figure 5(b), which has $2m$=42 parameters.

5. Further developments

The most obvious improvement needed would be a constrained linear least squares solver that could take advantage of the sparseness of the problem. The least squares matrix is $m \times 2n$ but has only $4m$ nonzero elements. If $ncon \leq 4(n-1)$ is the actual number of constraints, there will be fewer than $3ncon$ nonzero elements in the $ncon \times 2n$ constraint matrix. While the wasted space is tolerable in most of the problems on which PCHLS has been used to date, it will be prohibitive if the algorithm is to be extended to bivariate data.

The present automatic constraint setting algorithm is very primitive. It might

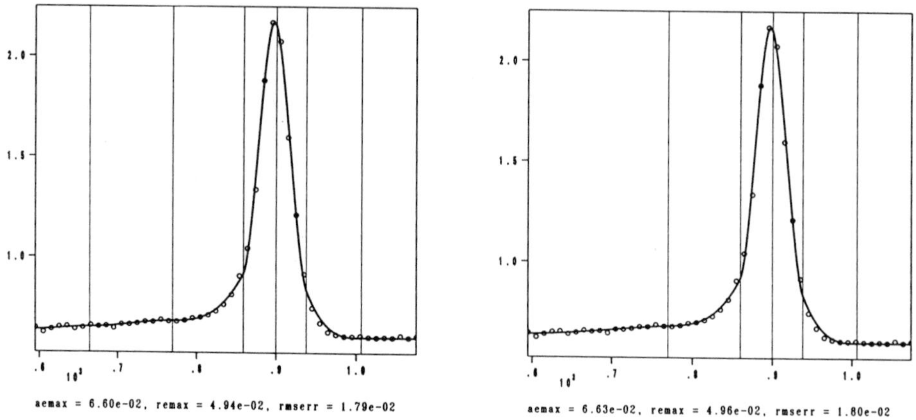

(a) (b)

Figure 3. Two minor adjustments. (a) Removed constraint in fourth segment. (b) Deleted first interior breakpoint.

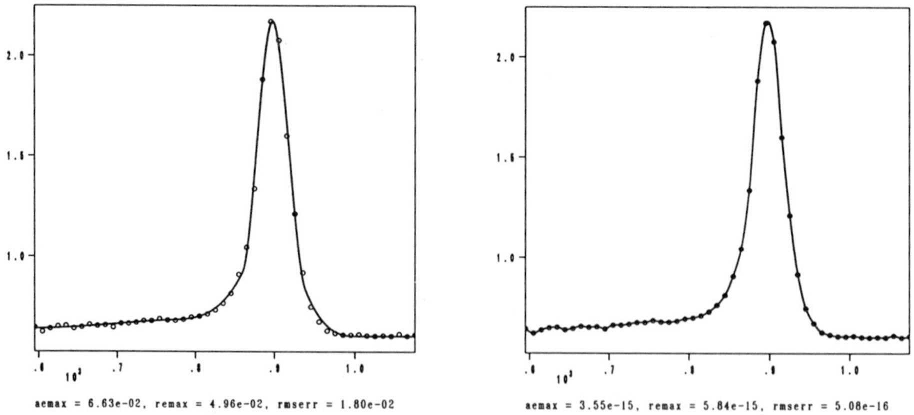

(a) (b)

Figure 4. Comparison of the latter curve (a) with the PCHIP interpolant (b) to the titanium data.

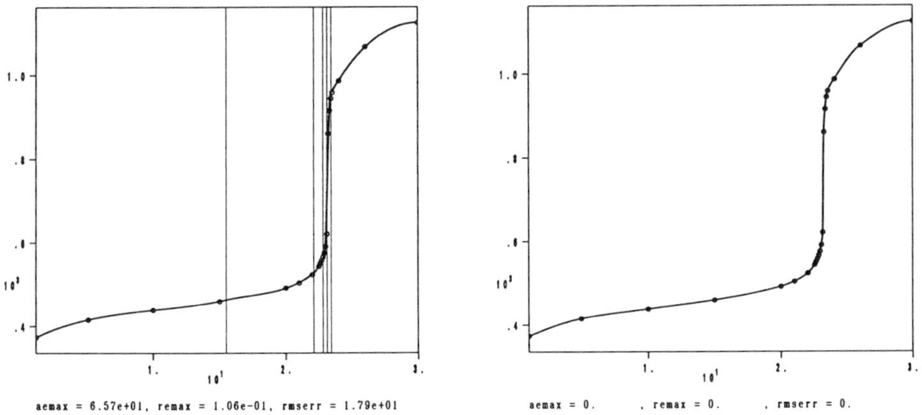

(a) (b)

Figure 5. Two approximations to potentiometric titration data. (a) The PCHLS fit with n=7 and equidistributed breakpoints. (b) The PCHIP interpolant.

be improved by doing local least squares fits of straight lines to the data in each breakpoint interval.

Because the most tedious part of the fitting process for a user is adjusting the locations of the breakpoints, work on automatic knot placement, such as that reported by Peter Harris at this conference [2], might be a valuable addition to the PCHLS driver.

The author has had good results on some problems by adding convexity constraints to the monotonicity constraints provided by PCHLS. This modification comes closer to providing a true shape preserving data fitting algorithm.

Another modification under consideration is the addition of a term to the objective function to reduce the size of the second derivative jumps at the data points.

Acknowledgments

This work was performed under the auspices of the U.S. Department of Energy by Lawrence Livermore Laboratory under contract W-7405-Eng-48.

References

1. R. E. Carlson and F. N. Fritsch. Monotone piecewise cubic interpolation. *SIAM J. Numer. Anal.*, 17:238–246, 1980.

2. M. G. Cox, P. M. Harris, and Helen M. Jones. A strategy for knot placement in least squares data fitting by splines. In *(These Proceedings)*, 1988.
3. Carl de Boor. *A Practical Guide to Splines*, page 222. Springer-Verlag, New York, 1978.
4. F. N. Fritsch and R. E. Carlson. *PCHIP Final Specifications.* LLNL Computer Documentation Report UCID-30194, Lawrence Livermore National Laboratory, Livermore, California, 1982.
5. Charles L. Lawson and Richard J. Hanson. *Solving Least Squares Problems*, chapter 23. Prentice-Hall, Englewood Cliffs, New Jersey, 1974.
6. Lawrence F. Shampine and Richard C. Allen, Jr. *Numerical Computing: an introduction*, page 40. Saunders, Philadelphia, 1973.

Direct and converse results on simultaneous approximation by the method of Bernstein–Durrmeyer operators

M. Heilmann and M. W. Müller

Lehrstuhl Mathematik VIII
Universität Dortmund, FRG

Abstract The n-th Bernstein-Durrmeyer operator M_n results from the n-th classical Bernstein operator with weights p_{nk} if the discrete values $f(\frac{k}{n})$ in its definition are replaced by an integral over the weighted function. For integrable functions the rate of simultaneous approximation will be related to the Ditzian-Totik modulus of smoothness. For the local rate of simultaneous approximation a generalized Voronovskaja theorem is proved.

Key words: Bernstein-Durrmeyer operators, Direct results, Inverse results, Local direct results, Simultaneous approximation.

1. Introduction

The n-th Bernstein-Durrmeyer operator M_n, $n \in \mathbb{N}$, results from the n-th classical Bernstein operator with weights $p_{nk}(x) = \binom{n}{k}x^k(1-x)^{n-k}$, $k \in \{0,\dots,n\}$, $x \in I = [0,1]$, if the discrete values $f(\frac{k}{n})$ in its definition are replaced by an integral over the weighted function. More precisely M_n assigns to a function $f \in L_1(I)$ the polynomial $M_n f$ of degree n defined by

$$(M_n f)(x) = (n+1)\sum_{k=0}^{n} p_{nk}(x)\int_0^1 p_{nk}(t)f(t)dt, \qquad x \in I.$$

These operators were introduced by Durrmeyer (1967) and first studied by

Derriennic (1979). For the rate of approximation in the L_p-metric global direct and inverse theorems have been proved by Ditzian and Ivanov. Their paper contains moreover a solution of the so-called nonoptimal approximation problem. The global saturation problem has been solved by Heilmann.

In this paper we will derive

a) global direct and inverse results for the rate of weighted simultaneous approximation $\|w(M_n f-f)^{(s)}\|_p$, $s \in \mathbb{N}_0$, of functions $f \in L_p^s(I) := \{g \mid g^{(s)} \in L_p(I)\}$, $1 \leq p < \infty$ (Theorems 3, 5, 6, 7),

b) a generalized Voronovskaja theorem for the local rate of (nonweighted) simultaneous approximation $(M_n f-f)^{(s)}(x)$, $f \in L_1^s(I)$ (Theorem 8).

If $f \in L_p^s(I)$, $s \in \mathbb{N}_0$, $1 \leq p \leq \infty$, $n > s$, $x \in I$, then

$$(M_n f)^{(s)}(x) = (n+1)\alpha(n,s) \sum_{k=0}^{n-s} p_{n-s,k}(x) \int_0^1 p_{n+s,k+s}(t) f^{(s)}(t)dt, \tag{1}$$

with $\alpha(n,s) = \frac{(n!)^2}{(n-s)!(n+s)!} < 1$, cf. Derriennic (1979).

Occasionally we will work for convenience with operators $M_{n,s}$ given for $h \in L_p(I)$, $1 \leq p \leq \infty$, $n > s$, $x \in I$ by

$$(M_{n,s}h)(x) = (n+1)\alpha(n,s) \sum_{k=0}^{n-s} p_{n-s,k}(x) \int_0^1 p_{n+s,k+s}(t)h(t)dt. \tag{2}$$

Evidently $(M_n f)^{(s)} = M_{n,s} f^{(s)}$ if $f \in L_p^s(I)$. (3)

Throughout this paper C_i will denote positive constants not necessarily the same at each occurance.

2. Basic properties

For every $n \in \mathbb{N}$, $k \in \{0,\dots, n\}$, $x \in I$, there holds

$$\sum_{k=0}^{n} p_{nk}(x) = 1, \quad \int_0^1 p_{nk}(t)dt = \frac{1}{n+1}, \tag{4}$$

$$\frac{k}{n} p_{nk}(x) = x \cdot p_{n-1,k-1}(x), \quad \int_0^1 t \cdot p_{n+s,k+s}(t)dt = \frac{k+s+1}{(n+s+1)(n+s+2)}, \tag{5}$$

$$p'_{nk}(x) = n(p_{n-1,k-1}(x) - p_{n-1,k}(x)), \tag{6}$$

$$\varphi(x)^2 p'_{nk}(x) = (k-nx)p_{nk}(x). \tag{7}$$

(Here and in the following we put formally $p_{nk}(x) = 0$ whenever $k < 0$ or $k > n$.)

Using (4) and the Riesz-Thorin theorem (cf. Bergh, Löfström, Theorem 1.1.1) it is easily proved that

$$\|M_{n,s}h\|_p \leq C\|h\|_p,\ h \in L_p(I),\ 1 \leq p \leq \infty,\ n > s, \tag{8}$$

with a constant C that is independent of n and p.

The following lemma is useful in connection with estimates for the moments $M_{n,s}(t-x)^m(x)$, $m \in \mathbb{N}_0$.

Lemma 1 Let $m \in \mathbb{N}_0$, $s \in \mathbb{N}_0$, $n > s$, $x \in I$ and

$$T_{n,s,m}(x) := \frac{n+s+1}{(n+1)\alpha(n,s)} M_{n,s}(t-x)^m(x) \tag{9}$$

$$= (n+s+1) \sum_{k=0}^{n-s} p_{n-s,k}(x) \int_0^1 p_{n+s,k+s}(t)(t-x)^m dt.$$

With $\varphi(x) = \sqrt{x(1-x)}$ we have the recursion formula

$$T_{n,s,0}(x) = 1, \quad T_{n,s,1}(x) = \frac{(s+1)(1-2x)}{n+s+2}, \tag{10}$$

$$T_{n,s,m+1}(x)$$

$$= \frac{1}{n+s+m+2}\left[\varphi(x)^2(T'_{n,s,m}(x)+2m\cdot T_{n,s,m-1}(x))+(r+m+1)(1-2x)T_{n,s,m}(x)\right],$$

$m \in \mathbb{N}$, and the representations

$$T_{n,s,2m}(x) = \sum_{i=0}^{m} \beta_{i,2m}(n)\left[\frac{\varphi(x)^2}{n}\right]^{m-i} \cdot n^{-2i}, \tag{11}$$

$$T_{n,s,2m+1}(x) = (1-2x)\sum_{i=0}^{m} \beta_{i,2m+1}(n)\left[\frac{\varphi(x)^2}{n}\right]^{m-i} \cdot n^{-2i-1},$$

where the $\beta_{i,2m}(n)$ and $\beta_{i,2m+1}(n)$ are independent of x and for fixed m uniformly bounded in n.

Proof $T_{n,s,0}(x)$ and $T_{n,s,1}(x)$ can be calculated directly. The recursion formula follows by an argument which is similar to Derriennic (1979), Proposition II.3. Then the representations (11) can easily be derived from (10) by induction. □

In section 3 the following corollary will be needed.

Corollary 2 For every $m \in \mathbb{N}_0$, $n > s$, $n \geq 3$ and $x \in I$ we have

$$|M_{n,s}(t-x)^{2m}(x)| \leq Cn^{-m}(\varphi(x)^2+n^{-1})^m, \tag{12}$$

$$|M_{n,s}(t-x)^{2m+1}(x)| \leq Cn^{-m-1}(\varphi(x)^2+n^{-1})^m,$$

where C is independent of n.

Proof For $x \in [0,\frac{1}{n}] \cup [1-\frac{1}{n},1]$ we have $\varphi(x)^2 \le \frac{n-1}{n^2}$. Thus by (11)

$$|M_{n,s}(t-x)^{2m}(x)| \le \frac{(n+1)\alpha(n,s)}{n+s+1} \sum_{i=0}^{m} |\beta_{i,2m}(n)| (\frac{n-1}{n^3})^{m-i} \cdot n^{-2i} \le Cn^{-2m}.$$

For $x \in [\frac{1}{n}, 1-\frac{1}{n}]$ we have $[n\varphi(x)^2]^{-1} \le \frac{n}{n-1} \le \frac{3}{2}$. Thus by (11)

$$|M_{n,r}(t-x)^{2m}(x)| \le \frac{(n+1)\alpha(n,s)}{n+s+1} n^{-m} \varphi(x)^{2m} \sum_{i=0}^{m} |\beta_{i,2m}(n)| [n\varphi(x)^2]^{-i}$$
$$\le Cn^{-m}\varphi(x)^{2m}.$$

The second estimate of (12) is proved analogously. □

For the two monomials e_0 and e_1 we obtain by direct computation

$$(M_{n,s}e_0)(x) = 1+O(n^{-1}), \quad (M_{n,s}e_1)(x) = x+O(n^{-1}) \tag{13}$$

uniformly for $x \in I$ and $n \to \infty$.

In our first global direct theorem the rate of (nonweighted) simultaneous approximation $\|(M_n f - f)^{(s)}\|_p$ will be estimated using the second order Ditzian-Totik modulus of smoothness. In our case the r-th order modulus of smoothness ($r \in \mathbb{N}$) is given by

$$\omega_\varphi^r(f,t)_p = \sup_{0<h\le t} \|\Delta_{h\varphi}^r f\|_p, \ f \in L_p(I), \ 1 \le p \le \infty, \ \varphi(x) = \sqrt{x(1-x)}, \text{ where}$$

$$\Delta_{h\varphi}^r f(x) = \sum_{k=0}^{r} (-1)^k \binom{r}{k} f(x+(\tfrac{r}{2}-k)h\varphi(x)), \text{ whenever } [x-\tfrac{r}{2}h\varphi(x), x+\tfrac{r}{2}h\varphi(x)] \subset I$$

and $\Delta_{h\varphi}^r f(x) = 0$ otherwise.

Ditzian and Totik proved in Chapter 3 the equivalence of this modulus with the modified K-functional

$$\overline{K}_\varphi^r(f,t^r)_p = \inf\{\|f-g\|_p + t^r\|\varphi^2 g^{(r)}\|_p + t^{2r}\|g^{(r)}\|_p \,;\, g \in L_p^r(I)\}.$$

3. Global results

Theorem 3 Let $\varphi(x) = \sqrt{x(1-x)}$, $n > s$. If $f \in L_p^s(I)$, $1 \le p < \infty$, then

$$\|(M_n f - f)^{(s)}\|_p \le C\{\omega_\varphi^2(f^{(s)}, n^{-1/2})_p + n^{-1}\|f^{(s)}\|_p\} \tag{14}$$

with a constant C independent of n.

Remarks (14) implies that $\lim_{n\to\infty} \|(M_n f-f)^{(s)}\|_p = 0$ for every $f \in L_p^s(I)$, $1 \le p < \infty$, since $\lim_{t\to 0} \omega_\varphi^2(f^{(s)},t)_p = 0$ holds true for every function of this class.

For $f \in L_p^{s+2}(I)$, $1 \le p < \infty$, (14) implies that the rate of simultaneous approximation is $\|(M_n f-f)^{(s)}\|_p = O(n^{-1})$ $(n\to\infty)$ because in this case $\omega_\varphi^2(f^{(s)},n^{-1/2})_p = O(n^{-1})$.

For the proof of Theorem 3 we still need the following lemma which can be proved in a similar way as Lemma 3.2 in Heilmann and Müller (1989).

Lemma 4 Let $s \in \mathbb{N}_0$, $n \in \mathbb{N}$, $n > s$, $u \in I$ and

$$H_n(u) = (n+1)\{\int_0^1\int_0^u - \int_0^u\int_0^1\}(u-t)\sum_{k=0}^{n-s} p_{n-s,k}(x)p_{n+s,k+s}(t)dtdx.$$

Then we have with $\varphi(u) = \sqrt{u(1-u)}$ the estimate $H_n(u) \le Cn^{-1}\varphi(u)^2$, where C denotes a constant independent of n and u.

Proof of Theorem 3 In order to prove (14) we make use of the equivalence

$$\omega_\varphi^2(f^{(s)},n^{-1/2})_p \sim \overline{K}_\varphi^2(f^{(s)},n^{-1})_p.$$

Using (3) and (8) we have for every $g \in L_p^2(I)$

$$\|(M_n f-f)^{(s)}\|_p \le C_1\|f^{(s)}-g\|_p + \|M_{n,s}g-g\|_p. \tag{15}$$

We expand g by the Taylor formula with integral remainder

$$g(t) = g(x)+(t-x)g'(x)+R_2(g,t,x), \quad R_2(g,t,x) = \int_x^t (t-u)g''(u)du.$$

(13) and (12) imply

$$\|M_{n,s}g-g\|_p \le C_2 n^{-1}[\|g\|_p+\|g'\|_p]+\|M_{n,s}R_2(g,\cdot,x)(x)\|_p$$
$$\le C_3 n^{-1}[\|g\|_p+\|\varphi^2 g''\|_p]+\|M_{n,s}R_2(g,\cdot,x)(x)\|_p, \tag{16}$$

where the last inequality results from the estimate (a) in the proof of Theorem 9.5.3 by Ditzian and Totik.

Next it will be shown that

$$\|M_{n,s}R_2(g,\cdot,x)(x)\|_p \le C_4 n^{-1}\|(\varphi^2+n^{-1})g''\|_p. \tag{17}$$

In fact it is enough to prove (17) for $p = 1$ and $p = \infty$. Then the cases $1 < p < \infty$ follow by the Riesz-Thorin theorem. Utilizing (12) with $m = 1$ the proof of (17) for

$p=\infty$ is the same as for (5.10) in the paper of Ditzian and Ivanov. For $p=1$ (17) is derived by applying Fubini's theorem twice, then the definition of $H_n(u)$ and Lemma 4. This gives

$$\int_0^1 |M_{n,s}(R_2(g,\cdot,x)(x)|dx$$
$$\leq (n+1)\alpha(n,s)\int_0^1 \sum_{k=0}^{n-s} p_{n-s,k}(x)\int_0^1 p_{n+s,k+s}(t)\left|\int_x^t (t-u)g''(u)du\right|dtdx$$
$$=(n+1)\alpha(n,s)\int_0^1 |g''(u)|\{\int_0^1\int_0^u - \int_0^u\int_0^1\}(u-t)\sum_{k=0}^{n-s} p_{n-s,k}(x)p_{n+s,k+s}(t)dtdxdu$$
$$\leq \alpha(n,s)\int_0^1 |g''(u)|H_n(u)du \leq C_5 n^{-1}\|\varphi^2 g''\|_1 \leq C_5 n^{-1}\|(\varphi^2+n^{-1})g''\|_1,$$

where C_5 is independent of n. Thus (17) is proved. Together with (15) and (16) this gives

$$\|(M_n f-f)^{(s)}\|_p$$
$$\leq C_1\|f^{(s)}-g\|_p + C_3 n^{-1}[\|f^{(s)}-g\|_p + \|f^{(s)}\|_p + \|\varphi^2 g''\|_p + \|(\varphi^2+n^{-1})g''\|_p]$$
$$\leq C\{\|f^{(s)}-g\|_p + n^{-1}\|\varphi^2 g''\|_p + n^{-2}\|g''\|_p + n^{-1}\|f^{(s)}\|_p\}.$$

Taking the infimum over all g on the right hand side leads to

$$\|(M_n f-f)^{(s)}\|_p \leq C\{\overline{K}^2_\varphi(f^{(s)},n^{-1})_p + n^{-1}\|f^{(s)}\|_p\},$$

which concludes the proof of Theorem 3. □

We did not succeed to invert Theorem 3. Therefore we studied weighted approximation. Our next theorem is again a global direct theorem, but now for the rate of weighted simultaneous approximation $\|\varphi^{2s}(M_n f-f)^{(2s)}\|_p$. This rate will be estimated using the second order weighted modulus of smoothness $\omega^2_\varphi(f,t)_{\varphi^{2s},p}$ on I, whose definition is given by Ditzian and Totik, Appendix (B.1). In the proof we use the equivalence of this modulus with the following modified weighted K-functional on I

$$\overline{K}^2_\varphi(f,t^2)_{\varphi^{2s},p}$$
$$=\inf\{\|\varphi^{2s}(f-g)\|_p + t^2\|\varphi^{2s}\varphi^2 g''\|_p + t^4\|\varphi^{2s}g''\|_p\,;\, g'\in AC_{loc}(I),\ \varphi^{2s}g''\in L_p(I)\}.$$

<u>Theorem 5</u> Let $\varphi(x)=\sqrt{x(1-x)}$, $n>2s$. If $\varphi^{2s}f^{(2s)}\in L_p(I)$, $1\leq p<\infty$, then
$$\|\varphi^{2s}(M_n f-f)^{(2s)}\|_p \leq C\{\omega^2_\varphi(f^{(2s)},n^{-1/2})_{\varphi^{2s},p} + n^{-1}\|\varphi^{2s}f^{(2s)}\|_p\}.$$

For the proof the representation (1) for $(M_n f)^{(2s)}$ has to be rewritten into

$$\varphi(x)^{2s}(M_n f)^{(2s)}(x) = (n+1)\sum_{k=0}^{n-2s} \beta(n,2s,k)p_{n,k+s}(x)\int_0^1 p_{n,k+s}(t)\varphi(t)^{2s}f^{(2s)}(t)dt$$

provided $\varphi^{2s}f^{(2s)} \in L_p(I)$, $1 \leq p < \infty$, $n > 2s$, where

$$\beta(n,2s,k) = \frac{[(k+s)!]^2}{k!(k+2s)!}\cdot\frac{[(n-k-s)!]^2}{(n-k)!(n-k-2s)!} < 1.$$

Again the most difficult step in the proof is to show that

$\|\varphi^{2s}M_{n,2s}R_2(g,\cdot,x)\|_p \leq Cn^{-1}\|\varphi^{2s}(\varphi^2+n^{-1})g''\|_p$. For details of the proof of this and the following two theorems see our forthcoming paper , where we derive similar results for the simultaneous approximation by the general method of Baskakov-Durrmeyer operators.

If $\varphi^{2s}f^{(2s)} \in L_p(I)$ and $\omega_\varphi^2(f^{(2s)},t)_{\varphi^{2s},p} = O(t^{2(\alpha-s)})$, $0 < \alpha - s \leq 1$ (i.e. $s < \alpha \leq s+1$), which is equivalent to $\omega_\varphi^{2(s+1)}(f,t)_p = O(t^{2\alpha})$ by Ditzian and Totik, Corollary 6.3.2, then Theorem 5 implies

$\|\varphi^{2s}(M_n f - f)^{(2s)}\|_p = O(n^{s-\alpha})$. For $s < \alpha < s+1$ this result can be inverted by using the Berens-Lorentz lemma.

<u>Theorem 6</u> Let $\varphi^{2s}f^{(2s)} \in L_p(I)$, $1 \leq p < \infty$, $\varphi(x) = \sqrt{x(1-x)}$, $n > 2s$, $s < \alpha < s+1$. Then $\|\varphi^{2s}(M_n f - f)^{(2s)}\|_p = O(n^{s-\alpha})$ implies $\omega_\varphi^{2(s+1)}(f,t)_p = O(t^{2\alpha})$.

Mainly as a corollary of Theorems 5 and 6 the following equivalence result is obtained.

<u>Theorem 7</u> Let $f \in L_p(I)$, $1 \leq p < \infty$, $\varphi(x) = \sqrt{x(1-x)}$, $n > 2s$, $s < \alpha < s+1$. Then the following statements are equivalent:

(i) $\omega_\varphi^{2(s+1)}(f,t)_p = O(t^{2\alpha})$

(ii) $\varphi^{2s}f^{(2s)} \in L_p(I)$ and $\|\varphi^{2s}(M_n f - f)^{(2s)}\|_p = O(n^{s-\alpha})$

Especially for $s = 0$ we obtain from Theorem 7 the equivalence

$\|M_n f - f\|_p = O(n^{-\alpha}) \Leftrightarrow \omega_\varphi^2(f,t)_p = O(t^{2\alpha})$, which has been proved by Ditzian and Ivanov, Theorem 7.4.

4. Local results

Derriennic (1979) proved the local convergence (cf. Théorème II.6)

$$\lim_{n\to\infty} (M_n f)^{(s)}(x) = f^{(s)}(x),$$

provided f is integrable and bounded on I and s-times differentiable at the point $x \in I$. Our final result is a generalization of the Voronovskaja theorem proved by Derriennic (1985).

Theorem 8 Let $f \in L_1^s(I)$, $s \in \mathbb{N}_0$, $f^{(s)}$ twice differentiable at a fixed point $x \in (0,1)$. Then $\lim_{n\to\infty} n(M_n f - f)^{(s)}(x) = \frac{d^{s+1}}{dx^{s+1}}\left[\varphi(x)^2 f'(x)\right]$, where $\varphi(x) = \sqrt{x(1-x)}$.

Proof Define $F(u) := \int_0^u f(t)dt$. Then $F'(u) = f^{(s)}(u)$ a.e. in I and

$$F'(x) = f^{(s)}(x),\ F''(x) = f^{(s+1)}(x),\ F'''(x) = f^{(s+2)}(x),$$

as $f^{(s)}$ is assumed to be twice differentiable in x. We consider the Taylor formula

$$F(t) = F(x)+(t-x)F'(x)+\tfrac{1}{2}(t-x)^2F''(x)+\tfrac{1}{3!}(t-x)^3F'''(x)+(t-x)^3R(t-x),$$

where $|R(t-x)| \leq C$ for $t \in I$ and $\lim_{t\to x} R(t-x) = 0$.

Differentiating this formula with respect to t leads to

$$f^{(s)}(t) = f^{(s)}(x)+(t-x)f^{(s+1)}(x)+\tfrac{1}{2}(t-x)^2f^{(s+2)}(x)+\tfrac{d}{dt}[(t-x)^3R(t-x)] \quad \text{a.e. in } I.$$

Together with Lemma 1 this gives

$$\begin{aligned} &n(M_n f-f)^{(s)}(x) \qquad (18)\\ &= n\Big\{f^{(s)}(x)\Big[\frac{\alpha(n,s)(n+1)}{(n+s+1)} - 1\Big]+f^{(s+1)}(x)\frac{\alpha(n,s)(n+1)(s+1)}{(n+s+1)(n+s+2)}\cdot(1-2x)\\ &\quad + \tfrac{1}{2}f^{(s+2)}(x)\frac{\alpha(n,s)(n+1)}{(n+s+1)(n+s+2)(n+s+3)}\\ &\quad \cdot\Big[\varphi(x)^2[2(n+1)-4(s+1)(s+2)]+(s+1)(s+2)\Big] + M_{n,s}\Big[\tfrac{d}{dt}[(t-x)^3R(t-x)]\Big](x)\Big\}. \end{aligned}$$

As

$$\begin{aligned} \frac{\alpha(n,s)(n+1)}{(n+s+1)} - 1 &= \Big\{\prod_{j=1}^{s}(n-s+j) - \prod_{j=1}^{s}(n+1+j)\Big\}\cdot \prod_{j=1}^{s}(n+1+j)^{-1}\\ &= -s(s+1)n^{s-1}\cdot\prod_{j=1}^{s}(n+1+j)^{-1}+O(n^{-2}) \end{aligned}$$

and $\lim_{n\to\infty}\frac{\alpha(n,s)(n+1)}{(n+s+1)} = 1$ we get from (18)

$$\lim_{n\to\infty} n(M_n f-f)^{(s)}(x) = -s(s+1)f^{(s)}(x)+(s+1)f^{(s+1)}(x)+\varphi(x)^2 f^{(s+2)}(x), \tag{19}$$

provided that

$$\lim_{n\to\infty} n\cdot M_{n,s}\left[\frac{d}{dt}[(t-x)^3 R(t-x)]\right](x) = 0 . \tag{20}$$

It is obvious that (19) can be rewritten in the form of the proposition of the theorem.

Now we look at the remainder and prove (20) where we consider the case $s > 0$. For $s = 0$ we refer to the paper by Derriennic (1985). Integration by parts leads to

$$R^*(x) := n\cdot M_{n,s}\left[\frac{d}{dt}[(t-x)^3 R(t-x)]\right](x)$$

$$= n(n+1)\alpha(n,s)\sum_{k=0}^{n-s} p_{n-s,k}(x)\int_0^1(-1)p'_{n+s,k+s}(t)\cdot(t-x)^3 R(t-x)dt .$$

Using (6) and changing the index of summation gives

$$R^*(x)=n(n+1)(n+s)\alpha(n,s)\left\{\sum_{k=0}^{n-s} p_{n-s,k}(x)\int_0^1 p_{n+s-1,k+s}(t)\cdot(t-x)^3 R(t-x)dt\right.$$
$$\left.+\sum_{k=-1}^{n-s-1} p_{n-s,k+1}(x)\int_0^1 p_{n+s-1,k+s}(t)\cdot(t-x)^3 R(t-x)dt\right\}$$
$$= n(n+1)(n+s)\alpha(n,s)\sum_{k=-1}^{n-s}(p_{n-s,k}(x)-p_{n-s,k+1}(x))$$
$$\times\int_0^1 p_{n+s-1,k+s}(t)\cdot(t-x)^3 R(t-x)dt ,$$

as we defined $p_{nk} = 0$ if $k < 0$ or $k > n$. Using again (6) and the relation (7) one obtains

$$\varphi(x)^2 R^*(x) = \frac{n(n+1)(n+s)}{(n-s+1)}\cdot\alpha(n,s)\sum_{k=-1}^{n-s} p_{n-s+1,k+1}(x)\cdot(k+1-(n-s+1)x)$$
$$\times\int_0^1 p_{n+s-1,k+s}(t)(t-x)^3 R(t-x)dt .$$

Applying the Cauchy-Schwarz inequality gives

$$[\varphi(x)^2 R^*(x)]^2 \le C\cdot n^4\left\{\sum_{k=-1}^{n-s} p_{n-s+1,k+1}(x)\cdot(k+1-(n-s+1)x)^2\right\}$$
$$\times\sum_{k=-1}^{n-s} p_{n-s+1,k+1}(x)\left\{\int_0^1 p_{n+s-1,k+s}(t)(t-x)^3 R(t-x)dt\right\}^2$$

The term in the first curly bracket equals $(n-s+1)\varphi(x)^2$ by Lorentz's formulas given in his book, p.14. Choosing $\epsilon > 0$ there exists a $\delta > 0$ such that $|R(t-x)| < \epsilon$ whenever $|t-x| < \delta$ and we obtain

$$[\varphi(x)^2R^*(x)]^2 \le C\cdot n^5\varphi(x)^2 \sum_{k=0}^{n-s+1} p_{n-s+1,k}(x)\frac{1}{n+s}\left\{\epsilon^2\int_0^1 p_{n+s-1,k+s-1}(t)(t-x)^6dt + \frac{C}{\delta^2}\int_0^1 p_{n+s-1,k+s-1}(t)(t-x)^8dt\right\}$$

$$\le C\cdot n^3\varphi(x)^2\left\{\epsilon^2\cdot T_{n,s-1,6}(x)+\frac{C}{\delta^2 n}\cdot T_{n,s-1,8}(x)\right\}$$

$$\le C\varphi(x)^2\left\{\epsilon^2+\frac{C}{\delta^2 n}\right\},$$

where we used Lemma 1 and Corollary 2. Thus for n big enough we get

$|\varphi(x)^2R^*(x)| \le C\epsilon$ which gives (20) as $x \in (0,1)$. □

References

Bergh J., Löfström J., 'Interpolation spaces, An introduction', Springer-Verlag, Berlin, Heidelberg, New York, 1976.

Derriennic M. M. (1979), 'Sur l'approximation des fonctions intégrables sur [0,1] par des polynômes de Bernstein modifiés', J. Approx. Theory, 26, 277-292.

Derriennic M. M. (1985), 'Additif au papier "Sur l'approximation de fonctions intégrables sur [0,1] par des polynômes de Bernstein modifiés" ', unpublished.

Ditzian Z., Ivanov K., 'Bernstein type operators and their derivates', to appear in J. Approx. Theory.

Ditzian Z., Totik V., 'Moduli of smoothness', Springer Series in Computational Mathematics 9, Springer-Verlag, Berlin, Heidelberg, New York, 1987.

Durrmeyer J. L. (1967), 'Une formule d'inversion de la transformée de Laplace: Applications à la théorie des moments', Thèse de 3e cycle, Faculté des Sciences de l'Université de Paris.

Heilmann M., 'L_p-saturation of some modified Bernstein operators', to appear in J. Approx. Theory.

Heilmann M., Müller M. W. (1989), 'On simultaneous approximation by the method of Baskakov-Durrmeyer operators', Numer. Funct. Anal. and Optimiz., 10, 127-138.

Heilmann M., Müller M. W., 'Direct and converse results on weighted simultaneous approximation by the method of operators of Baskakov-Durrmeyer type', in preparation.

Lorentz G. G., 'Bernstein polynomials', Chelsea Publishing Company New York, N. Y., 1986.

Orthogonality and approximation in a Sobolev space

A. Iserles

Department of Applied Mathematics and Theoretical Physics, University of Cambridge, UK

J. M. Sanz-Serna

Departamento de Matemática Aplicada y Computación, Universidad de Valladolid, Spain

P. E. Koch, S. P. Nørsett

Division of Mathematical Sciences, Norwegian Insitute of Technology, Trondheim, Norway

Abstract This paper explores polynomials orthogonal with respect to the Sobolev inner product

$$(f,g)_\lambda := \int_{-1}^{1} f(x)g(x)dx + \lambda \int_{-1}^{1} f'(x)g'(x)dx, \quad \lambda \geq 0.$$

We investigate expansions in Legendre polynomials—it transpires that their coefficients satisfy an explicitly known recurrence relation. Moreover, we re-interpret a result of Althammer [1962] and Gröbner [1967] on a differential relation which is obeyed by the Sobolev-orthogonal polynomials and exploit it to derive a useful expression for the corresponding Fourier coefficients. These results lead to an efficient algorithm for approximation by polynomials in the underlying Sobolev space. This algorithm is introduced and described in detail, accompanied by numerical examples.

Key words: Fourier coefficients, Legendre polynomials, Polynomial approximation, Recurrence relations, Sobolev norm, Ultraspherical polynomials.

1. Orthogonality in a Sobolev space

The theme of the present paper is orthogonality with respect to the *Sobolev inner product*

$$(f,g)_\lambda := \int_{-1}^{1} f(x)g(x)dx + \lambda \int_{-1}^{1} f'(x)g'(x)dx, \tag{1}$$

where λ is a non-negative real parameter, while f and g range across the Sobolev space W_2^1 (the set of real functions with L_2-integrable derivatives).

Each inner product that acts on polynomials generates (e.g. by the Gram-Schmidt process) a set of *orthogonal polynomials*. Disregarding normalization for the time being, we denote them by $p_0^{(\lambda)}, p_1^{(\lambda)}, p_2^{(\lambda)}, \ldots$:

$$\int_{-1}^{1} p_m^{(\lambda)}(x)p_n^{(\lambda)}(x)dx + \lambda \int_{-1}^{1} p_m^{(\lambda)\prime}(x)p_n^{(\lambda)\prime}(x)dx \begin{cases} = 0 & : m \neq n \\ > 0 & : m = n \end{cases}. \tag{2}$$

Sobolev orthogonality has been already introduced and debated by several authors, in particular Lewis [1947], Althammer [1962] and Gröbner [1967],[1] in a more general setting: Let $\varphi_0, \varphi_1, \ldots, \varphi_L$ be $L+1$ given distributions (i.e. real, right-continuous, monotonically non-decreasing functions with an infinite number of points of increase and with all moments bounded) on the real interval $\mathcal{R}$. One may consider polynomials orthogonal with respect to the inner product

$$\int_{\mathcal{R}} f(x)g(x)d\varphi_0(x) + \int_{\mathcal{R}} f'(x)g'(x)d\varphi_1(x) + \cdots + \int_{\mathcal{R}} f^{(L)}(x)g^{(L)}(x)d\varphi_L(x). \quad (3)$$

It is possible to prove easily that these *monic* polynomials solve the variational isoperimetric problem

$$\min\left\{\int_{\mathcal{R}} y^2(x)d\varphi_0(x) + \int_{\mathcal{R}} y'^2(x)d\varphi_1(x) + \cdots + \int_{\mathcal{R}} y^{(L)^2}(x)d\varphi_L(x)\right\},$$

where y ranges across all monic polynomials of given degree [Althammer, 1962]. Some of our present results extend to the inner product (3), and they will be subject of a forthcoming paper. In the present work we focus on the simpler (and the most useful!) form (1).

It follows readily from (2) that the underlying *Fourier coefficients* of a function f are

$$\hat{f}_n(\lambda) = \frac{(f, p_n^{(\lambda)})_\lambda}{(p_n^{(\lambda)}, p_n^{(\lambda)})_\lambda}, \; n = 0, 1, \ldots. \quad (4)$$

Thus, if f is an m-th degree polynomial then $f \equiv \sum_{n=0}^m \hat{f}_n(\lambda)p_n^{(\lambda)}$, whereas for any $f \in W_2^1$ it is true that $\lim_{m\to\infty} \|f - \sum_{n=0}^m \hat{f}_n(\lambda)p_n^{(\lambda)}\|_\lambda = 0$. Here $\|\cdot\|_\lambda$ is the norm induced by $(\cdot, \cdot)_\lambda$ [Lewis, 1947].

A mechanism to approximate functions by polynomials in Sobolev norm is useful in numerous applications, when derivatives, and not just function values, are important, e.g. in *spectral methods* for differential equations. The standard Legendre projection (i.e. $\lambda = 0$) produces poor approximation to the derivative, which might be pointwise worse by orders of magnitude than the underlying approximation to the function: an example to this effect features in §4.

In §2 we introduce a representation of $p_n^{(\lambda)}$ as a linear combination of Legendre polynomials. The coefficients, which depend on λ, obey a known three-term recurrence relation, hence can be obtained easily.

§3 is devoted to a differential relation that is satisfied by $p_n^{(\lambda)}$. Although it has been already debated, in a different form, by both Althammer [1962] and Gröbner [1967], we provide both easier derivation and a different interpretation of the result. Our approach leads to a relationship between the Fourier coefficients (4) and the quantities

$$\hat{f}_n^* := \int_{-1}^1 f(x)P_n^{(1,1)}(x)dx, \; n = 0, 1, \ldots$$

where $P_n^{(1,1)}$ are the $(1,1)$ *ultraspherical polynomials* [Chihara, 1978]. We produce recursively the coefficients that feature in that relationship. Results of §2 and §3 are assembled into a numerical algorithm to project functions into polynomials in W_2^1.

[1] We are grateful to Prof. W. Gautschi for drawing our attention to the above references.

Finally, in §4 we present a computational example demonstrating that, given a function with "awkward" derivative, the expansion in Legendre polynomials produces poor approximation to the derivative, while expansion in Sobolev-orthogonal polynomials brings about good approximation to both the function and its derivative.

Future papers will address themselves to the theory of Sobolev-orthogonal polynomials in a wider context.

2. Explicit representation of $p_n^{(\lambda)}$

As it calls for no extra effort, we consider polynomials orthogonal with respect to the inner product (3) with $L = 1$. We denote the polynomials orthogonal (in the conventional sense) with respect to φ_0 by $\{q_n\}$ and say that the distributions $\{\varphi_0, \varphi_1\}$ have *property* α if for every $m, n = 0, 1, \ldots$ it is true that $\int_{\mathcal{R}} q_n'(x)q_m'(x)d\varphi_1(x) = d_{\min\{m,n\}}$. It will be assumed henceforth that property α holds.

We seek $\gamma_{n,0}, \ldots, \gamma_{n,0}$ such that

$$p_n^{(\lambda)}(x) = \sum_{k=0}^{n} \gamma_{n,k}(\lambda)q_k(x). \tag{5}$$

Denoting $c_n := \int_{\mathcal{R}} q_n^2(x)d\varphi_0(x)$, $d_m := \int_{\mathcal{R}} q_n'(x)q_m'(x)d\varphi_1(x)$ (for $m \le n$) we have

$$0 = \left(p_n^{(\lambda)}, q_\ell\right)_\lambda = \gamma_{n,n}c_n + \lambda \sum_{k=0}^{n} \gamma_{n,k}d_{\min\{k,\ell\}}, \quad \ell = 0, 1, \ldots, n-1.$$

Thus, given that $C := \mathrm{diag}\{c_0, c_1, \ldots, c_n\}$, $D = \left(d_{\min\{k,\ell\}}\right)_{k,\ell=0}^{n}$ and e_n is the n-th unit vector, we have

$$(C + \lambda D)\gamma = \omega e_n, \tag{6}$$

where $\gamma_n = (\gamma_{n,0}, \gamma_{n,1}, \ldots, \gamma_{n,n})^T$ and $\omega \neq 0$ is a normalization constant, to be chosen later at our convenience. Solution of (6) follows easily by Cramer's rule:

$$p_n^{(\lambda)}(x) = \tilde{\omega} \det \begin{bmatrix} c_0 + \lambda d_0 & \lambda d_0 & \lambda d_0 & \cdots & \lambda d_0 & q_0(x) \\ \lambda d_0 & c_1 + \lambda d_1 & \lambda d_1 & \cdots & \lambda d_1 & q_1(x) \\ \lambda d_0 & \lambda d_1 & c_2 + \lambda d_2 & \cdots & \lambda d_2 & q_2(x) \\ \vdots & \vdots & \vdots & \ddots & \vdots & \vdots \\ \lambda d_0 & \lambda d_1 & \lambda d_2 & \cdots & c_{n-1} + \lambda d_{n-1} & q_{n-1}(x) \\ \lambda d_0 & \lambda d_1 & \lambda d_2 & \cdots & \lambda d_{n-1} & q_n(x) \end{bmatrix}, \tag{7}$$

where $\tilde{\omega}$ is, again, a non-zero constant.

We now subtract the bottom row of (7) from the remaining rows. This, in tandem with $\gamma_{n,0} = d_0 = 0$ yields

$$p_n^{(\lambda)}(x) = \tilde{\omega} \det \begin{bmatrix} c_1 & \lambda(d_1 - d_2) & \lambda(d_1 - d_3) & \cdots & \lambda(d_1 - d_{n-1}) & q_1(x) - q_n(x) \\ 0 & c_2 & \lambda(d_2 - d_3) & \cdots & \lambda(d_2 - d_{n-1}) & q_2(x) - q_n(x) \\ 0 & 0 & c_3 & \cdots & \lambda(d_3 - d_{n-1}) & q_3(x) - q_n(x) \\ \vdots & \vdots & \ddots & \ddots & \vdots & \vdots \\ 0 & 0 & 0 & \cdots & c_{n-1} & q_{n-1}(x) - q_n(x) \\ \lambda d_1 & \lambda d_2 & \lambda d_3 & \cdots & \lambda d_{n-1} & q_n(x) \end{bmatrix}.$$

Observe that, when performing *Gaussian elimination* to bring the matrix into upper triangular form, the coefficient r_ℓ used to eliminate the ℓ-th element, $\ell = 1, \ldots, n-1$, is independent of n. These coefficients satisfy

$$\lambda d_m - \lambda \sum_{k=1}^{m-1} (d_m - d_k) r_k + c_m r_m = 0, \quad m = 1, 2, \ldots, n-1. \tag{8}$$

Our contention is that the r_k's obey the three-term recurrence relation

$$\tau_\ell r_\ell = (\varsigma_\ell + \upsilon_\ell \lambda) r_{\ell-1} - \varepsilon_\ell r_{\ell-2}, \quad \ell = 3, 4, \ldots \tag{9}$$

where

$$\begin{aligned} \tau_\ell &= c_\ell (d_{\ell-1} - d_{\ell-2}); \\ \varsigma_\ell &= c_{\ell-1} (d_\ell - d_{\ell-2}); \\ \upsilon_\ell &= (d_\ell - d_{\ell-1})(d_{\ell-1} - d_{\ell-2}); \\ \varepsilon_\ell &= c_{\ell-2} (d_\ell - d_{\ell-1}); \end{aligned}$$

with the initial conditions

$$r_0 = 0; \quad r_1 = -\frac{d_1}{c_1}\lambda; \quad r_2 = -\frac{d_2}{c_2}\lambda - \frac{d_1(d_2 - d_1)}{c_1 c_2}\lambda^2.$$

Indeed, substituting (8) into (9) yields

$$\begin{aligned} &\tau_\ell r_\ell - (\varsigma_\ell + \upsilon_\ell \lambda) r_{\ell-1} + \varepsilon_\ell r_{\ell-2} \\ &= -\lambda \Bigg\{ (d_{\ell-1} - d_{\ell-2}) \left(d_\ell - \sum_1^{\ell-1} (d_\ell - d_k) r_k \right) - (d_\ell - d_{\ell-2}) \left(d_{\ell-1} - \sum_1^{\ell-2} (d_{\ell-1} - d_k) r_k \right) \\ &\quad + (d_\ell - d_{\ell-1}) \left(d_{\ell-2} - \sum_1^{\ell-3} (d_{\ell-2} - d_k) r_k \right) + (d_\ell - d_{\ell-1})(d_{\ell-1} - d_{\ell-2}) r_{\ell-1} \Bigg\} = 0 \end{aligned}$$

and, since both (8) and (9) possess unique solutions, our assertion is true.

A point of interest is that, subject to consecutive d_ℓ's being distinct and $\tau_\ell \varepsilon_\ell > 0$, (9) implies, via the Favard theorem [Chihara, 1978], that the shifted polynomials $\tilde{r}_\ell(\lambda) := -\frac{c_1}{d_1 \lambda} r_{\ell+1}(\lambda)$, $\ell = 0, 1, \ldots$ are orthogonal with respect to some real distribution.

Next, we evaluate for future use $\sum_1^{\ell-1} r_k$. The identity (8) implies that

$$\lambda \sum_{k=1}^{\ell-1} d_k r_k = -c_\ell r_\ell - d_\ell \lambda + d_\ell \lambda \sum_{k=1}^{\ell-1} r_k. \tag{10}$$

Shifting the index by one and adding $\lambda d_{\ell-1} r_{\ell-1}$ to both sides produces

$$\lambda \sum_{k=1}^{\ell-1} d_k r_k = -c_{\ell-1} r_{\ell-1} - d_{\ell-1} \lambda + d_{\ell-1} \lambda \sum_{k=1}^{\ell-1} r_k. \tag{11}$$

We now solve (10) and (11) for the unknowns $\sum_1^{\ell-1} r_k$ and $\sum_1^{\ell-1} d_k r_k$. Subject to $d_\ell \neq d_{\ell-1}$ this yields

$$\sum_{k=1}^{\ell-1} r_k = 1 + \frac{c_\ell r_\ell - c_{\ell-1} r_{\ell-1}}{\lambda (d_\ell - d_{\ell-1})} =: 1 + \sigma_\ell(\lambda). \tag{12}$$

We now proceed to evaluate the expansion of $p_n^{(\lambda)}$. To this end, we fix the normalization constant ω so that the coefficients are polynomials in λ and $p_n^{(0)} \equiv q_n$. Having eliminated the bottom row from the determinant, it follows at once that

$$p_n^{(\lambda)}(x) = q_n(x) + \sum_{\ell=1}^{n-1} r_\ell(\lambda)(q_\ell(x) - q_n(x)),$$

and (12) gives an explicit expansion of $p_n^{(\lambda)}$ in q_ℓ's,

$$p_n^{(\lambda)}(x) = -\sigma_n(\lambda)q_n(x) + \sum_{\ell=1}^{n-1} r_\ell(\lambda)q_\ell(x). \tag{13}$$

An interesting identity readily follows from the expansion: Subtract (13) from the corresponding expression for $p_{n+1}^{(\lambda)}$. This yields

$$p_{n+1}^{(\lambda)}(x) - p_n^{(\lambda)}(x) = -\sigma_{n+1}(\lambda)q_{n+1}(x) + (r_n(\lambda) + \sigma_n(\lambda))q_n(x).$$

However, since (12) implies that $\sigma_{n+1}(\lambda) = r_n(\lambda) + \sigma_n(\lambda)$, we obtain

$$p_{n+1}^{(\lambda)}(x) - p_n^{(\lambda)}(x) = -\sigma_{n+1}(\lambda)(q_{n+1}(x) - q_n(x)). \tag{14}$$

Identity (14) is useful in the explicit evaluation of the $p_n^{(\lambda)}$'s.

An example of distributions that obey property α is the Laguerre pair $d\varphi_0(x) \equiv d\varphi_1(x) = \frac{1}{\Gamma(1+\alpha)}x^\alpha e^{-x}dx$, where $0 < x < \infty$ and $\alpha > -1$. It can be proved that $d_\ell = \frac{(\alpha+2)_{\ell-1}}{(\ell-1)!}$ and that, for $\alpha = 0$, the r_ℓ's can be identified with Chebyshev polynomials of the second kind.

Legendre weights do not satisfy property α unamended. Fortunately, since, in that case, both $p_n^{(\lambda)}$ and $q_n \equiv P_n$ maintain the same parity as n, property α is "recovered" as long as attention is restricted to indices of the correct parity—the remaining coefficients vanish and are of no interest! Since, integrating by parts,

$$\int_{-1}^{1} P'_m(x)P'_n(x)dx = P'_m(1)P_n(1) - P'_m(-1)P_n(-1) - \int_{-1}^{1} P''_m(x)P_n(x)dx,$$

it follows readily that (subject to the aforementioned restriction on parity) $d_\ell = \ell(1+\ell)$. This, together with $c_\ell = \frac{2}{2\ell+1}$, (9) and (13) implies that

$$p_n^{(\lambda)} = -\sigma_n P_n + \sum_{\ell=1}^{[\frac{n-1}{2}]} r_{n-2\ell}P_{n-2\ell}, \tag{15}$$

where

$$\frac{2\ell-5}{2\ell+1}r_\ell = (2 + (2\ell-1)(2\ell-5)\lambda)r_{\ell-2} - \frac{2\ell-1}{2\ell-7}r_{\ell-4};$$

$$\sigma_n = \frac{\frac{r_n}{2n+1} - \frac{r_{n-2}}{2n-3}}{2(2n-1)\lambda};$$

with the initial conditions

$$r_0(\lambda) = 0; \quad r_1(\lambda) = -3\lambda; \quad r_2(\lambda) = -15\lambda; \quad r_3(\lambda) = -42\lambda - 105\lambda^2.$$

3. A differential equation

We restrict our attention in the present section to the Legendre distribution. Let t be an arbitrary polynomial of degree $\leq n-3$. Obviously, $\left(p_n^{(\lambda)},(1-x^2)t\right)_\lambda = 0$. Integration by parts produces

$$\int_{-1}^{1}(1-x^2)\left(p_n^{(\lambda)}(x)-\lambda p_n^{(\lambda)''}(x)\right)t(x)dx = 0.$$

Hence, $p_n^{(\lambda)} - \lambda p_n^{(\lambda)''}$ is orthogonal (with respect to the distribution $(1-x^2)dx$) to all polynomials of degree $\leq n-3$ and, taking parity into account, there exist $\alpha_n(\lambda)$ and $\beta_n(\lambda)$ so that the Sobolev-orthogonal polynomial obeys the ordinary differential equation

$$p_n^{(\lambda)}(x) - \lambda \frac{d^2}{dx^2}p_n^{(\lambda)}(x) = \alpha_n(\lambda)P_n^{(1,1)}(x) + \beta_n(\lambda)P_{n-2}^{(1,1)}(x), \tag{16}$$

where the $P_m^{(1,1)}$'s are *ultraspherical polynomials* [Chihara, 1978].

The importance of (16) becomes apparent upon the consideration of the Fourier coefficients. Integrating by parts we have

$$\begin{aligned}\hat{f}_n\|p_n^{(\lambda)}\|^2 &= \int_{-1}^{1} f(x)\left(p_n^{(\lambda)}(x)-\lambda p_n^{(\lambda)''}(x)\right)dx + \lambda\left\{p_n^{(\lambda)'}(1)f(1) - p_n^{(\lambda)'}(-1)f(-1)\right\}\\ &= \alpha_n\hat{f}_n^* + \beta_n\hat{f}_{n-2}^* + \lambda\left\{p_n^{(\lambda)'}(1)f(1) - p_n^{(\lambda)'}(-1)f(-1)\right\},\end{aligned} \tag{17}$$

where $\hat{f}_m^* := \int -1^1 f(x)P_n^{(1,1)}(x)dx$, $m = 0,1,\ldots$ Since the values of $p_n^{(\lambda)}$ and its derivative at the end-points can be easily derived from the results of §3, it remains to provide an explicit form of α_n and β_n.

Comparing coefficients of x^n in (15) and (16) yields at once

$$\alpha_n = -\frac{n+2}{2(2n+1)}\sigma_n.$$

It is an easy exercise to demonstrate that

$$\int_{-1}^{1} P_m^{(1,1)}(x)dx = \frac{2(1+(-1)^m)}{m+2}; \tag{18}$$

$$\int_{-1}^{1} xP_m^{(1,1)}(x)dx = \frac{2(1+(-1)^{m+1})}{m+2}. \tag{19}$$

Let n be even. Then integration of (16) from -1 to 1, in tandem with (18), yields

$$\frac{\alpha_n}{n+2} + \frac{\beta_n}{n} = -\frac{\lambda}{2}p_n^{(\lambda)'}(1).$$

Likewise, for odd n, we multiply (16) by x, integrate and employ (19): Since, by (17) and integration by parts respectively,

$$\begin{aligned}\int_{-1}^{1} xp_n^{(\lambda)}(x)dx &= \frac{2}{3}r_1;\\ \int_{-1}^{1} xp_n^{(\lambda)''}(x)dx &= 2\left(p_n^{(\lambda)'}(1) - p_n^{(\lambda)}(1)\right),\end{aligned}$$

we obtain

$$\frac{\alpha_n}{n+2} + \frac{\beta_n}{n} = \frac{1}{6} r_1 - \frac{\lambda}{2} \left(p_n^{(\lambda)\prime}(1) - p_n^{(\lambda)}(1) \right).$$

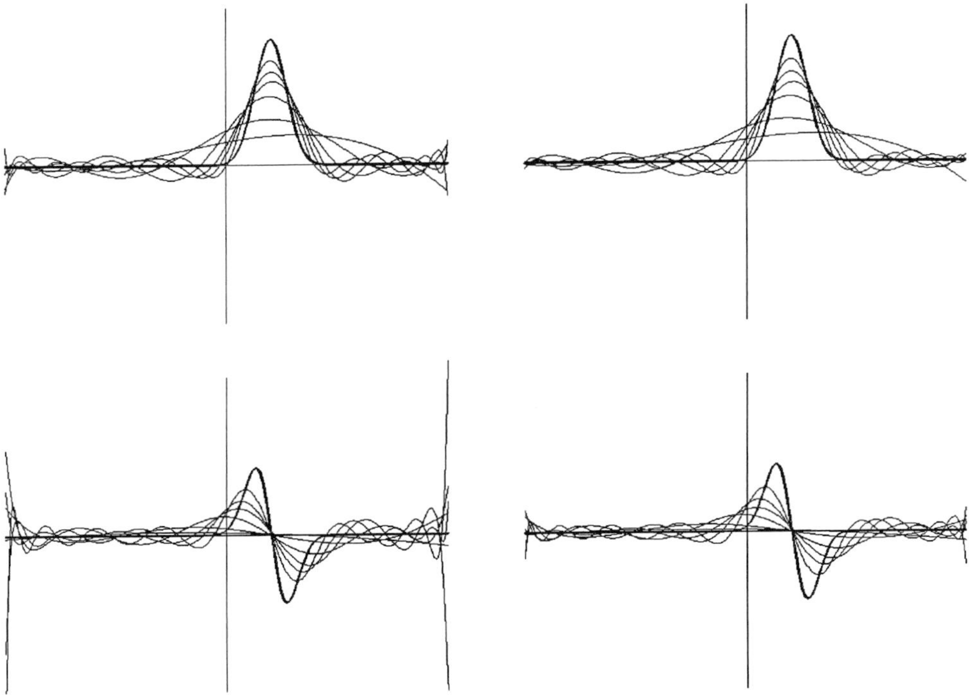

Legendre projection and its derivative   Sobolev projection and its derivative

Figure 1: *Legendre and Sobolev projections.*
The thicker line denotes the function f, whereas the thin lines stand for polynomials of degrees $3, 6, \ldots, 18$.

All the ingredients for an algorithm for the evaluation of a projection of f into n-th degree polynomials in W_2^1 are now in place: either (14) or (15) are used to evaluate $p_m^{(\lambda)}$'s, the quantities $\hat{f}_m^*$, which are independent of λ, are evaluated e.g. by numerical quadrature and substituted into (17), whereas α_n and β_n are also available from explicit formulae of this section. The norms $\|p_n^{(\lambda)}\|^2$ can be evaluated either directly from the explicit form or via a recurrence relation which is omitted here due to space constraints.

The satisfaction of an "interesting" ordinary differential equation by $p_n^{(\lambda)}$ has been already noticed by Althammer [1962], who obtained, in a considerably longer derivation, the right-hand side in terms od derivatives of Legendre polynomials, and by Gröbner [1967], who identified it with the n-th derivative of a Lagrange multiplier function of the isoperimetric variational problem from §1. However, none has used it to facilitate the evaluation of Fourier coefficients. Moreover, our method yields itself to other distributions as well. The characterization of all such distributions will be described in a forthcoming paper.

4. A numerical example

The approach of this paper becomes valuable when we wish to approximate a function f by its projection into polynomials and, simultaneously, to approximate its derivative by the derivative of the polynomial approximant. Given that the derivative of f is steep, it is only to be expected that the quality of the projection in the conventional L_2 norm (the *Legendre projection*) deteriorates. Several computational examples show that in this case the *Sobolev projection* is superior. Figure 1 displays approximants of both types (with $\lambda = 0$ and $\lambda = \frac{1}{100}$ respectively) to the function

$$f(x) = e^{-100\left(x-\frac{1}{5}\right)^2}, \quad -1 < x < 1,$$

by polynomials of degrees $n = 3, 6, 9, \ldots, 18$. It is evident that the Legendre projection is poor near the end-points, whereas the Sobolev projection displays reasonably good behaviour throughout the interval.

References

P. Althammer [1962], Eine Erweiterung des Orthogonalitätsbegriffes bei Polynomen und deren Anwendung auf die beste Approximation, *J. Reine Angew. Math. 211*, pp. 192–204.

T.S. Chihara [1978], *An Introduction to Orthogonal Polynomials*, Gordon & Breach, New York.

W. Gröbner [1967], Orthogonale Polynomsysteme die gleichzeitig mit $f(x)$ auch deren Ableitung $f'(x)$ approximieren, in *Funktionalanalysis, Approximationstheorie, Numerische Mathematik* (L. Collatz, G. Meinardus & H. Unger, eds.), Birkhäuser, Basel.

D.C. Lewis [1947], Polynomial least squares approximations, *Amer. J. Math. 69*, pp. 273–278.

Piecewise polynomial approximation of polynomial curves

M. A. Lachance

Department of Mathematics,
University of Michigan-Dearborn,
Michigan, USA

Abstract An algorithm based upon Chebyshev economization and constrained Chebyshev polynomials is discussed for producing low degree piecewise polynomial approximations for high degree polynomials. The method is shown to compare favorably to the Remez algorithm.
Key words Constrained Chebyshev polynomials, Chebyshev economization, Remez algorithm

1. Introduction

The majority of European and Asian computer-aided design (CAD) systems use parametric polynomials to represent 3D curves and surfaces. These systems differ in their internal representation of polynomials (power, Bernstein, and Hermite bases) and in their maximum degree (three, seven, nine, fifteen, and twenty-one). In order for these systems to exchange geometric data, representations must occasionally be altered, and/or degrees must be reduced. The numerical stability of altering representations has been recently investigated by Heybrock (1987) and Shearer

(1988). In this note we are concerned with the problem of degree reduction.

In general, a high degree polynomial will be approximated rather poorly by a lower degree polynomial, and hence piecewise or spline approximates are required. We shall describe a piecewise polynomial construction which is recursive, is uniformly near-best, controls geometric continuity, and is relatively inexpensive. The construction is compared with the Remez algorithm.

2. Constrained Chebyshev polynomials

Lachance, Saff and Varga (1979) introduced the collection of constrained Chebyshev polynomials as a solution to a minimax problem with Jacobi-type weights. These polynomials, denoted by $T_m^{(\alpha,\beta)}(x)$, are the unique monic polynomials which are extremal for

Problem 2.1 For each triple of nonnegative integers (m,α,β), determine

$$\underset{g\in\pi_{m-1}}{\text{minimum}} \quad \underset{-1\le x\le 1}{\text{maximum}} \quad (1-x)^{\alpha}(1+x)^{\beta}|x^m - g(x)|.$$

Here, π_m denotes the collection of real polynomials of degree at most m, and $\pi_{-1} = \{0\}$.

These constrained Chebyshev polynomials are a proper generalization of the classical Chebyshev polynomials (of the first kind) since $2^{m-1}T_m^{(0,0)}(x) = T_m(x)$. Most CAD systems parametrize curves on the interval $[0,1]$. We shall do the same, and introduce more convenient notation for the constrained Chebyshev polynomials on this interval.

Definition 2.1 For each pair of nonnegative integers (m,α), with $m \ge 2\alpha$, the constrained Chebyshev polynomials are defined by

$$C_m^\alpha(s) := 2^{-m}\, T_{m-2\alpha}^{(\alpha,\alpha)}(2s-1) \;=\; s^\alpha (s-1)^\alpha (s^{m-2\alpha} + \cdots);$$

$$E_m^\alpha := \max_{s\in[0,1]} |C_m^\alpha(s)|.$$

Thus m denotes the degree of the polynomial and α denotes the order zero at the endpoints 0 and 1.

3. The Construction

Let $\mathbf{P}(s) \in \pi_n^3$, $s\in[0,1]$, denote a vector-valued polynomial of degree n in R^3 and let E denote a prescribed tolerance. We wish to construct, within this tolerance, a visually smooth piecewise polynomial approximate to $\mathbf{P}(s)$ of degree m. We do so applying Chebyshev economization (CE) in a component-wise manner.

Let $k = n - m$, and define the sequence of polynomials $\{\mathbf{P}_i\}_{i=1}^k$ by

$$\mathbf{P}_{i+1}(s) := \mathbf{P}_i(s) - \frac{\mathbf{P}_i^{(n-i)}(0)}{(n-i)!}\, C_{n-i}^\alpha(s), \; i = 0,\ldots,k-1$$

where $\mathbf{P}_0(s) := \mathbf{P}(s)$. The use of the constrained Chebyshev polynomials was suggested by Lachance (1988) in order to control geometric continuity. A pointwise bound on the error between the original polynomial and the final approximate $\mathbf{P}_k(s)$ is given by

$$\|\mathbf{P}(s) - \mathbf{P}_k(s)\|_E \le \sum_{i=0}^{k-1} \|\mathbf{P}_{i+1}(s) - \mathbf{P}_i(s)\|_E$$

$$\le \sum_{i=0}^{k-1} \frac{\|\mathbf{P}_i^{(n-i)}(0)\|_E}{(n-i)!}\, E_{n-i}^\alpha,$$

where $\|\cdot\|_E$ denotes the Euclidean norm in R^3. Clearly, if the above error exceeds the prescribed tolerance E, then the approximation is unacceptable. In this event the original curve may be partitioned into two or more pieces, the pieces reparametrized to the interval $[0,1]$; and then the construction may be applied to each piece (see Shearer (1987)).

Since there is no a priori indication that the final error will be satisfactory, the difficulty that arises is how one should break up the original curve. Simple bisection leaves two polynomial curves which must be approximated, again with no prior knowledge that the reduction scheme will be successful, perhaps leading to a further refinement of $\mathbf{P}(s)$. A more efficient approach is to force the errors at each stage to fall below some threshold, say

$$\|\mathbf{P}_{i+1}(s) - \mathbf{P}_i(s)\|_E = \frac{\|\mathbf{P}_i^{(n-i)}(0)\|_E}{(n-i)!} E_{n-i}^{\alpha} \leq 2^{i-k}E,$$

for $0 \leq i \leq k-1$. With the stepwise errors so bounded, the accumulated error will always be less than E. If the stepwise error should exceed this bound at any step, then the polynomial $\mathbf{P}_i(s)$, instead of $\mathbf{P}(s)$, can be split and reparametrized into $\mathbf{P}_i(s/2)$ and $\mathbf{P}_i((1+s)/2)$, $s \in [0,1]$. This latter approach involves maintaining a stack of polynomials to be economized, and has the advantage of adaptively refining the original curve.

No matter which partitioning scheme is used, the geometric continuity of the final piecewise approximate is controlled by the value of the parameter α. The approximate will be continuous if $\alpha=1$, tangentially smooth if $\alpha=2$, and curvature continuous if $\alpha=3$. This is assured by the fact that each iterate $\mathbf{P}_i(s)$ matches tangent and acceleration directions where joined, depending on the value α. That is,

$$\mathbf{P}_i^{(j)}(\delta) = \mathbf{P}^{(j)}(\delta), \; j = 0,\ldots,\alpha-1; \; \delta = 0,1.$$

For a complete discussion of geometric versus parametric continuity we refer the interested reader to Bartels, Beatty and Barsky (1987).

4. Chebyshev Economization and the Remez Algorithm

At the heart of the preceding section is constrained Chebyshev economization. It is natural to ask how this construction compares with existing methods, and can it be improved upon. In this section we make some qualitative observations about this approximation process, and compare this method to an implementation of the Remez algorithm.

4.1 Chebyshev Economization

The Chebyshev expansion for a function on an interval has long been a tool of numerical analysts. The truncation of that series approximates the function almost uniformly, with an error on the order of the first neglected term. Thus it is not difficult to see that if an nth degree polynomial is approximated using CE by a constant polynomial, the coefficients in the degree reduction process are, in effect, the coefficients of the polynomial's Chebyshev expansion. Consequently, terminating the process at an intermediate stage produces a near uniform approximation to the original polynomial.

The cost in arithmetic operations to produce an mth degree polynomial approximate to an nth polynomial is (n^2-m^2+n-m). If a breakup of the original curve is required, then this cost times the number of pieces provides an upper bound on the total number of operations. This cost can, of course, be reduced if the original curve is refined adaptively, as suggested in the previous section.

4.2 The Remez Algorithm

The Remez algorithm is a standard method for estimating the best uniform polynomial approximate of a continuous function. It exploits the fact that the error between a function and its best m^{th} degree polynomial approximate attains its maximum (in absolute value) at least $m+2$ times, and does so while alternating signs. We will not discuss the method completely here, but only summarize some of the computational steps involved. Details can be found in Davis (1963).

Let $f(s)$ denote a real function to be approximated on an interval $[a,b]$, and let $\{s_i\}_{i=1}^{m+1}$ be an initial alternation set. A polynomial $p(s) \in \pi_m$ and an error E are determined so that $f(s_i)-p(s_i) = (-1)^i E$, $1 \le i \le m+2$. The location of a maximum of $|f(s)-p(s)|$, say s^*, is determined and exchanged with an appropriate element of $\{s_i\}_{i=1}^{m+1}$. The entire process is repeated with the new alternation set until the actual maximum is satisfactorily close to the computed value E.

Each iteration of the Remez algorithm requires a number of calculations. If $f(s) \in \pi_n$, then the initialization of the $m+2$ dimensional system of equations requires $(m+2)(2n+m(m-1)/2)$ operations, the solution of this system of equations requires on the order of $2(m+2)^3/3$ operations, and the difficulty of estimating the maxima of $|f(s)-p(s)|$ depends on the relative precision required. Some computational savings can be derived from the fact that the coefficient matrix need not be entirely redefined at each stage, but solution of the resulting system and the search for maxima can be costly.

To more directly compare CE with the Remez algorithm, consider an example where $n=15$ and $m=7$. A single pass through the CE process takes 184 arithmetic operations. The initialization of the system of equations for the Remez algorithm requires 459 arithmetic operations, to say nothing of the solution of the 9x9 system, or the search for the maxima.

5. Zolotareff Economization

Examination of the CE algorithm shows that it is a rather simple process involving only the coefficients of a set of fixed real polynomials, and that the degree of the polynomial is reduced by one in each iteration. It seems reasonable to seek a greater degree reduction with comparable arithmetic simplicity.

To investigate this possibility consider the real valued polynomial $p(s) = as^n + a\sigma s^{n-1} + \cdots \in \pi_n$. Conventional CE would have us define the first two approximates $p_1(s)$ and $p_2(s)$ by

$$p_1(s) = p(s) - a\ \overset{0}{C}_n(s) \in \pi_{n-1};$$

$$p_2(s) = p_1(s) - a(\sigma + n/2)\ \overset{0}{C}_{n-1}(s) \in \pi_{n-2}$$

Alternaltively, we could define an approximate $q(s)$ by $q_2(s) = p(s) - a\ Z_n(\sigma;s) \in \pi_{n-2}$, where

$$Z_n(\sigma;s) = s^n + \sigma s^{n-1} + \cdots \in \pi_n;$$

$$\max_{s\in[0,1]} |Z_n(\sigma;s)| = \min_{g\in\pi_{n-2}} \max_{s\in[0,1]} |s^n + \sigma s^{n-1} - g(s)|.$$

The extremal polynomials $Z_n(\sigma;s)$ were studied by Chebyshev's student Zolotareff (but with the problem phrased on the interval $[-1,1]$).

Zolotareff showed that $Z_n(\sigma;s)$ could be explicitly determined for only a small interval of σ values; and Kirchberger developed a first order approximation to $Z_n(\sigma;s)$ for values outside that interval (see Meinardus

(1967)). In the explicit case the extremal polynomial can be determined by a translation of the Chebyshev polynomials. In effect,

$$Z_n(\sigma;s) = s^n + \sigma s^{n-1} + \sum_{i=0}^{n-2} a_i(\sigma)\ s^i,$$

where the coefficients $a_i(\sigma)$ are polynomials in the variable σ. Otherwise $Z_n(\sigma;s)$ can be approximated by a linear combination of $T_n(s)$ and $U_n(s)$, the Chebyshev polynomials of the first and second kind, respectively.

The arithmetic complexity of using either of these representations in an economization scheme is at best no better than two iterates of conventional economization. As a result it would seem that CE is in some sense optimal relative to the quality of the result and the arithmetic computations required.

References

Bartels, R.H., Beatty, J.C., and Barsky, B.A. (1987), 'An Introduction to Splines for use in Computer Graphics and Geometric Modeling', Morgan Kaufmann Publishers, Los Altos.

Davis, P.J. (1963), 'Interpolation and Approximation', Dover Publications, New York.

Heybrock, P.C. (1987), 'Algorithms for conversion between alternative representations of parametric curves and surfaces', M.Sc. Thesis, Department of Applied Computing and Mathematics, Cranfield Institute of Technology.

Lachance, M.A. (1988), Chebyshev economization for parametric surfaces, Computer Aided Geometric Design 5, Elsevier Science Publishers, pp. 195-208.

Lachance, M.A., Saff, E.B., and Varga, R.S. (1979), Bounds

on incomplete polynomials vanishing at both end points of an interval, in: 'Constructive Approaches to Mathematical Models', Academic Press, New York, 421-437.

Meinardus, G. (1967), 'Approximation of Functions: Theory and Numerical Methods', Springer Verlag, New York.

Shearer, P.A. (1987), 'Chebyshev economization for parametric curves and surfaces', M.Sc. Thesis, Department of Applied Computing and Mathematics, Cranfield Institute of Technology.

Shearer, P.A. (1988), Conversion between different mathematical representations of the polynomial types of parametric curves and surfaces, a Report prepared for the Esprit CAD*I Project, WG3.CIT.0003.88.

Calculation of the energy of a piecewise polynomial surface

E. Quak and L. L. Schumaker

Department of Mathematics,
Vanderbilt University,
Nashville, Tennesee, USA

Abstract: This paper is related to the problem of constructing a smooth piecewise polynomial surface fit to data given at the vertices of a triangulation. We are interested in methods in which the surface is represented in Bernstein-Bézier form, and the coefficients are chosen so that the resulting surface minimizes some expression involving the energy of the surface (for example, see [1,3,5-9]). Here we do not go into the details of such methods themselves, but instead concentrate on the problem of representing the energy associated with a particular choice of coefficients. The formulae developed here will be applied elsewhere to specific interpolation and smoothing methods.

Keywords: multivariate splines, data fitting, interpolation, energy

1. Introduction

Let $\triangle$ be an arbitrary triangulation of a domain $\Omega \subset \mathbb{R}^2$. Given integers $0 \leq r < d$, we define the associated *space of polynomial splines of degree d and smoothness r* by

$$S_d^r(\triangle) = \{s \in C^r(\Omega) : s|_{T_i} \in \mathcal{P}_d, \quad i = 1, \ldots, N\}, \tag{1}$$

where $\mathcal{P}_d$ is the $\binom{d+2}{2}$ dimensional linear space of bivariate polynomials of total degree d, and where $T_i, i = 1, \ldots, N$, are the triangles of $\triangle$. While the constructive theory of these spaces of splines is not yet complete, they are clearly useful tools for fitting surfaces to data of the form

$$z_i = f(x_i, y_i) + \varepsilon_i, \quad i = 1, \ldots, V, \tag{2}$$

where $(x_i, y_i), i = 1, \ldots, V$, are given points scattered in the plane (and forming the vertices of the triangulation), and where $\varepsilon_i, i = 1, \ldots, V$, are measurement errors.

In this paper we are interested in two classes of surface fitting methods using the space $S_d^r(\Delta)$. The first class of methods involves minimizing some measure of smoothness $J(u)$ over the subset

$$U = \{u \in S_d^r(\Delta) \ : \ u(x_i, y_i) = z_i, \quad i = 1, \ldots, V\}. \tag{3}$$

The set U is the set of *interpolating splines.* We may call these kinds of methods *smoothest interpolation methods.*

The second class of methods of interest involve minimizing a combination of smoothness and goodness of fit such as

$$\rho_\lambda(s) = \lambda J(s) + E(s),$$

over $S_d^r(\Delta)$, where $\lambda \geq 0$ is a smoothing parameter, and where $E(s)$ measures the goodness of fit. For example, we might take

$$E(s) = \sum_{i=1}^{V} \left[s(x_i, y_i) - z_i\right]^2 .$$

This is an example of a *penalized least squares method* (see [6] in this volume).

While other energy expressions can be treated by similar methods, throughout the remainder of this paper we restrict our attention to the functional

$$J(u) = \sum_{i=1}^{N} J_{T_i}(u), \tag{4}$$

where

$$J_{T_i}(u) = \int\int_{T_i} [(u_{xx})^2 + 2(u_{xy})^2 + (u_{yy})^2] dx dy, \quad i = 1, \ldots, N,$$

where u_{xx} represents the second order derivative of u in the x direction, etc. This expression represents the energy of a "thin plate", and is a natural way of measuring the roughness of a surface (see [3,9]). We have defined J in (4) as a sum of integrals over the individual triangles of Δ in order to be able to apply it to spline spaces $S_d^r(\Delta)$ even if the global smoothness r is less than 2. (Indeed, in applications it is common to use C^1 splines, where $r = 1$).

The main purpose of this paper is to find formulae for representing $J(u)$ in terms of the coefficients of the Bernstein-Bézier representation of a spline u in $S_d^r(\Delta)$. This is an essential step in deriving numerical algorithms for calculating spline fits as discussed above. The remainder of this paper is organized as follows. In Section 2 we introduce the necessary notation for representing splines in $S_d^r(\Delta)$ in Bernstein-Bézier form, and introduce certain energy expressions. In Section 3 we derive convenient recursion formulae for these energy expressions. In Section 4 we apply our results to give explicit formulae for cubic and quartic splines.

2. The Bernstein-Bézier representation

In working with $S_d^r(\Delta)$ numerically, it will be useful to regard $S_d^r(\Delta)$ as the linear subspace of $S_d^0(\Delta)$ defined by

$$S_d^r(\Delta) = \{s \in S_d^0(\Delta) \ : \ s \in C^r(\Omega)\}. \tag{5}$$

Each spline $s \in S_d^0(\Delta)$ can be written in the form

$$s(x,y) = s_l(x,y) \quad \text{for} \quad (x,y) \in T_l, \quad l = 1, \ldots, N,$$

where s_l is a polynomial of degree d, $l = 1, \ldots, N$. Each of these polynomials can be written in Bernstein-Bézier form as

$$s_l(r,s,t) = \sum_{i+j+k=d} c_{ijk}^l \frac{d! r^i s^j t^k}{i! j! k!},$$

where (r,s,t) are the barycentric coordinates of a point (x,y) in the triangle T_l. Identifying the *Bézier-ordinates* c_{ijk}^l on common edges of triangles forces the continuity of the associated piecewise polynomial function.

With each Bézier–ordinate c_{ijk}^l we associate a *domain point*

$$P_{ijk}^\ell = \left(iV_1^\ell + jV_2^\ell + kV_3^\ell\right)/d,$$

where V_1^ℓ, V_2^ℓ, and V_3^ℓ denote the vertices of the triangle T_l. We omit the superscript l whenever this will cause no confusion. The set of all domain points is denoted by $\mathcal{B}_d(\Delta)$. On each triangle T of Δ there are precisely $(d+1)(d+2)/2$ points of $\mathcal{B}_d(\Delta)$ spaced uniformly over T. It is common to associate the coefficient c_{ijk}^ℓ with the domain point P_{ijk}^ℓ. Clearly, the set

$$\mathcal{C} = \{c_{ijk}^\ell\}, \quad \ell = 1, \ldots, N \quad \text{and} \quad i+j+k = d$$

of coefficients uniquely defines a spline in $S_d^0(\Delta)$. It is easy to see that

$$nc = \#(\mathcal{C}) = V + (d-1)E + \binom{d-1}{2} N,$$

where V is the number of vertices in Δ (which is also the number of data points), and where E is the number of edges and N the number of triangles of Δ.

It is well-known (cf. [2,4]) that a spline $s \in S_d^0(\Delta)$ lies in $S_d^r(\Delta)$ if and only if an appropriate set of continuity conditions, each of which can be described by a linear equation involving the Bernstein-Bézier coefficients, is satisfied. Thus, we can write

$$S_d^r(\Delta) = \{s \in S_d^0(\Delta) \ : \ Ac = b\}, \tag{6}$$

where c is the coefficient vector of length nc, A is an appropriate $m \times nc$ matrix (where m denotes the number of continuity conditions enforced), and b is an appropriate m vector. The exact nature of A and b is of no concern here, but is, of course, critical in implementing a specific surface fitting method. We remark that in general, $nc > V + m$, so that even after specifying both smoothness and interpolation conditions, there remain a certain number of free parameters to be used in the minimization process.

3. Formulae for the energy of a single patch

In this section we give formulae for the energy of a polynomial of degree d described in Bernstein-Bézier form on a triangle T. Throughout this section we suppose that the polynomial is written in the form

$$p(r,s,t) := \sum_{i=0}^{d} \sum_{j=0}^{i} b_{\frac{i(i+1)}{2}+j+1} \Phi_{d-i,i-j,j}(r,s,t), \tag{7}$$

where, in general

$$\Phi_{\alpha,\beta,\gamma}(r,s,t) = \frac{(\alpha+\beta+\gamma)!\, r^\alpha s^\beta t^\gamma}{\alpha!\,\beta!\,\gamma!}. \tag{8}$$

Now since $p \in \mathcal{P}_d$, it follows that p_{xx}, p_{xy}, and p_{yy} belong to $\mathcal{P}_{d-2}$ and that $(D_{xx}^2 + 2D_{xy}^2 + D_{yy}^2)p$ belongs to $\mathcal{P}_{2(d-2)}$. Introducing the notation $\mathbf{b} := (b_1, ..., b_n)^T$, where $n = \binom{d+2}{2}$, we find that the energy of p over the triangle T is given by the quadratic form

$$J_T(p) = \mathbf{b}^T E^{(d)} \mathbf{b},$$

where $E^{(d)}$ is an appropriate symmetric $n \times n$ matrix.

The remainder of this section will be devoted to finding the entries of $E^{(d)}$. We determine them recursively, starting with the case $d = 2$. As we are dealing with a rotation invariant energy expression, without loss of generality we can assume that the triangle T is in the canonical position shown in Figure 1 with vertices $V_1 = (b,c)$, $V_2 = (0,0)$, and $V_3 = (a,0)$. In this case, the barycentric coordinates (r,s,t) of a point (x,y) are given by the formulae $r = y/c$, $s = 1 - x/a + (b-a)y/ac$, and $t = x/a - by/ac$.

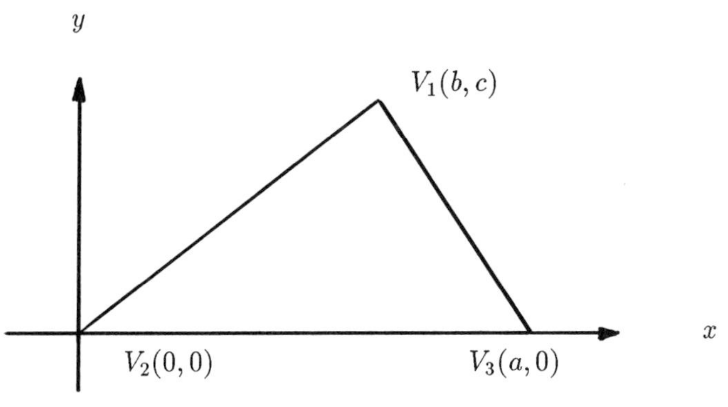

Figure 1. The canonical triangle

Now

$$p_{rr}(r,s,t) = d(d-1)\sum_{i=0}^{d-2}\sum_{j=0}^{i} b_{\frac{i(i+1)}{2}+j+1}\Phi_{d-2-i,i-j,j}(r,s,t),$$

with similar expressions for the other second-order partial derivatives. Using

$$p_{xx} = p_{rr}r_x^2 + 2p_{rs}r_x s_x + 2p_{rt}r_x t_x + p_{ss}s_x^2 + 2p_{st}s_x t_x + p_{tt}t_x^2 \tag{9}$$

and the analogous expressions for p_{xy} and p_{yy}, we find that $E^{(2)}/2ac$ is a symmetric 6×6 matrix whose entries in the upper triangular part are as follows:

$$\begin{pmatrix} (RR)^2 & 2RR\,RS & 2RR\,RT & (RS)^2 & 2RS\,RT & (RT)^2 \\ & 2RR\,SS + 2(RS)^2 & 2RR\,ST + 2RS\,RT & 2RS\,SS & 2SS\,RT + 2RS\,ST & 2RT\,ST \\ & & 2RR\,TT + 2(RT)^2 & 2RS\,ST & 2TT\,RS + 2RT\,ST & 2RT\,TT \\ & & & (SS)^2 & 2ST\,SS & (ST)^2 \\ & & & & 2SS\,TT + 2(ST)^2 & 2ST\,TT \\ & & & & & (TT)^2 \end{pmatrix}$$

where $R = (r_x, r_y)$, $S = (s_x, s_y)$, and $T = (t_x, t_y)$, each of which can be expressed in terms of the quantities a,b and c describing the locations of the vertices of T.

We now present the main result of this paper, a recursion relation for computing the entries in the matrix $E^{(d)}$ for arbitrary $d > 2$.

Theorem 1. *Let T be a canonical triangle as in Figure 1. Then for $i = 0, 1, \ldots, d$, $j = 0, 1, \ldots, i$, $k = 0, 1, \ldots, d$, and $l = 0, 1, \ldots, k$,*

$$E^{(d)}_{\frac{i(i+1)}{2}+j+1,\frac{k(k+1)}{2}+l+1} = d^2(d-1)^2 \int\int_T e^{(d)}_{i,j,k,l}(x,y)\,dx\,dy,$$

where

$$\begin{aligned} e^{(d)}_{i,j,k,l} &= r^2 e^{(d)}_{i,j,k,l} \\ &+ rs[e^{(d)}_{i,j,k-1,l} + e^{(d)}_{i-1,j,k,l}] \\ &+ rt[e^{(d)}_{i-1,j-1,k,l} + e^{(d)}_{i,j,k-1,l-1}] \\ &+ s^2 e^{(d)}_{i-1,j,k-1,l} \\ &+ st[e^{(d)}_{i-1,j,k-1,l-1} + e^{(d)}_{i-1,j-1,k-1,l}] \\ &+ t^2 e^{(d)}_{i-1,j-1,k-1,l-1} \end{aligned}$$

These recursions begin with the values

$$e^{(2)}_{i,j,k,l} = \frac{1}{2ac}E^{(2)}_{\frac{i(i+1)}{2}+j+1,\frac{k(k+1)}{2}+l+1},$$

contained in the matrix $E^{(2)}$ given above.

Proof: We proceed by induction. Since the area of T is $ac/2$, the assertion is clear for the case $d = 2$. To prove the result for $d > 2$, we use the easily verified recursion formula

$$\Phi_{\alpha,\beta,\gamma}(r,s,t) = r\Phi_{\alpha-1,\beta,\gamma}(r,s,t) + s\Phi_{\alpha,\beta-1,\gamma}(r,s,t) + t\Phi_{\alpha,\beta,\gamma-1}(r,s,t). \qquad (10)$$

Now by (9),

$$p_{xx}(x,y) = d(d-1)\sum_{i=0}^{d}\sum_{j=0}^{i} b_{\frac{i(i+1)}{2}+j+1} A_{ij}^{(d)}(x,y),$$

where

$$A_{i,j}^{(d)}(x,y) = r_x^2\Phi_{d-i-2,i-j,j} + 2r_xs_x\Phi_{d-i-1,i-j-1,j} + 2r_xt_x\Phi_{d-i-1,i-j,j-1}$$

$$+s_x^2\Phi_{d-i,i-j-2,j} + 2s_xt_x\Phi_{d-i,i-j-1,j-1} + t_x^2\Phi_{d-i,i-j,j-2}.$$

Using the recursion relation (10), we find that

$$A_{i,j}^{(d)}(x,y) = rA_{i,j}^{(d-1)}(x,y) + sA_{i-1,j}^{(d-1)}(x,y) + tA_{i-1,j-1}^{(d-1)}(x,y)$$

for $i = 0,1,\ldots,d$ and $j = 0,1,\ldots,i$, provided that we start with $A_{0,0}^{(2)} = r_x^2$, $A_{1,0}^{(2)} = 2r_xs_x$, $A_{1,1}^{(2)} = 2r_xt_x$, $A_{2,0}^{(2)} = s_x^2$, $A_{2,1}^{(2)} = 2s_xt_x$, and $A_{2,2}^{(2)} = t_x^2$. The polynomials p_{xy} and p_{yy} have similar expansions in terms of functions satisfying the same recursions, but with different starting values.

Using the above, we get

$$[p_{xx}]^2 = d^2(d-1)^2\sum_{i=0}^{d}\sum_{j=0}^{i}\sum_{k=0}^{d}\sum_{l=0}^{k} b_{\frac{i(i+1)}{2}+j+1} b_{\frac{k(k+1)}{2}+l+1} e_{xx,i,j,k,l}^{(d)},$$

where

$$e_{xx,i,j,k,l}^{(d)}(x,y) := A_{i,j}^{(d)}(x,y)\cdot A_{k,l}^{(d)}(x,y).$$

Let $e_{xy,i,j,k,l}^{(d)}$ and $e_{yy,i,j,k,l}^{(d)}$ be the relevant expressions involved in the expansions of $[p_{xy}]^2$ and $[p_{yy}]^2$, and let $e_{i,j,k,l}^{(d)} := e_{xx,i,j,k,l}^{(d)} + 2e_{xy,i,j,k,l}^{(d)} + e_{yy,i,j,k,l}^{(d)}$. Then it follows that $e_{i,j,k,l}^{(d)}$ satisfies the stated recursion relation, and

$$[D_{xx}^2 + 2D_{xy}^2 + D_{yy}^2]p = d^2(d-1)^2\sum_{i=0}^{d}\sum_{j=0}^{i}\sum_{k=0}^{d}\sum_{l=0}^{k} b_{\frac{i(i+1)}{2}+j+1} b_{\frac{k(k+1)}{2}+l+1} e_{i,j,k,l}^{(d)}.$$

The result follows since the entries of the energy matrix are obtained by integrating the expressions $e_{i,j,k,l}^{(d)}$ over the triangle T. ■

4. The cubic and quartic cases

In this section we consider cubic and quartic splines as they are the most likely to be useful in practice. In particular, we present the explicit entries in the energy matrices $E^{(3)}$ and $E^{(4)}$. Before proceeding, we observe that by the symmetries involved, in describing $E^{(d)}$, it is not necessary to give all of the $(d+1)^2(d+2)^2/4$ entries. First, by symmetry it suffices to work only with the upper-triangular part of the matrix. Moreover, it is clear that in the upper-triangular part, whenever a formula occurs involving R, S, there are always two others of the same form involving S, T and R, T (cf. the $E_{22}^{(2)}$, $E_{33}^{(2)}$, and $E_{55}^{(2)}$ entries of the matrix $E^{(2)}$ given above). In addition, whenever a formula appears involving either one or three of the letters, then there are always two other similar formulae obtained by cyclic permutation (cf. the $E_{11}^{(2)}$, $E_{44}^{(2)}$, and $E_{66}^{(2)}$, or the $E_{23}^{(2)}$, $E_{25}^{(2)}$, and $E_{35}^{(2)}$ entries of $E^{(2)}$).

The following theorem computes the number of formulae which must be given in order to completely specify the matrix $E^{(d)}$.

Theorem 2. *The number of formulae needed to specify the matrix $E^{(d)}$ is q_d, where $q_{-2} = q_{-1} = 0$, $q_0 = 1$, and where for $k > 0$,*

$$q_{2k} = 2k^3 + 4k + q_{2k-3}$$

$$q_{2k-1} = 2k^3 - 3k^2 + 5k - 2 + q_{2k-4}.$$

Proof: The result follows for $d = 2$ by inspection of the matrix $E^{(2)}$ given above. Now we can proceed by induction. We consider the recursion for q_{2k-1} first. In the first row it is easy to see that it suffices to give $2 + 4 + \cdots + 2k$ entries – this corresponds to the products of b_1 by the b_i associated with domain points in the left half of the triangle T. For example, in the cubic case (cf. Figure 2), we need the $(1,1), (1,2), (1,4), (1,5), (1,7), (1,8)$ entries. Next, for each $i = 1, \dots, k-1$, in the $\frac{i(i+1)}{2} + 1$-st row of the matrix we need $\frac{(n+1)(n+2)}{2} - 3(2i-1) - 1$ formulae. In the cubic case in Figure 2, these are the $(2,2), (2,3), (2,4), (2.5), (2,6), (2,9)$ entries. This accounts for all rows corresponding to coefficients on the boundary of T. To complete the proof we simply apply the induction hypothesis to the coefficients lying inside the triangle. In the cubic case this is just the $(5,5)$ entry.

The proof of the recursion for q_{2k} is similar. In this case, in the first row it suffices to give $1 + 3 + \cdots + (2k+1)$ entries – this corresponds to the products in the left half of the triangle T. Next, for each $i = 1, \dots, k-1$, in the $\frac{i(i+1)}{2} + 1$-st row of the matrix we again need $\frac{(n+1)(n+2)}{2} - 3(2i-1) - 1$ formulae. In this case we also need to do the $\frac{k(k+1)}{2} + 1$-st row, which requires $k(k-1)+2$ formulae. Finally, we apply the induction hypothesis to the coefficients lying inside the triangle. ■

Using Theorem 2 we see that for the cubic case we need 13 formulae, while for the quartic case we need 26 formulae.

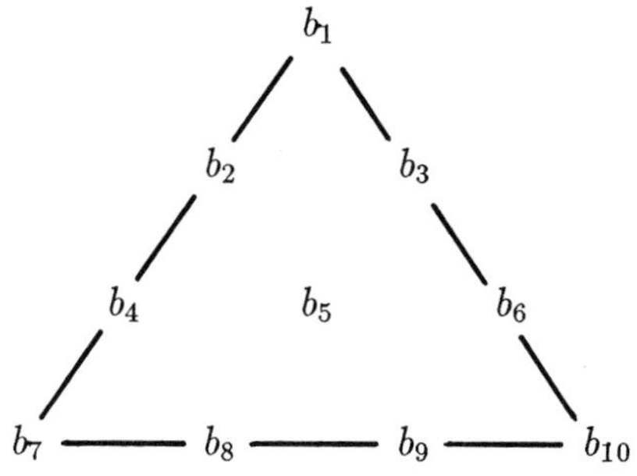

Figure 2. A typical triangle with associated coefficients

Example 3. *The thirteen essential formulae for the entries of* $G = 5E^{(3)}/9ac$ *are*

$$G^{(3)}_{1,1} = (RR)^2$$
$$G^{(3)}_{1,2} = 2RR\,RS + .5(RR)^2$$
$$G^{(3)}_{1,4} = (RS)^2 + RR\,RS$$
$$G^{(3)}_{1,5} = 2RS\,RT + RR\,RT + RR\,RS$$
$$G^{(3)}_{1,7} = .5(RS)^2$$
$$G^{(3)}_{1,8} = RS\,RT + .5(RS)^2$$
$$G^{(3)}_{2,2} = 2RR\,SS + 2(RS)^2 + 2RR\,RS + (RR)^2$$
$$G^{(3)}_{2,3} = 2RR\,ST + 2RS\,RT + RR\,RT + RR\,RS + .5(RR)^2$$
$$G^{(3)}_{2,4} = 2SS\,RS + RR\,SS + 1.5(RS)^2 + 2RR\,RS$$
$$G^{(3)}_{2,5} = 2SS\,RT + 2RS\,ST + RR\,ST + 2RS\,RT + RR\,SS + (RS)^2 + 2RR\,RT + RR\,RS$$
$$G^{(3)}_{2,6} = 2RT\,ST + .5(RT)^2 + RR\,ST + RS\,RT + RR\,RT$$
$$G^{(3)}_{2,9} = RT\,ST + SS\,RT + RS\,ST + (RT)^2 + RS\,RT$$
$$G^{(3)}_{5,5} = 2(SS\,TT + (ST)^2 + TT\,RS + RT\,ST + SS\,RT + RS\,ST + RR\,TT + (RT)^2$$
$$+ RR\,ST + RS\,RT + RR\,SS + (RS)^2)$$

Example 4. *The 26 essential formulae for the entries of* $H = 5E^{(4)}/4ac$ *are*

$$H^{(4)}_{1,1} = 6(RR)^2$$
$$H^{(4)}_{1,2} = 12RR\,RS + 3(RR)^2$$
$$H^{(4)}_{1,4} = 6(RS)^2 + 6RR\,RS + (RR)^2$$
$$H^{(4)}_{1,5} = 12RS\,RT + 6RR\,RT + 6RR\,RS + (RR)^2$$
$$H^{(4)}_{1,7} = 3(RS)^2 + 2RR\,RS$$
$$H^{(4)}_{1,8} = 6RS\,RT + 3(RS)^2 + 2RR\,RT + 2RR\,RS$$
$$H^{(4)}_{1,11} = (RS)^2$$
$$H^{(4)}_{1,12} = 2RS\,RT + (RS)^2$$
$$H^{(4)}_{1,13} = (RT)^2 + 2RS\,RT + (RS)^2$$
$$H^{(4)}_{2,2} = 12RR\,SS + 12(RS)^2 + 12RR\,RS + 4(RR)^2$$
$$H^{(4)}_{2,3} = 12RR\,ST + 12RS\,RT + 6RR\,RT + 6RR\,RS + 2(RR)^2$$
$$H^{(4)}_{2,4} = 12SS\,RS + 6RR\,SS + 9(RS)^2 + 10RR\,RS + 3(RR)^2$$
$$H^{(4)}_{2,5} = 12SS\,RT + 12RS\,ST + 6RR\,ST + 12RS\,RT + 6RR\,SS + 6(RS)^2 + 8RR\,RT + 6RR\,RS + 2(RR)^2$$
$$H^{(4)}_{2,6} = 12RT\,ST + 3(RT)^2 + 2RR\,RS + 6RR\,ST + 6RS\,RT + 4RR\,RT + (RR)^2$$
$$H^{(4)}_{2,7} = 6SS\,RS + 6(RS)^2 + 2RR\,SS + 6RR\,RS$$
$$H^{(4)}_{2,8} = 6SS\,RT + 6RS\,ST + 6SS\,RS + 10RS\,RT + 2RR\,ST + 2RR\,SS + 4(RS)^2 + 6RR\,RT + 4RR\,RS$$
$$H^{(4)}_{2,9} = 6RT\,ST + 6SS\,RT + 6RS\,ST + 4(RT)^2 + 2RR\,ST + 6RS\,RT + 2RR\,SS + 2(RS)^2 + 4RR\,RT + 2RR\,RS$$
$$H^{(4)}_{2,10} = 6RT\,ST + 2(RT)^2 + 2RR\,ST + 2RS\,RT + 2RR\,RT$$
$$H^{(4)}_{2,13} = 2RT\,ST + 2SS\,RT + 2RS\,ST + 2SS\,RS + 3(RT)^2 + 4RS\,RT + (RS)^2$$
$$H^{(4)}_{2,14} = 2RT\,ST + 2SS\,RT + 2RS\,ST + 2(RT)^2 + 2RS\,RT$$
$$H^{(4)}_{4,4} = 6(SS)^2 + 12SS\,RS + 8RR\,SS + 10(RS)^2 + 12RR\,RS + 6(RR)^2$$
$$H^{(4)}_{4,5} = 12SS\,ST + 6SS\,RT + 12RS\,ST + 6SS\,RS + 8RR\,ST + 10RS\,RT + 4RR\,SS + 5(RS)^2 + 6RR\,RT + 6RR\,RS + 3(RR)^2$$
$$H^{(4)}_{4,6} = 6(ST)^2 + 6RT\,ST + 6RS\,ST + (RT)^2 + 4RR\,ST + 4RS\,RT + (RS)^2 + 2RR\,RT + 2RR\,RS + (RR)^2$$
$$H^{(4)}_{4,9} = 3(ST)^2 + 6SS\,ST + 8RT\,ST + 4SS\,RT + 6RS\,ST + 2SS\,RS + 3(RT)^2 + 4RR\,ST + 6RS\,RT + 2RR\,SS + 2(RS)^2 + 6RR\,RT + 2RR\,RS$$
$$H^{(4)}_{5,5} = 12SS\,TT + 12(ST)^2 + 12TT\,RS + 12RT\,ST + 12SS\,RT + 12RS\,ST + 8RR\,TT + 8(RT)^2 + 8RR\,ST + 12RS\,RT + 8RR\,SS + 8(RS)^2 + 8RR\,RT + 8RR\,RS + 4(RR)^2$$
$$H^{(4)}_{5,8} = 6SS\,TT + 6(ST)^2 + 6SS\,ST + 10TT\,RS + 10RT\,ST + 6SS\,RT + 10RS\,ST + 8SS\,RS + 6RR\,TT + 6(RT)^2 + 6RR\,ST + 10RS\,RT + 4RR\,SS + 6(RS)^2 + 6RR\,RT + 8RR\,RS$$

Acknowledgment: The first author was supported in part by the Alexander von Humboldt Foundation, while the second author was supported in part by the National Science Foundation under Grant DMS-8602337.

References

1. Alfeld, P. Triangular extrapolation, University of Wisconsin–Madison Mathematics Research Center Technical Summary Report # 2707, 1984.
2. deBoor, B-form basics, in *Geometric Modeling: Algorithms and New Trends*, SIAM, Philadelphia, 1987, 131–148.
3. Duchon, J., Fonctions–spline homogenes plusieurs variables, Thèse, Grenoble, 1980.
4. Farin, G., Bézier polynomials over triangles and the construction of piecewise C^r polynomials, TR/91, Dept. of Mathematics, Brunel Univ., 1980.
5. Gmelig-Meyling, R., Approximation by cubic C^1 splines on arbitrary triangulations, Numer. Math. **51** (1987), 65–85.
6. von Golitschek, M. and L. L. Schumaker, Data fitting by penalized least squares, this volume.
7. Grandine, T., An iterative method for computing multivariate C^1 piecewise polynomial interpolants, CAGD **4** (1987), 307–319.
8. Schmidt, R., Eine Methode zur Konstruktion von C^1-Flächen zur Interpolation unregelmässig verteilter Daten, in *Multivariate Approximation Theory*, W. Schempp and K. Zeller, eds., Birkhäuser, Basel, 1982, 343–361.
9. Wahba, G., Spline bases, regularization and generalized cross-validation for solving approximation problems with large quantities of noisy data, in *Approximation Theory III*, E. Cheney (ed.), Academic Press, New York, 1980, 905–912.

3. Interpolation

Radial basis function interpolation on an infinite regular grid

M. D. Buhmann and M. J. D. Powell

Department of Applied Mathematics and Theoretical Physics, University of Cambridge, UK

Abstract A radial basis function approximation $s(\cdot)$ from $\mathcal{R}^n$ to $\mathcal{R}$ depends on a fixed set of points $\{x_i\}$ in $\mathcal{R}^n$ and on a fixed function $\phi(\cdot)$ from $\mathcal{R}^+$ to $\mathcal{R}$, as it has the form $\{s(x)=\sum_i \lambda_i \phi(\|x-x_i\|_2) \mid x\in\mathcal{R}^n\}$. We consider the case when $\{x_i\}$ is the infinite lattice $\mathcal{Z}^n$ and the coefficients $\{\lambda_i\}$ give interpolation on the lattice. Therefore we study the cardinal function $C(\cdot)$, which is in the space of approximations and satisfies $\{C(\ell)=\delta_{0\ell} \mid \ell\in\mathcal{Z}^n\}$. By employing Fourier transforms, we identify the rate of decay of $|C(x)|$ to zero as $\|x\| \to \infty$ for several useful choices of $\phi(\cdot)$. Further, an algorithm is presented for calculating $C(x)$ for moderate values of $\|x\|$. It provides some tabulations of the cardinal function that quantify the qualitative asymptotic theory. The results may be highly important to the development of new algorithms for multivariable interpolation, because they include some cases where the asymptotic decay to zero is very fast.

Keywords Cardinal functions. Fourier transforms. Gauss–Seidel iteration. Interpolation. Multivariable approximation. Radial basis functions.

1. Introduction

Radial basis functions provide a versatile family of functions from $\mathcal{R}^n$ to $\mathcal{R}$ that is highly promising for the approximation of functions of several variables in computer calculations. The generic radial basis function, $\phi(\cdot)$ say, is from $\mathcal{R}^+$ to $\mathcal{R}$, and a typical radial basis function approximation with n variables has the form

$$s(x) = \sum_{i=1}^{m} \lambda_i \, \phi(\|x-x_i\|_2), \quad x\in\mathcal{R}^n, \tag{1}$$

where $\{\lambda_i \mid i=1,2,\ldots,m\}$ and $\{x_i \mid i=1,2,\ldots,m\}$ are real parameters and fixed points in $\mathcal{R}^n$ respectively. When interpolating to values of a function $f(\cdot):\mathcal{R}^n\to\mathcal{R}$, it is usual for the $\{x_i \mid i = 1,2,\ldots,m\}$ to be the points at which the function values are given and for the parameters $\{\lambda_i \mid i = 1,2,\ldots,m\}$ to be defined by the interpolation conditions

$$s(x_i) = f(x_i), \quad i = 1,2,\ldots,m. \tag{2}$$

Of course the points $\{x_i \mid i = 1,2,\ldots,m\}$ should all be different, and then for many useful choices of $\phi(\cdot)$ the interpolation conditions are guaranteed to define the parameters uniquely (Micchelli, 1986). This important work was reviewed at the previous Shrivenham conference (Powell, 1987).

An obvious advantage of expression (1) over the methods that are employed usually for multivariable approximation is that this expression does not become more elaborate as the number of variables is increased. When m is large and $\phi(\cdot)$ does not have finite support, however, then the work of calculating $s(x)$ for any x can be much greater than the work of calculating a typical piecewise polynomial approximation. The purpose of the present paper is to consider another question that is highly important to the usefulness of radial basis function approximation, namely the localization properties of interpolation. In other words, assuming that $s(\cdot)$ can be expressed in the form

$$s(x) = \sum_{i=1}^{m} f(x_i)\, C_i(x), \quad x \in \mathcal{R}^n, \tag{3}$$

where the *cardinal functions* $\{C_i(\cdot) \mid i = 1,2,\ldots,m\}$ are independent of $f(\cdot)$ and satisfy $C_i(x_j) = \delta_{ij}$ in order that the interpolation conditions (2) hold, we ask how quickly $|C_i(x)|$ tends to zero as $\|x - x_i\|$ becomes large. Here $\|\cdot\|$ denotes any norm on $\mathcal{R}^n$. We see that this question is directly relevant to the effect on $s(\cdot)$ of perturbations to the given function values $\{f(x_i) \mid i = 1,2,\ldots,m\}$.

Perhaps the best known example of localization properties without finite support is cubic spline interpolation to values of a function of one variable at equally spaced points throughout the real line. In this case the only zeros of $C_i(\cdot)$ occur at the data points $\{x_j \mid j \neq i\}$, and the cardinal function satisfies the relation

$$C_i(x_i \pm \theta) = (\sqrt{3}-2)\, C_i(x_i \pm [\theta - h]), \quad \theta \geq 2h, \tag{4}$$

where h is the spacing between data points (see Powell, 1981, for instance). Thus, if k is the number of data points between x and x_i, then $|C(x)|$ is bounded above by a multiple of $(2-\sqrt{3})^k$. This example can be construed as radial basis function approximation with $n = 1$ and $\{\phi(r) = r^3 \mid r \in \mathcal{R}^+\}$.

We restrict attention to the case when the set of data points $\{x_i\}$ is the infinite lattice $\mathcal{Z}^n$ of points in $\mathcal{R}^n$ with integer components, in order that several decay properties of cardinal functions can be derived by Fourier techniques. It should be noted, however, that some of the main advantages of radial basis function interpolation occur when the data points are in general position. We let ℓ be the generic point of $\mathcal{Z}^n$, and, using translational symmetry, we write equation (3) in the form

$$s(x) = \sum_{\ell \in \mathcal{Z}^n} f(\ell)\, C(x-\ell), \quad x \in \mathcal{R}^n, \tag{5}$$

where now there is a single cardinal function that satisfies $\{C(j) = \delta_{0j} \mid j \in \mathcal{Z}^n\}$.

Provided that all the sums and integrals in this paragraph are uniformly convergent, it is suitable to let $C(\cdot)$ be the function

$$C(x) = \sum_{\ell \in \mathcal{Z}^n} c_\ell\, \phi(\|x-\ell\|_2), \quad x \in \mathcal{R}^n, \tag{6}$$

where the coefficients have the values

$$c_\ell = \frac{1}{(2\pi)^n} \int_{[-\pi,\pi]^n} \frac{e^{i\ell.\theta}}{\sum_{k \in \mathcal{Z}^n} e^{-ik.\theta} \phi(\|k\|_2)}\, d\theta, \quad \ell \in \mathcal{Z}^n. \tag{7}$$

Here i is $\sqrt{-1}$ and $\ell.\theta$ denotes the scalar product between $\ell \in \mathcal{Z}^n$ and $\theta \in \mathcal{R}^n$. We see that $C(\cdot)$ is in the correct linear space, and that the identity

$$\begin{aligned} C(j) &= \sum_{\ell \in \mathcal{Z}^n} c_\ell\, \phi(\|j-\ell\|_2) \\ &= \sum_{\ell \in \mathcal{Z}^n} c_{j-\ell}\, \phi(\|\ell\|_2) \\ &= \frac{1}{(2\pi)^n} \int_{[-\pi,\pi]^n} \frac{\sum_{\ell \in \mathcal{Z}^n} e^{i(j-\ell).\theta} \phi(\|\ell\|_2)}{\sum_{k \in \mathcal{Z}^n} e^{-ik.\theta} \phi(\|k\|_2)}\, d\theta \\ &= \frac{1}{(2\pi)^n} \int_{[-\pi,\pi]^n} e^{ij.\theta}\, d\theta = \delta_{0j}, \quad j \in \mathcal{Z}^n, \end{aligned} \tag{8}$$

is satisfied, which makes expression (6) plausible. This way of identifying the cardinal function is successful when $\{\phi(r) = e^{-r^2} \mid r \in \mathcal{R}^+\}$, for example, but, due to the conditions of absolute convergence, it cannot be applied directly unless $\sum_{\ell \in \mathcal{Z}^n} |\phi(\|\ell\|_2)|$ is finite. Therefore we employ Fourier methods to study the cardinal functions of the choices

$$\left.\begin{array}{lll} \phi(r) = r & & \text{(linear)} \\ \phi(r) = r^3 & & \text{(cubic)} \\ \phi(r) = \sqrt{r^2+c^2} & & \text{(multiquadric)} \\ \phi(r) = 1/\sqrt{r^2+c^2} & & \text{(inverse multiquadric)} \\ \phi(r) = r^2 \log r & & \text{(thin plate spline)} \end{array}\right\}. \tag{9}$$

We are interested in these radial basis functions because each one can give $s \equiv f$ when $f(\cdot)$ is a low order polynomial (Jackson, 1987; Buhmann, 1988b). Specifically, letting Π_q^n be the space of all polynomials from $\mathcal{R}^n$ to $\mathcal{R}$ of total degree at most q, the interpolation formula (5) reproduces each $f(\cdot) \in \Pi_q^n$, where $q = n$, $n+2$, n, $n-2$ and $n+1$ in the linear, cubic, multiquadric, inverse multiquadric (assuming $n \geq 2$) and thin plate spline cases respectively. In each of these cases $C(\cdot)$ has the form (6) for certain coefficients $\{c_\ell \mid \ell \in \mathcal{Z}^n\}$ that make the sum (6) absolutely convergent, so the cardinal function is well-defined. Further, the sum (5) is also absolutely convergent for every $x \in \mathcal{R}^n$ and every $f(\cdot) \in \Pi_q^n$, and it gives $\{s(x) = f(x) \mid x \in \mathcal{R}^n\}$ when $f(\cdot) \in \Pi_q^n$. It should be noted, however, that the double sum

$$\sum_{\ell \in \mathcal{Z}^n} f(\ell) \sum_{j \in \mathcal{Z}^n} c_j\, \phi(\|x - \ell - j\|_2), \tag{10}$$

which is obtained by substituting expression (6) in formula (5), need not be absolutely convergent. Indeed, if we were allowed to rearrange terms, then the factor $\sum_{\ell \in \mathcal{Z}^n} f(\ell)\, \phi(\|x - \ell - j\|_2)$ would multiply c_j, but when $f \equiv 1$ this factor is a divergent sum for all of the choices (9).

For example, consider the case when $n = 1$ and $\{\phi(r) = r \mid r \in \mathcal{R}^+\}$. The coefficients of the cardinal function (6) have the values $c_0 = -1$, $c_{-1} = c_1 = \frac{1}{2}$ and $\{c_\ell = 0 \mid |\ell| \geq 2\}$ in order that $C(\cdot)$ is the *hat function* of piecewise linear interpolation. Hence formula (5) reproduces all linear polynomials. Thus the space that is spanned by the radial basis functions $\{\, \{\|x - \ell\|_2 \mid x \in \mathcal{R}\} \mid \ell \in \mathcal{Z}\}$ includes Π_1^1, although no nonzero linear polynomial can be written as $\{\sum_{\ell \in \mathcal{Z}} \lambda_\ell \|x - \ell\|_2 \mid x \in \mathcal{R}\}$. Therefore, when studying the cases (9) we take the view that the approximation $s(\cdot)$ is to be in the linear space that is spanned by functions of the form (6), which admits the very powerful observation in the previous paragraph that q increases with n. Of course this point of view is unnecessary when m is finite in expression (1). It is also unnecessary when $\{x_i\} \equiv \mathcal{Z}^n$ and $\{\phi(r) = e^{-r^2} \mid r \in \mathcal{R}^+\}$ for example, but in this case the interpolation formula (5) does not give $s \equiv f$ when $f \equiv 1$ (Buhmann, 1988b). Therefore in practice the Gaussian radial basis function is usually far less successful than the nonintegrable functions (9).

We study the cardinal function $C(\cdot)$ theoretically in Section 2. Here we recall from Buhmann (1988b) an explicit expression for the Fourier transform of $C(\cdot)$ that is valid for all of the choices (9), and we also recall that the asymptotic decay rate of $|C(x)|$ to zero as $\|x\| \to \infty$ can be deduced from the differentiability properties of this transform. This analysis identifies the dominant term of $C(x)$ for large $\|x\|$ when $\phi(r) = r$ and $n = 2$.

In Section 3 we present a procedure for calculating values of $C(x)$. It employs a quasi-interpolating function

$$\psi(x) = \sum_{\ell \in \hat{\mathcal{Z}}} \gamma_\ell\, \phi(\|x - \ell\|_2), \quad x \in \mathcal{R}^n, \tag{11}$$

where $\hat{\mathcal{Z}}$ is a finite subset of $\mathcal{Z}^n$, that ideally possesses the diagonal dominance condition

$$|\psi(0)| > \sum_{\ell \in \mathcal{Z}^n}{}' |\psi(\ell)|. \tag{12}$$

Here and throughout this paper the prime on the summation indicates that the $\ell=0$ term is omitted. This function is used in an iteration of Gauss–Seidel type to obtain an approximation $C_b(\cdot)$ to $C(\cdot)$ that satisfies the cardinality equations

$$C_b(\ell) = \delta_{0\ell}, \quad \ell \in \mathcal{Z}^n, \quad \|\ell\|_\infty \le b, \tag{13}$$

for some fixed integer b. We find that this approximation is sufficiently accurate to give good estimates of the true cardinal function for $\|x\|_\infty \le \frac{1}{2}b$.

Thus several tables and graphs were calculated that show properties of $C(x)$ for $n=2$ and $n=3$. We find when $n=3$ and $\phi(r)=r$, for example, that $\{|C(x)| < 10^{-6} \mid \|x\|_\infty > 5\}$, which is superior to the decay rate (4) of cubic spline interpolation. These numerical results are given and discussed in Section 4, and we note that they suggest several topics for further research.

2. The Fourier transform and decay properties of the cardinal function

The Fourier transform of a continuous and absolutely integrable function $\{g(x) \mid x \in \mathcal{R}^n\}$ is defined by the formula

$$\hat{g}(t) = \int_{\mathcal{R}^n} e^{-ix.t}\, g(x)\, dx, \quad t \in \mathcal{R}^n, \tag{14}$$

and, when $\hat{g}(\cdot)$ is continuous and absolutely integrable too, the inverse formula is the relation

$$g(x) = \frac{1}{(2\pi)^n} \int_{\mathcal{R}^n} e^{ix.t}\, \hat{g}(t)\, dt, \quad x \in \mathcal{R}^n, \tag{15}$$

(see Stein and Weiss, 1971, for instance). In this section we derive some decay properties of the cardinal function $C(\cdot)$ from its Fourier transform $\hat{C}(\cdot)$ for each of the radial basis functions (9).

It is proved by Buhmann (1988b) that in all these cases $\hat{C}(\cdot)$ has the remarkably simple form

$$\hat{C}(t) = \frac{\hat{\phi}(\|t\|_2)}{\sum_{k \in \mathcal{Z}^n} \hat{\phi}(\|t - 2\pi k\|_2)}, \quad t \in \mathcal{R}^n, \tag{16}$$

where $\{\hat{\phi}(\|t\|_2) \mid t \in \mathcal{R}^n\}$ is the *generalized Fourier transform* of the radially symmetric function $\{\phi(\|x\|_2) \mid x \in \mathcal{R}^n\}$. Jones (1982) presents an excellent introduction to generalized transforms, and shows that the functions

$$\left.\begin{array}{lll}
\hat{\phi}(r) = r^{-n-1} & & \text{(linear)} \\
\hat{\phi}(r) = r^{-n-3} & & \text{(cubic)} \\
\hat{\phi}(r) = (c/r)^{(n+1)/2} K_{(n+1)/2}(cr) & & \text{(multiquadric)} \\
\hat{\phi}(r) = (c/r)^{(n-1)/2} K_{(n-1)/2}(cr) & & \text{(inverse multiquadric)} \\
\hat{\phi}(r) = r^{-n-2} & & \text{(thin plate spline)}
\end{array}\right\} \tag{17}$$

are appropriate to the radial basis functions (9), except that for simplicity we have suppressed some constant normalization factors because they are irrelevant to the ratio (16). Here $K_\nu(\cdot)$ is the ν-th modified Bessel function as defined in Abramowitz and Stegun (1970).

We see that in all these cases the generalized Fourier transform $\{\hat{\phi}(\|t\|_2) \mid t \in \mathcal{R}^n\}$ is unbounded at the origin, but otherwise it is positive, continuous and absolutely integrable over every closed part of $\mathcal{R}^n$ that excludes the origin. In view of equations (16) and (17), this unboundedness implies the values $\hat{C}(0)=1$ and $\{\hat{C}(2\pi k)=0 \mid k \in \mathcal{Z}^n \backslash \{0\}\}$, which are important to the fact that interpolation reproduces low order polynomials (Jackson, 1987). Further, when t is not in the set of lattice points $\{2\pi k \mid k \in \mathcal{Z}^n\}$, the denominator of expression (16) is an absolutely convergent sum that is periodic and bounded away from zero. It follows from all these remarks that $C(\cdot)$ is well defined when we substitute $\hat{g}(\cdot)=\hat{C}(\cdot)$ in the inverse Fourier transform formula (15).

Buhmann (1988b) uses some properties of generalized functions to prove that this $C(\cdot)$ is the required cardinal function, but the following simple analysis is sufficient when certain sums and integrals are absolutely convergent, including the conditions that $\{\phi(\|x\|_2) \mid x \in \mathcal{R}^n\}$ is the classical inverse Fourier transform of $\{\hat{\phi}(\|t\|_2) \mid t \in \mathcal{R}^n\}$ and $\sum_{k\in\mathcal{Z}^n} \hat{\phi}(\|t-2\pi k\|_2)$ is nonzero for all t. For example, these conditions hold for the Gaussian radial basis function $\{\phi(r)=e^{-r^2} \mid r \in \mathcal{R}^+\}$ because $\{\hat{\phi}(\|t\|_2) \mid t \in \mathcal{R}^n\}$ is a constant multiple of $\{e^{-\|t\|_2^2/4} \mid t \in \mathcal{R}^n\}$. Corresponding to the identity (8), equations (15) and (16) give the cardinality conditions

$$\begin{aligned} C(j) &= \frac{1}{(2\pi)^n} \int_{\mathcal{R}^n} \frac{e^{ij.t}\, \hat{\phi}(\|t\|_2)}{\sum_{k\in\mathcal{Z}^n} \hat{\phi}(\|t-2\pi k\|_2)}\, dt \\ &= \frac{1}{(2\pi)^n} \int_{[-\pi,\pi]^n} \frac{\sum_{\ell\in\mathcal{Z}^n} e^{ij.(t-2\pi\ell)}\, \hat{\phi}(\|t-2\pi\ell\|_2)}{\sum_{k\in\mathcal{Z}^n} \hat{\phi}(\|t-2\pi k\|_2)}\, dt \\ &= \frac{1}{(2\pi)^n} \int_{[-\pi,\pi]^n} e^{ij.t}\, dt \;=\; \delta_{0j}, \quad j \in \mathcal{Z}^n, \end{aligned} \tag{18}$$

where the middle line depends on the periodicity of the denominator and the last line on $e^{ij.(2\pi\ell)}=1$. Moreover, to show that $C(\cdot)$ has the form (6), we note that the continuous periodic function

$$\omega(t) = 1 \Big/ \sum_{k\in\mathcal{Z}^n} \hat{\phi}(\|t-2\pi k\|_2), \quad t \in \mathcal{R}^n, \tag{19}$$

is equal to its Fourier expansion, so we have the identity

$$\omega(t) = \sum_{\ell\in\mathcal{Z}^n} c_\ell\, e^{-i\ell.t}, \tag{20}$$

where the coefficients have the values

$$c_\ell = \frac{1}{(2\pi)^n} \int_{[-\pi,\pi]^n} e^{i\ell.t}\, \omega(t)\, dt, \quad \ell \in \mathcal{Z}^n. \tag{21}$$

Thus $C(\cdot)$ is the function

$$\begin{aligned} C(x) &= \frac{1}{(2\pi)^n} \int_{\mathcal{R}^n} e^{ix.t}\, \hat{C}(t)\, dt \\ &= \frac{1}{(2\pi)^n} \int_{\mathcal{R}^n} e^{ix.t}\, \omega(t)\, \hat{\phi}(\|t\|_2)\, dt \\ &= \sum_{\ell \in \mathcal{Z}^n} c_\ell \{ \frac{1}{(2\pi)^n} \int_{\mathcal{R}^n} e^{i(x-\ell).t}\, \hat{\phi}(\|t\|_2)\, dt \} \\ &= \sum_{\ell \in \mathcal{Z}^n} c_\ell\, \phi(\|x-\ell\|_2), \quad x \in \mathcal{R}^n, \end{aligned} \tag{22}$$

where the third and fourth lines depend on equations (20) and (15) respectively. This expression is identical to equation (6) as required. Further, it can be deduced from the *Poisson summation formula* (see Stein and Weiss, 1971, for instance) that the coefficients (7) and (21) are the same.

Next we consider the localization properties of interpolation that can be deduced from $\hat{C}(\cdot)$. If $\{\hat{C}(t) \mid t \in \mathcal{R}^n\}$ has absolutely integrable second derivatives and is sufficiently smooth for large $\|t\|$, then integration by parts applied to formula (15) gives the relation

$$\begin{aligned} C(x) &= \frac{1}{(2\pi)^n} \int_{\mathcal{R}^n} e^{ix.t}\, \hat{C}(t)\, dt \\ &= \frac{-1}{(2\pi)^n\,(ix_1)} \int_{\mathcal{R}^n} e^{ix.t}\, [\frac{d}{dt_1} \hat{C}(t)]\, dt \\ &= \frac{-1}{(2\pi)^n\, x_1^2} \int_{\mathcal{R}^n} e^{ix.t}\, [\frac{d^2}{dt_1^2} \hat{C}(t)]\, dt, \end{aligned} \tag{23}$$

which implies the identity

$$\|x\|_2^2\, C(x) = \frac{-1}{(2\pi)^n} \int_{\mathcal{R}^n} e^{ix.t}\, [\nabla^2 \hat{C}(t)]\, dt, \quad x \in \mathcal{R}^n, \tag{24}$$

and there is an upper bound on the modulus of the right hand side that is independent of x. Similarly, if $\hat{C}(\cdot)$ has absolutely integrable derivatives of order j and enough smoothness at infinity, this technique provides the inequality

$$|C(x)| \le \Lambda_j\, [1 + \|x\|_2]^{-j}, \quad x \in \mathcal{R}^n, \tag{25}$$

for some positive constant Λ_j. Thus the decay properties of the cardinal function $C(\cdot)$ are intimately related to the differentiability of its Fourier transform $\hat{C}(\cdot)$.

Therefore, as in Buhmann (1988b), we consider the differentiability of the function (16) for each of the choices (17) of $\hat{\phi}(\cdot)$. In the linear, thin plate spline and cubic cases we have the Fourier transform

$$\hat{C}(t) = [\, 1 + \|t\|_2^{n+m} \sum_{k \in \mathcal{Z}^n}{}' \|t - 2\pi k\|_2^{-n-m}\,]^{-1}, \quad t \in \mathcal{R}^n, \tag{26}$$

where $m=1,2$ and 3 respectively. We see that this function is infinitely differentiable throughout $\mathcal{R}^n$ when $n+m$ is even, which gives the important conclusion that a bound of the form (25) is satisfied for every positive j. In fact in these cases $|C(x)|$ decays exponentially as $\|x\| \to \infty$ (Madych and Nelson, 1987; Buhmann, 1988b), which means that the condition

$$|C(x)| \leq \Lambda\, \mu^{\|x\|}, \quad x \in \mathcal{R}^n, \tag{27}$$

is satisfied for some constants $\Lambda > 0$ and $\mu < 1$. One purpose of the numerical results of Section 4 is to indicate the magnitude of μ when $n=2$ and $n=3$.

When $n+m$ is odd in expression (26), however, then $\hat{C}(\cdot)$ has singularities at the grid points $t \in 2\pi\mathcal{Z}^n$ that restrict the value of j in inequality (25). The singularities at $t=0$ and when $t \in \{2\pi k \mid k \in \mathcal{Z}^n \backslash \{0\}\}$ are shown in the expansions

$$\hat{C}(t) = 1 - \|t\|_2^{n+m} \sum_{k\in\mathcal{Z}^n}{}' \|2\pi k\|_2^{-n-m} + O(\|t\|_2^{n+m+1}) \tag{28}$$

and

$$\hat{C}(t) = \|t - 2\pi k\|_2^{n+m}\, \|2\pi k\|_2^{-n-m} + O(\|t-2\pi k\|_2^{n+m+1}) \tag{29}$$

respectively, and it is the contribution from these singularities to $\{C(x) \mid x \in \mathcal{R}^n\}$ through the first line of equation (23) that dominates the magnitude of $|C(x)|$ for large $\|x\|$. Instead of integrating by parts, we pick out this contribution from the fact that the generalized inverse transform of $\{\|t\|_2^{n+m} \mid t \in \mathcal{R}^n\}$ when $n+m$ is odd is the function

$$\frac{2^{n+m}\,\Gamma(n+\frac{1}{2}m)}{\pi^{n/2}\,\Gamma(-\frac{1}{2}n-\frac{1}{2}m)}\, \|x\|_2^{-2n-m}, \quad x \in \mathcal{R}^n. \tag{30}$$

Specifically, the terms from the singularities of expressions (28) and (29) are given explicitly in the equation

$$\begin{aligned}
C(x) &= \frac{1}{(2\pi)^n} \int_{\mathcal{R}^n} e^{ix.t}\, \hat{C}(t)\, dt \\
&= \frac{1}{(2\pi)^n} \int_{\mathcal{R}^n} e^{ix.t} \Big[-\|t\|_2^{n+m} \sum_{k\in\mathcal{Z}^n}{}' \|2\pi k\|_2^{-n-m} + \cdots \\
&\quad + \sum_{k\in\mathcal{Z}^n}{}' \|t-2\pi k\|_2^{n+m}\, \|2\pi k\|_2^{-n-m} + \cdots \Big]\, dt \\
&= \frac{1}{(2\pi)^n} \int_{\mathcal{R}^n} e^{ix.t} \Big[-\|t\|_2^{n+m} \sum_{k\in\mathcal{Z}^n}{}' \|2\pi k\|_2^{-n-m} + \cdots \Big] \\
&\quad + \Big[\sum_{k\in\mathcal{Z}^n}{}' e^{ix.(t+2\pi k)}\, \|t\|_2^{n+m}\, \|2\pi k\|_2^{-n-m} + \cdots \Big]\, dt \\
&= \frac{1}{(2\pi)^n} \int_{\mathcal{R}^n} e^{ix.t}\, \|t\|_2^{n+m} \sum_{k\in\mathcal{Z}^n}{}' [-1 + e^{2\pi i\, x.k}]\, \|2\pi k\|_2^{-n-m} + \cdots\, dt \\
&= \frac{2^{n+m}\,\Gamma(n+\frac{1}{2}m)}{\pi^{n/2}\,\Gamma(-\frac{1}{2}n-\frac{1}{2}m)} \Big\{ \sum_{k\in\mathcal{Z}^n}{}' [-1 + e^{2\pi i\, x.k}]\, \|2\pi k\|_2^{-n-m} \Big\}\, \|x\|_2^{-2n-m} \\
&\quad + o(\|x\|_2^{-2n-m}),
\end{aligned} \tag{31}$$

which is valid for all large $\|x\|$. We see that the sum in the last line of this expression is absolutely convergent, real, nonpositive and periodic in $x \in \mathcal{R}^n$. Further, it vanishes at the grid points $x \in \mathcal{Z}^n$ in accordance with the cardinality conditions $\{C(\ell) = \delta_{0\ell} \mid \ell \in \mathcal{Z}^n\}$.

Later we give particular attention to the linear radial basis function $\phi(r) = r$ when $n=2$ and $m=1$. Therefore we note that in this case expression (30) has the value $9/(2\pi\|x\|_2^5)$, and for any x it is easy to work out numerically the sum in the last line of equation (31). Thus we find the relations

$$\left.\begin{array}{ll} C(x) \approx -0.05747\,\|x\|_2^{-5}, & [x + (\tfrac{1}{2}, 0)] \in \mathcal{Z}^2 \\ C(x) \approx -0.06734\,\|x\|_2^{-5}, & [x + (\tfrac{1}{2}, \tfrac{1}{2})] \in \mathcal{Z}^2 \end{array}\right\} \tag{32}$$

for the values of the cardinal function at the midpoints of edges of grid squares and at the midpoints of grid squares respectively. These asymptotic results agree well with the calculations of Section 4 for moderate values of $\|x\|$.

The following argument shows that, for even n, the asymptotic decay properties of $C(\cdot)$ in the multiquadric and inverse multiquadric cases are similar to expression (31). The dominant singularity of $\{K_\nu(\theta) \mid \theta \in \mathcal{R}^+\}$ at the origin is a multiple of $\theta^{-\nu}$, so we let $m=1$ and $m=-1$ for multiquadrics and inverse multiquadrics respectively, in order that lines 3 and 4 of expression (17) give the relation

$$\hat{\phi}(r) = \mu_m\, r^{-n-m} + O(r^{-n-m+1}), \quad r \in \mathcal{R}^+, \tag{33}$$

where μ_m is a constant. It follows from formula (16) that, instead of equations (28) and (29), we now have the expansions

$$\hat{C}(t) = 1 - \mu_m^{-1}\,\|t\|_2^{n+m} \sum_{k\in\mathcal{Z}^n}{}' \hat{\phi}(\|2\pi k\|_2) + O(\|t\|_2^{n+m+1}) \tag{34}$$

and

$$\hat{C}(t) = \mu_m^{-1}\,\|t - 2\pi k\|_2^{n+m}\,\hat{\phi}(\|2\pi k\|_2) + O(\|t - 2\pi k\|_2^{n+m+1}). \tag{35}$$

Thus expression (31) is still valid except that the sum inside the braces is replaced by the term

$$\mu_m^{-1} \sum_{k\in\mathcal{Z}^n}{}' [-1 + e^{2\pi i\, x.k}]\,\hat{\phi}(\|2\pi k\|_2). \tag{36}$$

In order to be more explicit, we recall from Abramowitz and Stegun (1970) that, when $\nu + \frac{1}{2}$ is a positive integer, the ν-th modified Bessel function is the finite sum

$$K_\nu(\theta) = \sqrt{\pi}\, e^{-\theta} \sum_{q=0}^{\nu-\frac{1}{2}} \frac{(2\nu - q - 1)!}{q!\,(\nu - q - \frac{1}{2})!}\,(2\theta)^{q-\nu}, \quad \theta \in \mathcal{R}^+, \tag{37}$$

which after some manipulation gives the value

$$\frac{\hat{\phi}(\|2\pi k\|_2)}{\mu_m} = \|2\pi k\|_2^{-n-m}\, e^{-\|2\pi c k\|_2} \sum_{q=0}^{\frac{1}{2}(n+m-1)} \frac{(\frac{1}{2}[n+m-1])!\,(n+m-q-1)!\,\|4\pi c k\|_2^q}{(\frac{1}{2}[n+m-1]-q)!\,(n+m-1)!\,q!}. \tag{38}$$

When this expression replaces $\|2\pi k\|_2^{-n-m}$ in equation (31), we have the dominant asymptotic term of the cardinal function $C(\cdot)$ for the multiquadric ($m = 1$) and inverse multiquadric ($m=-1$) radial basis functions, provided that n is even.

In particular, taking $n=2$, $m=1$, $c=\frac{1}{2}$ and x at the midpoint of an edge of a grid square, this analysis gives the relation

$$\begin{aligned} C(x) &= \frac{9}{2\pi}\sum_{k\in\mathcal{Z}^2}{}'[-1+e^{2\pi i x.k}]\,[\frac{1}{\|2\pi k\|_2^3}+\frac{c}{\|2\pi k\|_2^2}]\,e^{-c\|2\pi k\|_2}\|x\|_2^{-5}+o(\|x\|_2^{-5}) \\ &= -0.005212\,\|x\|_2^{-5}+o(\|x\|_2^{-5}),\quad [x+(\tfrac{1}{2},0)]\in\mathcal{Z}^2, \end{aligned} \tag{39}$$

when $\|x\|$ is large. One unexpected conclusion from equations (32) and (39) is that in two dimensions multiquadric interpolation can have better localization properties than linear interpolation ($\phi(r)=r$).

If n is odd, however, then neither multiquadrics nor inverse multiquadrics give the exponential decay (27). In these cases $n+m$ is even in equation (33), so the r^{-n-m} term does not cause a singularity in the Fourier transform (16) of the cardinal function, but there is some loss of differentiability due to the expansion of the modified Bessel function $\{K_{(n+m)/2}(cr)\mid r\in\mathcal{R}^+\}$ at $r=0$. Specifically, the dominant term of $\hat{\phi}(r)$ for small r and the first term that gives a singularity are $\mu_m r^{-n-m}$ and $\hat{\mu}_m \log r$ respectively, where μ_m and $\hat{\mu}_m$ are constants. Now equation (16) implies the relation

$$\begin{aligned} \hat{C}(t) &\approx 1-[\hat{\phi}(\|t\|_2)]^{-1}\sum_{k\in\mathcal{Z}^n}{}'\hat{\phi}(\|2\pi k\|_2) \\ &\approx 1-\mu_m^{-1}\|t\|_2^{n+m}[1+\cdots+\frac{\hat{\mu}_m}{\mu_m}\|t\|_2^{n+m}\log\|t\|_2+\cdots]^{-1}\sum_{k\in\mathcal{Z}^n}{}'\hat{\phi}(\|2\pi k\|_2) \end{aligned} \tag{40}$$

for small t. Thus the leading singularity in $\hat{C}(\cdot)$ is proportional to $\{\|t\|_2^{2n+2m}\log\|t\|_2 \mid t\in\mathcal{R}^n\}$. The generalized inverse Fourier transform of this expression is a multiple of $\{\|x\|_2^{-3n-2m}\mid x\in\mathcal{R}^n\}$. Hence, for odd n and large $\|x\|$, $|C(x)|$ is $O(\|x\|_2^{-3n-2})$ and $O(\|x\|_2^{-3n+2})$ for the multiquadric and inverse multiquadric radial basis functions respectively.

3. The calculation of cardinal functions

The results of Jackson (1988) suggest that, when $n=3$ and $\{\phi(r)=r\mid r\in\mathcal{R}^+\}$, it is possible for a quasi-interpolating function

$$\psi(x)=\sum_{\ell\in\hat{\mathcal{Z}}}\gamma_\ell\,\phi(\|x-\ell\|_2),\quad x\in\mathcal{R}^n, \tag{41}$$

where $\hat{\mathcal{Z}}$ is a finite subset of $\mathcal{Z}^n$ and $\{\gamma_\ell \mid \ell \in \hat{\mathcal{Z}}\}$ is a set of constant coefficients, to possess the diagonal dominance condition

$$|\psi(0)| > \sum_{\ell \in \mathcal{Z}^n}{}' |\psi(\ell)|. \tag{42}$$

In this case, with the normalization $\psi(0)=1$, one can deduce from the Gauss–Seidel iteration

$$C^{[k+1]}(x) = C^{[k]}(x) - \{C^{[k]}(j) - \delta_{0j}\}\, \psi(x-j), \quad x \in \mathcal{R}^n, \tag{43}$$

j being a vector from $\mathcal{Z}^n$ that depends on k, that a cardinal function of interpolation exists. One uses this formula to generate a convergent sequence of approximating functions $\{C^{[k+1]}(\cdot) \mid k = 1,2,3,\ldots\}$ whose limit satisfies $\{C^{[\infty]}(\ell) = \delta_{0\ell} \mid \ell \in \mathcal{Z}^n\}$. This technique is mentioned because in this section we present a procedure for the calculation of approximations to cardinal functions that includes an iteration of Gauss–Seidel type that is derived from equation (43).

Our procedure begins by generating a quasi-interpolating function of the form (41) in a way that is described later. In order to take advantage of symmetry, we let $\hat{\mathcal{Z}} = \mathcal{Z}_q^n$ for some small integer q (usually $q=3$ or $q=4$), where we define $\mathcal{Z}_q^n$ to be the set $\mathcal{Z}^n \cap [-q,q]^n$. Thus the coefficients of expression (41) are chosen to satisfy $\gamma_{P\ell} = \gamma_\ell$, where $P\ell$ is any vector in $\mathcal{Z}^n$ whose components have moduli that are a permutation of the moduli of the components of ℓ. In this case we say that $P\ell$ and ℓ are *equivalent*. Each function $C^{[k]}(\cdot)$ has the form

$$C^{[k]}(x) = \sum_{\ell \in \mathcal{Z}_b^n} \mu_\ell^{[k]}\, \psi(x-\ell), \quad x \in \mathcal{R}^n, \tag{44}$$

a typical value of the integer b being $b=20$, and each integer vector that is analogous to j in equation (43) is restricted to the set $\mathcal{Z}_b^n$. Therefore in the limit $k \to \infty$ we try to achieve the conditions

$$C^{[\infty]}(\ell) = \delta_{0\ell}, \quad \ell \in \mathcal{Z}_b^n, \tag{45}$$

which correspond to the equations (13) with $C_b(\cdot) \equiv C^{[\infty]}(\cdot)$. We see that expressions (44) and (45) give a $(2b+1)^n \times (2b+1)^n$ system of linear equations in the coefficients $\{\mu_\ell^{[\infty]} \mid \ell \in \mathcal{Z}_b^n\}$ that is to be solved by a Gauss–Seidel iteration. We take advantage of the equivalence symmetry $\mu_{P\ell}^{[\infty]} = \mu_\ell^{[\infty]}$ to reduce the number of unknowns to $\frac{1}{2}(b+1)(b+2)$ and $\frac{1}{6}(b+1)(b+2)(b+3)$ when $n=2$ and $n=3$ respectively, which are the only values of n in the numerical results of Section 4.

To begin the iteration we set $C^{[1]}(\cdot) = \psi(\cdot)$ and $k=1$. For each k the vector $j \in \mathcal{Z}_b^n$ is chosen to maximize the residual $\{|C^{[k]}(j) - \delta_{0j}| \mid j \in \mathcal{Z}_b^n\}$. If this maximum residual is less than a preset tolerance the calculation ends, and from now on we use the notation $C_b(\cdot)$ to denote the final $C^{[k]}(\cdot)$. Otherwise a step of Gauss–Seidel type is applied to give $C^{[k+1]}(\cdot)$, but equation (43) is not used as it stands because it would lose the equivalence symmetry. Instead we define $\mathcal{P}(j)$ to be the subset of $\mathcal{Z}_b^n$ that is composed of j and all other vectors that are equivalent to j, and we let $C^{[k+1]}(\cdot)$ be the function

$$C^{[k+1]}(x) = C^{[k]}(x) - \{C^{[k]}(j) - \delta_{0j}\} \sum_{m\in\mathcal{P}(j)} \psi(x-m), \quad x\in\mathcal{R}^n. \tag{46}$$

Because the symmetry implies $\{C^{[k]}(m)=C^{[k]}(j) \mid m\in\mathcal{P}(j)\}$ and because $\psi(0)=1$, it can be shown that this iteration gives the inequality

$$\begin{aligned}\sum_{\ell\in\mathcal{Z}_b^n} |C^{[k+1]}(\ell) - \delta_{0\ell}| \;\le\; & \sum_{\ell\in\mathcal{Z}_b^n\setminus\mathcal{P}(j)} |C^{[k]}(\ell) - \delta_{0\ell}| \\ & + \sum_{m\in\mathcal{P}(j)} |C^{[k]}(m) - \delta_{0m}| \sum_{\ell\in\mathcal{Z}_b^n\setminus\{m\}} |\psi(\ell-m)|. \end{aligned} \tag{47}$$

Therefore convergence is guaranteed if condition (42) holds. The iterative procedure terminates quite efficiently in all of the calculations that are mentioned in the next section, but the diagonal dominance (42) is not obtained in some of these cases.

Several of these calculations require more than 24 hours of running time on a Sun 3/50 workstation, because of the number of terms that occur. For example, when $n=3$, $b=20$ and $\hat{\mathcal{Z}}=\mathcal{Z}_3^3$, the computation of $C_b(x)$ for general x requires 103,823 different values of $\phi(r)$ to be determined and each one involves a square root or a logarithm. It is therefore important to plan the details of the algorithm carefully. The following features are included in our iterative procedure.

Before beginning the iterations we calculate the values $\{\psi(\ell) \mid \ell\in\mathcal{Z}_{2b}^n\}$ of the quasi-interpolating function (41), and for each k we have available the coefficients $\{\mu_\ell^{[k]} \mid \ell\in\mathcal{Z}_b^n\}$, the residuals $\{C^{[k]}(\ell) - \delta_{0\ell} \mid \ell\in\mathcal{Z}_b^n\}$ and the value of ℓ that gives the residual of maximum modulus, which is the j of formula (46). In view of this formula, the new coefficients $\{\mu_\ell^{[k+1]} \mid \ell\in\mathcal{Z}_b^n\setminus\mathcal{P}(j)\}$ are the same as the old ones $\{\mu_\ell^{[k]} \mid \ell\in\mathcal{Z}_b^n\setminus\mathcal{P}(j)\}$ but the $\{\mu_\ell^{[k]} \mid \ell\in\mathcal{P}(j)\}$ are overwritten by $\{\mu_\ell^{[k]} - \rho^{[k]} \mid \ell\in\mathcal{P}(j)\}$ where $\rho^{[k]} = C^{[k]}(j) - \delta_{0j}$. Further, the residuals are overwritten by the numbers

$$\{C^{[k+1]}(\ell) - \delta_{0\ell}\} = \{C^{[k]}(\ell) - \delta_{0\ell}\} - \rho^{[k]} \sum_{m\in\mathcal{P}(j)} \psi(\ell-m), \quad \ell\in\mathcal{Z}_b^n, \tag{48}$$

and during this process the next value of j is determined. Because the values of $\psi(\cdot)$ in equation (48) are available explicitly, there are no calculations of radial basis functions during the iterations and the amount of work per iteration is independent of the number of terms in $\hat{\mathcal{Z}}=\mathcal{Z}_q^n$. In all of these operations we take advantage of the equivalence symmetry. The computer program stores and updates only one of the coefficients $\{\mu_\ell^{[k]}\}$ and one of the residuals $\{C^{[k]}(\ell) - \delta_{0\ell}\}$ for each of the equivalence sets $\mathcal{P}(\ell)$, but, in order to save some work when equation (48) is applied, it stores $\psi(\ell)$ for every ℓ in $\mathcal{Z}^n\cap[0,2b]^n$.

Another feature of this program is that it allows a range of values of b. They are treated in ascending order, and we set $C^{[1]}(\cdot)\equiv\psi(\cdot)$ only when b is least. For each b the iterations continue until the termination condition gives $C_b(\cdot)$, and this approximation to the cardinal function is chosen as the $C^{[1]}(\cdot)$ of the next b. This feature is useful because comparing tabulations of $C_b(\cdot)$ for several values of b provides a good indication of the accuracy of these approximations to the cardinal function. An example of such a comparison is considered later.

The coefficients $\{\gamma_\ell \mid \ell \in \mathcal{Z}_q^n\}$ of the quasi-interpolating function (41) are determined by a linear programming procedure that tries to achieve inequality (42) subject to the normalization condition

$$\psi(0) = 1. \tag{49}$$

Because the sum (42) is infinite, a finite integer a satisfying $a \geq q$ has to be chosen. Then the coefficients are calculated to minimize the objective function

$$\sum_{j\in\mathcal{Z}_a^n} |\psi(j)| = \sum_{j\in\mathcal{Z}_a^n} |\sum_{\ell\in\mathcal{Z}_q^n} \gamma_\ell\, \phi(\|j-\ell\|_2)\,|. \tag{50}$$

Except in the $n=2$ inverse multiquadric case, condition (49) is augmented by further linear equality constraints that help $|\psi(x)|$ to become small as $\|x\|\rightarrow\infty$.

Most of these constraints are contained in the statement that, except in the $n=2$ inverse multiquadric case, $\psi(\cdot)$ shall have the form

$$\psi(x) = \sum_{\ell\in\mathcal{Z}_{q-1}^n} \hat{\gamma}_\ell \{\sum_{i=1}^{n} \delta_i^2 \phi(\|x-\ell\|_2)\}, \quad x\in\mathcal{R}^n, \tag{51}$$

where δ_i^2 is the central difference operator

$$\delta_i^2\phi(\|x-\ell\|_2) = \phi(\|x-e_i-\ell\|_2) - 2\phi(\|x-\ell\|_2) + \phi(\|x+e_i-\ell\|_2), \tag{52}$$

e_i being the unit vector along the i-th coordinate direction of $\mathcal{R}^n$. In addition to helping the boundedness of $\psi(\cdot)$, expression (51) has the strong advantage in the linear, cubic and multiquadric cases that some cancellation can be done analytically when $\psi(x)$ is computed for large $\|x\|$. For example, the identity

$$\begin{aligned}&\{(x_1-1)^2+\alpha^2\}^{\frac{1}{2}} - 2\{x_1^2+\alpha^2\}^{\frac{1}{2}} + \{(x_1+1)^2+\alpha^2\}^{\frac{1}{2}}\\ &= 8\,\alpha^2\,/\,(\,[\,\{(x_1-1)^2+\alpha^2\}^{\frac{1}{2}}\,\{(x_1+1)^2+\alpha^2\}^{\frac{1}{2}} + x_1^2 + \alpha^2 - 1\,]\\ &\quad\times[\,\{(x_1-1)^2+\alpha^2\}^{\frac{1}{2}} + 2\,\{x_1^2+\alpha^2\}^{\frac{1}{2}} + \{(x_1+1)^2+\alpha^2\}^{\frac{1}{2}}\,]\,)\end{aligned} \tag{53}$$

is highly useful for large x_1 when $\phi(r)=r$ or $\surd(r^2+c^2)$. Such identities improve the accuracy of many of the numerical results of Section 4, particularly when $\phi(r)=r$ and $n=3$, because in this case some values of the cardinal function are given whose moduli are less than 10^{-10}.

We also include the constraints

$$\sum_{\ell\in\mathcal{Z}^n_{q-1}} \hat{\gamma}_\ell = \sum_{\ell\in\mathcal{Z}^n_{q-1}} \hat{\gamma}_\ell\,(\ell_1^4+\ell_2^4+\ell_1^2+\ell_2^2-6\,\ell_1^2\ell_2^2)=0, \tag{54}$$

$$\sum_{\ell\in\mathcal{Z}^n_{q-1}} \hat{\gamma}_\ell = \sum_{\ell\in\mathcal{Z}^n_{q-1}} \hat{\gamma}_\ell\,(\ell_1^4+\ell_2^4+\ell_3^4+\ell_1^2+\ell_2^2+\ell_3^2-3\,[\ell_1^2\ell_2^2+\ell_2^2\ell_3^2+\ell_3^2\ell_1^2])=0 \tag{55}$$

and

$$\begin{aligned}\sum_{\ell\in\mathcal{Z}^n_{q-1}} \hat{\gamma}_\ell &= \sum_{\ell\in\mathcal{Z}^n_{q-1}} \hat{\gamma}_\ell\,(\ell_1^2+\ell_2^2+\ell_3^2) = \sum_{\ell\in\mathcal{Z}^n_{q-1}} \hat{\gamma}_\ell\,(\ell_1^4+\ell_2^4+\ell_3^4-3\,[\ell_1^2\ell_2^2+\ell_2^2\ell_3^2+\ell_3^2\ell_1^2])\\
&= \sum_{\ell\in\mathcal{Z}^n_{q-1}} \hat{\gamma}_\ell\,(\ell_1^4\ell_2^2+\ell_1^4\ell_3^2+\ell_2^4\ell_3^2+\ell_2^4\ell_1^2+\ell_3^4\ell_1^2+\ell_3^4\ell_2^2+\ell_1^2\ell_2^2+\ell_2^2\ell_3^2+\ell_3^2\ell_1^2-18\,\ell_1^2\ell_2^2\ell_3^2)\\
&= \sum_{\ell\in\mathcal{Z}^n_{q-1}} \hat{\gamma}_\ell\,(2\,[\ell_1^6+\ell_2^6+\ell_3^6]+15\,[\ell_1^2\ell_2^2+\ell_2^2\ell_3^2+\ell_3^2\ell_1^2]-90\,\ell_1^2\ell_2^2\ell_3^2)=0\end{aligned} \tag{56}$$

for the thin plate spline when $n=2$, the linear and multiquadric functions when $n=3$, and the cubic radial basis function when $n=3$ respectively. We see that the objective function (50) and all the constraints of each linear programming problem provide the required symmetry $\{\gamma_{P\ell}=\gamma_\ell \mid P\ell\in\mathcal{P}(\ell)\}$, so we reduce the number of variables of this calculation to the number of equivalence sets $\mathcal{P}(\ell)$ in $\mathcal{Z}^n_q$ or in $\mathcal{Z}^n_{q-1}$, the choice depending on whether we work with $\{\gamma_\ell\}$ or with $\{\hat{\gamma}_\ell\}$. We set $q=3$ in the $n=2$ inverse multiquadric case and $q=4$ otherwise. In all cases we let $a=6$ or $a=4$ for $n=2$ or $n=3$ respectively. No difficulties occurred in the computations of the quasi-interpolating functions $\psi(\cdot)$ by some software that was developed by one of the authors (MJDP) for the solution of general, linearly constrained, optimization problems.

The purpose of each of the constraints (54), (55) and (56) is to optimize the asymptotic decay properties of $\psi(\cdot)$. Specifically, they ensure that the integer j in the bound

$$|\psi(x)| = O(\|x\|^{-j}),\quad \|x\|\to\infty, \tag{57}$$

is as large as possible, subject to $\psi(\cdot)$ having the form (51) and subject to the *integral condition* that enough freedom remains in the coefficients $\{\hat{\gamma}_\ell \mid \ell\in\mathcal{Z}^n_{q-1}\}$ for it to be possible for $\int_{\mathcal{R}^n}\psi(x)\,dx$ to be nonzero. The details of equations (54)–(56) can be derived from the Taylor series expansion of expression (51) for large $\|x\|$ or from the relations between asymptotic decay rates and Fourier transforms that are mentioned in Section 2. Thus one finds the values $j=6$, $j=7$ and $j=7$ in the bound (57) for the $n=2$ thin plate spline, the $n=3$ linear and multiquadric cases and the $n=3$ cubic radial basis function respectively. Hence the right hand side of the diagonal dominance condition (42) is finite. Further, when q and a have the values given in the previous paragraph, it seems that in each of these cases the $\psi(\cdot)$ from the linear programming calculation actually satisfies inequality (42), provided that c is not too large in the multiquadric radial basis function.

The integral condition that has just been mentioned keeps j finite in inequality (57) by ruling out functions $\psi(\cdot)$ that are formed by applying high order divided difference operators to $\{\phi(\|x\|_2) \mid x \in \mathcal{R}^n\}$, and, more importantly, it is a necessary condition for inequality (42) to hold in each of the cases (9). To justify this statement we deduce a contradiction if we have the diagonal dominance (42), $\psi(\cdot)$ being an absolutely integrable function with $\int_{\mathcal{R}^n} \psi(x)\, dx = 0$. Inequality (42) allows the Gauss–Seidel method, as mentioned in the first paragraph of this section, to construct a function

$$C^{[\infty]}(x) = \sum_{\ell \in \mathcal{Z}^n} \mu_\ell^{[\infty]} \psi(x-\ell), \quad x \in \mathcal{R}^n, \tag{58}$$

that satisfies $\{C^{[\infty]}(\ell) = \delta_{0\ell} \mid \ell \in \mathcal{Z}^n\}$, and the sum $\sum_{\ell \in \mathcal{Z}^n} |\mu_\ell^{[\infty]}|$ of moduli of coefficients is finite, which implies $\int_{\mathcal{R}^n} C^{[\infty]}(x)\, dx = 0$. The cardinal functions $C(\cdot)$ that are studied in Section 2 for each of the cases (9), however, have the property that the interpolation formula (5) gives $s \equiv f$ when $f \equiv 1$, which implies $\int_{\mathcal{R}^n} C(x)\, dx = 1$. Therefore $C^{[\infty]}(\cdot)$ is different from $C(\cdot)$. Due to the asymptotic decay properties of $C(\cdot)$ and $C^{[\infty]}(\cdot)$, given in Section 2 and deducible from the smoothness and absolute integrability of $\psi(\cdot)$, we now have a contradiction to the uniqueness result in Theorem 17 of Buhmann (1988b).

It follows from this argument that inequality (42) is satisfied only if $\psi(\cdot)$ is not absolutely integrable or if $\int_{\mathcal{R}^n} \psi(x)\, dx \neq 0$. Remembering that the number of terms of $\hat{\mathcal{Z}}$ in equation (41) is finite, and that $\phi(\cdot)$ is one of the functions (9), the first alternative would imply that the right hand side of expression (42) is a divergent sum. Therefore diagonal dominance occurs only if $\psi(\cdot)$ is absolutely integrable with a nonzero integral. It is proved in Jackson (1988), however, that no such function exists in the linear case $\phi(r) = r$ when n is even. It also follows from his analysis that condition (42) cannot hold for the linear, cubic, multiquadric and inverse multiquadric functions when n is even nor for the thin plate spline when n is odd. Fortunately this lack of diagonal dominance does not prevent satisfactory convergence of the Gauss–Seidel iteration in all the calculations of Section 4.

For example, in the inverse multiquadric case with $c = \frac{1}{2}$ and $n = 2$, the only constraint in the linear programming calculation is the normalization condition (49). With $q = 3$ and $a = 6$, the final value of the objective function (50) is 1.166, which suggests that inequality (42) might hold, but the right hand side of this expression is infinite because $\sum_{\ell \in \mathcal{Z}_3^2} \gamma_\ell = 0.0148$. One can deduce from the bound (47), however, that the Gauss–Seidel iteration converges for values of b up to 15, because the calculated $\psi(\cdot)$ satisfies the condition

$$\sum_{\ell \in \mathcal{Z}_{15}^2 \setminus \{m\}} |\psi(\ell - m)| \leq 0.919, \quad m \in \mathcal{Z}_{15}^2, \tag{59}$$

the inequality being an equation when $m = 0$. In fact the largest value of b in these calculations is $b = 24$, and then $\sum_{\ell \in \mathcal{Z}_b^2 \setminus \{0\}} |\psi(\ell)| = 1.810$. Even in this case

Table 1 Some values of $C_b(x)$ when $\phi(r)=r$ and $n=3$

x	$b=4$	$b=5$	$b=6$	$b=7$	$b=8$
(0.5,0,0)	0.399676	0.399676	0.399676	0.399676	0.399676
(1.5,0,0)	-0.021868	-0.021868	-0.021868	-0.021868	-0.021868
(2.5,0,0)	8.7094_{-4}	8.7095_{-4}	8.7095_{-4}	8.7095_{-4}	8.7095_{-4}
(3.5,0,0)	2.7883_{-6}	1.6869_{-6}	1.6721_{-6}	1.6733_{-6}	1.6733_{-6}
(4.5,0,0)	-2.4931_{-5}	-4.8588_{-6}	-5.1298_{-6}	-5.1386_{-6}	-5.1381_{-6}
(5.5,0,0)	-2.6420_{-5}	-6.6890_{-6}	7.3171_{-7}	6.9654_{-7}	6.9279_{-7}
(6.5,0,0)	-6.8362_{-6}	-8.6200_{-6}	-1.3422_{-6}	-5.7800_{-8}	-5.9653_{-8}
(7.5,0,0)	-5.0118_{-7}	-1.0026_{-6}	-1.5452_{-6}	-3.0887_{-7}	-3.3524_{-10}
(8.5,0,0)	1.6935_{-7}	-2.9567_{-8}	-2.3408_{-7}	-3.6663_{-7}	-7.4295_{-8}

the convergence of the Gauss–Seidel iteration is entirely satisfactory, the number of iterations to reduce the maximum residual by a factor of 10 on the 49×49 grid being about 300, where each iteration adjusts at most 8 of the coefficients $\{\mu_\ell^{[k]}\}$ because the set $\mathcal{P}(j)$ contains at most 8 elements. If necessary we could have increased the values of q and a to ensure the convergence of our iterative procedure, but in all cases the original values of q and a give an adequate quasi-interpolating function $\psi(\cdot)$.

The dependence of $C_b(\cdot)$ on b is illustrated in Table 1 for the linear radial basis function in three dimensions, where p_q denotes $p\times 10^q$. In this case $C(\cdot)$ has an exponential asymptotic rate of decay of the form (27), while $\psi(\cdot)$ satisfies condition (57) with $j=7$. Looking across the rows of the table, we see that $C_b(x)$ settles down very quickly as b increases. Therefore much of the last column reflects the decay of the true cardinal function $C(\cdot)$. The later entries of the $b=4$ column, however, show the algebraic rate of decay of $C_b(\cdot)$ that is inherited from $\psi(\cdot)$.

To conclude this section we address the goodness of the approximation $C_b(\cdot)\approx C(\cdot)$, keeping in mind the results of Table 1. We argue that the decay rate of the true cardinal function provides excellent accuracy in $C_b(x)\approx C(x)$ for $0\le\|x\|_\infty\le\frac{1}{2}b$, although $C_b(\cdot)$ may decay much more slowly than $C(\cdot)$. This argument depends on a conjecture, which in it strongest form asserts that, in each of the cases (9), every bounded function has a unique bounded interpolant on $\mathcal{Z}^n$, namely the function (5), where $C(\cdot)$ is the inverse Fourier transform of expression (16). An equivalent claim is that the zero function is the only bounded function in our space of approximations that vanishes on $\mathcal{Z}^n$.

Our argument requires the choice of $\psi(\cdot)$ and the convergence of the Gauss–Seidel iteration to be such that $\{C_b(x)\mid x\in\mathcal{R}^n\}$ is a bounded function, and we also require the test for termination of the iterative procedure to be so fine that we can suppose that $C_b(\cdot)$ satisfies the equations

$$C_b(\ell)=\delta_{0\ell},\quad \ell\in\mathcal{Z}_b^n. \tag{60}$$

Then the error function

$$\Gamma_b(x) = C(x) - C_b(x), \quad x \in \mathcal{R}^n, \tag{61}$$

is uniformly bounded, and, because it is in the space of approximations, the conjecture of the previous paragraph allows $\Gamma_b(\cdot)$ to be equated to its interpolant, which is the identity

$$\Gamma_b(x) = \sum_{\ell \in \mathcal{Z}^n} \Gamma_b(\ell)\, C(x-\ell), \quad x \in \mathcal{R}^n, \tag{62}$$

the right hand side being absolutely convergent because of the decay properties of $C(\cdot)$ that are given in Section 2. Equations (60)–(62) and $\{C(\ell) = \delta_{0\ell} \mid \ell \in \mathcal{Z}_b^n\}$ imply the bound

$$|\Gamma_b(x)| \le \|\Gamma_b(\cdot)\|_\infty \sum_{\ell \in \mathcal{Z}^n \setminus \mathcal{Z}_b^n} |C(x-\ell)|, \quad x \in \mathcal{R}^n, \tag{63}$$

and we draw our conclusions from the fact that $\ell \notin \mathcal{Z}_b^n$ on the right hand side. The main points are that only small values of $|C(x-\ell)|$ occur in this absolutely convergent sum when x is well inside the box $\|x\|_\infty \le b$, and the magnitudes of these small terms depend not on the decay rate of $C_b(\cdot)$ but on the decay rate of $C(\cdot)$. Hence the true cardinal function dominates the approximation $C_b(x) = C(x) - \Gamma_b(x)$ to $C(x)$ when $0 \le \|x\|_\infty \le \frac{1}{2}b$, but $\Gamma_b(x)$ becomes important for larger values of $\|x\|_\infty$. For example, $\Gamma_b(\cdot)$ causes the final increases in $|C_b(x)|$ in the $b=7$ column of Table 1. Moreover, because the top half of the $b=8$ column gives excellent approximations to $C(\cdot)$, we see that the contributions from $\Gamma_b(\cdot)$ to the first row of Table 1 are scaled by factors of at most 10^{-5}, which provides the consistency across this row of the table.

In the calculations of the next section the tolerance that stops the Gauss–Seidel iteration is set to such a small number, typically 10^{-16}, that the discrepancies in equation (60) due to $C_b(\cdot) \not\equiv C^{[\infty]}(\cdot)$ are below the rounding errors of the given tables. Further, when estimating each value of $C(x)$ we increased b until the changes to $C_b(x)$ were less than the displayed accuracy. We believe, therefore, that the tabulations of $C(\cdot)$ in Section 4 are close to the true cardinal functions of interpolation, and that there is no need to refer back to the details of the approximations of the algorithm of this section.

4. Numerical results and discussion

Table 2 presents some values of cardinal functions when $n = 2$ in the linear, multiquadric, inverse multiquadric and thin plate spline cases. The lower precision in the penultimate column is due to the rather slow convergence of the sequence of approximations $\{C_b(\cdot) \mid b = 1, 2, 3, \ldots\}$ when $\phi(r) = (r^2 + \frac{1}{4})^{-1/2}$. From a practical

Table 2 Some values of $C(x)$ when $n=2$

x	$\phi(r)=r$	$(r^2+\frac{1}{4})^{1/2}$	$(r^2+\frac{1}{4})^{-1/2}$	$r^2\log r$
(0.5,0)	0.434190	0.560007	0.473146	0.535938
(1.5,0)	-0.018506	-0.090066	-0.035897	-0.074666
(2.5,0)	-1.0044_{-3}	0.018726	3.4454_{-3}	0.013094
(3.5,0)	-4.6723_{-5}	-4.0942_{-3}	-1.7455_{-4}	-2.4549_{-3}
(4.5,0)	-2.7134_{-5}	9.2458_{-4}	8.525_{-5}	4.8375_{-4}
(5.5,0)	-1.0850_{-5}	-2.1614_{-4}	3.095_{-5}	-9.8539_{-5}
(6.5,0)	-4.7241_{-6}	5.0800_{-5}	2.114_{-5}	2.0522_{-5}
(7.5,0)	-2.3377_{-6}	-1.2491_{-5}	1.382_{-5}	-4.3399_{-6}
(8.5,0)	-1.2608_{-6}	2.9014_{-6}	9.64_{-6}	9.2796_{-7}
(9.5,0)	-7.2707_{-7}	-7.9513_{-7}	6.97_{-6}	-2.0006_{-7}
(10.5,0)	-4.4266_{-7}	1.4590_{-7}	5.20_{-6}	4.3408_{-8}

point of view the most important question is often 'what is the greatest value of $\|x\|$ such that $|C(x)|\geq 10^{-6}$', because for all larger values of $\|x\|$ one may be able to treat $C(x)$ as negligible. Table 2 shows that, according to this criterion, the thin plate spline is marginally better than the linear and multiquadric radial basis functions, while the inverse multiquadric is least good. If larger values of $|C(x)|$ can be neglected, then $\phi(r)=r$ seems to be the best of the four given cases, but $c=\frac{1}{2}$ is the only c of this tabulation of multiquadric and inverse multiquadric cardinal functions. If one preferred $\phi(r)=(r^2+\frac{1}{100})^{1/2}$, for example, then one would have localization properties that are similar to those of $\phi(r)=r$ and one would have differentiability too.

Because every x in Table 2 is on the first coordinate axis, Figures 1–4 present some properties of the cardinal functions on $\mathcal{R}^2$. Each figure is constructed from the values of $C(\cdot)$ on the 201×201 square grid with mesh size 0.1 on the square $\|x\|_\infty \leq 10$. For each of the heights $\{h=10^{-k} \mid k=1,2,3,\ldots\}$, we calculated the convex hull of the points of this mesh at which $|C(x)|\geq h$. The piecewise linear curves in the figures are the boundaries of these convex hulls in the fourth quadrant, the range of x being $\|x\|_\infty \leq 8$. The shaded regions of the figures are the spaces between the $h=10^{-5}$ and $h=10^{-6}$ curves, which show clearly that the inverse multiquadric radial basis function is the least successful of the four cases.

For each radial basis function $\phi(\cdot)$, the spacings between the convex hull boundaries of the heights $\{h=10^{-k} \mid k=1,2,3,\ldots\}$ indicate the asymptotic decay rate of the cardinal function of interpolation, but, particularly in Figure 2, the range $\|x\|_\infty \leq 8$ is rather small for this purpose. When the cardinal function decays exponentially, which implies a bound of the form (27), then the *average* spacing tends to a constant, and this feature is shown well in Figure 4. In the other three cases, however, the asymptotic decay rate is algebraic as suggested by inequality (25), so the spacings between convex hull boundaries tend to diverge. This property is clear

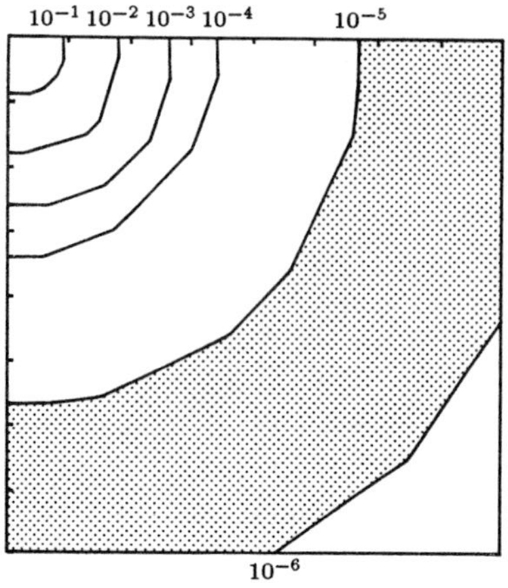

Figure 1
Convex hulls: $\phi(r) = r$

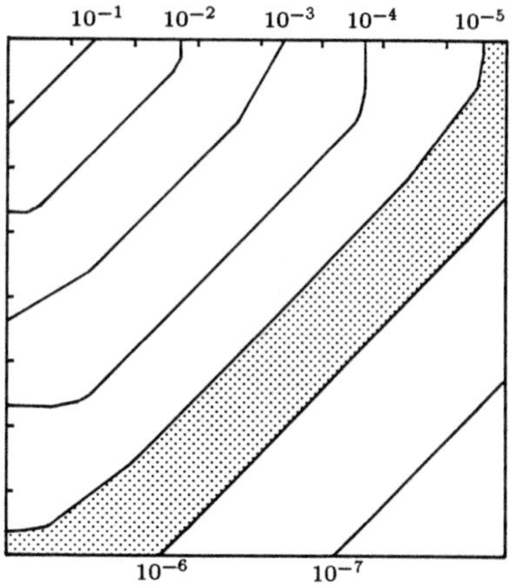

Figure 2
Convex hulls: $\phi(r) = (r^2 + \frac{1}{4})^{1/2}$

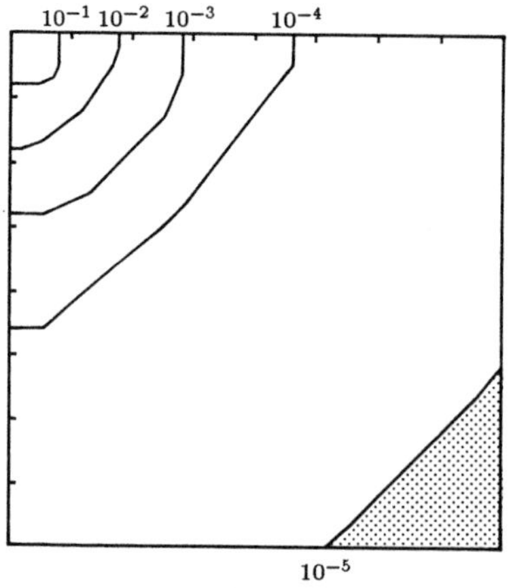

Figure 3
Convex hulls: $\phi(r) = (r^2 + \frac{1}{4})^{-1/2}$

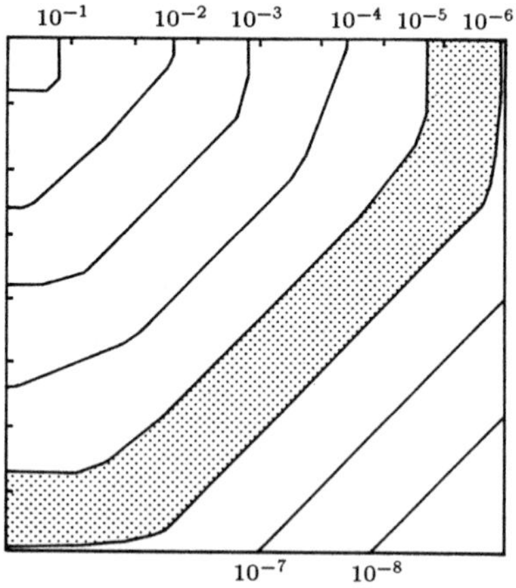

Figure 4
Convex hulls: $\phi(r) = r^2 \log r$

Table 3 Further values of $C(x)$ when $n=2$

	$\phi(r)=(r^2+\frac{1}{4})^{1/2}$		$\phi(r)=r$	
x	$C(x)$	$\|x\|_2^5\,C(x)$	$C(x)$	$\|x\|_2^5\,C(x)$
(10.5,0)	1.4590_{-7}	0.01862	-4.4266_{-7}	-0.05650
(11.5,0)	-6.8753_{-8}	-0.01382	-2.8178_{-7}	-0.05668
(12.5,0)	-4.396_{-9}	-0.00134	-1.8618_{-7}	-0.05682
(13.5,0)	-1.365_{-8}	-0.00612	-1.2698_{-7}	-0.05694
(14.5,0)	-6.928_{-9}	-0.00444	-8.899_{-8}	-0.05704
(15.5,0)	-5.698_{-9}	-0.00510	-6.386_{-8}	-0.05713

in Figures 1 and 3, but the multiquadric figure is more interesting because of its similarity to the exponential decay of the thin plate spline for $h \geq 10^{-6}$.

The shapes of the convex hull boundaries also deserve comment. We know from equation (31) that they tend to be circular in the linear and multiquadric cases, and we see this feature in Figure 1. Asymptotic circularity also occurs for inverse multiquadrics (Buhmann, 1988b), but the range $\|x\|_\infty \leq 8$ is too small for it to appear in Figures 2 and 3. In Figure 4, however, where the asymptotic decay rate of $C(\cdot)$ is exponential, it seems that the convex hull boundaries tend to be diamond shaped ($\|x\|_1 = \text{constant}$). This property is obtained in cubic spline interpolation on $\mathcal{Z}^n$, because the exponential rate of decay along any line that is parallel to a coordinate direction is independent of the displacement of the line from the origin. Perhaps a similar property holds for radial basis function interpolation methods with exponential decay, in which case it would be most appropriate to let $\|x\| = \|x\|_1$ in the bound (27).

The figures so far fail to reflect the asymptotic behaviour of $C(\cdot)$ in the multiquadric case $\phi(r)=(r^2+\frac{1}{4})^{1/2}$. Indeed the signs of the entries in the middle column of Table 2 alternate, but the theory of Section 2, in particular expressions (31) and (38), show that $C(x)$ is negative whenever $\|x\|$ is large and x is away from the points of the grid $\mathcal{Z}^n$. Therefore Table 3 provides some more values of this cardinal function. Now we see that the dominant part of $C(x)$ can be a negative multiple of $\|x\|_2^{-5}$, and that there is fair agreement with the factor -0.005212 in equation (39). It is easier, however, to use tabulated values of $C(\cdot)$ to estimate the constants of expression (32) in the $\phi(r)=r$ case, partly because the first derivative discontinuities of the radial basis function allow $C(x)$ to be nonpositive for all moderate values of $\|x\|$. Thus the last column of Table 3 shows good agreement with the number -0.05747. In order to support the other constant of expression (32), we note that at $x=(9.5,9.5)$, for example, we have $C(x)=-1.54\times10^{-7}$ and $\|x\|_2^5\,C(x)=-0.0672$.

Some values of cardinal functions in three dimensions are given in Table 4. The asymptotic decay rate of $C(\cdot)$ when $\phi(r)=r$ is particularly impressive, being about a factor of 10 per data point. We recall from the last paragraph of Section 2 that

Table 4 Some values of $C(x)$ when $n=3$

x	$\phi(r)=r$	$(r^2+\frac{1}{100})^{1/2}$	$(r^2+\frac{1}{4})^{1/2}$	$\phi(r)=r^3$
(0.5,0,0)	0.399676	0.439796	0.545232	0.560447
(1.5,0,0)	-0.021868	-0.034425	-0.082773	-0.099110
(2.5,0,0)	8.7095_{-4}	3.2618_{-3}	0.017282	0.025761
(3.5,0,0)	1.6733_{-6}	-3.5220_{-4}	-3.9703_{-3}	7.5461_{-3}
(4.5,0,0)	-5.1381_{-6}	4.3009_{-5}	9.8001_{-4}	2.3896_{-3}
(5.5,0,0)	6.9300_{-7}	-5.7089_{-6}	-2.5541_{-4}	-7.9975_{-4}
(6.5,0,0)	-6.1395_{-8}	7.9934_{-7}	6.9290_{-5}	2.7841_{-4}
(7.5,0,0)	3.5619_{-9}	-1.1589_{-7}	-1.9363_{-5}	-9.9717_{-5}
(8.5,0,0)	-1.9969_{-11}	1.7210_{-8}	5.5319_{-6}	3.6474_{-5}
(9.5,0,0)	-2.8859_{-11}	-2.6008_{-9}	-1.6073_{-6}	-1.3556_{-5}
(10.5,0,0)	4.8041_{-12}	3.9834_{-10}	4.7315_{-7}	5.1017_{-6}

the multiquadric cardinal function decays like $\|x\|_2^{-11}$ as $\|x\| \to \infty$, and we see in the table that this algebraic rate is sufficiently strong for the localization properties of multiquadric interpolation with moderate c to compare favourably with the exponential decay of the $\phi(r)=r^3$ case. Therefore, if one cannot tolerate the first derivative discontinuities in the interpolant when $\phi(r)=r$, it may be better to turn to multiquadric radial basis functions instead of $\phi(r)=r^3$, but the obvious choice for a regular grid is cubic spline interpolation. In large calculations the amount of work of cubic spline methods is orders of magnitude less than that of radial basis function techniques, but perhaps the introduction of transputers and further research will alleviate this gap. Therefore we note that the cubic spline decay rate of $(2-\sqrt{3})$ per data point is less good than the localization properties when $\phi(r)=r$, and that the comparison with multiquadrics depends on c. All of these methods reproduce cubic polynomials in the theoretical case of interpolation on an infinite grid, but only the cubic spline has this property when the grid is finite.

The relevance of approximation on an infinite grid to finite grids has been studied by Dyn and Jackson (private communication, 1988). It is important not only because finiteness is necessary in practical calculations but also because of the gains in efficiency that may be achievable by partitioning very large finite grids into smaller ones. Therefore we give it some attention now. As at the end of Section 3, we relate an interpolant $s_b(\cdot)$ on the finite grid $\mathcal{Z}_b^n \equiv \mathcal{Z}^n \cap [-b,b]^n$ to our interpolant (5) on the infinite grid $\mathcal{Z}^n$. The equation

$$s_b(x) = s(x) - \eta(x), \quad x \in \mathcal{R}^n, \tag{64}$$

defines a function $\eta(\cdot)$ that is in the space of approximations. Therefore, assuming a suitable uniqueness condition for interpolation on $\mathcal{Z}^n$, we may equate $\eta(\cdot)$ to its interpolant, which yields the identity

$$s(x) - s_b(x) = \sum_{\ell \in \mathcal{Z}^n \backslash \mathcal{Z}_b^n} \{f(\ell) - s_b(\ell)\}\, C(x-\ell), \quad x \in \mathcal{R}^n. \tag{65}$$

Even if $f(\cdot)$ is bounded, the magnitude of $s_b(\ell)$ for large $\|\ell\|$ depends on the growth of $\phi(r)$ as $r \to \infty$. Thus, corresponding to the derivation of inequality (63) from equation (62), we find a bound of the form

$$|s(x) - s_b(x)| \le \beta \sum_{\ell \in \mathcal{Z}^n \backslash \mathcal{Z}_b^n} \{1 + \phi(\|\ell\|_2)\}\, |C(x-\ell)|, \quad x \in \mathcal{R}^n, \tag{66}$$

where β is a constant, the sum being absolutely convergent for all of the cardinal functions that are considered in Section 2. Again all the factors $|C(x-\ell)|$ are small when x is well inside the box $\|x\|_\infty \le b$. Hence the extent to which $s_b(x)$ enjoys the properties of $s(x)$ depends on the distance from x to the set $\mathcal{Z}^n \backslash \mathcal{Z}_n^b$. If $n=3$ and $\phi(r)=r$, for example, Table 4 suggests the strong result that $|s(x) - s_b(x)|$ is $O(10^{-d})$, where $d = \min\{\|x-\ell\|_\infty \mid \ell \in \mathcal{Z}^n \backslash \mathcal{Z}_b^n\}$, so there is rapid decay in the perturbations to the approximation from the finiteness of the grid.

Tables 2–4 give excellent support to the assumptions on uniqueness of interpolation that are crucial to equations (62) and (65). Here we have in mind that, for each of the radial basis functions (9), the inverse Fourier transform of the function (16) provides a unique cardinal function of the form (6) whose coefficients $\{c_\ell \mid \ell \in \mathcal{Z}^n\}$ satisfy the asymptotic conditions of Theorem 17 of Buhmann (1988b). These conditions correspond to the decay of $|C(x)|$ as $\|x\| \to \infty$ that is studied in Section 2. If, however, the equations $\{C(\ell) = \delta_{0\ell} \mid \ell \in \mathcal{Z}^n\}$ failed to define a unique bounded $C(\cdot)$ of the form (6), then different attempts to calculate $C(\cdot)$ might yield different cardinal functions. Now the method of the algorithm of Section 3 has nothing to do with the inversion of Fourier transforms, and we even apply the Gauss–Seidel iteration in some cases when the diagonal dominance (42) is not obtained. Thus it is likely that our algorithm would find a cardinal function that is different from the inverse transform of $\hat{C}(\cdot)$ if it were possible to do so. Tables 2–4, however, seem to confirm the asymptotic properties that are established in Section 2. Further, when $\psi(\cdot)$ satisfies the diagonal dominance condition (42), the proof of Lemma 10 of Buhmann (1988a) shows that the Gauss–Seidel iteration of Section 3 gives the cardinal function that is the inverse Fourier transform of $\hat{C}(\cdot)$. Therefore we believe that the uniqueness assumptions are true.

Although the algorithm of Section 3 is suitable for calculating the cardinal functions that we have studied, it is not recommended for interpolation to a general function on a finite grid. Our reasons include the loss of equivalence symmetry and the need for a range of quasi-interpolating functions $\psi(\cdot)$ in order to take account of the effects of edges of the grid. Therefore some other algorithms are currently under investigation at Cambridge, particularly the use of preconditioned conjugate gradient methods for solving the interpolation equations. Such methods have already been applied successfully by Dyn, Levin and Rippa (1986) when the data are not confined to a regular grid.

In conclusion we note that the asymptotic properties of cardinal functions are so encouraging that many more questions deserve attention. Here are three examples. If we estimate the interpolant $s(x)$ in a way that ignores the function values $\{f(\ell) \mid \ell \in \mathcal{Z}^n\}$ for large $\|\ell - x\|$, is there a good way of making the estimate a continuous function of x? If we extend the space of approximating functions to include low order polynomials explicitly, can the interpolant on $\mathcal{Z}_b^n$ provide not only reproduction of these low order polynomials but also the localization properties that are noted in the paragraph that includes expressions (64)–(66)? Although the analysis of Section 2 is an application of Fourier techniques, do the main conclusions of this section require interpolation on a *regular* grid? Clearly there is much practical and theoretical work to be done, and it seems probable that it will yield some very powerful new algorithms for multivariable approximation.

References

Abramowitz, M. and Stegun, I.A. (1970), *Handbook of Mathematical Functions*, Dover Publications (New York).

Buhmann, M.D. (1988a), "Multivariate interpolation in odd dimensional Euclidean spaces using multiquadrics", Technical Report DAMTP 1988/NA6, University of Cambridge.

Buhmann, M.D. (1988b), "Multivariate interpolation with radial basis functions", Technical Report DAMTP 1988/NA8, University of Cambridge.

Dyn, N., Levin, D. and Rippa, S. (1986), "Numerical procedures for surface fitting of scattered data by radial functions", *SIAM J. Sci. Stat. Comput.* **7**, 639–659.

Jackson, I.R.H. (1987), "An order of convergence for radial basis functions", Technical Report DAMTP 1987/NA11, University of Cambridge.

Jackson, I.R.H. (1988), "Convergence properties of radial basis functions", *Constructive Approximation* **4**, 243–264.

Jones, D.S. (1982), *The Theory of Generalised Functions*, Cambridge University Press (Cambridge).

Madych, W.R. and Nelson, S.A. (1987), "Polyharmonic cardinal splines", preprint.

Micchelli, C.A. (1986), "Interpolation of scattered data: distance matrices and conditionally positive functions", *Constructive Approximation* **2**, 11–22.

Powell, M.J.D. (1981), *Approximation Theory and Methods*, Cambridge University Press (Cambridge).

Powell, M.J.D. (1987), "Radial basis functions for multivariable interpolation: a review", in *Algorithms for Approximation*, eds. Mason, J.C. and Cox, M.G., Oxford University Press (Oxford), pp. 143–167.

Stein, E.M. and Weiss, G. (1971), *Introduction to Fourier Analysis on Euclidean Spaces*, Princeton University Press (Princeton).

The Fourier operator of even order and its application to an extremum problem in interpolation

L. Brutman

Department of Mathematics and Computer Science, University of Haifa, Israel

Abstract The Fourier operator of even order $\tilde{S}_n$ is defined as a projection of the family of continuous 2π-antiperiodic functions onto the trigonometric polynomials of half-angles. It is shown that this operator is a natural extension of the classical Fourier operator and corresponds to the case where the dimension of the projection subspace is even. The operator $\tilde{S}_n$ is used to extend a result of Szabados, concerning optimal choice of nodes for trigonometric interpolation, to the case of an even number of nodes as well as to the case of complex interpolation by algebraic polynomials on the unit circle.

Key words: Fourier operator, Minimum norm projection, Trigonometric interpolation.

1. The Fourier operator of even order

Various problems of approximation theory involve the trigonometric polynomials of half-angles

$$\tilde{h}_n(\theta) = \sum_{k=1}^{n} [c_k \cos(2k-1)\frac{\theta}{2} + d_k \sin(2k-1)\frac{\theta}{2}]$$

Properties of these polynomials and their application to interpolation and to the theory of quadrature were considered by Turečkiĭ (1960,1968). Recently (see Brutman and Pinkus (1980)) the polynomials $\tilde{h}_n(\theta)$ were found to be useful in proving the Erdös conjecture concerning minimal norm interpolation on the unit circle for even number of points. In the following the trigonometric polynomials of half-angles are used in order to define an even-order analogue of the classical Fourier operator.

Let $\tilde{C}_{2\pi}$ be the Banach space of functions continuous and antiperiodic on $[0,2\pi)$ (i.e. satisfying $f(0) = -f(2\pi)$), equipped with the uniform norm, and denote by $\tilde{H}_n$ the following 2n-dimensional subspace of $\tilde{C}_{2\pi}$

$$\tilde{H}_n = \text{span}\{\sin\frac{\theta}{2}, \cos\frac{\theta}{2}, \ldots, \sin\frac{(2n-1)\theta}{2}, \cos\frac{(2n-1)\theta}{2}\}$$

It is known that $\tilde{H}_n$ is an orthogonal system on $[0,2\pi)$. The Fourier-type operator $\tilde{S}_n : \tilde{C}_{2\pi} \to \tilde{H}_n$ is defined by

$$(\tilde{S}_n f)(\theta) = \sum_{k=1}^{n} \{c_k \cos(2k-1)\frac{\theta}{2} + d_k \sin(2k-1)\frac{\theta}{2}\}, \tag{1}$$

where

$$c_k = \frac{1}{\pi}\int_0^{2\pi} f(\theta)\cos(2k-1)\frac{\theta}{2}\, d\theta,$$
$$d_k = \frac{1}{\pi}\int_0^{2\pi} f(\theta)\sin(2k-1)\frac{\theta}{2}\, d\theta. \tag{2}$$

The following integral representation of $\tilde{S}_n$ holds

Theorem 1

$$(\tilde{S}_n f)(\theta) = \frac{1}{\pi}\int_0^{2\pi} f(\tau)\tilde{D}_n(\tau-\theta)d\tau, \tag{3}$$

where

$$\tilde{D}_n(\theta) = \sum_{k=1}^{n} \cos(2k-1)\frac{\theta}{2} = \frac{\sin n\theta}{2\sin(\theta/2)} . \tag{4}$$

Note that the kernel $\tilde{D}_n(\theta)$ in (3) is a natural extension of the classical Dirichlet kernel $D_m(\theta) = \sin(m\theta/2)/2\sin(\theta/2)$, $m = 2n+1$, and corresponds to the case of even m. As a direct consequence of the representation (3) we get

Corollary 1

$$\|\tilde{S}_n\| = \frac{2}{\pi}\int_0^{\pi/2} \frac{|\sin 2n\theta|}{\sin\theta}d\theta \equiv \tilde{\rho}_n . \tag{5}$$

Remark To the best of our knowledge, the quantities $\tilde{\rho}_n$ appear for the first time in Szegö (1921): "Mann kann die Konstanten

$$r_m = \frac{2}{\pi}\int_0^{\pi/2} \frac{|\sin m\theta|}{\sin} d\theta , \quad m = 0,1,2,...$$

betrachten, welche sich für ungerade m auf die Lebesgueschen Konstanten reduzieren." Szegö has proved the following representation

$$r_m = \frac{16}{\pi^2}\sum_{\nu=1}^{\infty}\left[\frac{1}{1} + \frac{1}{3} + \frac{1}{5} + ... + \frac{1}{2m\nu-1}\right]\frac{1}{4\nu^2-1} , \tag{6}$$

which was used by Galkin (1971) to obtain the precise estimate

$$0.9894 < r_m - \frac{4}{\pi^2}\log m \leq 1. \tag{7}$$

The quantities $\tilde{\rho}_n$ were also used by Geddes and Mason (1975) as an upper bound for the norm of the Taylor projection of even order. They gave the following formula

$$\tilde{\rho}_n = \frac{4}{\pi} \sum_{k=1}^{n} \frac{1}{2k-1} \tan \frac{2k-1}{4n} \pi \tag{8}$$

which is analogous to the representation of the classical Lebesgue constants due to Fejér (1910).

Continuing the analogy with the classical case we define the Fejér-type operator $\tilde{F}_n : \tilde{C}_{2\pi} \to \tilde{H}_n$ to be the arithmetic mean:

$$\tilde{F}_n = \frac{1}{n}\{\tilde{S}_1 + \tilde{S}_2 + \ldots + \tilde{S}_n\} \tag{9}$$

Then we have

Theorem 2

$$(\tilde{F}_n f)(\theta) = \frac{1}{n\pi} \int_0^{2\pi} f(\tau)\tilde{\sigma}_n(\tau-\theta)\,d\tau, \tag{10}$$

where

$$\tilde{\sigma}_n(\theta) = \frac{\sin \frac{n+1}{2}\theta \sin \frac{n\,\theta}{2}}{2\sin^2 \frac{\theta}{2}} \quad . \tag{11}$$

Note that in contrast to the classical case, the kernel $\tilde{\sigma}_n(\theta)$ is not positive. Yet this is not surprising, since the operator $\tilde{F}_n$ is defined on the space $\tilde{C}_{2\pi}$ which does not contain positive functions. This observation raises the natural question of whether every $f \in \tilde{C}_{2\pi}$ can be uniformly approximated by the Fejér-type polynomials $\tilde{F}_n(\theta)$. It can be shown by applying the classical result of Fejér (1904) that the answer to this question is affirmative, namely

Theorem 3 The set $\tilde{H} = \{\cos(\theta/2), \sin(\theta/2), \cos(3\theta/2), \sin(3\theta/2), \ldots\}$ is dense in $\tilde{C}_{2\pi}$.

Applying Theorem 3 and arguments analogous to those used for the classical trigonometric polynomials, the following analog of the Marcinkiewicz-Zygmund-Berman identity can be proven.

Theorem 4 Let $\tilde{P}_n$ be any projection from $\tilde{C}_{2\pi}$ onto $\tilde{H}_n$. Then

$$\frac{1}{2\pi} \int_0^{2\pi} \tilde{P}_n(f(x-\tau), \theta+\tau)\,d\tau = (\tilde{S}_n f)(\theta) \tag{12}$$

As a consequence of the above theorem we have the following minimal norm property of $\tilde{S}_n$ (For the corresponding classical results see Losinskiĭ (1948) and Cheney et al. (1969)).

Corollary 2

$$\|\tilde{S}_n\| \leq \|\tilde{P}_n\| \tag{13}$$

Moreover, the equality in (13) holds only if $\tilde{P}_n = \tilde{S}_n$.

2. An interpolation-theoretical extremum problem.

Turečkiĭ (1968) posed the following problem: Let $0 \leq \theta_0 < \theta_1 < \ldots < \theta_{2n} < 2\pi$ and let $t_k(\theta) = t_{k,n}(\theta)$, $k=0,.,\ldots,2n$ be the fundamental trigonometric polynomials of degree n, i.e. such that $t_k(\theta_j) = \delta_{k,j}$, $k,j=0,1,\ldots,2n$. For what system of nodes $\{\theta_k\}_{k=0}^{2n}$ will

$$I_p = I_p(\theta_0,\theta_1,\ldots,\theta_{2n}) = \int_0^{2\pi} \{ \sum_{k=0}^{2n} |t_k(\theta)|^p \} d\theta, \quad (0<p<\infty) \tag{14}$$

be minimal? It was conjectured by Turečkiĭ that the optimal solution corresponds to the equidistant nodes. Schumacher (1976) proved this conjecture in case p=1, as a corollary of a much more general theorem. Szabados (1980) found a direct solution to the problem in the case $1 \leq p < \infty$, specifically he proved the following

Theorem 5 Let $p \geq 1$. The integral $I_p(\theta_0,\theta_1,\ldots,\theta_{2n})$ is minimal iff the nodes are identical with the equidistant nodes $\theta_k = 2k\pi/(2n+1)$, $k=0,1,\ldots,2n$, or with their translation. Moreover,

$$\min_{\theta_k} I_p(\theta_0,\theta_1,\ldots,\theta_{2n}) = \frac{2^p}{(2n+1)^{p-1}} \int_0^{\pi} |D_n(\theta)|^p d\theta \tag{15}$$

where $D_n(\theta)$ is the classical Dirichlet kernel.

In the following we extend the above result to the case where the number of interpolation nodes is even. Let $0 \leq \theta_1 < \theta_2 < \ldots < \theta_{2n} < 2\pi$ and denote by $\tilde{t}_k(\theta) = \tilde{t}_{k,n}(\theta)$, $k=1,2,\ldots,2n$, the fundamental trigonometric polynomials of half-angles, i.e. $\tilde{t}_k \in \tilde{H}_n$ and $\tilde{t}_k(\theta_j) = \delta_{kj}$. Let

$$\tilde{I}_p = \tilde{I}_p(\theta_1, \theta_2, \ldots, \theta_{2n}) = \int_0^{2\pi} \{ \sum_{k=1}^{2n} |\tilde{t}_k(\theta)|^p \} d\theta \tag{16}$$

The following result holds

Theorem 6 The integral $\tilde{I}_p(\theta_1, \theta_2, \ldots, \theta_{2n})$, $p \geq 1$, is minimal iff the nodes are identical with $\theta_k = k\pi/n$, $k=0,1,\ldots,2n-1$, or with their translation. Moreover

$$\min_{\theta_k} \tilde{I}_p(\theta_1, \theta_2, \ldots, \theta_{2n}) = \frac{2^p}{(2n)^{p-1}} \int_0^{2\pi} |\tilde{D}_n(\theta)|^p d\theta, \tag{17}$$

where $\tilde{D}_n(\theta)$ is the Dirichlet kernel corresonding to the trigonometric polynomials of half-angles. The proof of this theorem, which is based on (12), follows the same reasoning as the proof of Szabados.

Our final result concerns complex interpolation by algebraic polynomials on the unit circle. Let $z_k = \exp(i\theta_k)$, $k=0,1,\ldots,m$ be m+1 distinct points and let $l_{k,m}(z)$, $k=0,.,\ldots,m$ be the corresponding fundamental polynomials. It was proved by Brutman (1980) that

$$\begin{aligned} |\ell_{k,m}(e^{i\theta})| &= |t_{k,n}(\theta)|, \quad m=2n \\ &= |\tilde{t}_{k,n}(\theta)|, \quad m=2n-1 \end{aligned} \tag{18}$$

Combining (18) with Theorems 4 and 5, we arrive at the following

Theorem 7 The integral

$$I_p(z_0, z_1, \ldots, z_m) = \int_0^{2\pi} \{ \sum_{k=0}^{m} |\ell_{k,m}(\ell^{i\theta})|^p \} d\theta, \; p \geq 1 \tag{19}$$

is minimal iff $z_k = \exp[ik\pi/(m+1)+\alpha]$, $k=0,1,\ldots,m$, some α. Moreover

$$\min_{z_k} I_p(z_0, z_1, \ldots, z_m) = \frac{2^p}{(m+1)^{p-1}} \int_0^{2\pi} |D_m(\theta)|^p d\theta \tag{20}$$

where $D_m(\theta)=\sin(m\theta/2)/2\sin(\theta/2)$ is a generalized Dirichlet kernel.

References

Brutman, L. (1980), On the polynomial and rational projections in the complex plane, SIAM J. Numer. Anal. 17, 366-372

Brutman, L. and Pinkus, A. (1980), On the Erdös conjecture concerning minimal norm interpolation on the unit circle, SIAM J. Numer. Anal. 17, 373-375

Cheney, C.W., Hobby, C.R., Morris, P.D., Shurer, F. Wulbert, D.E. (1969) On the minimal property of the Fourier projection, Trans. Amer. Math. Soc. 143, 249-258

Fejér, L. (1904), Untersuchungen über Fouriersche Reihen, Math. Ann. 58, 51-69.

Fejér, L. (1910), Lebesguesche Konstanten und divergente Fourierreihen, J. Reine Angew. Math. 138, 22-53

Galkin, P.V. (1971), Estimates for the Lebesgue constants, Proc. Steklov Inst. Math. 109, 1-3

Geddes, K.O. and Mason, J.C. (1975), Polynomial approximation by projections on the unit circle, SIAM J. Numer. Anal. 1, 111-120

Lozinskiǐ, S.M. (1948), On a class of linear operators, Dokl. Akad. Nauk SSSR 61, 193-196

Schumacher, R. (1976), Zur Minimalität trigonometrischer Polynomoperatoren, Manuscripta Math. 19, 133-142

Szabados, J. (1980), On an interpolation-theoretic extremum problem, Periodica Math. Hungarica 11, 145-150

Szegö, G. (1921), Über die Lebesgueschen Konstanten bei den Fourierschen Reihen, Math. Z. 9, 163-166.

Turečkiǐ, A.H. (1960), On the quadrature formulas with even number of nodes accurate for the trigonometric polynomials, Dokl. Akad. Nauk BSSR 9, 365-366 (In Russian)

Turečkiǐ, A.H. (1968), 'Theory of Interpolation in Problem Form', Vyšeišaya Škola, Minsk, (In Russian); MR 41, #5840

On multivariate polynomial interpolation

N. Dyn

School of Mathematical Science, Tel Aviv University, Israel

A. Ron

Department of Computer Science, University of Wisconsin, Madison, Wisconsin, USA

Abstract A class of spaces of multivariate polynomials, closed under differentiation, is studied and corresponding classes of well posed Hermite-type interpolation problems are presented. All Hermite-type problems are limits of well posed Lagrange problems.

The results are based on a duality between certain spaces of multivariate exponential-polynomials $\mathcal{H}$ and corresponding spaces of multivariate polynomials $\mathcal{P}$, used by Dyn and Ron (1988) to establish the approximation order of the span of translates of exponential box splines. In the interpolation theory $\mathcal{P}$ is the space of interpolating polynomials and $\mathcal{H}$ characterizes the interpolation points and the interpolation conditions, both spaces being defined in terms of a set of hyperplanes in $\mathbb{R}^s$.

This geometric approach extends the work of Chung and Yao (1977) on Lagrange interpolation, and also a subset of the Hermite-type problems considered via the Newton scheme, by several authors (see Gasca and Maetzu (1982) and references therein). For a different approach to the interpolation problem see Chui and Lai (1988).

It is the systematic and unified analysis of a wide class of interpolation problems which is the main contribution of this paper to the study of multivariate polynomial interpolation.

Keywords: Multivariate interpolation, multivariate polynomials, Hermite-type interpolation.

1. The Interpolating Polynomial Spaces

The spaces of interpolating polynomials we consider here are more general than the total degree polynomials π_m (polynomials of degree $\leq m$), and are still closed under differentiation.

Given a set of directions $A = \{\mathbf{a}^1, \ldots, \mathbf{a}^n\} \subset \mathbb{R}^s$, with the property $\operatorname{span} A = \mathbb{R}^s$, consider the space of polynomials

$$\mathcal{P}(A) = \operatorname{span}\Big\{ \prod_{i \in I} (\mathbf{a}^i \cdot \mathbf{x}) \;\Big|\; I \in S(A) \Big\} \tag{1}$$

where $S(A)$ consists of index sets corresponding to "small enough" subsets of A, namely

$$S(A) = \big\{ I \subset \{1, \ldots, n\} \;\big|\; \operatorname{span}\{\mathbf{a}^i \mid i \notin I\} = \mathbb{R}^s \big\} \ . \tag{2}$$

By choosing $I \in S(A)$ such that $\{1, \ldots, n\} \backslash I$ is a basis of $\mathbb{R}^s$, we conclude that

$$\mathcal{P}(A) \subset \pi_{n-s} \ . \tag{3}$$

To see that $\mathcal{P}(A)$ is closed under differentiation, observe that

$$\frac{\partial}{\partial x_j} \prod_{i \in I} (\mathbf{a}^i \cdot \mathbf{x}) = \sum_{\ell \in I} a_j^\ell \prod_{\substack{i \in I \\ i \neq \ell}} (\mathbf{a}^i \cdot \mathbf{x}) \ , \tag{4}$$

and that if $I \in S(A)$ then any subset of I is in $S(A)$.

A more involved analysis is required in order to show the following two properties of $\mathcal{P}(A)$, demonstrated in Dyn and Ron (1988):

(a) Let $d = d(A) = \min \big\{ |I| \;\big|\; I \subset \{1, \ldots, n\} \ , \ I \notin S(A) \big\}$. Then $\pi_{d-1} \subset \mathcal{P}(A)$.

(b) The dimension of $\mathcal{P}(A)$ equals the number of bases that can be formed from A.

Combining (a) and (3) we conclude that

$$\pi_{d-1} \subset \mathcal{P}(A) \subset \pi_{n-s} \ . \tag{5}$$

If $\mathbf{a}^1, \ldots, \mathbf{a}^n$ are in "general position", namely any s vectors among $\mathbf{a}^1, \ldots, \mathbf{a}^n$ form a basis of $\mathbb{R}^s$, then it is easy to see that $d = n - s + 1$. Hence $\mathcal{P}(A) = \pi_{n-s}$.

To introduce a basis of $\mathcal{P}(A)$, consider n hyperplanes

$$H_i = \{ x \in \mathbb{R}^s \mid \mathbf{a}^i \cdot \mathbf{x} = \gamma_i \} \ , \qquad i = 1, \ldots, n \ , \tag{6}$$

determined by $\Gamma = (\gamma_1, \ldots, \gamma_n) \in \mathbb{R}^n$. For each $\mathbf{v} \in \mathbb{R}^s$ define

$$I_{\mathbf{v}} = \big\{ i \in \{1, \ldots, n\} \;\big|\; \mathbf{v} \in H_i \big\} \ . \tag{7}$$

and consider the set of intersection points of $H_1, \dots, H_n$,

$$V(A,\Gamma) = \{\mathbf{v} \in I\!R^s \mid \operatorname{span} A_\mathbf{v} = I\!R^s\} , \qquad A_\mathbf{v} = \{\mathbf{a}^i \mid i \in I_\mathbf{v}\} . \tag{8}$$

Choosing Γ so that $|I_\mathbf{v}| = s$ for $\mathbf{v} \in V(A,\Gamma)$, where $|I_\mathbf{v}|$ denotes the cardinality of $I_\mathbf{v}$, we conclude from (b) that

$$\dim \mathcal{P}(A) = |V(A,\Gamma)| . \tag{9}$$

Furthermore, the following polynomials

$$p_\mathbf{v}(\mathbf{x}) = \prod_{i \notin I_\mathbf{v}} \frac{(\mathbf{a}^i \cdot \mathbf{x} - \gamma_i)}{(\mathbf{a}^i \cdot \mathbf{v} - \gamma_i)} , \qquad \mathbf{v} \in V(A,\Gamma) , \tag{10}$$

are linearly independent, since

$$p_\mathbf{v}(\mathbf{u}) = \begin{cases} 0 & \mathbf{u} \neq \mathbf{v} , \\ 1 & \mathbf{u} = \mathbf{v} , \end{cases} \qquad \mathbf{v}, \mathbf{u} \in V(A,\Gamma) , \tag{11}$$

and hence constitute a basis of $\mathcal{P}(A)$.

The pair (A,Γ) is termed "simple" (for simple intersection points as opposed to multiple ones) if $|I_\mathbf{v}| = s$ for all $\mathbf{v} \in V(A,\Gamma)$.

Remark 1 It is shown by Ron (1988) that for fixed A the set of all $\Gamma \in I\!R^n$ such that (A,Γ) is simple, is dense in $I\!R^n$.

The explicit form (10) of a basis of $\mathcal{P}(A)$ indicates that the following result holds.

Proposition 1 $\mathcal{P}(A)$ consists of polynomials of degree $\leq n - s$, which are of degree $\leq n - s - \big|\{i \in \{1,\dots,n\} \mid \mathbf{a}^i \in \operatorname{span}\{\mathbf{y}\}\}\big| + 1$ along hyperplanes of the form $\mathbf{y} \cdot \mathbf{x} = \lambda$, $\mathbf{y} \in A$, $\lambda \in I\!R$. Furthermore, let $Y = \{\mathbf{y}^1, \dots, \mathbf{y}^k\}$, be $k < s$ pairwise distinct directions in A. Then the degree of any $p \in \mathcal{P}(A)$ along the intersection of k hyperplanes of the form

$$\mathbf{y}^j \cdot \mathbf{x} = \mu_j \quad , \quad j = 1,\dots,k , \qquad \mu_1,\dots,\mu_k \in R , \tag{12}$$

is at most

$$n - s - \big|\{i \in \{1,\dots,n\} \;\big|\; \mathbf{a}^i \in \langle Y \rangle \}\big| + \dim \langle Y \rangle , \tag{13}$$

where $\langle Y \rangle = \operatorname{span} Y$.

Proof Since for (A,Γ) simple, and $\mathbf{v} \in V(A,\Gamma)$, $\{\mathbf{a}^i \mid i \in I_\mathbf{v}\}$ is a basis of $I\!R^s$, each $p_\mathbf{v}$ in (10) consists of at least $\left|\{i \in \{1,\ldots,n\} \mid \mathbf{a}^i \in \text{span}\{\mathbf{y}\}\}\right| - 1$ factors which are constant along $\mathbf{y} \cdot \mathbf{x} = \lambda$, $\lambda \in I\!R$. Similarly, one can count the constant factors in $p_\mathbf{v}$ of (10) along the intersection of the hyperplanes (12), to conclude (13).

Remark 2 The space $\mathcal{P}(A)$ consists of all polynomials over $I\!R^s$ with the properties stated in Proposition 1. This will be shown elsewhere.

2. The Interpolation Problems

In this section we present a class of interpolation problems which are unisolvent in $\mathcal{P}(A)$ for fixed A. The interpolation points and the data at each point, which is of Hermite type, are determined by the choice of $\Gamma = (\gamma_1, \ldots, \gamma_n) \in I\!R^n$. The set of interpolation points consists of all points of intersection of at least s of the hyperplanes (6), namely, it is the set denoted by $V(A,\Gamma)$. To define the interpolation conditions at each $\mathbf{v} \in V(A,\Gamma)$, we consider the set of directions related to $\mathbf{v}$

$$A_\mathbf{v} = \{\mathbf{a}^i \mid i \in I_\mathbf{v}\}\ , \tag{14}$$

and a corresponding polynomial space defined by

$$\mathcal{K}(A_\mathbf{v}) = \{p \in \pi \mid [\prod_{i \in I}(\mathbf{a}^i \cdot D)]p \equiv 0\ ,\ I \notin S(A_\mathbf{v})\}\ , \tag{15}$$

where $D = \left(\frac{\partial}{\partial x_1}, \ldots, \frac{\partial}{\partial x_s}\right)$. Since each I in (15) satisfies $|I| \geq d_\mathbf{v} = d(A_\mathbf{v})$, it is clear that $\pi_{d_\mathbf{v}-1} \subset \mathcal{K}(A_\mathbf{v})$. The space $\mathcal{K}(A_\mathbf{v})$ is closed under differentiation since $D^\mathbf{m}$ commutes with any polynomial in D. In terms of $\mathcal{K}(A_\mathbf{v})$ the interpolation conditions at $\mathbf{v}$ are

$$[q(D)p](\mathbf{v}) = [q(D)f](\mathbf{v})\ , \qquad q \in \mathcal{K}(A_\mathbf{v})\ , \tag{16}$$

where f is smooth enough, and $q(D)$ is obtained from the polynomial $q(\mathbf{x})$ by replacing the vector $\mathbf{x}$ by the vector D. With these definitions we can introduce the interpolation problem determined by A and Γ:

$$\text{Find } p \in \mathcal{P}(A) \text{ satisfying (16) for all } \mathbf{v} \in V(A,\Gamma). \tag{17}$$

The solvability of (17) is due to the following result from Dyn and Ron (1988):

Theorem 1 The spaces $\mathcal{P}(A)$ and the space

$$\mathcal{H}(A,\Gamma) = \bigoplus_{\mathbf{v}\in V(A,\Gamma)} \{e^{\mathbf{v}\cdot\mathbf{x}}q(\mathbf{x}) \mid q \in \mathcal{K}(A_{\mathbf{v}})\} , \tag{18}$$

are dual to each other under the pairing

$$[p(D)h](0) = [q(D)p](\mathbf{v}) , \quad p \in \mathcal{P}(A) , \quad h(\mathbf{x}) = e^{\mathbf{v}\cdot\mathbf{x}}q(\mathbf{x}) \in \mathcal{H}(A,\Gamma) . \tag{19}$$

Corollary 1 There exists a unique $p \in \mathcal{P}(A)$ solving the interpolation problem (17).

It follows from Theorem 1 that $\mathcal{K}(A_{\mathbf{v}})$ is dual to $\mathcal{P}(A_{\mathbf{v}})$ in the sense of (19), and by (b) $\dim \mathcal{K}(A_{\mathbf{v}}) = \#$ of bases in $A_{\mathbf{v}}$. Furthermore, since $\mathcal{P}(A_{\mathbf{v}}) \subset \pi_{|A_{\mathbf{v}}|-s}$ we conclude that $\mathcal{K}(A_{\mathbf{v}}) \subset \pi_{|A_{\mathbf{v}}|-s}$. Hence

$$\pi_{d_{\mathbf{v}}-1} \subset \mathcal{K}(A_{\mathbf{v}}) \subset \pi_{|A_{\mathbf{v}}|-s} , \tag{20}$$

in analogy to (5). Moreover, if the directions in $A_{\mathbf{v}}$ are in general position then $\mathcal{K}(A_{\mathbf{v}}) = \pi_{|A_{\mathbf{v}}|-s}$.

Corollary 2 Let Γ be such that for each $\mathbf{v} \in V(A,\Gamma)$ the directions in $A_{\mathbf{v}}$ are in general position. Then the interpolation conditions in (16) are pure Hermite of the form

$$D^{\mathbf{m}}p(\mathbf{v}) = D^{\mathbf{m}}f(\mathbf{v}) , \qquad |\mathbf{m}| = \sum_{i=1}^{s} m_i \le |A_{\mathbf{v}}| - s , \ m_i \ge 0 \ , \ i = 1,\ldots,s . \tag{21}$$

In case A consists of directions in general position, then so does each $A_{\mathbf{v}}$, $\mathbf{v} \in V(A,\Gamma)$, and the interpolation problem becomes: Find $p \in \pi_{n-s}$ satisfying (21) for each $\mathbf{v} \in V(A,\Gamma)$. In R^2 the conditions on Γ in Corollary 2 are satisfied if $\gamma_i \ne \lambda\gamma_j$ whenever $\mathbf{a}^i = \lambda\mathbf{a}^j$, $\lambda \in I\!R$, $i \ne j$, $i,j \in \{1,\ldots,n\}$, namely if the hyperplanes $H_1,\ldots,H_n$ in (6) are pairwise disjoint.

An especially interesting interpolation problem is the Lagrange interpolation, obtained when (A,Γ) is simple. In this case $|A_{\mathbf{v}}| = s$, $\mathcal{K}(A_{\mathbf{v}}) = \pi_0$, and p satisfies $p(\mathbf{v}) = f(\mathbf{v})$, $\mathbf{v} \in V(A,\Gamma)$. The solution is given explicitly, in terms of the basis (10), as

$$p(x) = \sum_{\mathbf{v}\in V(A,\Gamma)} f(\mathbf{v})p_{\mathbf{v}}(x) . \tag{22}$$

This together with Remark 1 implies that the interpolation problem (17) is a limit of a sequence of Lagrange interpolation problems.

For general (A, Γ) the interpolation conditions (16) at $\mathbf{v} \in V(A, \Gamma)$ are determined by the structure of a chosen basis of $\mathcal{K}(A_{\mathbf{v}})$. The construction of such bases is discussed by Dahmen (this volume) and by deBoor and Ron (1988).

3. Examples

The first two examples are in R^2 and can be displayed graphically. We consider two Lagrange interpolation problems, for the same set of directions A, and then two Hermite-type problems, obtained as limits of the Lagrange problems.

Example 1 Let $A = \{\mathbf{a}^1, \ldots, \mathbf{a}^6\}$ with $\mathbf{a}^1 = \mathbf{a}^4 = (1,0)$, $\mathbf{a}^2 = \mathbf{a}^5 = (0,1)$, $\mathbf{a}^3 = \mathbf{a}^6 = (1,1)$, and let $\Gamma = \left(0, 0, 1, \frac{1}{2} + \varepsilon, \frac{1}{2} + \varepsilon, \frac{1}{2}\right)$ for $\varepsilon > 0$. The space $\mathcal{P}(A)$ is of dimension 12 and consists of quartic polynomials which reduce to cubics along hyperplanes of the form $\mathbf{a}^i \cdot \mathbf{x} = \text{const}$. The hyperplanes $\mathbf{a}^i \cdot \mathbf{x} = \gamma_i$, $i = 1, \ldots, 6$ are depicted in Figure 1, together with the twelve interpolation points. Since each interpolation point belongs to exactly two hyperplanes, (A, Γ) is simple, and the data at each point is just the function value.

For $\varepsilon = 0$ the three interpolation points $\mathbf{v}^1 = (1,0)$, $\mathbf{v}^2 = (0,1)$, $\mathbf{v}^3 = (0,0)$ remain unchanged together with the corresponding $A_{\mathbf{v}^i}$. Hence also in this problem only function values are required at $\mathbf{v}^i$, $i = 1,2,3$. Each of the other three interpolation points $\mathbf{v}^4 = \left(\frac{1}{2}, 0\right)$, $\mathbf{v}^5 = \left(0, \frac{1}{2}\right)$, $\mathbf{v}^6 = \left(\frac{1}{2}, \frac{1}{2}\right)$ is the limit of three interpolation points in the case $\varepsilon > 0$, with $A_{\mathbf{v}^i} = \{\mathbf{a}^1, \mathbf{a}^2, \mathbf{a}^3\}$, $i = 4,5,6$. Thus $\mathcal{K}(A_{\mathbf{v}^i}) = \pi_1$, $i = 4,5,6$, and the Hermite conditions are of the form

$$(f-p)(\mathbf{v}^i) = 0 \quad , \quad \frac{\partial}{\partial x_1}(f-p)(\mathbf{v}^i) = 0 \quad , \quad \frac{\partial}{\partial x_2}(f-p)(\mathbf{v}^i) = 0 \quad , \quad i = 4,5,6 \; .$$

Example 2 Let A be as in Example 1 and let $\Gamma = (0, 0, 1, \varepsilon, \varepsilon, 1 - \varepsilon)$ for $\varepsilon > 0$. The space $\mathcal{P}(A)$ is as in Example 1. The hyperplanes $\mathbf{a}^i \cdot \mathbf{x} = \gamma_i$, $i = 1, \ldots, 6$ are depicted in Figure 2. These hyperplanes have twelve intersection points, each belonging to exactly two hyperplanes. Thus for $\varepsilon > 0$, (A, Γ) is simple and the interpolation is of Lagrange type.

In the limit $\varepsilon \to 0$, there are only three interpolation points: $\mathbf{v}^1 = (1,0)$, $\mathbf{v}^2 = (0,1)$, $\mathbf{v}^3 = (0,0)$, each being the limit of four interpolation points in the case $\varepsilon > 0$. The

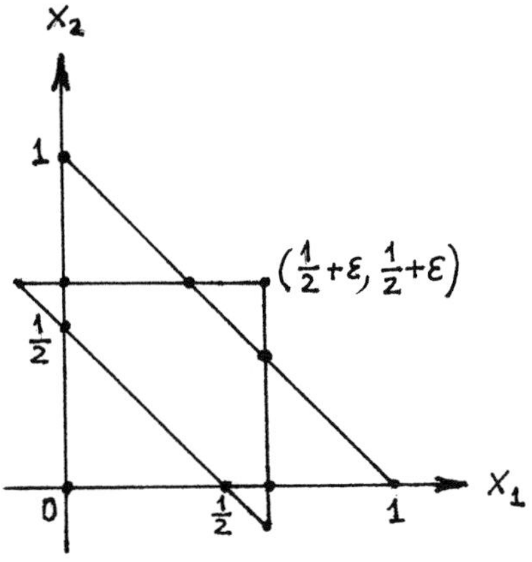

Figure 1

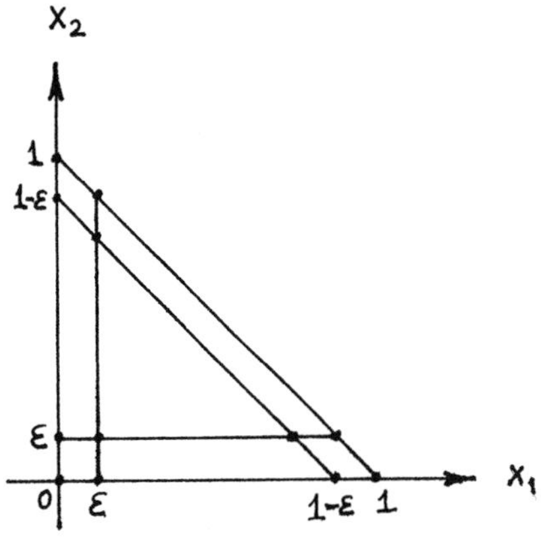

Figure 2

interpolation conditions at $\mathbf{v}^i$ are determined by $\mathcal{K}(A_{\mathbf{v}^i})$ where

$$A_{\mathbf{v}^i} = \{\mathbf{a}^j, \mathbf{a}^{j+3} \mid j \neq i \ , \ j = 1,2,3\} \ .$$

By (15)

$$\begin{aligned}\mathcal{K}(A_{\mathbf{v}^i}) &= \left\{p \in \pi \,\middle|\, (\mathbf{a}^j \cdot D)^2 p = 0 \ , \ j \neq i \ , \ j = 1,2,3\right\} \\ &= \operatorname{span}\left\{1, x_1, x_2, \prod_{\substack{j=1 \\ j \neq i}}^{3} (\mathbf{n}^j \cdot \mathbf{x})\right\} \ ,\end{aligned}$$

where $\mathbf{n}^i \cdot \mathbf{a}^i = 0$, $i = 1,2,3$. Hence the interpolation conditions are

$$(f-p)(\mathbf{v}^i) = 0 \ , \ \frac{\partial}{\partial x_1}(f-p)(\mathbf{v}^i) = 0 \ , \ \frac{\partial}{\partial x_2}(f-p)(\mathbf{v}^i) = 0 \ , \quad i = 1,2,3$$

$$\left(\frac{\partial^2}{\partial x_1^2} - \frac{\partial^2}{\partial x_1 \, \partial x_2}\right)(f-p)(\mathbf{v}^1) = 0 \quad , \quad \left(\frac{\partial^2}{\partial x_2^2} - \frac{\partial^2}{\partial x_1 \, \partial x_2}\right)(f-p)(\mathbf{v}^2) = 0 \ ,$$

$$\frac{\partial^2}{\partial x_1 \, \partial x_2}(f-p)(\mathbf{v}^3) = 0 \ .$$

This interpolation problem is a special case of the one solved by Gregory (1985), where the interpolation points are the vertices of a simplex in $I\!R^s$, and A consists of $s+1$ directions in general position each repeated $N \geq 2$ times. The next example deals with an extended version of this case in terms of our analysis.

Example 3 Let $B = \{\mathbf{b}^1, \ldots, \mathbf{b}^{s+1}\} \subset I\!R^s$ be in general position and let (B, Δ) be simple, with $\Delta = (\delta_1, \ldots, \delta_{s+1}) \in I\!R^{s+1}$. Given $s+1$ positive integers $m_1, \ldots, m_{s+1}$, $n = \sum_{i=1}^{s+1} m_i$, consider $A = \{\mathbf{a}^1, \ldots, \mathbf{a}^n\}$ consisting of $\mathbf{b}^i$ repeated m_i times, and $\Gamma =$

$(\gamma_1, \ldots, \gamma_n)$ consisting of δ_i repeated m_i times, $i = 1, \ldots, s+1$. The hyperplanes $H_i = \{\mathbf{x} \mid \mathbf{b}^i \cdot \mathbf{x} = \delta_i\}$, $i = 1, \ldots, s+1$, intersect at $s+1$ points $\mathbf{v}^1, \ldots, \mathbf{v}^{s+1}$, forming the vertices of a simplex. Let $\mathbf{v}^i$ denote the intersection of the s hyperplanes H_j, $j \neq i$, $j = 1, \ldots, s+1$. Then $A_{\mathbf{v}^i}$ consists of $\mathbf{b}^j$ repeated m_j times $j \neq i$, $j = 1, \ldots, s+1$, and by (15)

$$\mathcal{K}(A_{\mathbf{v}^i}) = \left\{p \in \pi \;\middle|\; (\mathbf{b}^j \cdot D)^{m_j} p \equiv 0 \;,\; j \neq i \;,\; j = 1, \ldots, s+1\right\} .$$

The dimension of $\mathcal{K}(A_{\mathbf{v}^i})$ is the number of bases in $A_{\mathbf{v}^i}$ given by $M_i = \prod_{j=1, j\neq i}^{s+1} m_j$.

Now the edge of the simplex connecting vertices $\mathbf{v}^i$ and $\mathbf{v}^\ell$ belongs to the intersection of the hyperplanes H_j, $j \neq i, \ell$, $j = 1, \ldots, s+1$. Hence $(\mathbf{v}^i - \mathbf{v}^\ell) \cdot \mathbf{b}^j = 0$, $j \neq i, \ell$, $j = 1, \ldots, s+1$, from which we conclude that any polynomial of the form

$$\prod_{\substack{\ell=1 \\ \ell \neq i}}^{s+1} [(\mathbf{v}^i - \mathbf{v}^\ell) \cdot \mathbf{x}]^{\alpha_\ell} \;, \qquad 0 \leq \alpha_\ell < m_\ell \;,\; \ell = 1, \ldots, s+1 \;,$$

is annihilated by $(\mathbf{b}^j \cdot D)^{m_j}$, $j \neq i$, and therefore belongs to $\mathcal{K}(A_{\mathbf{v}^i})$. The number of these polynomials is M_i and they are linearly independent, thus forming a basis of $\mathcal{K}(A_{\mathbf{v}})$. In terms of this basis the Hermite type conditions at $\mathbf{v}^i$ are

$$\prod_{\substack{j=1 \\ j \neq i}}^{s+1} [(\mathbf{v}^i - \mathbf{v}^j) \cdot D]^{\alpha_j} (f - p)(\mathbf{v}^i) = 0 \;, \quad 0 \leq \alpha_j < m_j \;,\; j \neq i \;,\; j = 1, \ldots, s+1 \;.$$

References

de Boor, C. and Ron, A. (1988), On polynomials ideals of finite codimension with applications to box spline thoery, CMS TSR # 89-21 Univ of Wisconsin-Madison.

Chui, C.K. and Lai, M.J. (1988), Vandermonde determinants and Lagrange interpolation in $I\!R^s$, in nonlinear and convex analysis, ed. by B.L. Lin, Marcel Dekker, N.Y.

Chung, K.C. and Yao, T.H. (1977), On lattices admitting unique Lagrange interpolation, SIAM J. Num. Anal. **14**, 735-743.

Dahmen, W., A basis of certain spaces of multivariate polynomials and exponentials, this volume.

Dyn, N. and Ron, A. (1988), Local approximation by certain spaces of exponential polynomials, appoximation order of exponential box splines and related interpolation problems, to appear in Trans. AMS.

Gasca, M. and Maeztu, J.I. (1982), On Lagrange and Hermite interpolation in $I\!R^n$, Num. Math. **39**, 1-14.

Gregory, J.A. (1985), Interpolation to boundary data on the simplex, CAGD **2**, 43-52.

Ron, A. (1988), Exponential box splines, to appear in Const. Approx.

Algorithms for the construction of data dependent triangulations

N. Dyn, D. Levin and S. Rippa
School of Mathematical Sciences,
Tel-Aviv University, Israel

Abstract Given a set of data points in R^2 and corresponding data values it is clear that the quality of a piecewise linear interpolation over triangles depends on the specific triangulation of the data points. While conventional triangulation methods depend only on the distribution of the data points in R^2, we suggested in [1] the construction of triangulations which depend on the data values as well. In this paper we present and compare some algorithms for the construction of such data dependent triangulations.
Key words : Triangulation, Data dependent triangulation, Piecewise linear interpolation.

1 Introduction

Let $V = \{v_i = (x_i, y_i) \in R^2 , i = 1, \dots, N\}$ be a set of distinct and non-collinear data points and $F = (F_1, \dots, F_N)$ a (real) data vector. Suppose furthermore that $\Omega \supset V$ is a region with a polygonal boundary $\partial\Omega$ with all vertices in V.

Definition 1 *A set $T = \{T_i\}_1^t$ of non-degenerate, open, triangles is a triangulation of Ω if :*

- *V is the set of all vertices of triangles in T.*
- *Every edge of a triangle in T contains only two points from V, namely its endpoints,*
- $\overline{\Omega} = \bigcup_{i=1}^{t} \overline{T_i}$, $T_i \bigcap T_j = \emptyset$, $i \neq j$.

Given a triangulation T of Ω we consider the space $S_1^0(T)$ of piecewise linear polynomials defined over T, i.e.

$$S_1^0(T) = \{g \in C^0(\Omega) \mid \; g|_{T_i} \in \Pi_1\},$$

where Π_1 is the three dimensional space of linear polynomials. Finally we denote by f_T the unique function from $S_1^0(T)$ interpolating F, i.e.

$$f_T(x_i, y_i) = F_i \;, \; i = 1, \ldots, N.$$

A linear function is uniquely defined by its values at the three vertices of a triangle and thus a Piecewise Linear Interpolating Surface (PLIS) is uniquely determined by the choice of a specific triangulation of Ω. It is clear that the quality of approximation by a PLIS depends on the particular choice of the triangulation and naturally we look for an optimal triangulation. The classical theory says that long, thin, triangles should be avoided and that triangles should be as equiangular as possible ([4]). A popular choice of triangulation for interpolation schemes is the well known Delaunay triangulation (see e.g. [4]) which, among other nice properties, is a MaxMin triangulation, i.e. it is a triangulation T^* maximizing the quantity

$$\alpha(T) = \min_{T_i \in T}(\text{ smallest angle in } T_i).$$

The Delaunay triangulation, as various others in use, depends only on the set V and not on the data vector F. In [1] we suggested the use of *data dependent* criteria for measuring the quality of a triangulation. These criteria depend on the set V of data points and on the data vector F as well. Given a data dependent criterion and an initial triangulation, a data *dependent triangulation* T' may be constructed by the familiar procedure of swapping diagonals of convex quadrilaterals in order to get better triangulations where "better" should be interpreted as better with respect to the given data dependent criterion. Numerical tests, reported in [1], demonstrate very clearly that PLISes defined over data dependent triangulations provide better approximation, to various test functions, than the PLIS defined over the Delaunay triangulation of the same set of data points.

In the present paper we discuss algorithms for constructing data dependent triangulations. In §2 the concept of data dependent triangulations is presented as well as two algorithms for their construction. In order to simplify the presentation we use an example rather than a more detailed definition to illustrate the ideas involved. It is straightforward to extend the results to the more general setting of [1]. The most basic algorithm for the construction of data dependent triangulations is Lawson's LOP algorithm which swaps the diagonals of convex quadrilaterals in order to decrease a certain cost function. The Modified LOP (MLOP) suggested here defines the specific order of swapping edges. Two strategies for swapping edges are presented in §3 and numerical experiments comparing the LOP and MLOP algorithms are reviewed in §4.

2 Data dependent triangulations

Let V be a fixed set of data points, F a data vector, T a triangulation of Ω and suppose that f_T is the piecewise linear interpolant to F. For each interior edge e of T a real cost function $S(f_T, e)$ is assigned. and the index vector N_T, of length q, containing the cost functions of all interior edges is constructed:

$$N_T = (S(f_T, e_1), \ldots, S(f_T, e_{\mathrm{q}})).$$

The cost function of a triangulation is defined to be:

$$c(f_T) = \sum_{i=1}^{\mathrm{q}} |S(f_T, e_i)|.$$

Two examples for cost functions are : (a) Jump in Normal Derivative (JND) $S_1(f_T, e)$, the (magnitude of the) jump in the normal derivative of f_T across the edge e and (b) Angle Between Normals (ABN) $S_2(f_T, e)$, the angle between the normal vectors to the two facets of the surface f_T on both sides of the edge e. These cost functions, and others, are discussed in more detail in [1].

Definition 2 *A triangulation T' of Ω is called optimal if $c(f_{T'}) \leq c(f_T)$ for every triangulation T of Ω.*

In most practical situations it is very difficult to obtain a globally optimal triangulation so here we consider only locally optimal triangulations. Let T be a triangulation of Ω, e an interior edge of T and Q a quadrilateral formed from the two triangles having e as a common edge. If Q is strictly convex then there are two possible ways of triangulating Q (see Figure 1).

Definition 3 *An edge e is called locally optimal if one of the following conditions holds: (a) Q is not strictly convex or (b) Q is strictly convex and $c(f_T) \leq c(f_{T'})$ where T' is obtained from T by replacing e by the other diagonal of Q.*

Figure 1: Two triangulations of a convex quadrilateral

Definition 4 *A locally optimal triangulation of* Ω *is a triangulation* T' *in which all edges are locally optimal.*

The basic algorithm for constructing locally optimal triangulations is the Local Optimization Procedure (LOP) suggested by Lawson [3]:

Algorithm 1 LOP

1. Construct an initial triangulation $T^{(0)}$ of Ω and set $T \leftarrow T^{(0)}$.

2. If T is locally optimal, end the procedure; else go to step 3.

3. Let e be an interior edge of T which is not locally optimal and let Q be the strictly convex quadrilateral formed from the two triangles in T having e as a common edge.

   (a) Swap diagonals of Q: Replace e by the other diagonal of Q, therefore transforming T into a triangulation T'

   (b) Set $T \leftarrow T'$ and go to step 2.

Each time an edge swap occurs, the cost function of the resulting triangulation is strictly smaller than that of the previous one. Since the number of triangulations of Ω is finite, the LOP converges, after a finite number of edge swaps, to a locally optimal triangulation.

The above LOP seems to work very well in the numerical experiments reported in [1], but it has a major drawback: the resulting locally optimal triangulation depends on the labelling of the data points and on the software implementation of the LOP. We would like to control more closely the order in which edges are swapped during the LOP iterations in order to obtain a better defined algorithm. Given a triangulation T

we consider the set $E(T)$ of all the interior edges e of T which are not locally optimal. Suppose that $E(T) \neq \emptyset$ and assume that the edges are labeled such that:

$$E(T) = \{e_1, e_2, \ldots, e_{k(T)}\}.$$

A swap of some edge e_j , $1 \leq j \leq k(T)$, will transform T into a triangulation $T^{(j)}$. The Modified LOP (MLOP) selects, at each step from the set $E(T)$, the next edge to be swapped according to a predetermined swapping strategy. In §3 we discuss and compare some selection strategies. Suppose that the edge $e_p \in E(T)$ is swapped. Then $E(T^{(p)})$ has to be computed for the next iteration. This can be done efficiently since $E(T^{(p)})$ and $E(T)$ differ only in edges belonging either to the two triangles T_i and T_j which have e_p as a common edge or in edges belonging to triangles sharing an edge with T_i or T_j. Thus at most 12 edges, which may be in one of the sets and not in the other, need to be checked in each MLOP iteration (after the edge swap, e_p becomes locally optimal and thus is excluded from the set $E(T^{(p)})$).

3 Selecting the edges to swap

The first selection strategy is the *Maximal Reduction (MR)* strategy. Since a data dependent criterion selects triangulations which minimize the cost function $c(f_T)$, a natural strategy for the MLOP is to swap, in each iteration, the edge $e_p \in E(T)$ for which the $c(f_{T^{(p)}})$ is minimal, i.e.

$$c(f_{T^{(p)}}) \leq c(f_{T^{(j)}}), \quad 1 \leq j \leq k(T).$$

The second strategy is based on the observation that often the LOP/MLOP terminates in a poor local minimum since many of the edges become interior to non-convex quadrilaterals. To avoid this we would like to swap the edges in a way which will leave the maximum possible number of convex quadrilaterals for the next MLOP iteration. Let us divide the set $E(T)$ into classes of edges according to the value of

$$m_j = I(T^{(j)}) - I(T), \quad j = 1, \ldots, k(T),$$

where $I(T)$ is the number of convex quadrilaterals in the triangulation T. Let

$$E'(T) \subseteq E(T)$$

be the set of edges $e_j \in E(T)$ for which m_j is greatest. The *Maximal Opportunity (MO)* strategy chooses the next edge e_p to swap from the set $E'(T)$ such that $c(f_{T^{(p)}})$

is minimal, i.e.

$$c(f_{T(p)}) \leq c(f_{T(j)}), \quad e_j \in E'(T).$$

For comparison we also generate the "worst possible" sequence of edge swaps. This is obtained by the *Minimal Reduction (MinR)* strategy which is just the opposite of the Maximal Reduction strategy, i.e. we choose the next edge to swap so that the reduction in the value of the cost function is minimized.

4 Numerical experiments

In our numerical experiments two sets of data points taken from Franke ([2]) were used. The first set contains 100 data points distributed more or less uniformly over the unit square, while the second set, with 33 data points, was designed with larger variations in the density of the data points. The data vectors $F = (F_1, \ldots, F_N)$ were obtained by evaluating various test functions, also used in [1], at the data points.

Several data dependent criteria were tested including those relating to the JND and ABN cost functions mentioned in §2 as well as others presented in [1]. For each data set and data dependent criterion, several data dependent triangulations were generated by using different strategies for swapping the edges:

- LOP - No strategy, edges are swapped according to their labelling in the edge list.
- MLOP-MR - Edges are swapped according to the Maximal Reduction strategy.
- MLOP-MO - Edges are swapped according to the Maximal Opportunity strategy.
- MLOP-MinR - Edges are swapped according to the Minimal Reduction strategy.

The Delaunay triangulation of the data points was used in all cases as an initial triangulation.

On each triangulation, the piecewise linear function f_T, interpolating the data vector F, was constructed, and the error between f_T and the test function which generates F was computed on a grid of 33×33 nodes. The mean, root mean square and maximum of these errors were tabulated along with the value of the cost function and the number of edge swaps needed for convergence of the LOP/MLOP to the data dependent triangulation.

In the numerical experiments it became clear that the order in which edges are swapped during LOP/MLOP iterations may have a large influence on the final locally optimal triangulation in terms of the quality of approximation to the test functions and the value of the cost function of the locally optimal triangulation. It was found that the value of the JND and ABN cost functions of a triangulation is usually a good indicator to the quality of approximation to the test functions: a PLIS defined over a low cost triangulation (for which these cost functions have small values) usually provides a better approximation than a PLIS defined on a high cost triangulation.

The MLOP with the MR edge swapping strategy performed quite well in general. The resulting data dependent triangulations were, in most cases, comparable or better than data dependent triangulations resulting from other strategies of edge swapping. When we say "better" we mean that a better approximation to the test functions was achieved. The MLOP-MR triangulation was in most cases the triangulation which attains the smallest value of the cost function. Also the MLOP-MR converges in the fewest number of edge swaps and thus is the most efficient. In view of these reasons this is the scheme of our choice.

The MLOP-MO strategy was comparable to the MLOP-MR strategy but is more difficult to program and thus we do not recommend the use of it.

The MLOP with the MinR strategy did produce in most cases the worst data dependent triangulation of all and used the largest number of edge swaps. The MLOP-MinR triangulation demonstrates that there is a sequence of edge swaps which can lead to very bad data dependent triangulations.

As can be expected from the above observations, the LOP swapping according to labelling resulted in good and bad triangulations depending on the labelling used, the triangulation criterion and the test function. The results are usually quite acceptable but often worse than the MLOP-MR triangulation and sometimes much worse. There are advantages, however, in using the LOP since its programming is simpler and it requires less computer storage than the MLOP for which the list $E(T)$ of edges has to be stored and maintained.

We note that none of the above strategies performed well in all cases. For any of these strategies there are examples of poor data dependent triangulations generated by it. It may be interesting to look for other heuristic strategies for edge swapping.

An example of data dependent triangulation, taken from [1], concludes this paper. The cliff function $F = (\tanh(9y - 9x) + 1)/9$ (see Figure 2), was sampled on a set of 33 data points. The Delaunay triangulation of the set of data points and the related PLIS are displayed in Figure 3. A data dependent triangulation, based on the JND cost function of §2, and the PLIS defined over it are displayed in Figure 4. These pictures demonstrate very clearly the advantage in using data dependent triangulations.

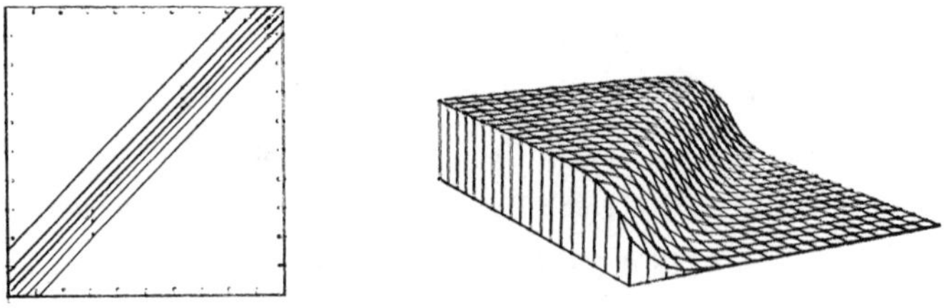

Figure 2: The cliff function

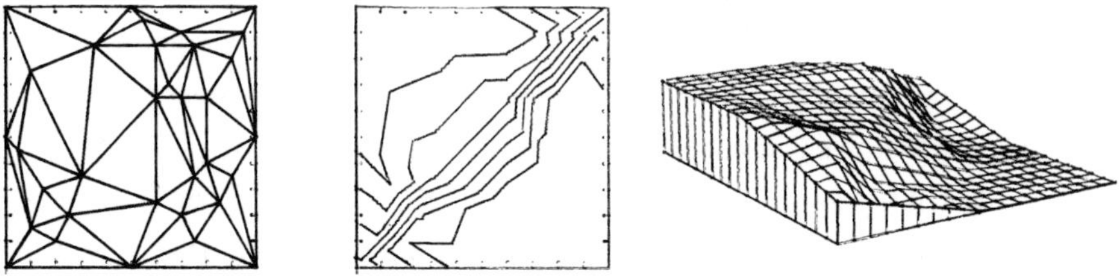

Figure 3: The Delaunay triangulation and the PLIS defined over it

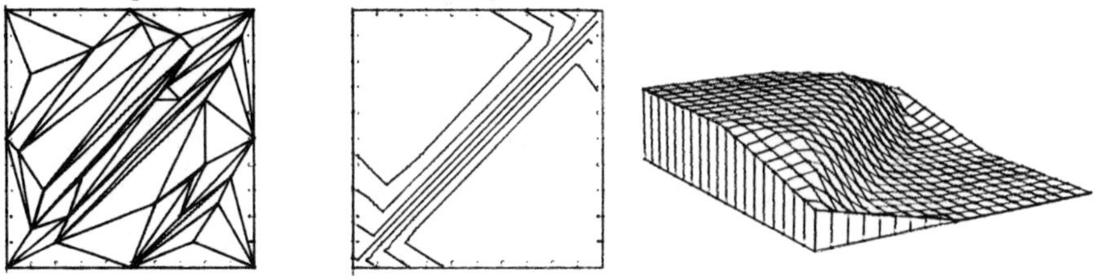

Figure 4: A data dependent triangulation and the PLIS defined over it

References

[1] Dyn, N., Levin, D. and Rippa, S. *Data dependent triangulations for piecewise linear interpolation.* To appear in IMA J. Numer. Anal.

[2] Franke, R., *Scattered data interpolation: tests of some methods,* Math Comp. **38** (1982), 181-200.

[3] Lawson, C. L., *Software for C^1 interpolation,* in Mathematical Software III, J. R. Rice (ed.), Academic Press, New York, 1977, 161-194.

[4] Schumaker, L. L., *Triangulation Methods,* in Topics in Multivariate Approximation, C. K. Chui, L. L. Schumaker and F. I. Utreras (eds.), Academic Press, 1987, 219-232.

Algorithms for computing best parametric cubic interpolation

C. Rademacher and K. Scherer

Institut für Angewandte Mathematik, Universität Bonn, FRG

Abstract We consider the problem of finding a curve passing through prescribed points in $\mathbb{R}^d$ such that the kinetic energy of particle moving along this curve is minimized. Recently uniqueness of the solution of this problem has been established by the second author. In this contribution algorithms for computing the solution are devised and their (global) convergence properties studied.

Key words: Best interpolation, cubic spline curves, constrained minimization, stationary points, projected gradient method, Newton Method.

1. Introduction. The problem of best parametric interpolation of given data $\{\underline{y}_i\}_{i=1}^n$ in $\mathbb{R}^d$ is the following: find a function $\underline{s}(t) \in \mathcal{L}_2^k(0,1)$, the space of $\mathbb{R}^d$-valued functions on [0,1] with components in $L_2^k(a,b)$, which attains

$$\inf_{\underline{t}} \inf \{\int_0^1 ||\underline{f}^{(k)}(t)||^2 dt \; : \; f \in \mathcal{L}_2^k(0,1), \underline{f}(t_i)=\underline{y}_i \, , \; 1 \le i \le n\}. \tag{1}$$

The first infimum runs over all sequences $\underline{t}$ of nodes satisfying

$$0 = t_1 < t_2 < \ldots < t_n = 1, \tag{2}$$

and the norm $\| \ \|$ in (1) denotes the usual Euclidean norm in $\mathbb{R}^d$. If one considers only the second infimum in (1) the problem reduces to the classical problem of best interpolation in spline theory. For the motivation of the general problem we refer to [5], [3] and [4]. In [3] and [4] the uniqueness of the solution of (1) in the cubic case $k=2$ is proved ($d=1$ in [3] and general d in [4]) under the condition

$$\underline{y}_i \neq \underline{y}_{i+1}, \qquad 1 \leq i \leq n-1, \tag{3}$$

on the data.

In this note we describe two types of convergent algorithm for computing the solution $\underline{s}^*(t)$ of (1) in the case $k=2$. First we summarize some well-known facts and the relevant results of [4]. For fixed $\underline{t}$ the inner infimum in (1) is attained by a cubic spline function $\underline{s}(t)$ whose second derivative is

$$\underline{s}''(t) = \sum_{i=1}^{n-2} \underline{a}_i \, N_{i,2}(t), \tag{4}$$

where the $N_{i,2}(t)$ denote the piecewise linear B-splines with knots in $\underline{t}$ satisfying

$$N_{i,2}(t_j) = \delta_{j,i+1}, \qquad 1 \leq j,i \leq n-2 .$$

The coefficients $\underline{a}_i \in \mathbb{R}^d$ in (4) are therefore equal to $\underline{s}''(t_{i+1})$. They are determined by the tridiagonal linear system

$$\sum_{i=1}^{n-2} G_{ij} \, \underline{a}_i = \underline{z}_j := \frac{\underline{w}_{j+1}}{h_{j+1}} - \frac{\underline{w}_j}{h_j}, \qquad \underline{w}_j := \underline{y}_{j+1} - \underline{y}_j , \tag{5}$$

where

$$G_{ij} = \begin{cases} 2(h_i + h_{i+1}), & j = i, \\ h_{\max(i,j)}, & |i-j| = 1, \quad h_i := t_{i+1} - t_i, \\ 0, & \text{otherwise}. \end{cases} \tag{6}$$

We can write (5) in matrix form as

$$G\,\underline{A} = \underline{Z} \tag{7}$$

if we introduce the vectors

$$\underline{A} := (\underline{a}_1,\ldots,\underline{a}_{n-2}), \qquad \underline{Z} = (\underline{z}_1,\ldots,\underline{z}_{n-2}).$$

Furthermore one can easily verify

$$\int_o^1 ||\underline{s}''(t)||^2 dt = <\underline{A}\,,\,G\underline{A}> \;=\; <\underline{A}\,,\,\underline{Z}>, \tag{8}$$

where $<\ ,\ >$ denotes the obvious scalar product. From this one concludes that the infimum problem (1) is equivalent to

$$\inf_{\underline{h}\in K} \{F(\underline{h}) : F(\underline{h}) := <\underline{A}\,,\,\underline{Z}>, \quad G\underline{A} = \underline{Z}>, \tag{9}$$

where

$$K := \{\underline{h} \in \mathbb{R}^{n-1} : \underline{h} = (h_1,\ldots,h_{n-1}),\ h_i > 0, \quad \sum_{i=1}^{n-1} h_i = 1\}. \tag{10}$$

Hence a solution of (1) is given by a pair $(\underline{h}\,,\,\underline{A}(\underline{h}))$, where $\underline{A} = A(\underline{h})$ satisfies (7) and $\underline{h}$ the equations for a critical point of $F(\underline{h})$ in K. These equations read

$$0 = \lambda + \frac{\partial F}{\partial h_j} \equiv \lambda - \frac{2T_j}{h_j^2} - \alpha_j, \qquad 1 \le j \le n-1, \tag{11a}$$

$$1 = \sum_{j=1}^{n-1} h_j, \tag{11b}$$

with Lagrangian parameter $\lambda \in \mathbb{R}$. For the quantities in (11) one has the explicit expressions (with $\underline{a}_o = \underline{a}_{n-1} \equiv 0$)

$$T_j := -\,h_j^2 <\underline{A}\,,\,\frac{\partial Z}{\partial h_j}> = (\underline{a}_{j-1} - \underline{a}_j\,,\,\underline{w}_j), \tag{12a}$$

$$\alpha_j := < \underline{A}, \frac{\partial G}{\partial h_j} \underline{A} > = \frac{1}{3} [(\underline{a}_j, \underline{a}_j) + (\underline{a}_{j-1}, \underline{a}_{j-1}) + (\underline{a}_j, \underline{a}_{j-1})]. \quad (12b)$$

Here (,) denotes the usual scalar product in $\mathbb{R}^d$.

The equations (11) together with (7) form a nonlinear system of equations for $\underline{h} \in K$. In [4] the following was proved:

Theorem 1: There exists only one minimum $\underline{h}^*$ of $F(h)$ in K (which is then the unique, global, solution of (1)). The corresponding pair $\underline{h}^*$, $\underline{A}^* = A(\underline{h}^*)$ is a solution of the saddle point problem

$$\inf_{\underline{h} \in K} \sup_{\underline{A}} \phi(\underline{A},\underline{h}) = \sup_{\underline{A}} \inf_{\underline{h} \in K} \phi(\underline{A},\underline{h}), \quad (13)$$

where $\phi(\underline{A},\underline{h}) := - < \underline{A}, G\underline{A} > + 2 < \underline{A}, \underline{Z} >$. A critical point of $F(\underline{h})$, i.e. a solution of (11) and (7), is a solution of (1) if and only if the corresponding A^* lies in

$$B := \{\underline{A} \in \mathbb{R}^{(n-2)d} : T_j := (\underline{a}_{j-1} - \underline{a}_j, \underline{w}_j) > 0, \ 1 \leq j \leq n-1\}. \quad (14)$$

2. Descent Methods. In the following we describe how some descent methods for constrained minimization problems may be adapted and modified to problem (9) so as to guarantee global convergence. A general method for minimization with linear equality constraints has the form:

Given the approximation $\underline{h}^{(\nu)}$ for the solution $\underline{h}^* \in \mathbb{R}^{n-1}$, one computes an (improved) approximation $\underline{h}^{(\nu-1)}$ for $\underline{h}^*$ by

$$\underline{h}^{(\nu+1)} := \underline{h}^{(\nu)} + \delta_\nu P\underline{d}^{(\nu)}, \qquad \nu = 0,1,\ldots, \quad (15)$$

where $\underline{d}^{(\nu)}$ is the direction of search ($||\underline{d}^\nu|| = 1$), the positive scalar δ_ν the step length, and P the linear projection of $\underline{d}^{(\nu)}$ to ker R, where the constraints are given in the form $R\underline{h} = \underline{b}$ with matrix R. In the case of problem (9) a simple choice is to set

$$(P\underline{d})_j := d_j - \frac{1}{(n-1)} \sum_{i=1}^{n-1} d_i \quad (16)$$

for the vector $\underline{d} = (d_1,\ldots,d_{n-1})$.

In order that the above method is a descent method we assume that there is a number $\beta > 0$, independent of $\nu = 0,1,2,\ldots$, such that (note that $P = P^T$ with the choice (16))

$$\left(\underline{g}^{(\nu)}, P\underline{d}^{(\nu)}\right) \le -\ \beta\,||P^T g^{(\nu)}||, \tag{17}$$

where $g^{(\nu)}$ denotes the gradient of F at $\underline{h}^{(\nu)}$, i.e.

$$\underline{g}^{(\nu)} := \text{grad } F\left(\underline{h}^{(\nu)}\right).$$

(Note that here (,) stands for the scalar product in $\mathbb{R}^{n-1}$.) Then Taylor expansion gives the decrease of $F(\underline{h}^{(\nu+1)})$ for δ_ν sufficiently small. The following step-length procedure realizes this.

Choose $\delta_\nu = 2^{-j}$, where $j = j(\nu)$ is the smallest non-negative number such that

$$F\left(\underline{h}^{(\nu)} + 2^{-j}\underline{d}^{(\nu)}\right) < F\left(\underline{h}^{(\nu)}\right) + 2^{-j-1}\left(\underline{g}^{(\nu)}, P\,\underline{d}^{(\nu)}\right) \tag{18a}$$

is valid. In this procedure the constraints $0 < h_i^{(\nu+1)} < 1$, $1 \le i \le n-1$, implied by $\underline{h}^{(\nu+1)} \in K$ can also be incorporated. To this end we require additionally that $\delta_\nu = 2^{-j}$ shall satisfy

$$-h_i^{(\nu)} < 2^{-j}\left(P\,\underline{d}^{(\nu)}\right)_i < 1 - h_i^{(\nu)}, \qquad 1 \le i \le n-1\,. \tag{18b}$$

Both (18a) and (18b) are always satisfied for $\delta_\nu = 2^{-j}$ sufficiently small. This is guaranteed by the fact that for any $\underline{h} \in K$ the inequality

$$\min_j h_j \ge \min_i ||\underline{y}_{i+1} - \underline{y}_i||^2 / [(n-1)\max_j ||\underline{y}_{j+1} - y_j||^2 + F(\underline{h})] \tag{19}$$

holds (see [4]).

The above considerations only ensure convergence to a critical point. In order to improve this we introduce so called "T steps" for $\underline{h}^{(\nu)} \in K$ such that not all corresponding T_i's defind by (12a) are positive. In this case a critical point $\underline{h}^*$ of F(h) cannot be the solution of (1) according to (14). Suppose it is not a strict local minimum too. Then

$$\eta \equiv \min_{\|\underline{d}\|=1} \left(P\underline{d} \,,\, H(\underline{h}^*) P\underline{d} \right) \leq 0 \tag{20}$$

must hold, where $H(\underline{h})$ denotes the Hessian of $F(\underline{h})$ at a point $\underline{h}$. If $\underline{d}^*$ is a direction attaining η in (20), the decrease of $F(\underline{h})$ along $\underline{d}^*$ satisfies, for $0 < \delta \leq \delta_o$,

$$F(\underline{h}^* + \delta P\underline{d}^*) \leq F(\underline{h}^*) + \max\,(-C\delta^{k_o},\ \eta\,\delta^2) \tag{21}$$

with some constant $C > 0$ and a natural number k_o. This can be seen from Taylor expansion about $\underline{h}^*$.

Therefore we define a "T-step" as follows: for an estimate $\underline{h}^{(\nu)}$ compute a direction $\tilde{\underline{d}}^{(\nu)}$ for which

$$\eta^{(\nu)} = \left(P\tilde{\underline{d}}^{(\nu)} \,,\, H(\underline{h}^{(\nu)})\, P\tilde{\underline{d}}^{(\nu)} \right) = \min_{\|\underline{d}\|=1} \left(P\underline{d} \,,\, H(\underline{h}^{(\nu)}) P\underline{d} \right). \tag{22}$$

The step length $\tilde{\delta}_\nu$ for this direction is chosen as $\tilde{\delta}_\nu = 2^{-\ell_o}$ and ℓ_o by

$$\varphi(2^{-\ell_o}) = \min\,\{\varphi(2^{-\ell}) \;:\; \varphi(2^{-\ell}) \leq F(\underline{h}^{(\nu)} + \delta_\nu P\underline{d}^{(\nu)})\}. \tag{18c}$$

Here each j has to satisfy (18b), and $\varphi(t)$ is defined by $\varphi(t) := F(\underline{h}^{(\nu)} + t\,P\tilde{\underline{d}}^{(\nu)})$. We remark that (18c) can be satisfied only if some T_i are non-positive. Therefore we combine the "T-steps" with the "normal steps" to obtain the following algorithm:

Given an estimate $\underline{h}^{(\nu)}$ for the solution $\underline{h}$, compute

(i) the vector $A^{(\nu)} \equiv \underline{A}(\underline{h}^{(\nu)})$ from (7) with $G = G(\underline{h}^{(\nu)})$, $Z = Z(\underline{h}^{(\nu)})$,

(ii) a direction of search $\underline{d}^{(\nu)}$ and its projection $P\underline{d}^{(\nu)}$ according to (16) such that (17) is satisfied, and a step length σ_ν according to (18a), (18b),

(iii) the new estimate $\underline{h}^{(\nu+1)}$ via formula (15), if all T_i formed in (12a) with respect to $\underline{A}^{(\nu)}$ are positive, otherwise

iv) a direction of search $\tilde{\underline{d}}^{(\nu)}$ according to (22), and the test (18c),

v) the new estimate $\underline{h}^{(\nu+1)}$ by setting $\underline{h}^{(\nu+1)} := \underline{h}^{(\nu)} + \tilde{\sigma}_\nu \tilde{\underline{d}}^{(\nu)}$

if the test is positive, otherwise define $\underline{h}^{(\nu+1)}$ as in iii).

We can now prove

Theorem 2: The above algorithm converges under the assumotion (18) for any starting vector $\underline{h}^{(o)} \in K$ to a strict local minimum of problem (9).

Proof By (19) all the iterates $\underline{h}^{(\nu)}$ lie in a compact subset of K. Hence there is a subsequence converging to some $\underline{h}^* \in [0,1]^n$ which we denote by $\{\underline{h}^{(j\nu)}\}_{\nu=1}^{\infty}$.

Following standard lines (cf. [2])it can be established that $\underline{h}^*$ is a critical point of $F(\underline{h})$. But we have still to show that $\underline{h}^*$ is a strict local minimum of (9). To this end we choose for $\varepsilon > 0$ suitable $\nu_o = \nu_o(\varepsilon)$ so large that for all $\varepsilon \geq \nu_o$

$$||\underline{h}^{(j\nu)} - \underline{h}^*|| \leq \varepsilon, \qquad |||H(\underline{h}^{(j\nu)}) - H(\underline{h}^*)||| \leq \gamma_1(\varepsilon), \quad \nu \geq \nu_o,$$

where $|||\cdot|||$ denotes the matrix norm on $\mathbb{R}^{n-1}$ and $\gamma_1(\varepsilon)$ is a constant tending to zero for $\varepsilon \to 0$. Then we can find $\gamma_2(\varepsilon)$ with $\gamma_2(\varepsilon) \to 0$ for $\varepsilon \to 0$ such that, for $\nu \geq \nu_o$,

$$F\left(\underline{h}^{(j\nu)} + \delta P \tilde{\underline{d}}^{(j\nu)}\right) \leq F\left(\underline{h}^* + \delta P \underline{d}^*\right) + \gamma_2(\varepsilon), \tag{23}$$

where $\underline{d}^*$ is defined via (20). This is a consequence of the fact that η and $\underline{d}^*$ depend continously on $\underline{h}^*$.

We assume then that $\underline{h}^*$ is not a strict local minimum of (9), i.e. (20) holds. Then it follows from (23) and (21) (for $\tilde{\sigma}_{j\nu} \leq \delta_o$) that

$$\varphi(2^{-\ell}) \equiv F\left(\underline{h}^{(j\nu)} + 2^{-\ell}\tilde{\underline{d}}^{(j\nu)}\right) \leq F(\underline{h}^*) + \max\left(-C2^{-\ell k_o}, \eta 4^{-\ell}\right) + \gamma_2(\varepsilon)$$

$$\leq F\left(\underline{h}^{(j\nu)} + \sigma_{j\nu} P\underline{d}^{(j\nu)}\right) + \max\left(-C2^{-\ell k_o}, \eta 4^{-\ell}\right) + \gamma_2(\varepsilon).$$

This means that in (18c) a finite step length $\tilde{\sigma}_{j\nu} = 2^{-\ell_o} \leq \delta_o/2$ will be chosen provided ε - and hence $\nu_o = \nu_o(\varepsilon)$ - are chosen such that $\gamma_2(\varepsilon) \leq (1/2)\min(C,-\eta)$. Then we obtain a contradiction since ℓ_o is such that

$$F\left(\underline{h}^{(j\nu)+1}\right) = \varphi(2^{-\ell_o}) \leq F(\underline{h}^*) + (1/2)\max(-c\,,\,\eta).$$

As an application of this theorem we state

Corollary: The above algorithm converges to a strict local minimum if one takes as direction of search $\underline{d}^{(\nu)}$ the direction of the projected gradient $Pg^{(\nu)}$.

3. Further methods. The proposed modification of the projected gradient method is not yet practical, since a "T step" requires too much work. This is due to the fact that the Hessian of $F(\underline{h})$ is given only implicitly, namely by (cf. [4])

$$\frac{\partial^2 F}{\partial h_i \partial h_j} = \left\langle \frac{\partial A}{\partial h_i}\,,\, G\,\frac{\partial A}{\partial h_j} \right\rangle + \delta_{ij}\, T_i h_i^{-3}\,,$$

where the vectors $\{\partial A/\partial h_i\}_{i=1}^{n-1}$ have to be determined by the systems

$$G(\partial \underline{A}/\partial h_i) = (\partial \underline{Z}/\partial h_i) - (\partial \underline{G}/\partial h_i)\,\underline{A}. \tag{24}$$

Thus the amount of work of a "T step" is n-times larger than that of a usual one if one assumes that the essential work is solving the $(n-2) \times (n-2)$ tridiagonal systems (with always the same matrix G). Namely, a "normal" step requires the solution of d such systems (see (5), (9) for the evaluation F(h)) whereas a "T step" needs (n-1)d additional solutions.

In the actual computation we used a simplified version of a "T step" based on the use of directions $\underline{e}^{(\ell)}$ defined via

$$e_i^{(\ell)} := \begin{cases} -1, & i = \ell, \\ 1, & i = -\ell, \\ 0 & \text{otherwise}, \end{cases}$$

and the following definition of $\tilde{d}^{(\nu)}$ in the above algorithm:

$$\underline{\tilde{d}}^{(\nu)} := \sum_{\ell \in I^{(\nu)}} \underline{e}^{(\ell)}, \quad I^{(\nu)} := \{\ell \in \{2,\ldots,n-1\} : T_\ell^{(\nu)} \le 0, \; T_{\ell-1}^{(\nu)} > 0\}, \tag{25}$$

where the $T_\ell^{(\nu)}$ are defined via (12a) with respect to $\underline{A}^{(\nu)} \equiv \underline{A}(\underline{h}^{(\nu)})$. The reason for the restriction $\ell \in I^{(\nu)}$ is that in the first order expansion $(h_\ell \equiv h_\ell^{(\nu)})$,

$$F\left(\underline{h}^{(\nu)} + \delta \underline{\tilde{d}}^{(\nu)}\right) \approx F\left(\underline{h}^{(\nu)}\right) + \delta \sum_{\ell \in I^{(\nu)}} \left[\alpha_\ell^{(\nu)} - \alpha_{\ell-1}^{(\nu)} + 2T_\ell^{(\nu)} h_\ell^{-2} - 2T_{\ell-1}^{(\nu)} h_{\ell-1}^{-2}\right],$$

the negative part with the $T_\ell^{(\nu)}$, $T_{\ell-1}^{(\nu)}$ should dominate.

We used direction (25) combined with the step length procedure (18b), (18c) in the above algorithm successfully despite the fact that we could not prove that it is a direction of descent near a critical point other than the solution of (1). Our experience was that it usually produced a larger decrease of F(h) than the (projected) direction of steepest descent. The effectivness of this technique is highlighted by the fact that only with its help the solution of (1) could be found in presence of another critical point (see example 2 below). Here one should remark that this statement is not quite precise in that even Theorem 2 only guarantees convergence to a strict local minimum of $F(\underline{h})$. However it is very likely (with homotypy arguments) that there is only one such minimum, namely the global one.

Another idea is to use Newton's method. We omit the details since each iteration would require the solution of (n-1)d tridiagonal systems for the computation of the Hessian (cf. (24)). Instead we used a less expensive variant based on a different parametrization of $F(\underline{h})$ which will be described shortly. Introducing

$$\underline{s}_i := s'(t_i), \quad \underline{S} := (\underline{s}_1, \dots, \underline{s}_n),$$

we obtain after some computation, instead of (8), the representation

$$\int_0^1 ||\underline{s}''(t)||^2 dt = 12\left[\sum_{i=1}^n ||\underline{w}_i||^2 h_i^{-3} - < \underline{S}, \underline{R} > + < \underline{S}, \Gamma \underline{S} >\right] \equiv \Psi(\underline{h}, \underline{S}), \tag{26}$$

where $\underline{R} := (\underline{r}_1, \dots, \underline{r}_n)$ is given by

$$\underline{r}_j := \begin{cases} \underline{w}_1 h_1^{-2} &, \quad j = 1 \\ \underline{w}_j h_j^{-2} + \underline{w}_{j-1} h_{j-1}^{-2}, & 2 \le j \le n-1 \\ \underline{w}_{n-1} h_{n-1}^{-2} &, \quad j = n \end{cases}$$

and $\Gamma = (\Gamma_{ij})$ denotes the symmetric $n \times n$ matrix

$$\Gamma := \frac{1}{3}\begin{pmatrix} 2h_1^{-1} & h_1^{-1} & & & \\ h_1^{-1} & \ddots & \ddots & & \\ & h_{i-1}^{-1} & 2(h_i^{-1} + h_{i-1}^{-1}) & h_i^{-1} & \\ & & \ddots & \ddots & \\ & & & h_{n-1}^{-1} & 2h_{n-1}^{-1} \end{pmatrix}. \tag{27}$$

Since any cubic spline function with $\underline{s}(t_i)$, $\underline{s}'(t_i)$ prescribed for $1 \le i \le n$ is feasible in the infimum of (1), this problem is equivalent to

$$\inf_{\underline{h} \in K, \underline{S} \in \mathbb{R}^{nd}} \Psi(\underline{h}, \underline{S}).$$

The equations for a critical point $\underline{h} \in K, \underline{S}$ of $\Psi(\underline{h}, \underline{S})$ read

$$\begin{aligned} 0 = \mu + \frac{\partial \Psi}{\partial h_j} \equiv \mu - \frac{36}{h_j^4}||\underline{w}_j||^2 + \frac{24}{h_j^3}(\underline{w}_j, \underline{s}_{j+1} + \underline{s}_j) - \\ - \frac{4}{h_j^2}\left[||\underline{s}_j||^2 + ||\underline{s}_{j+1}||^2 + (\underline{s}_j, \underline{s}_{j+1})\right], \end{aligned} \tag{28}$$

$$1 = \sum_{j=1}^{n-1} h_j, \tag{28b}$$

$$0 = \frac{\partial \Psi}{\partial \underline{S}} \equiv -12\underline{R} + 24\,\Gamma\,\underline{S}, \tag{28c}$$

with Lagrangian parameter μ. The notation $\partial\Psi/\partial\underline{S}$ stands for the gradient with respect to the variables in $\underline{S}$. It is well known ([1, chapter 4]) that (28c) is a defining system for the natural cubic interpolating spline curve in $\mathbb{R}^d$. Hence the solutions of equations (8), (11) in $\underline{h}$, $\underline{A}$ are same as those in (28).

In order to compute the Jacobian of (28) the $n \times d$ unknowns of these $n \times d$ equations are ordered as

$$h_1,\ldots,h_{n-1},h_n := \mu\,;\ \underline{S} = (\underline{S}_1,\ldots,\underline{S}_d)\ ;\ \underline{S}_\ell := (s_i^{(\ell)},\ldots,s_n^{(\ell)}) \in \mathbb{R}^n.$$

The associated $n \times d$ mappings are ($\underline{h}$ denotes now the old $\underline{h}$ enlarged by h_n)

$$\begin{aligned} \varphi_i(\underline{h},\underline{S}) &:= h_n + \partial\Psi/\partial h_i, \quad 1 \le i \le n-1, \\ \varphi_n(\underline{h},\underline{S}) &:= \sum_{i=1}^{n-1} h_i - 1, \quad i = n, \\ \phi_{i,\ell}(\underline{h},\underline{S}) &:= \partial\Psi/\partial s_i^{(\ell)}, \quad 1 \le i \le n,\ 1 \le \ell \le d. \end{aligned} \tag{29}$$

The Jacobian J of this mapping on $\mathbb{R}^{nd}$ into $\mathbb{R}^{nd}$ then has the block form

$$J = \begin{pmatrix} E & B_1 & \cdots & B_d \\ B_1^T & \Gamma & & \\ \vdots & & \ddots & \\ B_d^T & & & \Gamma \end{pmatrix}, \qquad E = \begin{pmatrix} \eta_1 & & & 1 \\ & \ddots & & \vdots \\ & & \eta_{n-1} & 1 \\ 1 & \cdots & 1 & 0 \end{pmatrix},$$

where the numbers η_j are given by

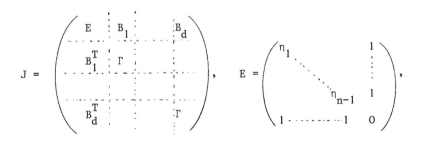

$$\eta_j = 144||\underline{w}_j||^2 h_j^{-5} - 72(\underline{w}_j, \underline{s}_j + \underline{s}_{j+1}) h_j^{-4} + 8h_j^{-3}\Big[||\underline{s}_j||^2 + ||\underline{s}_{j+1}|| + $$

$$+ \quad (\underline{s}_j, \underline{s}_{j+1}) \Big]. \tag{30}$$

The only nonzero entries of the matrices B_ℓ are

$$b_{ii}^{(\ell)} = 24 w_i^{(\ell)} h_i^{-3} - 8 s_i^{(\ell)} h_i^{-2} - 4 s_{i+1}^{(\ell)} h_i^{-2}, \qquad 1 \le i \le n-1,$$

$$b_{i,i+1}^{(\ell)} = 24 w_i^{(\ell)} h_i^{-3} - 8 s_{i+1}^{(\ell)} h_i^{-2} - 4 s_i^{(\ell)} h_i^{-2}, \qquad 1 \le i \le n-1,$$

and the matrix Γ is defined by (27). With the knowledge of J the Newton method for solving the system (30) can be set up in the standard manner. We omit the details due to the lack of space. We only mention that by using the block structure of J the solution of the linear equations with matrix J can be reduced to the solution of d systems with matrix Γ and one system with matrix $E - \sum_{\nu=1}^{d} B_\nu \Gamma^{-1} B_\nu^T$.

In the actual computation we simplified J by setting the matrices $B_1^T, \ldots, B_d^T$ equal to zero. Then the solution of the resulting system requires the solution of d systems with Γ and one system with E. Thus the amount of work for computing a new direction of search is comparable to that of the projected gradient method above. We then formed a descent method with this direction of search and step length procedure as above together with "T steps" in the version of (25). We could not prove global convergence of this Quasi-Newton method. But we have at least local convergence in view of the relation

$$\eta_j = 12 T_j h_j^{-3} + 8 h_j^{-3}\Big[||\underline{s}_j||^2 + ||\underline{s}_{j+1}||^2 + (\underline{s}_j, \underline{s}_{j+1}) \Big],$$

which follows from (30) and (12a). Specifically in view of (14), all these numbers are strictly positive near the solution and hence E has a bounded inverse.

4. Numerical results. Many examples were computed in order to

compare the behaviour of the projected gradient method (PG) with that of the Quasi-Newton method (QN). As starting points for the algorithms we chose either the uniform parametrization (U) with knots $h_i = 1/(n-1)$ or the normalized accumulated chord length parametrization (CL).

It was observed that for the PG method a start with CL parametrization often took fewer iterations for convergence, whereas for the QN method this happened for the U parametrization. Concerning the dependence of the data it turned out that for "smooth" data (cf. the data in S-form in [3]) the QN method often needed more iterations than the PG method but even fewer function evaluations (which correspond to the number of solutions of tridiagonal systems). For data with isolated corners or peaks the QN method seems to be generally superior. The following table illustrates this (# It denote the number of iterations, # F the number of function evaluations, the tolerance for the error was chosen as 10^{-4}):

	CL		U	
QN method	# It	41	# It	114
	# F	130	# F	674
PG method	# It	332	# It	555
	# F	2737	# F	4481

Example 2

The following two examples (figures 1, 2 and 3) are chosen to illustrate the qualitative behaviour of the different parametric spline interpolants. The curved marked by ... uses the uniform, - .. - the CL and - the "optimal" parametrization. The small circles mark the data points (in figures 1 and 2 the four lower corner points are nearly identical).

The first example is particularly interesting since it exhibits the existence of a critical (or stationary) point other than the global minimum. The PG method without "T steps" terminated at a point with T_i's alternating in sign (figure 2) and at a point with

FIGURE 1 : THE MINIMUM

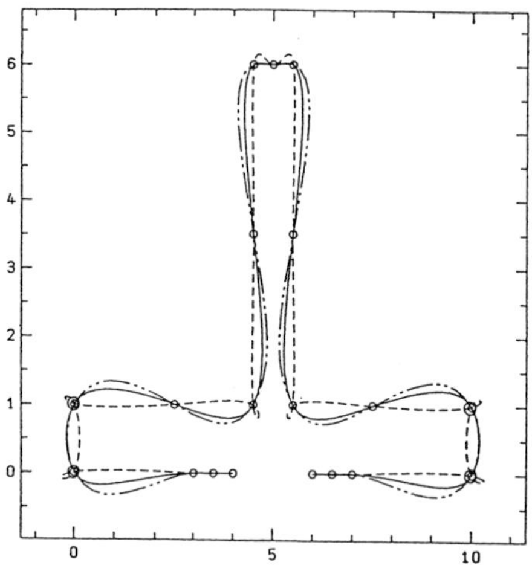

FIGURE 2 : A STATIONARY POINT

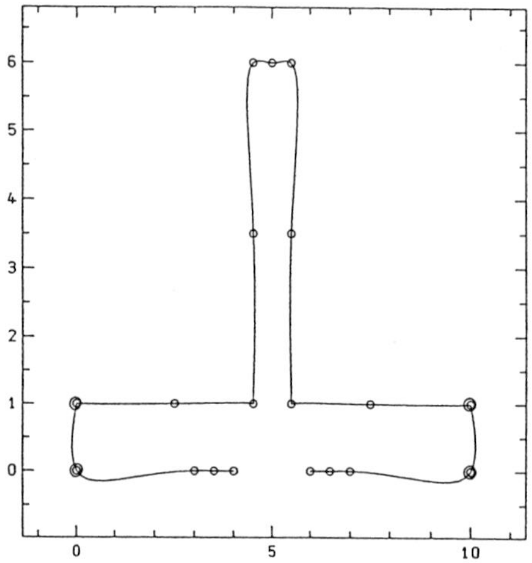

FIGURE 3

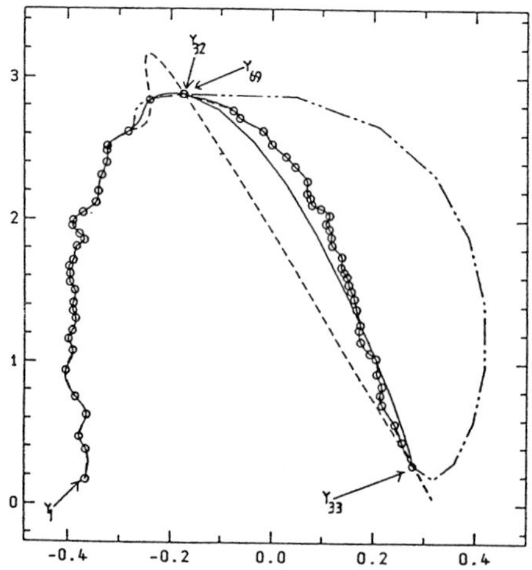

Table 1

Data of Figure 1 and Figure 2

i	y_i
1	(-1.0 , 0.0)
2	(-1.5 , 0.0)
3	(-2.0 , 0.0)
4	(-4.999 , 0.0)
5	(-5.0 , 0.001)
6	(-5.0 , 0.999)
7	(-4.999 , 1.0)
8	(-0.501 , 1.0)
9	(-0.5 , 1.001)
10	(-0.5 , 3.5)
11	(-0.5 , 6.0)
12	(0.0 , 6.0)
:	:

positive T_i if "T steps" were used (figure 1). According to (14) only the latter point is the global minimum.

The second example (figure 3) illustrates very clearly the influence of the parametrization on the resulting spline interpolant. The data points $y_1,\dots,y_{32}$ and $y_{33},\dots,y_{69}$ follow closely each other so that the curve is well described by them. Consequently the parametrization has nearly no effect; all three curves are essentially identical. However, between y_{32} and y_{33} there is a large gap in the data which provide therefore only poor information. Here the use of sound parametrization turns out to be essential.

References

[1] De Boor, C. (1978),'A practical guide to splines', Springer, New York.

[2] Gill, P.E., Murray, W., Wright, M.H. (1981), 'Practical Optimization', Academic Press, London-New York.

[3] Martin, S. (1984), 'An approach to data parametrization in parametric cubic spline-interpolation problems', J. Approx. Theory 41 (1984), 64-86.

[4] Scherer, K. (1988), 'Uniqueness of best parametric interpolation by cubic spline curves', Preprint, University of Bonn.

[5] Toepfer, H.J. (1981), 'Models for curve fitting', in: Numer. Methoden der Approximationstheorie 6 (L. Collatz ed.) ISNM 59, Birkhäuser, Basel.

4. Smoothing and Constraint Methods

Data fitting by penalized least squares

M. Von Golitschek

Institut für Angewandte Mathematik, Universität Würzburg, FRG

L. L. Schumaker

Department of Mathematics, Vanderbilt University, Nashville, Tennesee, USA

Abstract: In this paper we deal with the problem of approximating an unknown smooth function, given a set of data consisting of measurements on the function. We are especially interested in the case where the number of data is fairly large, and is subject to considerable noise. We focus on a method for solving this problem, called the method of *penalized least squares*, which, for some reason, does not seem to have received much attention in the Approximation Theory literature, despite the fact that it often seems to work much better than several other more commonly used methods. Our aim is to give a complete and self-contained treatment of penalized least squares, including a discussion of how to choose the smoothing parameter.

Key words: Data fitting, smoothing, least squares, penalized least squares

1. Introduction

Suppose f is an unknown function of one or more variables, and that we are given a set of measurements

$$z_i = \lambda_i f + \epsilon_i, \quad i = 1, \ldots, n, \tag{1}$$

where $\lambda_1, \ldots, \lambda_n$ are linear functionals defined on some linear space $\mathcal{F}$ containing f, and where $\epsilon_1, \ldots, \epsilon_n$ are measurement errors. Typically the λ_i will be point-evaluation of the function or one of its derivatives, although other functionals are also of interest.

Our aim is to approximate f using a finite dimensional subspace $\mathcal{S}$ of $\mathcal{F}$ spanned by functions $B_1, \ldots, B_k$, where $k \le n$. For each $c := (c_1, \ldots, c_k)^T$, let

$$s_c = \sum_{i=1}^{k} c_i B_i. \tag{2}$$

The problem is to devise some reasonable means of choosing the coefficients. Perhaps the best-known approach is the classical method of *least-squares* whereby we minimize

$$E_z(c) = \frac{1}{n}\sum_{i=1}^{n}[\lambda_i s_c - z_i]^2 \tag{3}$$

over all choices of c.

While straightforward least-squares fitting is often appropriate, for some data fitting problems it produces a function which is not sufficiently smooth (for an example, see [11]). In such cases, it may be better to look for a function in $\mathcal{S}$ which minimizes a combination of the goodness of fit (3) with an appropriate measure of smoothness of the fitting function.

Many reasonable measures of smoothness of the function s_c can be expressed in the form

$$J(c) = c^T E c, \tag{4}$$

where E is a symmetric nonnegative-definite $k \times k$ matrix. For typical examples in univariate and bivariate data fitting, see Section 5 and Remark 5, respectively.

Suppose now that $\lambda \geq 0$ is given and that $E \neq 0$. Then for each $c \in \mathbb{R}^k$, let

$$\rho_\lambda(c) = \lambda c^T E c + \frac{1}{n}\sum_{i=1}^{n}[\lambda_i s_c - z_i]^2. \tag{5}$$

The second term in (5) is just the mean square error in approximating the data by s_c. The first term can be thought of as a *penalty term* which, with the appropriate choice of E, measures the smoothness of s_c. The parameter λ can be regarded as a *smoothing parameter*.

Definition 1. *The penalized least-squares fit of the function f based on data z is the function $s_{\lambda,z} := s_{c(\lambda)}$ corresponding to the choice of $c(\lambda)$ which minimizes the functional $\rho_\lambda(c)$ defined in (5).*

While the use of such penalty terms seems to be well established practice in statistical data fitting as well as in regularization of ill-posed problems, the method seems to have received much less attention in Approximation Theory circles. For more on the history of penalized least squares methods, see Remark 1 in Section 6 below.

The remainder of this paper is organized as follows. In the following section we establish the existence and uniqueness of solutions to the penalized least squares problem, and also discuss the special cases when $\lambda = 0$ and when λ approaches ∞. In Section 3 we discuss the connection between the choice of the smoothing parameter λ and the goodness of fit. Section 4 is devoted to an automatic procedure for selecting the smoothing parameter λ called *generalized cross validation*. In Section 5 we give details of how penalized least squares can be applied to fit data in the univariate case using splines. The last section is devoted to remarks.

2. Existence and uniqueness

Throughout the remainder of this paper we assume that the linear functionals $\lambda_1, \ldots, \lambda_n$ and basis functions $B_1, \ldots, B_k$ are such that for some $1 \le \nu_1 < \cdots < \nu_k \le n$,

$$\det\,(\lambda_{\nu_i} B_j)_{i,j=1}^k \neq 0. \tag{6}$$

Under this basic hypothesis, we can now establish the following existence and uniqueness theorem for penalized least-squares fitting.

Theorem 2. *For any $\lambda \ge 0$, there exists a unique vector $c(\lambda)$ minimizing the functional $\rho_\lambda(c)$ in (5). In particular, $c(\lambda)$ is the unique solution of the system*

$$(B^T B + n\lambda E)c(\lambda) = B^T z, \tag{7}$$

where $z = (z_1, \ldots, z_n)^T$ and B is the $n \times k$ matrix with entries

$$B_{ij} = \lambda_i B_j, \quad i = 1, \ldots, n; \; j = 1, \ldots, k.$$

Proof: Setting the gradient of the function $\rho_\lambda(c)$ equal to 0 leads immediately to the system of equations (7). The basic hypothesis (6) assures that the matrix $G := B^T B$ is symmetric, positive definite, and nonsingular. Since E is assumed to be symmetric nonnegative-definite, we conclude that for every $\lambda \ge 0$, $G + n\lambda E$ is also symmetric positive definite. ■

As in the case of the classical smoothing splines, clearly the size of the smoothing parameter λ has a major impact on how well the function $s_{\lambda,z}$ fits the function f which generated the data. In Sections 3 and 4 below we shall discuss how to select λ effectively. In the remainder of this section we discuss the nature of the fit in the extreme cases when $\lambda = 0$ and when λ approaches ∞.

When $\lambda = 0$, it is clear that the functional $\rho_\lambda(c)$ is just the mean square error when fitting the data using functions of the form (2). Hence this case corresponds to classical least-squares fitting using the space $\mathcal{S} = \text{span}\,\{B_i\}_1^k$, and the system (7) is just the usual set of normal equations. If $k = n$, then the least-squares fit is simply the function in $\mathcal{S}$ which interpolates the data.

The situation as λ approaches ∞ is somewhat more complicated. First we prove a lemma concerning the behavior of the inverse of the matrix $G + n\lambda E$, where $G = B^T B$.

Lemma 3. *For all $\lambda \ge 0$, the entries of $(G + n\lambda E)^{-1}$ are uniformly bounded by $1/\alpha$, where α is the smallest eigenvalue of G.*

Proof: Let x be the rth column of $(G + n\lambda E)^{-1}$. Then $(G + n\lambda E)x = e_r$, where e_r is the vector in R^k with all zero entries except for the rth. It follows that $x^T(G + n\lambda E)x = x^T e_r = x_r$. Now by the nonnegative definiteness of E, we conclude that $x^T G x \le x_r$. On the other hand, by the positive definiteness of G, the smallest eigenvalue α of G satisfies

$$0 < \alpha = \min_{y \neq 0} \frac{y^T G y}{y^T y}.$$

It follows that $\alpha x^T x \leq x^T G x \leq x_r$, and thus $\alpha x_j^2 \leq x_r$, for all $j = 1, \ldots, k$. This inequality asserts in particular that $0 \leq x_r \leq 1/\alpha$, and it immediately follows that $|x_j| \leq 1/\alpha$, for all $j = 1, \ldots, k$. Since r was an arbitrary integer satisfying $1 \leq r \leq k$, this completes the proof. ■

We can now examine the behavior as $\lambda \to \infty$ of the coefficient vector $c(\lambda)$ which minimizes the functional $\rho_\lambda(c)$ in (5).

Lemma 4. *There exists a vector $c(\infty)$ in R^k such that*

$$\lim_{\lambda \to \infty} c(\lambda) = c(\infty).$$

Moreover, $Ec(\infty) = 0$.

Proof: Clearly, $c(\lambda) = (G + n\lambda E)^{-1} B^T z$. Since the entries of $(G + n\lambda E)^{-1}$ are rational functionals of λ and are uniformly bounded for all $\lambda \geq 0$, it follows that $(G + n\lambda E)^{-1}$ converges to some matrix L as $\lambda \to \infty$. But then $c(\lambda)$ approaches $LB^T z$ as $\lambda \to \infty$. Finally, since

$$\left(\frac{G}{n\lambda} + E \right) c(\lambda) = \frac{B^T z}{n\lambda},$$

it follows that $Ec(\infty) = 0$. ■

Theorem 5. *The function $s_{c(\infty)}$ is the least-squares fit of the data from the following subspace of $\mathcal{S}$:*

$$\mathcal{S}_E = \left\{ \sum_{i=1}^{k} c_i B_i \ : \ Ec = 0 \right\}. \tag{8}$$

Proof: Since

$$Gc(\lambda) + n\lambda Ec(\lambda) = B^T z,$$

and the first term converges as $\lambda \to \infty$, we conclude that $n\lambda Ec(\lambda)$ converges to $u := B^T z - Gc(\infty)$. The matrices E and (E, u) have the same rank, and it follows that $u = Ex$ for some element Ex with x in R^k. Thus we have

$$\begin{bmatrix} G & E \\ E & 0 \end{bmatrix} \begin{bmatrix} c(\infty) \\ x \end{bmatrix} = \begin{bmatrix} B^T z \\ 0 \end{bmatrix}.$$

But this is precisely the system of equations which uniquely determines the least-squares fit of the data from the subspace $\mathcal{S}_E$. ■

3. Properties of the fit

In this section we discuss the behavior of the fit as we vary the smoothing parameter λ. Given measurements as in (1), let $s_{\lambda,z}$ be the penalized least-squares fit of f constructed from the noisy data vector z. Then the mean-square error using this fit is given by

$$T_z(\lambda) = \frac{1}{n}\sum_{i=1}^{n}[\lambda_i(s_{\lambda,z} - f)]^2. \tag{9}$$

We begin by considering the case where we have exact measurements; i.e., where the errors $\epsilon_1, \ldots, \epsilon_n$ in (1) are all zero. In this case we want to consider $T_{\mathbf{f}}(\lambda)$, where $\mathbf{f} = (\lambda_1 f, \ldots, \lambda_n f)^T$.

Theorem 6. *The function $T_{\mathbf{f}}(\lambda)$ is monotone increasing for $\lambda \geq 0$ with $\dot{T}_{\mathbf{f}}(0) = 0$ and $\lim_{\lambda\to\infty} \dot{T}_{\mathbf{f}}(\lambda) = 0$.*

Proof: First we establish the monotonicity of $T_{\mathbf{f}}(\lambda)$. For exact measurements $z = \mathbf{f}$, the minimum of the expression ρ_λ in (5) is attained at $c_{\mathbf{f}}(\lambda)$, and is given by

$$\rho_\lambda(c_{\mathbf{f}}(\lambda)) = \lambda c_{\mathbf{f}}(\lambda)^T E c_{\mathbf{f}}(\lambda) + T_{\mathbf{f}}(\lambda).$$

Similarly, for all $\tilde{\lambda} > \lambda > 0$,

$$\rho_{\tilde{\lambda}}(c_{\mathbf{f}}(\tilde{\lambda})) = \tilde{\lambda} c_{\mathbf{f}}(\tilde{\lambda})^T E c_{\mathbf{f}}(\tilde{\lambda}) + T_{\mathbf{f}}(\tilde{\lambda}).$$

For ease of notation, we now define $\alpha = c_{\mathbf{f}}(\lambda)^T E c_{\mathbf{f}}(\lambda)$, $\tilde{\alpha} = c_{\mathbf{f}}(\tilde{\lambda})^T E c_{\mathbf{f}}(\tilde{\lambda})$, $\beta = T_{\mathbf{f}}(\lambda)$, and $\tilde{\beta} = T_{\mathbf{f}}(\tilde{\lambda})$. Then the minimality of $c_{\mathbf{f}}(\lambda)$ and $c_{\mathbf{f}}(\tilde{\lambda})$, respectively, imply that

$$\lambda\alpha + \beta \leq \lambda\tilde{\alpha} + \tilde{\beta} \quad \text{and} \quad \tilde{\lambda}\tilde{\alpha} + \tilde{\beta} \leq \tilde{\lambda}\alpha + \beta.$$

It follows that

$$\lambda\alpha + \tilde{\lambda}\tilde{\alpha} \leq \lambda\tilde{\alpha} + \tilde{\lambda}\alpha \quad \text{and} \quad \tilde{\lambda}(\tilde{\alpha} - \alpha) \leq \lambda(\tilde{\alpha} - \alpha),$$

and thus that $\tilde{\alpha} \leq \alpha$. Since $\lambda(\alpha - \tilde{\alpha}) \leq \tilde{\beta} - \beta$, we conclude that $\tilde{\beta} \geq \beta$ which asserts that $T_{\mathbf{f}}(\lambda)$ is monotone increasing.

To establish the assertions about the derivative of $T_{\mathbf{f}}(\lambda)$, we need an explict formula for it. To this end we introduce the $n \times n$ *influence matrix*

$$A(\lambda) = B(G + n\lambda E)^{-1}B^T. \tag{10}$$

By the properties of the matrices appearing in (10), it follows immediately that $A(\lambda)$ is symmetric positive definite. In addition, we claim that $A(\lambda)$ provides the connection between the data vector z, and the associated values of the penalized least squares fit; i.e.,

$$\begin{bmatrix} \lambda_1 s_{\lambda,z} \\ \vdots \\ \lambda_n s_{\lambda,z} \end{bmatrix} = A(\lambda)z. \tag{11}$$

Indeed, $(\lambda_1 s_{\lambda,z}, \ldots, \lambda_n s_{\lambda,z})^T = Bc(\lambda) = B(G + n\lambda E)^{-1} B^T z = A(\lambda)z$.

Now it is clear that

$$T_{\mathbf{f}}(\lambda) = \frac{1}{n}[A(\lambda)\mathbf{f} - \mathbf{f}]^T [A(\lambda)\mathbf{f} - \mathbf{f}].$$

From this we compute that

$$n[T_{\mathbf{f}}(\lambda) - T_{\mathbf{f}}(0)] = \mathbf{f}^T [A(\lambda)^T A(\lambda) - 2A(\lambda) - A(0)^T A(0) + 2A(0)]\mathbf{f}.$$

Using (10), we arrrive at the formula

$$T_{\mathbf{f}}(\lambda) = T_{\mathbf{f}}(0) + n\lambda^2 c_{\mathbf{f}}(\lambda)^T E G^{-1} E c_{\mathbf{f}}(\lambda). \tag{12}$$

Forming the difference quotient of $T_{\mathbf{f}}(\lambda)$ at 0 and using the continuity of $c_{\mathbf{f}}(\lambda)$, (as shown in the proof of Lemma 4), it immediately follows that the derivative at $\lambda = 0$ is zero as asserted. The assertion about the derivative as $\lambda \to \infty$ follows from the boundedness of $T_{\mathbf{f}}(\lambda)$. ■

It is now clear that to minimize the error in the approximation when fitting exact data, we should choose the value of the smoothing parameter λ to be zero. We should note, however, that for a certain class of functions $\mathcal{F}_E$ which we are about to define, $T_{\mathbf{f}}(\lambda)$ is constant, and hence all values of λ are equally as good. The class is

$$\mathcal{F}_E = \{f \ : \ d(f, \mathcal{S}) = d(f, \mathcal{S}_E)\}, \tag{13}$$

where

$$d(f, \mathcal{S}) = \inf_{s \in \mathcal{S}} \frac{1}{n} \sum_{i=1}^{n} [\lambda_i (s - f)]^2, \tag{14}$$

and $d(f, \mathcal{S}_E)$ is defined similarly. Indeed, in this case, $T_{\mathbf{f}}(\lambda)$ has the same value for both $\lambda = 0$ and for $\lambda = \infty$. Thus, since $T_{\mathbf{f}}(\lambda)$ is monotone increasing, it must be a constant.

In practice we almost always have noisy measurements, and as we shall see shortly, in this case it will be more advantageous to take $\lambda > 0$. To analyze the noisy case, we need to make some basic assumption about the nature of the noise. Throughout the remainder of this section we suppose that the errors $\epsilon_1, \ldots, \epsilon_n$ in the measurements (1) are independent, identically distributed random variables with mean 0 and variance σ^2.

It is easy to see that

$$nT_z(\lambda) = nT_{\mathbf{f}}(\lambda) + 2\epsilon^T A(\lambda)^T [A(\lambda)\mathbf{f} - \mathbf{f}] + \epsilon^T A(\lambda)^T A(\lambda)\epsilon.$$

Taking the *expected value* of this expression gives

$$\mathcal{E}T_z(\lambda) = T_{\mathbf{f}}(\lambda) + \frac{\sigma^2 \text{trace}\ (A^2(\lambda))}{n}. \tag{15}$$

Before analyzing this expression further, we need some information on the trace of the influence matrix $A(\lambda)$. Let $0 \le \kappa_1 \le \kappa_2 \le \cdots \le \kappa_k$ be the eigenvalues of $G^{-1}E$, and suppose that $\{v_1, \ldots, v_k\}$ are the corresponding eigenvectors. We suppose that $\kappa_1 = \cdots = \kappa_d = 0 < \kappa_{d+1}$. This is equivalent to assuming that the dimension of the space $\mathcal{S}_E$ defined in (8) is d.

Lemma 7. *For $0 \leq \lambda < \infty$ the matrix $A(\lambda)$ has exactly k non-zero eigenvalues*

$$\mu_j = \frac{1}{1+n\lambda\kappa_j}, \quad j = 1, \ldots, k.$$

For $\lambda = \infty$, the eigenvalues of $A(\lambda)$ consist of 0 with multiplicity $n-d$ and 1 with multiplicity d.

Proof: We first consider the case where $0 \leq \lambda < \infty$. Since for all z, the vector $A(\lambda)z$ lies in a k-dimensional subspace of $\mathbb{R}^n$, the first $n-k$ eigenvalues of A are all 0. Suppose now that $\lambda = 0$, and fix $1 \leq j \leq k$. Since the least squares fit of the function B_j is B_j itself, it follows that $(\lambda_1 B_j, \ldots, \lambda_n B_j)^T$ is an eigenvector of $A(0)$ corresponding to eigenvalue 1. This proves that the k nonzero eigenvalues of $A(0)$ are all 1.

We now treat the case where $0 < \lambda < \infty$. Suppose $v \in \mathbb{R}^k$ is an eigenvector of $G^{-1}E$ corresponding to $\kappa > 0$; i.e., $G^{-1}Ev = \kappa v$. Let $\mu = \frac{1}{1+n\lambda\kappa}$ and suppose $u \in \mathbb{R}^n$ is such that $G^{-1}B^T u = v$. Let $w = Au$. Since B is of rank k,

$$Au = 0 \qquad \text{iff} \qquad B^T u = 0. \tag{16}$$

The fact that $v \neq 0$ coupled with (16) assures that $w \neq 0$. Now it is easy to see that the following equalities are equivalent:

$$\kappa v = G^{-1}Ev.$$

$$Gv = \mu Gv + n\lambda\mu Ev$$

$$Gv = \mu(G + n\lambda E)v$$

$$(G + n\lambda E)^{-1}B^T u = \mu G^{-1}B^T u$$

$$G(G + n\lambda E)^{-1}B^T u = \mu B^T u$$

$$B^T(Au - \mu u) = 0$$

$$A(Au - \mu u) = 0$$

$$Aw = \mu w.$$

This establishes that μ is an eigenvector of A. To complete the proof we have to consider the possibility that κ is an eigenvalue of $G^{-1}E$ of multiplicity m. Suppose $v_1, \ldots, v_m$ are linear independent eigenvectors of $G^{-1}E$ associated with κ. For each $i = 1, \ldots, m$, let $u_i \in \mathbb{R}^n$ be such that $G^{-1}B^T u_i = v_i$, and let $w_i = Au_i$. Now we claim that the w_i are linearly independent eigenvectors of A. Indeed, if $\alpha_1 w_1 + \cdots + \alpha_m w_m = 0$, then $\alpha_1 Au_1 + \cdots + \alpha_m Au_m = 0$, and by (16), it follows that $\alpha_1 G^{-1}B^T u_1 + \cdots + \alpha_m G^{-1}B^T u_m = \alpha_1 v_1 + \cdots + \alpha_m v_m = 0$. But by the linear independence of the v's, this implies $\alpha_1 = \cdots = \alpha_m = 0$, and we have established the asserted linear independence of the w_i's.

It remains to deal with the case $\lambda = \infty$. In this case, we are performing least squares in $\mathcal{S}_E$, and hence s_λ lies in a d dimensional subspace of $\mathbb{R}^n$. It follows that $n-d$ of the eigenvalues of $A(\infty)$ must be zero. On the other hand, since each of the d basis functions of $\mathcal{S}_E$ is approximated exactly, just as in the case $\lambda = 0$ we see that 1 is an eigenvalue of $A(\infty)$ of multiplicity d. ■

For later use, we note that since the trace of a matrix is equal to the sum of its eigenvalues, for $0 \leq \lambda < \infty$,

$$t(\lambda) := \text{trace } A(\lambda) = \sum_{i=1}^{k} \left(\frac{1}{1+n\lambda\kappa_i} \right), \tag{17}$$

and, since the eigenvalues of $A^2(\lambda)$ are the squares of the eigenvalues of $A(\lambda)$,

$$t_2(\lambda) := \text{trace } A^2(\lambda) = \sum_{i=1}^{k} \left(\frac{1}{1+n\lambda\kappa_i} \right)^2. \tag{18}$$

In particular, $t(0) = t_2(0) = k$. We also note that $t(\infty) = t_2(\infty) = d$.

Theorem 8. *The function $\phi(\lambda) := \mathcal{E}T_z(\lambda)$ has the value $\mathcal{E}T_z(0) = T_{\mathbf{f}}(0) + k\sigma^2/n$, and asymptotically approaches the value $T_{\mathbf{f}}(\infty) + d\sigma^2/n$ as $\lambda \to \infty$. Its derivative at $\lambda = 0$ is negative. Moreover, for all $\lambda \geq 0$,*

$$\mathcal{E}T_z(\lambda) - \mathcal{E}T_z(0) \geq \frac{\sigma^2}{n}[t_2(\lambda) - t_2(0)], \tag{19}$$

where $t_2(\lambda)$ is defined in (18).

Proof: Putting $t_2(0) = k$ and $t_2(\infty) = d$ in (15), we immediately obtain the asserted values of ϕ at $\lambda = 0$ and $\lambda = \infty$. By (15),

$$\dot{\phi}(\lambda) = \dot{T}_{\mathbf{f}}(\lambda) + \frac{\sigma^2 \dot{t}_2(\lambda)}{n}. \tag{20}$$

Now since

$$\dot{t}_2(\lambda) = -2\sum_{i=1}^{k} \frac{n\kappa_i}{(1+n\lambda\kappa_i)^3}$$

and $\dot{T}_{\mathbf{f}}(0) = 0$ by Theorem 6, we immediately deduce that $\dot{\phi}(0) < 0$ as asserted.

Finally, to prove (19), we note that by (12) and (15),

$$\mathcal{E}T_z(\lambda) - \mathcal{E}T_z(0) = n\lambda^2 c_{\mathbf{f}}(\lambda)^T E G^{-1} E c_{\mathbf{f}}(\lambda) + \frac{\sigma^2[t_2(\lambda) - t_2(0)]}{n}.$$

Since the first term on the right is nonnegative, the result follows. ■

It is clear from (20) that in the case where $f \in \mathcal{F}_E$, the function $\mathcal{E}T_z(\lambda)$ is monotone decreasing since, as observed earlier, $T_{\mathbf{f}}(\lambda)$ is constant in this case while $\dot{t}_2(\lambda) < 0$.

4. Choosing the smoothing parameter λ

In the previous section, we have examined the connection between the choice of the smoothing parameter and the size of the true error or expected value of the true error. We can summarize our observations as follows:

Case 1: (The noise is zero) In this case, in order to minimize the true error of fit, we should take $\lambda = 0$; i.e., we should perform least squares using the space $\mathcal{S}$.

Case 2: (Noisy data) In this case it is reasonable to choose λ to minimize the expected value $\mathcal{E}T_z(\lambda)$ of the true error. There are two subcases. If $\mathcal{E}T_z(\lambda)$ is monotone decreasing on $[0, \infty)$, then we should take $\lambda = \infty$. On the other hand, if $\mathcal{E}T_z(\lambda)$ is not monotone decreasing on $[0, \infty)$, then since by Theorem 8, $\mathcal{E}T_z(\lambda)$ has a negative derivative at $\lambda = 0$, there is at least one value of λ in $(0, \infty)$ which minimizes $\mathcal{E}T_z(\lambda)$. Let λ_n^T be the smallest choice of λ which provides a minimum.

In the remainder of this section we restrict our attention to the case of noisy data. Our immediate aim is to discuss an automatic method, called *generalized cross validation*, for computing a reasonable estimate for λ_n^T. Given a data vector z as in (1), let $s_{\lambda,z}$ be the associated penalized least-squares fit, and let $A(\lambda)$ be the influence matrix defined in (10). Then we define the associated *generalized cross validation function* as

$$V(\lambda) = \frac{E_z(\lambda)}{[1 - t(\lambda)/n]^2}, \tag{21}$$

where

$$E_z(\lambda) = \frac{1}{n}\sum_{i=1}^{n}(\lambda_i s_{\lambda,z} - z_i)^2, \tag{22}$$

and $t(\lambda)$ is the trace of $A(\lambda)$ as given in (17).

Lemma 9. *The function $E_z(\lambda)$ is monotone increasing for $\lambda \geq 0$ with $\dot{E}_z(0) = 0$ and $\lim_{\lambda \to \infty} \dot{E}_z(\lambda) = 0$.*

Proof: The proof of this lemma is virtually the same as the proof of Theorem 6, and we omit the details. ∎

Lemma 9 does not assert that $E_z(\lambda)$ is *strictly* monotone increasing. For some z it can happen that $E_z(\lambda)$ is constant for all $0 \leq \lambda < \infty$. We now show that $V(\lambda)$ has properties very similar to those of $\mathcal{E}T_z(\lambda)$.

Theorem 10. *The function $V(\lambda)$ has value $E_z(0)/[1 - k/n]^2$ at 0 and asymptotically approaches the value $E_z(\infty)/[1 - d/n]^2$ as $\lambda \to \infty$. Moreover, $V(\lambda)$ has a negative derivative at $\lambda = 0$.*

Proof: The first assertions follow immediately from the properties of $t(\lambda)$. To prove the last assertion, we note that

$$\dot{V}(\lambda) = \frac{n^2(n - t(\lambda))\dot{E}_z(\lambda) + 2n^2 E_z(\lambda)\dot{t}(\lambda)}{[n - t(\lambda)]^3}.$$

The assertions now follow from the properties of $E_z(\lambda)$ and the fact that

$$\dot{t}(\lambda) = -\sum_{i=1}^{k} \frac{n\kappa_i}{(1+n\lambda\kappa_i)^2}. \quad \blacksquare$$

Theorem 10 implies that either $V(\lambda)$ is monotone decreasing on $[0,\infty)$ in which case we can think of its minimum as occuring at $\lambda_n^V := \infty$, or there is at least one value of λ in $(0,\infty)$ where E_z has an absolute minimum, in which case we define λ_n^V to be the smallest choice of λ which works.

The process of computing the quantity λ_n^V as an approximation to λ_n^T is called the *method of generalized cross validation.* To implement it numerically, we have to find the minimum of the validation function $V(\lambda)$. The standard approach to doing this is to compute $V(\lambda)$ at several choices of λ, and then use a search procedure. For each choice of λ, the bulk of the computational effort typically goes into finding $t(\lambda)$ = trace $A(\lambda)$. In those cases where the matrix B has a band structure (as happens for example when using univariate splines – see Section 5 below), the trace can be computed efficiently using the LU decomposition of the matrix $(G + n\lambda E)$, see [13,21]. In the general case it may be more efficient to resort to finding the singular value decomposition of $(G + n\lambda E)$, which then leads to an explicit expression for $t(\lambda)$, see [2].

From a theoretical standpoint, it is of interest to consider the expected value of the validation function V. Since

$$\mathcal{E}E_z(\lambda) = T_{\mathbf{f}}(\lambda) + \frac{\sigma^2}{n} t_3(\lambda),$$

where

$$t_3(\lambda) = \text{trace } (A - I)^2,$$

we obtain

$$\mathcal{E}V(\lambda) = \frac{T_{\mathbf{f}}(\lambda) + \frac{\sigma^2 t_3(\lambda)}{n}}{[1 - \frac{t(\lambda)}{n}]^2}.$$

The following theorem, whose proof is straightforward, shows that $\mathcal{E}V(\lambda)$ has properties very similar to those of both $\mathcal{E}T_z(\lambda)$ and $V(\lambda)$.

Theorem 11. *The function $\mathcal{E}V(\lambda)$ has the value $\frac{T_{\mathbf{f}}(0)}{[1-\frac{k}{n}]^2}$ at 0, and asymptotically approaches the value $\frac{T_{\mathbf{f}}(\infty)}{[1-\frac{d}{n}]^2}$ as $\lambda \to \infty$. Its derivative at $\lambda = 0$ is negative.*

Theorem 11 shows that either the function $\mathcal{E}V(\lambda)$ is monotone decreasing for all λ, or there is some value of $0 < \lambda < \infty$ where it has a minimum. We denote the first such point by λ_n^{EV}. In practice, we cannot compute this value because we have no way of computing either $T_{\mathbf{f}}(\lambda)$ or σ. (Indeed, if we could compute these quantities, we could find the minimum of $\mathcal{E}T_z(\lambda)$ directly). We have the following interesting theorem concerning λ_n^{EV}.

Theorem 12. *The quantity* λ_n^{EV} *is an asymptotically optimal estimate of the best smoothing parameter* λ_n^T *in the sense that*

$$1 \leq \frac{\mathcal{E}T_z(\lambda_n^{EV})}{\mathcal{E}T_z(\lambda_n^T)} \to 1 \tag{23}$$

as $n \to \infty$.

Proof: As in [23], it is easy to show that for $\lambda > 0$,

$$1 \leq \frac{\mathcal{E}T_z(\lambda_n^{EV})}{\mathcal{E}T_z(\lambda_n^T)} \leq \frac{1 + \Delta_n(\lambda_n^T)}{1 - \Delta_n(\lambda_n^{EV})}, \tag{24}$$

where

$$\Delta_n(\lambda) = \frac{\left|t(\lambda)^2 - 2nt(\lambda) + nt(\lambda)^2/t_2(\lambda)\right|}{[n - t(\lambda)]^2},$$

and where $t(\lambda)$ and $t_2(\lambda)$ are defined in (17) and (18), respectively. Now since $d \leq t(\lambda) \leq k$ and $d \leq t_2(\lambda) \leq k$, it follows that the quotient in (24) converges to zero at a rate $\mathcal{O}(\frac{1}{n})$ as $n \to \infty$. ■

For smoothing splines, the only known way of producing sequences λ_n which are asymptotically optimal is via generalized cross validation. The following result shows, however, that for penalized least squares fitting, the situation is different. Here there is a range of values of λ_n which are asymptotically optimal.

Theorem 13. *Suppose that for each* $n > 1$, *the quantity* $\lambda_n \geq 0$ *is such that* $\mathcal{E}T_z(\lambda_n) \leq \mathcal{E}T_z(0)$. *Then*

$$\lim_{n \to \infty} \frac{\mathcal{E}T_z(\lambda_n)}{\mathcal{E}T_z(\lambda_n^T)} = 1. \tag{25}$$

Proof: By (19) and the fact that $t_2(0) - t_2(\lambda) \leq k - d$, we have

$$1 \leq \frac{\mathcal{E}T_z(\lambda_n)}{\mathcal{E}T_z(\lambda_n^T)} \leq \frac{\mathcal{E}T_z(0)}{\mathcal{E}T_z(0) - (k-d)\sigma^2/n}.$$

Clearly this expression converges to 1 at a rate of $\mathcal{O}(\frac{1}{n})$. ■

Theorem 13 asserts in particular that $\lambda_n = 0$ provides an asymptotically optimal sequence of smoothing parameters. In fact, since $\mathcal{E}T_z(\lambda)$ has a negative derivative at $\lambda = 0$, any sequence of "sufficiently small" numbers λ_n would work.

The fact that $\mathcal{E}T_z(0) - (k-d)\sigma^2/n \leq \mathcal{E}T_z(\lambda) \leq \mathcal{E}T_z(0)$ for all sufficiently small λ suggests that for fairly large values of n, there is little to gain in terms of goodness of fit by striving for the optimal value of λ; it is enough to make sure that λ is sufficiently small. On the other hand, for noisy data, in practice it is often necessary to take $\lambda > 0$ in order to get a smooth fit. Generalized cross validation can be a useful automatic procedure for selecting a good value of λ.

5. Penalized Least Squares Fitting of Curves Using Splines

In this section we show how the penalized least squares method can be applied to fit a spline to the data

$$\lambda_i f = f(t_i) + \varepsilon_i, \quad i = 1, \ldots, n,$$

where $f \in C[a,b]$, and

$$a = t_1 < \cdots < t_n = b.$$

Let $m, k > 0$ be prescribed integers, and suppose that

$$B_i(x) = N_i^{2m}(x), \tag{26}$$

are the normalized B-splines of order m associated with an extended knot vector $y_1 \le \cdots \le y_{2m+k}$, with $y_i < y_{i+2m}$ for all i (cf. [20]). We may assume that the knots have been chosen in such a way that $y_{2m} \le a$ and $b \le y_{k+1}$, and so that (6) is satisfied. By the Schoenberg-Whitney Theorem (cf. Theorem 4.64 of [20]), (6) will be satisfied provided we choose the knots so that there exist $1 < \nu_2 < \cdots < \nu_{k-1} < n$ so that $y_i < t_{\nu_i} < y_{i+m}$ for all $i = 2, \ldots, k-1$.

When using splines, a natural measure for the smoothness of a fit s would be

$$J(s) = \int_a^b [D^m g(t)]^2 dt. \tag{27}$$

In this case, if s is a spline of the form

$$s(x) = \sum_{i=1}^{k} c_i B_i(x),$$

then $J(g) = c^T E c$ with

$$E_{ij} = \int_a^b D^m B_i(t) D^m B_j(t)\, dt, \quad i,j = 1, \ldots, k. \tag{28}$$

We now discuss the problem of computing the quantities appearing in (28). First we show how to represent the derivative of a B-spline in terms of lower order ones.

Lemma 14. *Let $N_i^\ell(x)$ be the normalized B-splines of order ℓ associated with the knot sequence $y_i \le \cdots \le y_{i+\ell}$ with $y_i < y_{i+\ell}$ (cf. [20]). Define $h_{i,\nu} = y_{i+\nu} - y_i$. Then for any $q > 0$,*

$$D^q N_i^\ell(x) = \sum_{\nu=0}^{q} a_{i,\nu}^q Q_{i+\nu}^{\ell-q}, \tag{29}$$

where the Q's are unnormalized B-splines on the same knots, and where the $a^q_{i,\nu}$'s can be computed recursively as follows:

1. Set

$$A_\nu = \begin{cases} 1, & \nu = 2 \\ -1, & \nu = 3 \\ 0, & \nu = 1, 4, \ldots q+2. \end{cases}$$

2. Do $\nu = 2$ to q

   Do $\mu = \nu + 2$ step -1 to 2
   $A_\mu = (\frac{A_\mu}{h_{i+\mu-2,\ell+1-\nu}} - \frac{A_{\mu-1}}{h_{i+\mu-3,\ell+1-\nu}})$

3. Do $\nu = 0$ to q
   $a_{i,\nu} = fac * A_{\nu+2}$,

where fac $= (\ell-1)(\ell-2)\cdots(\ell-q)$. If any of the h's is zero in step 2, the corresponding term should be omitted.

Proof: This result follows easily from the basic formula (cf. Theorem 4.16 in [20])

$$DQ_i^\ell(x) = (\ell-1)\frac{(Q_i^{\ell-1}(x) - Q_{i+1}^{\ell-1}(x)}{h_{i,\ell}},$$

and the fact that $N_i^\ell(x) = h_{i,\ell}Q_i^\ell(x)$. ■

Theorem 15. *For all i, j we have*

$$E_{i,j} = \int_a^b D^m Q_i^{2m}(x) D^m Q_j^{2m}(x)dx = \sum_{\nu=0}^{m}\sum_{\mu=0}^{m} a^m_{i,\nu} a^m_{i,\mu} \int_a^b Q^m_{i+\nu}(x) Q^m_{i+\mu}(x)dx.$$

Proof: The result follows easily from Lemma 14. ■

Theorem 15 can easily be converted into an algorithm for computing the entries of the matrix E. Let

$$P = \left(\int_a^b Q_i^m(x) Q_j^m(x)dx\right)_{i,j=1}^k. \tag{30}$$

This is a $2m-1$ banded symmetric matrix which can be computed accurately and efficiently using Gauss quadrature (see [4,20]). Then

$$E_{i,j} = a_i^T P(i,j) a_j,$$

where $P(i,j)$ is the $(m+1)\times(m+1)$ minor of P obtained by taking rows $i, \ldots, i+m$ and columns $j, \ldots, j+m$, and where $a_i = [a^m_{i,0}, \ldots, a^m_{i,m}]^T$ for all i.

Penalized least squares using cubic splines with uniform knots has been used in [11] to fit some medical data, and in [15] to fit some mechanical data.

6. Remarks

Remark 1

The idea of minimizing a combination of goodness of fit and some measure of smoothness is well-known in the approximation literature. Such expressions arise, for example, in the definition of smoothing splines (see [5,13-14,16-25,31]) and of thin plate splines (see [7-10,25,28-30]). In these methods the approximating function turns out to be a linear combination of n basis functions (see Remark 3 below). The idea of working with a smaller number k of basis functions was explicitly mentioned in Wahba [26-28], but seems to have been carried out explicitly in only a few papers. In curve fitting case, cubic splines with equally spaced knots were used in [11] to fit some medical data, and in [15] to fit some mechanical data. Penalized least squares methods have also been used to fit surfaces to scattered data using tensor-product splines [6] and finite elements defined on rectangles [1].

Remark 2

The method of penalized least squares as described here is closely related to the method of ridge regression as studied by statisticians (see, for example, [12]). The ridge regression problem is to minimize the expression $\rho_\lambda(c)$ in (5), with E replaced by the identity matrix. It has been shown in [2] that the general penalized least squares problem can be reduced to the ridge regression problem by appropriate matrix manipulations.

Remark 3

Suppose $a = t_1 < \cdots < t_n = b$ as in Section 5. Then if we minimize

$$\lambda \int_a^b [f^m(x)]^2 dx + \frac{1}{n} \sum_{i=1}^{n} [f(t_i) - z_i]^2$$

over all functions with square integrable m-th derivative, we get the classical natural smoothing spline (see [5,13-14,16-25,31]). The theory of smoothing splines can be regarded as a special case of the penalized least squares method treated here. Indeed, if we set $k = n$ and take the natural B-splines (cf. [14,20]) as basis functions, then the solution of the penalized least squares problem is precisely the smoothing spline. Almost all of what we have done here is valid for the case where $k = n$ with the exception of the assertions about asymptotic optimality (see Remark 4).

Remark 4

The asymptotic assertion of Theorems 12 and 13 have been proved here only for the case where k is fixed. However. it is not hard to see that the results are also valid for a sequence $k(n)$ with the property that $k(n)/n \to 0$ as $n \to \infty$. The situation when $k(n)/n$ does not go to zero as $n \to \infty$ is more delicate. For example, to prove the analog of Theorem 12 for natural smoothing splines (see [22]) or for complete smoothing splines (cf. [21]), it was necessary to give precise estimates on the eigenvalues of E.

Remark 5

In Section 5 we showed how to use univariate splines to fit noisy data in one dimension by penalized least squares. Clearly, the method can also be used to fit surfaces to noisy data. For example, suppose f is defined on some subset $\Omega \subseteq \mathbb{R}^2$. Then we may measure the smoothness of possible fits g of f by

$$J(g) = (g, g)_m, \tag{31}$$

where, in general,

$$(\phi, \psi)_m := \sum_{\nu+\mu=m} \int D_x^\nu D_y^\mu \phi(x,y) D_x^\nu D_y^\mu \psi(x,y)\, dxdy. \tag{32}$$

In this case, if s is as in (2), then $J(s) = c^T E c$ with

$$E_{ij} = (B_i, B_j)_m.$$

Penalized least squares of surfaces can be performed using a variety of bases including tensor product splines (cf. [6]), radial basis functions (cf. [26]), finite elements (cf. [1]), or multivariate splines on triangulations.

The energy expression (31) arises in the definition of thin plate splines (see [7-10,25,28-30] and references therein). Thin plate splines can be regarded as arising from penalized least squares using $k = n$ and appropriate basis functions.

Remark 6

Generalized cross validation methods have been heavily studied in the case of spline smoothing (cf. [5,13-14,16-25,31]) as well as for ridge regression [12] and for regularization of ill-posed problems [27]. In connection with general penalized least squares problems, generalized cross validation is treated in [2]. It was also used in [15] in connection with a cubic spline method. The basic statistical hypothesis that the errors are independent identically distributed random variables is standard in discussing such methods. We should note that we are *not* assuming that the distributions are Gaussian, only that they are all the same. However, even this hypothesis may well not be satisfied in some practical situations.

Remark 7

Error bounds for spline smoothing can be found in [16,19] and for thin plate splines in [25]. Error bounds for penalized least squares fitting of surfaces using finite elements can be found in [1]. We intend to treat error bounds for penalized least squares in a separate paper.

Remark 8

In some practical problems, in forming the error expression (3), it may be desirable to weigh the i-th measurement with a weight factor $w_i > 0$ for $i = 1, \ldots, n$. In this case, the results of the paper remain valid with only minor adjustments.

Acknowledgment: The work of the second author was supported in part by the Deutsche Forschungsgemeinschaft, and by the National Science Foundation under Grant DMS-8602337.

References

1. Apprato, D., R. Arcangeli, and R. Manzanilla, On the construction of surfaces of class C^k, preprint, 1987.
2. Bates, D. M., M. J. Lindstrom, G. Wahba, and B. S. Zandell, GCVPACK–Routines for generalized cross validation, Tech. Rpt. 775, Statistics Dept., Univ. Wisconsin, Madison, 1986.
3. Bates, D. and G. Wahba, Computational methods for generalized cross-validation with large data sets, in *Treatment of Integral Equations by Numerical Methods*, C. T. Baker and G. Miller (eds.), Academic Press, New York, 1982, 283–296.
4. de Boor, C., T. Lyche, and L. L. Schumaker, On calculating with B-splines II: Integration, in *Numerische Methoden der Approximationstheorie*, L. Collatz, G. Meinardus, and H. Werner, eds., Birkhäuser Verlag, Basel, 1976, 123–146.
5. Craven, P. and G. Wahba, Smoothing noisy data with spline functions: estimating the correct degree of smoothing by the method of generalized cross-validation, Numer. Math. **31** (1979), 377–403.
6. Dierckx, Paul, A fast algorithm for smoothing data on a rectangular grid using spline functions, SIAM J. Numer. Anal **19** (1982), 1286–1304.
7. Dyn, N. and D. Levin, Bell shaped basis functions for surface fitting, in *Approximation Theory and Applications*, Z. Ziegler (ed.), Academic Press, New York, 1981, 113–129.
8. Dyn, N., D. Levin, and S. Rippa, Surface interpolation and smoothing by thin plate splines, in *Approximation Theory IV*, C. Chui, L. Schumaker, and J. Ward (eds.), Academic Press, New York, 1983, 445–449.
9. Dyn, N., D. Levin, and S. Rippa, Numerical procedures for global surface fitting of scattered data by radial functions, SIAM J. Sci. Stat. Comp. **7** (1986), 639–659.
10. Dyn, N. and G. Wahba, On the estimation of functions of several variables from aggregated data, SIAM J. Math. Anal. **13** (1982), 134–152.
11. von Golitschek, M., F. Schardt, and M. Wiegand, Mathematische Auswertung ergospirometrischer Messungen, in *Numerical Methods of Approximation Theory*, L. Collatz, G. Meinardus, and G. Nürnberger (eds.), Birkhäuser, Basel, 1987,

12. Golub, G., M. Heath, and G. Wahba, Generalized cross-validation as a method for choosing a good ridge parameter, Technometrics **21** (1979), 215–222.
13. Hutchinson, M. and F. DeHoog, Smoothing noisy data with spline functions, Numer. Math. **47** (1985), 99–106.
14. Lyche, T. and L. L. Schumaker, Computation of smoothing and interpolating splines via local bases, SIAM J. Numer. Anal. **10** (1973), 1027–1038.
15. Pope, S. B. and R. Gadh, Fitting noisy data using cross-validated cubic smoothing splines, Tech. Rpt. FDA-87-2, Cornell Univ., 1987.
16. Ragozin, D., Error bounds for derivative estimates based on spline smoothing of exact or noisy data, J. Approx. Th. **37** (1983), 335–355.
17. Reinsch, C. H., Smoothing by spline functions, Numer. Math. **10** (1967), 177–183.
18. Reinsch, C. H., Smoothing by spline functions II., Numer. Math. **16** (1971), 451–454.
19. Rice, J. and M. Rosenblatt, Integrated mean square error of a smoothing spline, J. Approx. Th. **33** (1981), 353–369.
20. Schumaker, L. L., *Spline Functions: Basic Theory*, Wiley Interscience, New York, 1981.
21. Schumaker, L. and F. Utreras, Asymptotic properties of complete smoothing splines and appications, SIAM J. Sci. Stat. Comp. **9** (1988), 24–31.
22. Utreras, F., Natural spline functions: their associated eigenvalue problem, Numer. Math. **42** (1983), 107–117.
23. Utreras, F., On generalized cross validation for multivariate smoothing spline functions, SIAM J. Sci. Stat. Comp.
24. Wahba, G., Smoothing noisy data with spline functions, Numer. Math. **24** (1975), 383–393.
25. Wahba, G., Convergence rates of thin plate smoothing splines when the data are noisy, in *Smoothing Techniques in Curve Estimation*, M. Rosenblatt and T. Glasser (eds.), Lecture Notes 757, Springer-Verlag, Heidelberg, 1979,
26. Wahba, G., Spline bases, regularization and generalized cross-validation for solving approximation problems with large quantities of noisy data, in *Approximation Theory IV*, E. Cheney (ed.), Academic Press, New York, 1980, 905–912.
27. Wahba, G., Ill-posed problems: numerical and statistical methods for mildly, moderately, and severely ill-posed problems with noisy data, in *Proceedings of the International Conference on Ill-posed Problems*, M. Nashed (ed.).
28. Wahba, G., Surface fitting with scattered, noisy data on Euclidean d-spaces and on the sphere, Rocky Mt. J. **14** (1984), 281–299.
29. Wahba, G., Cross-validated methods for the estimation of multivariate functions from data on functionals, in *Statistics: An Appraisal*, H. A. David and H. T. David (eds.), Iowa State Univ. Press, 1984, 205–235.
30. Wahba, G. and J. Wendelberger, Some new mathematical methods for variational objective analysis using splines and cross validation, Mon. Wea. Rev. **108** (1980), 36–57.

31. Wahba, G. and S. Wold, A completely automatic French curve: Fitting spline functions by cross-validation, Comm. Stat. 4 (1975), 1–17.
32. Wahba, G. and S. Wold, Periodic splines for spectral density estimation: The use of cross-validation for determining the degree of smoothing, Comm. Stat. 4 (1975), 125–141.

A semiinfinite programming algorithm for constrained best approximation

K. W. Bosworth

Department of Mathematics and Statistics, Utah State University, Logan, Utah, USA

Abstract The general linearly constrained best approximation problem with arbitrary norm is cast as a semiinfinite linear programming problem, using a device of A. Haar (1918) . The resulting optimization problem - that of maximizing a linear form over a convex set described by the intersection of its supporting halfspaces - can be solved numerically by a 2-phase hybrid "interior point / cutting plane" algorithm. A brief outline of both phases of the algorithm is given. Phase 1 is concerned with an "activity" analysis of the constraints posed on the desired fit, and with the construction of an initial feasible solution. Phase 2 produces a sequence of feasible (interior) *and* infeasible (exterior) solutions, both sequences converging to the optimal constrained solution. The method is competitive in *non-*L_2 settings, especially when the norm is not smooth and/or the constraints posed are particularly active.

Key words: Constrained approximations, Semiinfinite programming, Interior point algorithm.

1. Introduction

1.1 Rationale

Researchers in the empirical sciences are faced with the general problem of "fitting" data in a pleasing , rational, and efficient manner. One means of attack is to cast the data fitting problem as a constrained best approximation problem in a normed linear space. The constraints should express various shape qualities and interpolation conditions that the data fit must possess to be "pleasing" or useful. The choice of norm is dictated by both the form of the hypothesized or suspected errors on the data and by the application intended for the fit after it has been produced. The choice of approximating space is made on the basis of one or several of the following criteria: computational ease, model predictions, degree of approximation, and manipulative ease in later applications.

Although algorithms exist for solving constrained best approximation problems for certain specific combinations of constraints, norm, and approximating family, no general all purpose algorithm has been available. This is the motivation for the present work; framing the constrained approximation problem as a semiinfinite convex programming problem posessing special structure, and developing a numerical algorithm that could then be applied in any situation that may obtain.

1.2 The linearly constrained best approximation problem

In this section the data fitting problem is cast in a mathematical framework, flexible enough to handle most practical situations. Let $\{f_1, f_2, \ldots, f_n\}$ be n linearly independent, real valued functions defined and continuous on the compact set $\mathbf{I} \subset R^k$. The linear space $\mathcal{V} := \text{lin}\{f_1, f_2, \ldots, f_n\}$ will be referred to as the approximating subspace, and the set $\{f_1, f_2, \ldots, f_n\}$ as the approximating family. Let f denote real data, defined and continuous on the set $\mathbf{I}$; i.e., $f \in C[\mathbf{I}]$. If one wishes to fit the data f using elements from $\mathcal{V}$, one needs a criterion for measuring the goodness of the fit; this measure is usually a norm or seminorm defined on the linear space $\mathcal{W}$ of functions gotten by adjoining f to $\mathcal{V}$; and is denoted by $||h||$, where $h \in \mathcal{W}$. The problem of finding $g_{ba} \in \mathcal{V}$ such that:

$$||f - g_{ba}|| = \inf_{g \in \mathcal{V}} ||f - g||$$

is the classical problem of best approximation from a linear subspace. The element g_{ba} is called the best "$||\cdot||$ - approximation" to f from $\mathcal{V}$. Using standard compactness and continuity arguments, one can easily prove the existence of such an element $g_{ba} \in \mathcal{V}$ for each $f \in C[\mathbf{I}]$.

However, one often desires that the best fit to the data also possess other shape or structural qualities. In a large majority of cases, such constraints can be formulated in terms of an indexed collection of linear constraints of the form:

$$a(p) \leq L(p; f, g) \leq b(p), \ \forall p \in \mathbf{J}$$

where:

$$a(p) : \mathbf{J} \to R^e,$$
$$b(p) : \mathbf{J} \to R^e,$$

with R^e the extended reals; $L(p; f, g)$ is a continuous linear functional acting on f or $g \in \mathcal{V}$ or both, indexed by and continuous in $p \in \mathbf{J}$; and $\mathbf{J}$ is a compact topological space, called the *index set*, often identical with $\mathbf{I}$ or a subset of $\mathbf{I}$. The function a (resp. b) is required to be u.s.c. (resp. l.s.c.) on $\mathbf{J}$, continuous on $\mathbf{J} - a^{-1}\{-\infty\}$ (resp. $\mathbf{J} - b^{-1}\{\infty\}$), and to avoid vacuous problems, $a(p) \leq b(p)$, $\forall p \in \mathbf{J}$.

In the case where $\mathbf{I} \subset R^1$, one has as common examples of such linear constraints (assuming $g \in \mathcal{V}$ and f are sufficiently smooth):

1. **Interpolation Conditions:** the approximant g is to exactly agree with the data f at the specified points $p \in \mathbf{J} \subseteq \mathbf{I}$. Then take $a(p) = b(p) = 0$, and

$$L(p; f, g) := f(p) - g(p), \ \forall p \in \mathbf{J}.$$

2. **Non-negativity Constraints:** the approximant g is to remain non-negative over some subset $\mathbf{J} \subseteq \mathbf{I}$. Set $a(p) = 0$, $b(p) = \infty$, and

$$L(p; f, g) := g(p), \ \forall p \in \mathbf{J}.$$

3. **Specified Moments:** the approximant g is to have its pth moments, $p \in \mathbf{J} \subset Z$, specified as either the same as that of the data, f, or as some value m_p. In the first instance, take $a(p) = b(p) = 0$ for $p \in \mathbf{J}$, and

$$L(p; f, g) := \int_{\mathbf{I}} x^p (f(x) - g(x)) dx.$$

In the second instance, take $a(p) = b(p) = m_p$ for $p \in \mathbf{J}$, and

$$L(p; f, g) := \int_{\mathbf{I}} x^p g(x) dx.$$

Notice that in examples **1** and **2**, it happened that $\mathbf{J} \subseteq \mathbf{I}$, whereas example **3** shows that $\mathbf{J}$ may just as well be an arbitrary index set. One is free to impose several of the above conditions simultaneously in any approximation problem, and then the index set $\mathbf{J}$ might contain several "copies" of $\mathbf{I}$ or subsets of $\mathbf{I}$.

1.3 Haar's transcription

To transcribe the constrained best approximation problem:

$$\mathcal{P}_c : \begin{cases} \min_{g \in \mathcal{V}} \|f - g\| \\ a(p) \leq L(p; f, g) \leq b(p), \ \forall p \in \mathbf{J} \end{cases}$$

into a semiinfinite linear program, one introduces homogeneous coordinates $\vec{x} := (x_1, \ldots, x_n, x_{n+1})$, $x_{n+1} > 0$, as follows. For each $g \in \mathcal{V}$, $g = \sum_{i=1}^{n} \alpha_i f_i$, and $f - g = f - \sum_{i=1}^{n} \alpha_i f_i$, so $e := f - g \in \mathcal{W}$. In $\mathcal{W}$, consider now all linear combinations of the form:

$$w = \sum_{i=1}^{n} x_i f_i + x_{n+1} f, \qquad x_{n+1} > 0,$$

which we write as:

$$w = \vec{x} \cdot \vec{f}, \qquad \vec{e}_{n+1} \cdot \vec{x} > 0$$

where:

$$\vec{x} := (x_1, \ldots, x_n, x_{n+1}),$$
$$\vec{f} := (f_1, \ldots, f_n, f),$$
$$\vec{e}_{n+1} := (0, \ldots, 0, 1),$$

and the dot product is the standard scalar product on R^{n+1} . Homogeneous is a suitable adjective for the $\vec{x}$ coordinates, as each ray $\vec{r}_{\vec{u}} := \{\vec{x} \mid \vec{x} = t\vec{u},\ t > 0\} \subset H$, where $H := \{\vec{x} \mid \vec{e}_{n+1} \cdot \vec{x} > 0\}$, corresponds to precisely one error function $e := f - g$, $g = \sum_{i=1}^{n} \alpha_i f_i$, via the well defined map: $\alpha_i = -x_i/x_{n+1} = -u_i/u_{n+1}$, $i = 1, \ldots, n$.

Now, defining $\beta_i(p) := L(p; 0, f_i)$ for $i = 1, \ldots, n$ and $\beta_{n+1}(p) := L(p; f, 0)$, the collection of constraints in $\mathcal{P}_c$ becomes, in homogeneous coordinates:

$$a(p)x_{n+1} \leq -\sum_{i=1}^{n} \beta_i(p)x_i + \beta_{n+1}(p)x_{n+1} \leq b(p)x_{n+1}, \qquad \forall p \in \mathbf{J}.$$

Defining the vectors:

$$\vec{n}_+(p) := (\beta_1(p), \ldots, \beta_n(p), \beta_{n+1}(p) - a(p)),$$
$$\text{for all } p \in \mathbf{J} - a^{-1}\{-\infty\},$$
$$\vec{n}_-(p) := (-\beta_1(p), \ldots, -\beta_n(p), -\beta_{n+1}(p) + b(p)),$$
$$\text{for all } p \in \mathbf{J} - b^{-1}\{\infty\},$$

with $\vec{n}_+(p) := \vec{0}_{n+1}$ on $a^{-1}\{-\infty\}$ and $\vec{n}_-(p) := \vec{0}_{n+1}$ on $b^{-1}\{\infty\}$, the homogeneous variable transcription of the constraints is the closed convex cone:

$$\mathcal{C} := \{\vec{x} \in R^{n+1} \mid \vec{n}_+(p) \cdot \vec{x} \leq 0,\ \vec{n}_-(p) \cdot \vec{x} \leq 0,\ \forall p \in \mathbf{J}\}.$$

That is to say, any vector $\vec{x} \in \mathcal{C} \cap H$ corresponds under the above rule of correspondence to a function $g \in \mathcal{V}$ satisfying the totality of constraints imposed on $\mathcal{P}_c$. Moreover, any $g \in \mathcal{V}$ *not* satisfying some constraint has homogeneous coordinates *not* in $\mathcal{C} \cap H$.

Lastly, define the closed convex body (called by Haar the "Eichkörper"):

$$K := \{\vec{x} \in R^{n+1} \mid \|\vec{x} \cdot \vec{f}\| \leq 1\},$$

and consider the semiinfinite linear program:

$$\mathcal{P}_h : \begin{cases} \max\ \vec{e}_{n+1} \cdot \vec{x} \\ \vec{x} \in K \cap \mathcal{C} \end{cases}$$

With the above notation and definitions, one has the simple result:

Result 1. Problem $\mathcal{P}_c$ and $\mathcal{P}_h$ are equivalent in the following sense:

1. $\mathcal{P}_c$ is inconsistent if and only if $\mathcal{P}_h$ has optimal solution $\vec{x}_{opt} = \vec{0}_{n+1}$.

2. $\mathcal{P}_c$ is consistent and has $\|f - g_{opt}\| = 0$ if and only if $\mathcal{P}_h$ has unbounded feasible solutions and unbounded objective values; that is, there exists a ray $r_{\vec{u}} := \{\vec{x} = t\vec{u},\ t \geq 0, \vec{e}_{n+1} \cdot \vec{u} > 0\} \subset K \cap \mathcal{C}$. In this case, any g_{opt} corresponds to some such $\vec{u}$ under the map $g = \sum_{i=1}^{n} \alpha_i f_i$ with $\alpha_i := -u_i/u_{n+1}$ and vice versa: any such ray $r_{\vec{u}}$ yields an optimal solution g_{opt} by the same map.
3. $\mathcal{P}_c$ is consistent and has $\|f - g_{opt}\| = E > 0$ if and only if $\mathcal{P}_h$ has a finite optimal solution $\vec{x}_{opt} = \vec{v}$ with $\vec{e}_{n+1} \cdot \vec{v} = v_{n+1} > 0$. In this case, the error in the constrained best approximation is $E = 1/v_{n+1}$. Any constrained best approximation g_{opt} corresponds to some optimal vector $\vec{v}$ under the map $g = \sum_{i=1}^{n} \alpha_i f_i$ with $\alpha_i := -v_i/v_{n+1}$ and vice versa: any such optimal vector $\vec{v}$ yields an optimal solution g_{opt} by the same map.

The proof of this result is straightforward, and is in fact a direct application of a minimum norm duality theorem of Deutsch and Maserick (1967) and found in Luenberger (1969), page 119, and thus will not be given here.

2. A 2 phase algorithm

For convenience in what follows, the supporting normal vectors appearing in $\mathcal{C}$ are renamed and reindexed, omitting any that are $\vec{0}_{n+1}$, i.e., $\mathcal{C} = \{\vec{x} \in R^{n+1} \mid \vec{n}(p) \cdot \vec{x} \leq 0,\ p \in \mathbf{J}'\}$. Also, set $\mathcal{C} \cap \overline{H} := \mathcal{C}_H$.

2.1 Phase 1

Problem $\mathcal{P}_h$ can be efficiently solved by an interior point algorithm to be described in the following section *provided* $K \cap \mathcal{C}_H$ has interior in R^{n+1}. As K always has interior, int$(K \cap H)$ is empty iff int$(\mathcal{C}_H) = \emptyset$. In terms of the constraints in $\mathcal{P}_c$, this situation obtains when either interpolatory conditions are imposed (not satisfied by *all* $g \in \mathcal{V}$), *or* when a collection of "one-sided" constraints has "ganged up" to form a " generalized" interpolation condition or conditions. As the solution of general problems is the goal, Phase 2 is preceded by an analysis of the cone $\mathcal{C}_H$ determining the dimension of its relative interior, followed, if necessary, by a suitable introduction of a reduced set of basis functions $\{\tilde{g}_1, \ldots, \tilde{g}_{\tilde{n}}\}, \tilde{n} < n$, with linear span $\tilde{\mathcal{V}} \subset \mathcal{C}$, such that *all* interpolation conditions are automatically satisfied by any $\tilde{g} \in \tilde{\mathcal{V}}$. Hence, by adopting $\tilde{\mathcal{V}}$ as the approximating space, the resulting cone $\tilde{\mathcal{C}}_H$ will have interior in the reduced dimension space $R^{\tilde{n}+1}$.

It turns out that the key to Phase 1 is determining whether $\vec{0}_{n+1}$ is in the convex hull of $\vec{e}_{n+1} \cup \{\vec{n}(p)\}_{p \in \mathbf{J}'}$. If so, one determines the set L of all such vectors which are involved in nontrivial convex combinations yielding $\vec{0}_{n+1}$. The resulting set L has orthogonal compliment $L^{\perp}$ containing $\mathcal{C}_H$. In essence, the set L corresponds precisely to the constraints (if any) which have "ganged up" in $\mathcal{P}_c$ to reduce the degrees of freedom in the approximating space. That is, a basis for $L^{\perp}$ yields, under the inverse of the homogeneous coordinate transcription, a reduced basis $\{\tilde{g}_1, \ldots, \tilde{g}_{\tilde{n}}\}$ to work in, with $\dim(L^{\perp}) = \tilde{n}$, or shows, in case $\tilde{n} = 0$, that the problem $\mathcal{P}_c$ is overconstrained. Fortunately, such a task is efficiently undertaken; see Bosworth (1988) for a detailed algorithm description. Moreover, if L is empty, then the just cited algorithm provides one with a vector $\vec{u} \in C_H$, separating $\vec{0}_{n+1}$ from the convex hull of $\vec{e}_{n+1} \cup \{\vec{n}(p)\}_{p \in \mathbf{J}'}$.

2.2 Phase 2: the tunnelling algorithm

A description is now given of the upper level structure of the actual Phase 2 semi-infinite programming algorithm solving the problem $\mathcal{P}_h$. A detailed description of Phase 2 can be found in Bosworth (1984) . Note that on successful completion of Phase 1, a vector $\vec{u}$ pointing into the interior of $K \cap C_H$ is available, and that $\vec{0}_{n+1} \in K \cap C$. Note also that in the course of Phase 1, the dimension n of the problem has been suitably redefined if necessary. Initialize $\vec{x}_{cm} := \vec{0}_{n+1}$.

2.2.1 Shooting

From the point $\vec{x}_{cm}$, "shoot" in the positive $\vec{u}$ direction to $s\vec{x} \in \partial(K \cap C)$. If no such point exists, then $\vec{u}$ satisfies Result 1, condition 2, and the algorithm terminates. If $s\vec{x} \notin \partial K$, then scale $s\vec{x}$ positively until it reaches ∂K. Again, if no such point exists, then the direction vector $s\vec{x}$ satisfies Result 1, condition 2, and the algorithm terminates.

At $s\vec{x}_1 := s\vec{x}$ compute a unit outer supporting normal $s\vec{n}_1$ to ∂K, using the subgradient $\partial||\vec{x} \cdot \vec{f}||$ at $s\vec{x}_1$. If more than one normal exists (nonsmooth norm), select one with maximal $n+1^{\text{st}}$ component. If the normal $s\vec{n}_1 = \vec{e}_{n+1}$, then $s\vec{x}_1$ is the optimal solution to $\mathcal{P}_h$, Result 1, case 3, and the algorithm terminates.

2.2.2 Webbing

Set $M := \{\vec{x} \in R^{n+1} | \ \vec{e}_{n+1} \cdot \vec{x} = \vec{e}_{n+1} \cdot s\vec{x}_1\}$ and $N := \{\vec{x} \in R^{n+1} | \ \vec{e}_{n+1} \cdot \vec{x} = 0\}$. I.e., $M = N + s\vec{x}_1$.

A collection of $n+1$ supporting halfspaces to the set $K \cap C$ with support points on $\partial(K \cap C) \cap M$ is constructed by means of a n stage recursion (using the shooting technique of the previous subsection), the aim being to "equidistribute" the support points on $\partial(K \cap C)$. These halfspaces will be used in the following stage of the algorithm to define a *minimal* linear program approximating $\mathcal{P}_h$.

The kth stage of the recursion ($k = 2, \ldots, n+1$) is "roughly" the following:

$$\begin{aligned} \vec{x}_{cm,k-1} &:= \tfrac{1}{k-1} \textstyle\sum_{i=1}^{k-1} s\vec{x}_i \\ \vec{n}_i &:= \text{proj}_N \, s\vec{n}_i \, , \ i = 1, \ldots, k-1, \end{aligned}$$

define the kth search direction $\vec{u}_k$ by:

$$\begin{aligned} \vec{u}_k \cdot \vec{n}_i &= -1 \, , \ i = 1, \ldots, k-1 \\ \vec{u}_k &\in \text{lin}\{\vec{n}_1, \ldots, \vec{n}_{k-1}\}, \end{aligned}$$

and then the supporting vector $s\vec{x}_k$ and outer supporting normal $s\vec{n}_k$ to $K \cap C$ are found by "shooting" from $\vec{x}_{cm,k-1}$ in the positive $\vec{u}_k$ direction. However, here

$\vec{sn}_i$ is allowed to be a supporting normal to the cone C. (If the vectors $\{\vec{n}_i\}_{i=1}^{k-1}$ are dependent, randomly chosen vectors in N are added to the set until a well defined direction $\vec{u}_k$ can be obtained from the above relations, hence the adjective "roughly".)

By construction, $K \cap C \subseteq \{\vec{x} \in R^{n+1} | \; \vec{sn}_i \cdot \vec{x} \leq \vec{sx}_i \cdot \vec{sx}_i\}$ for $i = 1, \cdots, k$. At the conclusion of the recursion, set $\vec{x}_{cm} := \frac{1}{n+1} \sum_{i=1}^{n+1} \vec{sx}_i$.

2.2.3 Analysis of a minimal linear program

Consider the simple linear program:

$$\mathcal{P}_{mlp} : \begin{cases} \max & \vec{e}_{n+1} \cdot \vec{x} \\ \vec{sn}_i \cdot \vec{x} & \leq \vec{sn}_i \cdot \vec{sx}_i \qquad i = 1, \ldots, n+1 \end{cases}$$

By construction, $K \cap C$ is within the feasible region defined by $\mathcal{P}_{mlp}$, and hence an optimal value of $\mathcal{P}_{mlp}$ provides an upper bound on the optimal value of $\mathcal{P}_h$. Moreover, $\mathcal{P}_{mlp}$ is trivial to analyze (one "Simplex Pivot"). Since $\mathcal{P}_{mlp}$ can be considered to be a "coarse" discretization of the semiinfinite program $\mathcal{P}_h$, it will be reasonable to use any information obtained from the optimality analysis of $\mathcal{P}_{mlp}$ as approximating the situation in $\mathcal{P}_h$.

Four cases may obtain in $\mathcal{P}_{mlp}$:

1. There exists a unique optimal solution $\vec{v}_{mlp}$. Then the feasible region in $\mathcal{P}_{mlp}$ intersected with $\overline{H}$ is a bounded polytope containing $K \cap C_H$. The solution $\vec{v}_{mlp}$ can be considered an approximate solution to $\mathcal{P}_h$. Define a new search direction for the "shooting" stage, $\vec{u} := \vec{v}_{mlp} - \vec{x}_{cm}$.
2. There exists infinitely many optimal solutions $\vec{w}_{mlp}$, all with the same $n+1^{\text{st}}$ component. Define a new search direction $\vec{u} := \vec{v}_{mlp} - \vec{x}_{cm}$, where $\vec{v}_{mlp}$ is the closest optimal solution to $\vec{x}_{cm}$. It can be easily shown that this selection of $\vec{u}$ has nonnegative $n+1^{\text{st}}$ component.
3. The system of equalities $A\vec{x} = \vec{b}$ with $\text{row}_i A := \vec{sn}_i$, $b_i := \vec{sn}_i \cdot \vec{sx}_i$, $i = 1, \ldots, n+1$ is singular, but has vectors $\vec{w} \in \text{Kern}(A)$ with positive $n+1^{\text{st}}$ component. Select the unit vector $\vec{u} \in \text{Kern}(A)$ with maximal $n+1^{\text{st}}$ component as the new shooting direction.
4. The system $A\vec{x} = \vec{b}$ is nonsingular, but $\mathcal{P}_{mlp}$ has unbounded objective values. Then it can be shown that there exists a ray $r_{\vec{u}}$ of unbounded feasible solutions to $\mathcal{P}_{mlp}$ with unit direction vector $\vec{u}$ having maximal positive $n+1^{\text{st}}$ component. This is the new shooting direction.

In each of the 4 cases, a vector $\vec{u}$ is obtained, which heuristically points from $\vec{x}_{cm} \in K \cap C$ in a direction of "ascent" for $\mathcal{P}_h$.

2.2.4 Convergence

If, in case 1 or 2 above, $\vec{v}_{mlp} - \vec{x}_{cm}$ is "sufficiently small", then one can conclude convergence. If $\vec{e}_{n+1} \cdot \vec{x}_{cm} < (1/\text{tol})$, where tol is an upper bound on the error needed in a near-best approximation, then one concludes convergence. In either case, to obtain the final $\vec{v}$ to use in computing g_{ba}, shoot one last time from $\vec{x}_{cm}$ in the $\vec{u}$ direction to $\vec{v} \in \partial(K \cap C)$.

Else, one returns to the shooting stage, and continues with the algorithm.

3. Conclusions

Linear convergence of the algorithm can be concluded provided Phase 1 is successfully completed; see Bosworth (1984). That is, the sequence of points $\{\vec{x}_{cm}\}$ converges in objective value, from below, to the optimal value for $\mathcal{P}_h$ at least at a linear rate. If $\mathcal{P}_h$ is in situation 3 of Result 1, then one is also guaranteed the existence of subsequences of $\{\vec{x}_{cm}\}$ and $\{\vec{v}_{mlp}\}$ having convergence in R^{n+1} to an optimal solution $\vec{v}$ of $\mathcal{P}_h$ (different subsequences can have different limits; however, all tend to the same objective value). In polyhedral norm situations, with a finite set of constraints, the algorithm converges in a finite number of steps, often outperforming the Dual Simplex Algorithm of linear programming, due to its ability to bypass in one step several pivots taken by the Dual Simplex Algorithm. (The Dual Simplex Algorithm computes a sequence of infeasible solutions, converging to the optimal solution of $\mathcal{P}_h$, similar to the solutions $\{\vec{v}_{mlp}\}$ computed here.)

Numerical results for the case of uniform norm shape constrained approximation problems have appeared in Bosworth (1987) . Applications to the l_1 and other specially tailored norms will appear shortly. One final observation is that the viability of the algorithm appears most strongly in *non* l_2 settings. In constrained l_2 settings, the code of Lawson and Hanson (1974) is to be recommended.

References

1. Ken W. Bosworth. *A general method for the computation of best uniform norm approximations.* PhD thesis, Rensselaer Polytechnic Institute, Troy, N.Y., 1984.
2. Ken W. Bosworth. *A numerical algorithm for the determination of the linear span of a cone given in polar form.* Technical Report, Utah State Univ. , Dept. of Mathematics and Statistics, 1988.
3. Ken W. Bosworth. Shape constrained curve and surface fitting. In G. E. Farin, editor, *Geometric Modeling: Algorithms and New Trends*, pages 247–263, SIAM, 1987.
4. F.R. Deutsch and P.H. Maserick. Applications of the Hahn-Banach theorem in approximation theory. *SIAM Review*, 9(3):516–530, July 1967.
5. A. Haar. Über die Minkowskische Geometrie und die Annäherung an stetige Funktionen. *Math. Ann.*, 78:294–311, 1918.
6. C. L. Lawson and R. J. Hanson. *Solving Least Square Problems.* Prentice Hall, 1974.
7. D.L. Luenberger. *Optimization by Vector Space Methods.* John Wiley, 1969.

Inference region for a method of local approximation by using the residuals

M. Bozzini

Dipartimento di Matematica
Università di Lecce, Italy

L. Lenarduzzi

Istituto Applicazioni Matematica ed
Informatica C.N.R.,
Milano, Italy

Abstract We consider a formula for smoothing noisy data in which the variances are different. The formula is obtained by a local weighted approximation. The weight is a function of a smoothing parameter α, which varies with the point. Then a diagnostic band for data analysis is obtained : this determines whether an initial and computationally quick choice of α constant for all the data allows a homogeneous accuracy of approximation on the whole domain. In the case when it is necessary for α to be variable, a fast algorithm to evaluate α is proposed.

Key words: Inference region, Weighted local approximation, Smoothing, Variable smoothing parameter, Data analysis.

1. Introduction

In the problem of approximating a function from a sample of function values, it is of interest to know the reliability achieved with the chosen approximation method. In other words what is the probability that the approximating function fits the unknown function to a specified accuracy?

On this subject we refer to the work by Wahba (1983), who, on the basis of a Bayesian model (that is to say the region is that within which the approximating function can fall with that sampling) for spline functions, assuming that the noise variance is constant determines an inference region with constant band. A later work by Silverman (1985) generalizes the work by Wahba by considering more general data configurations and noise characterized by variance $\sigma^2 = \sigma^2(x)$ as a function of the abscissa x. The band obtained varies according to the data location density and variance.

Müller and Stadtmüller (1987) study a method with a localized smoothing parameter based on estimates of the higher derivatives of $f(x)$.

In this note, on the basis of an approximation method with constant window, we construct an inference region for the unknown function $f(x)$ (in other words $f(x)$ falls in the region derived from that sampling with given probability) by using samples of random variables $\tilde{f}_r(x)$ $r = 1, \ldots, l$ $l \geq 1$ and finite, with $E(\tilde{f}_r(x)) = f(x)$ and covariance matrix $\sigma^2_{rs} = \delta_{rs}\sigma^2_r$.

The method, which is computationally simple, has variable band and indicates the regions where more accuracy is needed when constructing the approximating function.

Then a computationally straightforward method is suggested to improve the approximation locally, by modifying the local smoothing parameter.

2. Approximation formula

Let a set $J = \{x_i, \tilde{f}(x_i)\}_1^N$ be given on a domain $D = [a, b] \subset R$. An approximation to $f(x)$ is required.

We assume that the function values $\tilde{f}(x_i)$ are the result of sampling from l populations with sample sizes $n_1, n_2, \ldots, n_l$, respectively, each one according to a uniform distribution (for simplicity) and $\sum_1^l n_r = N$ and such that

$$E(\tilde{f}_r(x)) = f(x),$$
$$\sigma^2_{rs} = \delta_{rs}\sigma^2_r.$$

For simplicity, in the following it is assumed that $n_r = n$ for $r = 1, \ldots, l$.

Let us consider the method studied in Bozzini and Lenarduzzi (1988a) to approximate the unknown function $f(x)$.

Precisely, after indicating with x_i a generic assigned point , take a neighbourhood $I = [x_i - R, x_i + R]$ and determine the constant function $\tilde{c}_i$ by the weighted least squares method:

$$min \sum_{x_j \in I} w_{j\alpha}(c - \tilde{f}(x_j))^2$$

where the weights $w_{j\alpha}$ have the following expression:

$$w_{j\alpha} = \begin{cases} \frac{1}{(d^2+\alpha)\sigma_r^2} & if\ d < R, \\ 0 & if\ d \geq R, \end{cases}$$

with $d^2 = dist^2(x_i, x_j)$ the Euclidean distance and α a positive smoothing parameter.

Then the approximating function

$$\tilde{m}(x) = \frac{\sum_1^N \tilde{c}_i \phi_i(x)}{\sum_1^N \phi_i(x)} \tag{1}$$

is constructed.

Here the $\phi_i(x)$ are smooth functions with support $(x_i - R_1, x_i + R_1)$ with $R_1 < R$ and such that $\tilde{m}(x_i) = \tilde{c}_i$.

This approximating function depends on α; in particular, as $\alpha \rightarrow 0$ the function $\tilde{m}(x)$ tends to an interpolating function, while, as $\alpha \rightarrow \infty$ $\tilde{m}(x)$ tends to the function in which the constants $\tilde{c}_i$ are solution of the weighted least squares problem with weights $w_{ir} = \frac{1}{\sigma_r^2}$.

3. Inference region

Let us assume we work in asymptotic conditions (that is to say with $N \rightarrow \infty, R \rightarrow 0, NR \rightarrow \infty$) and that we also have information beyond the boundary of D.

After constructing the function $\tilde{m}(x)$ on the basis of (1) we determine a computable inference region.

From the Markov inequality (see Ross (1980)), one gets for $i = 1, \ldots, N$:

$$Prob\{|\ f(x_i) - \tilde{m}(x_i)\ | < k\{E_e[f(x_i) - \tilde{m}(x_i)]^2\}^{\frac{1}{2}}\} > 1 - \frac{1}{k^2}.$$

The expected mean squared error $E_e[f(x_i) - \tilde{m}\ (x_i)]^2$, which will be called $EMSE_i$ in what follows, has the expression:

$$EMSE_i = \{E_e[f(x_i) - \tilde{m}(x_i)]\}^2 + var(\tilde{m}(x_i))$$

and, of course, it is a function of the parameter α.

In Bozzini and Lenarduzzi (1988a) it was proved that there is one and only one value $\alpha_i^\star$ of α that minimizes $EMSE_i(\alpha)$:

$$EMSE_i(\alpha_i^\star) = min_{\alpha > 0} EMSE_i(\alpha)$$

and such a value falls in$(0, R^2)$ or in $[R^2, \infty)$ depending on whether the ratio

$$\frac{R^4 f_i^{(2)2}}{[\frac{\delta}{2nR}]}$$

is greater or less than one, (here $\delta = (\sum^l \frac{1}{\sigma_r^2})^{-1}$).

Furthermore, using the hypothesis that $f(x)$ can be developed as a power series, the principal part of $EMSE_i(\alpha_i^\star)$ has the following expression:

$$EMSE_i(\alpha_i^\star\ ) \simeq \frac{f_i^{(2)2}}{\pi^2} R^2 \alpha_i^\star + \frac{\delta}{2n\sqrt{\alpha_i^\star}\pi}$$

if $\alpha_i^\star \in (0, R^2)$ and

$$EMSE_i(\alpha_i^\star) \simeq \frac{f_i^{(2)2} R^4 (\frac{1}{36} - \frac{R^2}{30\alpha_i^\star}) + \frac{\delta}{2nR}(1 - \frac{2R^2}{3\alpha_i^\star})}{(1 - \frac{R^2}{3\alpha_i^\star})^2}$$

if $\alpha_i^\star \in [R^2, \infty)$

It can be verified analytically that an upper bound to the principal part of $EMSE_i(\alpha_i^\star)$ is:

$$1.5\delta p_{ii}(\alpha_i^\star), \tag{2}$$

where $p_{ii}(\alpha)$ is the asymptotic value of the weight $w_{i\alpha}/\sum w_{j\alpha}$ (index of influence of the i-th datum) divided by $\frac{\delta}{\sigma_i^2}$ and associated to the value $\tilde{c}_i$ at the point x_i.

Consider the random variable

$$\theta_{\alpha_i} = \frac{\sum^{N_I} \beta_i^2 (\tilde{f}_i - \tilde{m}_{\alpha_i}(x_i))^2}{\sum^{N_I}(1 - p_{ii}(\alpha_i))} \tag{3}$$

with $\beta_i^2 = \frac{\delta}{\sigma_i^2}$;

then:

$$E_e(\theta) = \delta + \frac{\sum^{N_I} \beta_i^2 EMSE_i}{\sum^{N_I}(1 - p_{ii})} - \delta \frac{\sum^{N_I} p_{ii}}{\sum^{N_I}(1 - p_{ii})}.$$

If one assumes that the second derivative does not change too much on I then

$$AMSE_I(\alpha_i) = E_e \int_I (f(t) - \tilde{m}(t))^2 dt = 2R\, EMSE_i(\alpha_i) + O(R^6)$$

In optimal conditions (that is to say $R \asymp n^{-\frac{1}{5}}, \alpha_i^\star \asymp R^2$), it has been proved in Bozzini and Lenarduzzi (1988a)that $EMSE_i(\alpha_i^\star) = O(n^{-\frac{4}{5}})$ and $p_{ii} = O(n^{-\frac{4}{5}})$; so:

$$E_e(\theta(\alpha_i^\star)) = \delta + O(n^{-\frac{4}{5}}). \tag{4}$$

As can be seen by developing the calculations, one has:

$$var(\theta(\alpha_i^\star)) = O(n^{-\frac{4}{5}})$$

so that $\theta(\alpha_i^\star)$ is consistent estimator for δ.

From (4) it follows that the principal part of $1.5 p_{ii}(\alpha_i^\star) E_e(\theta(\alpha_i^\star))$ is given by $1.5\delta p_{ii}(\alpha_i^\star)$. On the basis of (2) and (3) it follows that an inference region which is computable for $f(x)$ at the probability level $(1 - \frac{1}{k^2})$ is given by:

$$\tilde{m}_{\alpha_i^\star}(x) - k(1.5\theta_{\alpha_i^\star} p_{ii}(\alpha_i^\star))^{\frac{1}{2}} \le f(x) \le \tilde{m}_{\alpha_i^\star}(x) + k(1.5\theta_{\alpha_i^\star} p_{ii}(\alpha_i^\star))^{\frac{1}{2}}. \tag{5}$$

In order to make the method more efficient, in the case when the second derivative does not change too much on the domain, it is preferable to calculate a constant value of α for all $x_i \in J$.

This is possible by calculating the value α_{opt} which minimizes $IMSE = E_e \int_D (f(t) - \tilde{m}(t))^2 dt$ (see Bozzini and Lenarduzzi (1988a)).

In this case (5) becomes:

$$\tilde{m}\ \alpha_{opt}\ (x) - k(1.5\theta_{\alpha_{opt}}\ p_{ii}(\alpha_{opt}\ ))^{\frac{1}{2}} \leq f(x) \leq \tilde{m}\ \alpha_{opt}\ (x) + k(1.5\theta_{\alpha_{opt}}\ p_{ii}(\alpha_{opt}\ )))^{\frac{1}{2}}. \tag{6}$$

In the case in which, for some x_i, $(f_i^{(2)})^2 \gg (f_M^{(2)})^2 = \int_a^b \frac{f^{(2)2}}{b-a}(t)dt$ then, as derived in Bozzini and Lenarduzzi (1988a), both $EMSE_i(\alpha_i^\star\ )$ and $E(\theta(\alpha_{opt}\ ))p_{ii}(\alpha_{opt}\ )$ increase because of the bias; therefore this band is wider than that for the other points.

Remark:In the case when there is only one variance, that is to say $\sigma_r^2 = \sigma^2$, $r = 1, \ldots, l$, the above remains true provided that one substitutes $w_{j\alpha} = \frac{1}{(d^2+\alpha)\sigma_r^2}$ with $w_{j\alpha} = \frac{1}{d^2+\alpha}$ and δ with σ^2.

4. Variable smoothing parameter

Let us consider a smooth function $f(x)$ which has peaks at some points. Such a function therefore has a second derivative $f_M^{(2)2}$ which is almost constant and given by $\int_a^b \frac{f^{(2)2}}{b-a} dx$ but at the point x_i, $f_i^{(2)2} \gg f_M^{(2)2}$.

The parameter α,which is optimal on the whole interval $[a, b]$, falls in $[R^2, \infty)$ as the method constructs an approximation function in the least squares sense which removes the noise and provides a smooth function. On the contrary, the optimal value of α at the point x_i falls in the interval $(0, R^2)$ in order to give a function which is locally nearer the interpolating function.

Consider the random variable:

$$Z = \frac{\sum_1^N \beta_i^2 (\tilde{f}_i - \tilde{m}_i)^2}{\sum_1^N (1 - p_{ii})}.$$

Its expected value is

$$E(Z) = \delta + \frac{IMSE}{\Sigma(1 - p_{ii})} - \delta \frac{\sum p_{ii}}{\Sigma(1 - p_{ii})}.$$

In the case we are studying, the optimal value of α for $IMSE$ is α_{opt} and therefore:

$$E(Z_{opt}) = \delta(1 + k_1 n^{-\frac{4}{5}}).$$

Assuming $\alpha_i^\star \asymp n^{-\frac{2}{5-\epsilon}}$ one gets:

$$E(\theta(\alpha_i^\star\ )) = \delta(1 + k_2 n^{-\frac{4-\epsilon}{5-\epsilon}}),$$

from which

$$Ratio = \frac{E(Z_{opt})}{E_e(\theta(\alpha_i^\star))} \simeq 1 - k_2 n^{-\frac{4-\epsilon}{5-\epsilon}}.$$

On the contrary, when one constructs locally the function $\tilde{m}(x)$ using the value α_{opt} instead of $\alpha_i^\star$:

$$E_e(\theta(\alpha_{opt})) \simeq \delta(1 + k_3 n^{-\frac{20-7\epsilon}{5(5-\epsilon)}}),$$

from which

$$Ratio = \frac{E(Z_{opt})}{E_e(\theta(\alpha_{opt}))} \simeq 1 - k_3 n^{-\frac{20-7\epsilon}{5(5-\epsilon)}}$$

(here the k_j are positive constants).

It is now evident that the band, calculated by using $\theta(\alpha_{opt})$, is no longer an inference band but only a tool for data analysis : in fact on one hand $E(\theta(\alpha_i^\star))p_{ii}(\alpha_i^\star)$ and $EMSE_i(\alpha_i^\star)$ are of the order of $n^{-\frac{4-\epsilon}{5-\epsilon}}$, but on the other $E(\theta(\alpha_{opt}))p_{ii}(\alpha_{opt})$ is of the order of $n^{-\frac{4}{5}}$.

In order to obtain a better estimate of the function on that interval and more reliability it is appropriate to modify locally the value of α.

This can be done according to the following law:

$$\alpha = \alpha_{opt} * g\{\frac{Z_{opt}}{\theta(\alpha_{opt})}\},$$

where g is an increasing function of its argument and $g(1) = 1$.

5. Numerical results

We refer to two cases from the references.

For the estimates of α_{opt} and σ_r^2 see Bozzini and Lenarduzzi (1988b) .

Case 1: test function from Müller and Stadtmüller (1987):

$$g_2(x) = h(0.25, 0.05) + h(0.5, 0.1),$$

where h is a Gaussian density $N = 100$, $k = 4.48$, $R = .039$, $\sigma_1 = 0.01$, $\sigma_2 = 0.02$.

The function is smooth, so a good approximation accuracy is reached and the confidence band is uniform enough, by using (6). The results are presented in figure 1.

Case 2: test function number 4 in Wahba (1983):

$$f(x) = \begin{cases} 0, & 0 \le x \le \frac{1}{3}, \\ 36(t - \frac{1}{3}), & \frac{1}{3} \le t \le \frac{1}{2}, \\ 36(\frac{2}{3} - t), & \frac{1}{2} \le t \le \frac{2}{3}, \\ 0, & \frac{2}{3} \le t \le 1, \end{cases}$$

normal noise $\sigma = 0.02$, $N = 100$, $k = 4.48$, $R = .049$, $\alpha = 0.01$.

The function presents discontinuity points in the first derivative.

When using (6) to construct the inference band, one observes an enlargement of the band near the discontinuity points (see figure 2).

In figure 3 can be seen the approximation with the global α and the data.

To obtain better accuracy it is possible to use the method described in section 4, with the following function:

$$g(ratio) = (ratio)^{10}.$$

The resulting band is presented in figure 4.

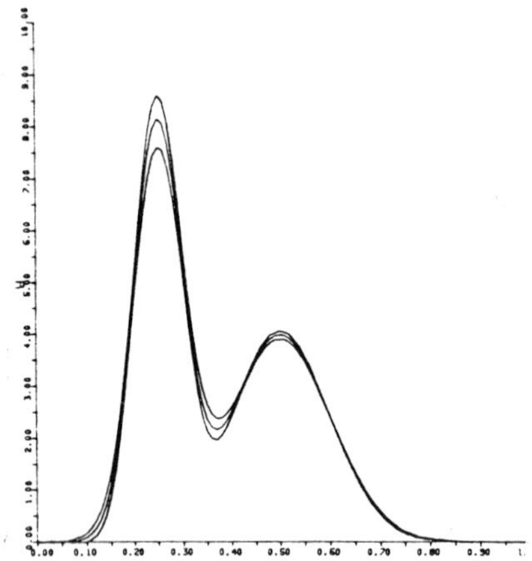

Figure 1

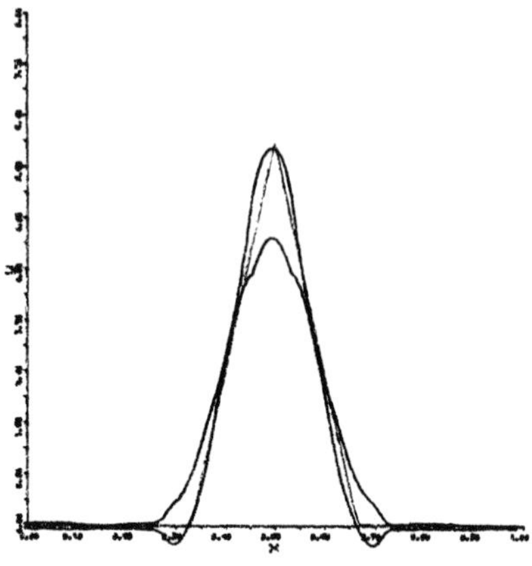

Figure 2

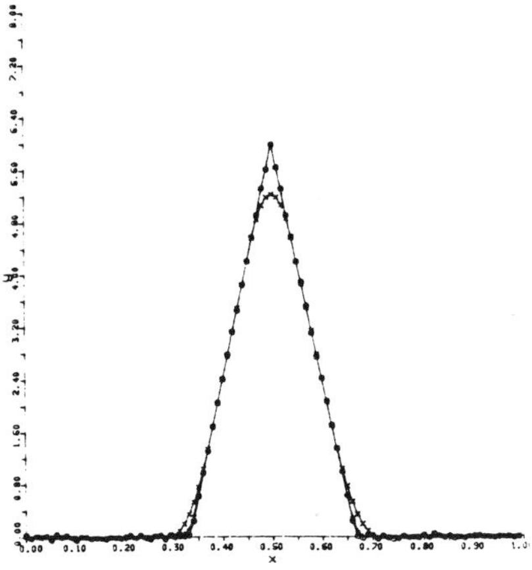

Figure 3

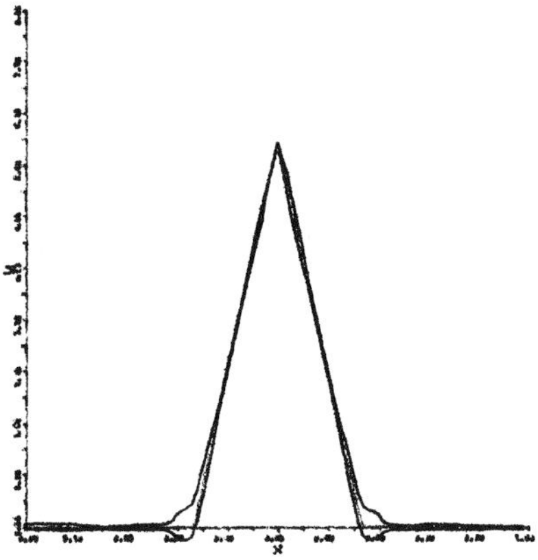

Figure 4

References

Bozzini, M. and Lenarduzzi, L. (1988a), Smoothing with weight variable both in shape and support. To be submitted

Bozzini, M. and Lenarduzzi, L. (1988b), Smoothing with weight variable both in shape and support: numerical aspects. To be submitted

Müller, H. G. and Stadtmüller, U. (1987), Variable bandwidth kernel estimators of regression curves. Ann. Statist. 15, 182-201

Ross, S. M. (1980), 'Introduction to Probability Models', Academic Press, New York, pp 60-61

Silverman, B. W. (1985), Some aspects of the spline smoothing approach to non-parametric regression curve fitting. J. Roy. Statist. Soc. B, 47, 1-52

Wahba, G. (1983), Bayesian confidence intervals for the cross validated smooth ing spline. J. Roy. Statist. Soc. B, 45, 135-150

5. Complex Approximation

Numerical methods for Chebyshev approximation of complex-valued functions

G. A. Watson

Department of Mathematics and Computer Science, University of Dundee, UK

Abstract This paper is concerned with the problem of best Chebyshev approximation of a given complex function on a contour of the complex plane. The emphasis is on the provision of methods which can combine global convergence with rapid local convergence, and some different approaches are examined, with particular attention being paid to the special case when the approximating function is linear. Examples are used to illustrate various situations which can arise, and numerical results are presented to demonstrate the perfomance of the methods on a variety of linear problems.

Keywords: Chebyshev approximation, complex functions, algorithms, global convergence.

1. Introduction

Let Γ be a contour of the complex plane C, and let $r:C^n \times \Gamma \rightarrow C$ be an analytic function. Then of interest here is the problem:

$$\text{find } \mathbf{z} \in C^n \text{ to minimise } \|r(\mathbf{z},.)\| , \tag{1}$$

where $\|r(\mathbf{z},.)\| = \max_{w \in \Gamma} |r(\mathbf{z},w)|$, and $\mathbf{z} = (z_1, z_2, \ldots, z_n)^T \in C^n$. Such problems frequently arise from best approximation problems defined on a simply connected region of the complex plane with boundary Γ, for the use of the maximum modulus theorem means that attention may be restricted to the boundary. For $j = 1,2,\ldots,n$, let $\phi_j(\mathbf{z},w)$ denote the partial derivative of r with respect to z_j. By considering the real problem equivalent to (1) (i.e. with $2n$ real variables formed by taking the real and imaginary parts of z), the following result may be established (see, for example, Ben-Tal et al (1979)), giving necessary conditions for a solution to (1).

Theorem 1 Let $\mathbf{z}\in C^n$ solve (1). Then there exist $m\leq 2n+1$ points $w_1,w_2,\ldots,w_m$ in Γ, with

$$|r(\mathbf{z},w_j)| = \|r(\mathbf{z},.)\|, \quad j=1,2,\ldots,m,$$

and λ_j, $j=1,2,\ldots,m$ real nonnegative numbers (not all zero) such that

$$\sum_{j=1}^{m}\lambda_j\overline{r(\mathbf{z},w_j)}\phi_k(\mathbf{z},w_j) = 0, \ k=1,2,\ldots,n.$$

Remark For the special case when r is an affine function of z, it follows that the conditions of the theorem are also sufficient for a solution, and therefore represent characterization conditions (see also Lorentz(1966)). Otherwise the conditions may be said to define a stationary point.

Assume that Γ can be parameterized by the real number $t\in[\alpha,\beta]$ so that $w:[\alpha,\beta]\rightarrow\Gamma$ and $w_j = w(t_j)$, $j=1,2,\ldots,m$. Define $f:R^{2n}\times R\rightarrow R$ by

$$f(\mathbf{x},t)=|r(\mathbf{z},w(t))|^2, \quad t\in[\alpha,\beta],\mathbf{z}\in C^n,$$

where $\mathbf{x} = \begin{bmatrix}\mathbf{z}^R\\ \mathbf{z}^I\end{bmatrix}\in R^{2n}$, and the superscripts R and I denote real and imaginary parts respectively. Without loss of generality, assume that $t_1,t_2,\ldots,t_p$ are internal to $[\alpha,\beta]$. Then if m and p are known, the conditions of the theorem may be written as the nonlinear system of equations

$$\sum_{j=1}^{m}\lambda_j\nabla f(\mathbf{x},t_j) = 0, \tag{2a}$$

$$1-\sum_{j=1}^{m}\lambda_j = 0, \tag{2b}$$

$$f'(\mathbf{x},t_j) = 0, \quad j=1,2,\ldots,p, \tag{2c}$$

$$f(\mathbf{x},t_j)-h = 0 \ , \ j=1,2,\ldots,m, \tag{2d}$$

where ∇ denotes differentiation with respect to $\mathbf{x}\in R^{2n}$, the dash denotes differentiation with respect to t and the fact is used that

$$\nabla f(\mathbf{x},t)=2\begin{bmatrix}Re(\overline{r}\boldsymbol{\phi})\\ -Im(\overline{r}\boldsymbol{\phi})\end{bmatrix}\in R^{2n},$$

with $\boldsymbol{\phi}=(\phi_1,\phi_2,\cdots,\phi_n)^T$. This is a system of $2n+m+p+1$ real equations for the $2n+m+p+1$ real unknowns $\mathbf{x} \in R^{2n}, \lambda_j \in R$, $j=1,2,..,m$, $t_j \in R$, $j=1,2,...,p$, and $h \in R$, where $h=\|r(\mathbf{z},.)\|^2$ at a solution to (1).

2. A locally convergent method

The Jacobian matrix of the system (2) may be written in the form

$$\begin{bmatrix} H & 0 & A^T\Lambda & G^T \\ 0 & 0 & 0^T & -\mathbf{e}^T \\ A & 0 & D & 0 \\ G & -\mathbf{e} & \begin{bmatrix} D_1 \\ 0 \end{bmatrix} & 0 \end{bmatrix} \tag{3}$$

where

$$H = \sum_{j=1}^{m} \lambda_j \nabla^2 f(\mathbf{x}, t_j) \in R^{2n \times 2n},$$

$$\Lambda = \text{diag}\ \{\lambda_1, \lambda_2, \ldots, \lambda_p\},$$

$$A_{jk} = \frac{\partial^2 f(\mathbf{x}, t_j)}{\partial t\, \partial x_k}, \quad j=1,2,...,p, \quad k=1,2,...,2n,$$

$$G_{jk} = \frac{\partial f(\mathbf{x}, t_j)}{\partial x_k}, \quad j=1,2,...,m, \quad k=1,2,...,2n,$$

$$D = \text{diag}\ \{f''(\mathbf{x}, t_j),\ j=1,2,..,p\},$$

$$D_1 = \text{diag}\ \{f'(\mathbf{x}, t_j),\ j=1,2,..,p\},$$

and $\mathbf{e}$ denotes the vector in R^m all of whose components are 1. Notice that D is negative semi-definite at a solution to (2) which is also a solution to (1). Also

$$\nabla^2 f(\mathbf{x}, t) = 2\begin{bmatrix} Re(\phi\overline{\phi}^T) & Im(\phi\overline{\phi}^T) \\ -Im(\phi\overline{\phi}^T) & Re(\phi\overline{\phi}^T) \end{bmatrix} + 2\begin{bmatrix} Re(r\overline{T}) & Im(r\overline{T}) \\ Im(r\overline{T}) & -Re(r\overline{T}) \end{bmatrix}, \tag{4}$$

where T is the Hessian matrix of r with respect to $\mathbf{z}$.

The system of equations defining the Newton step is therefore an unsymmetric $2n+m+p+1$ by $2n+m+p+1$ system of equations. Considerable improvement may be obtained if use is made of the fact that D_1 is zero at a solution to (1), so that it may be replaced by zero in (3) without inhibiting a second order convergence rate. The system of equations defining the increment vector may then be written

$$\begin{bmatrix} H & 0 & A^T\Lambda & G^T \\ 0 & 0 & 0 & -\mathbf{e}^T \\ A & 0 & D & 0 \\ G & -\mathbf{e} & 0 & 0 \end{bmatrix} \begin{bmatrix} \delta\mathbf{x} \\ \delta h \\ \delta\mathbf{t} \\ \delta\boldsymbol{\lambda} \end{bmatrix} = \begin{bmatrix} -G^T\boldsymbol{\lambda} \\ \boldsymbol{\lambda}^T\mathbf{e}-1 \\ -\mathbf{f}' \\ h\mathbf{e}-\mathbf{f} \end{bmatrix}, \tag{5}$$

where $\mathbf{f}' \in R^p$ denotes the vector with jth component $f'(\mathbf{x}, t_j)$, $j=1,2,\ldots,p$, and $\mathbf{f} \in R^m$ denotes the vector with jth component $f(\mathbf{x}, t_j)$, $j=1,2,..,m$. This system of equations may readily be broken down into smaller systems. Indeed the following algorithm may be used which first computes a first order estimate of $\boldsymbol{\lambda}$ (an approximation to $\boldsymbol{\lambda}+\delta\boldsymbol{\lambda}$ satisfying (5)) and then provides approximations to $\delta\mathbf{x}$, δh, $\delta\mathbf{t}$ satisfying (5). When used as increment vectors, a second order convergence rate is not inhibited (see, for example Watson(1985)).

Algorithm 1

Step 1 Let $C^T \equiv \begin{bmatrix} G^T \\ -\mathbf{e}^T \end{bmatrix} = [\ Y : Z\ ] \begin{bmatrix} R \\ 0 \end{bmatrix} = Y\ R,$

where $[Y\ :\ Z]$ is $(2n+1)\times(2n+1)$ orthogonal and R is $m\times m$ upper triangular, nonsingular if C has full rank (a necessary condition for the system (5) to be nonsingular).

Step 2 Determine $\boldsymbol{\lambda}$ satisfying $R\boldsymbol{\lambda} = -Y^T\mathbf{e}_{2n+1}$ ($\mathbf{e}_{2n+1}$ is the $(2n+1)^{th}$ co-ordinate vector) and scale λ so that $\boldsymbol{\lambda}^T\mathbf{e} = 1$.

Step 3 Solve $R^T\mathbf{d}_1 = h\mathbf{e}-\mathbf{f}$.

Step 4 Form $W = H - A^T\Lambda D^{-1}A$, and solve

$$Z_1^T W Z_1 \mathbf{d}_2 = Z_1^T(\ A^T\Lambda D^{-1}\mathbf{f}' - WY_1\mathbf{d}_1\ ) - \mathbf{z}_2,$$

where Y_1,Z_1 denote the first $2n$ rows of Y,Z respectively, and $\mathbf{z}_2^T$ denotes the last row of Z.

Step 5 Form $\begin{bmatrix} \delta\mathbf{x} \\ \delta h \end{bmatrix} = Y\mathbf{d}_1 + Z\mathbf{d}_2$ and set $\mathbf{x} = \mathbf{x}+\delta\mathbf{x}$, $h = h+\delta h$.

Step 6 Form $\delta\mathbf{t} = -D^{-1}(\mathbf{f}'+A\,\delta\mathbf{x})$ and (unless convergence) set $\mathbf{t} = \mathbf{t}+\delta\mathbf{t}$ and return to Step 1.

This algorithm will be dominated by the calculation of the QR factorization of C^T in Step 1. It can converge to a solution of (1) if provided with good initial approximations, and with the correct information about m and p. Of course even if convergence is obtained it has still to be confirmed that $h = \|f\|$, and this is a nontrivial problem in itself. Its solution is connected with a modification of this method in which the vector $\mathbf{t}$ at each iteration is defined to be the appropriate set of m local maxima of the current f in $[\alpha,\beta]$: this calculation can be done as a preliminary step, with Step 6 of course no longer required. Provided that D is nonsingular, and $f'(\mathbf{x},t_j)\neq 0, j=p+1,..,m$ the implicit function theorem can be used to give $\mathbf{t}$ as a differentiable function of $\mathbf{x}$, and the Newton step (in the variables $\mathbf{x},h,\lambda$) derived. The analogue of Algorithm 1 is then obtained by introducing Step 0 as described, dropping Step 6 (testing for convergence in Step 5) and setting $\mathbf{f}'$ to zero in Step 4: a second order convergence rate can again be established.

The following examples all use Algorithm 1 in its original form.

Example 1 (Glashoff and Roleff(1981)) Consider the approximation of w^8 by a complex polynomial of degree 7 on the rectangle $[-2,2] \times [-1,1]$ in the complex plane. Because of symmetry, it is sufficient to consider the approximation of w^8 by a *real* linear combination of $1,w^2,w^4,w^6$ on

$$w(t) = \begin{cases} 2+it, & 0\leq t\leq 1 \\ (3-t)+i, & 1\leq t\leq 3 \end{cases} .$$

The fact that the coefficients are real means that H,A and G may all be reduced in size with consequent reduction in the computation involved in Algorithm 1. Taking the initial approximation $\mathbf{z} = \{(-15.19822,0), (46.28396,0), (-31.88876,0), (7.748537,0)\}$, $\mathbf{t} =$ (0.325, 1.8, 1.0, 3.0), with $m = 4$ and $p = 2$ (interpreting $t_3 = 1.0$ as an end point), and $h = 10428.39$, the performance of the algorithm is summarized in Table 1. The column headed k gives the iteration number and δ gives the maximum modulus component of $\delta\mathbf{x},\delta h$ and $\delta\mathbf{t}$. The CPU time (DEC 10 single precision) was 0.78 seconds. (Notice that the size of the matrix (3) in this case is 11×11, whereas the corresponding Jacobian matrix in the method used by Glashoff and Roleff is 15×15).

k	δ	t_1	t_2	h
1	0.149281	0.315838	1.794251	10428.54
2	0.479269	0.316292	1.795387	10429.02
3	0.008201	0.316291	1.795388	10429.03
4	0.000025	0.316291	1.795388	10429.03

Table 1

Example 2 Consider the approximation of $1/(w - (2+i))$ by a polynomial of degree 7 on the part of the unit circle defined by $\theta \in [0,3\pi/2]$. In this case m=9 and p=7. Taking the initial approximation **z** = {(–0.4,0.2), (–0.1202,0.1602) , (–0.01653,0.08797), (0.01073,0.03786), , (0.001226,0.01225), (0.008144,0.002251), (0.004343,–0.00308), (0.001954,–0.000459)}, **t** = (0,0.3436 ,0.8467, 1.399, 2.0985, 2.815, 3.607, 4.3196, $3\pi/2$) and h=0, the algorithm performed as shown in Table 2. The CPU time was 2.71 seconds.

k	δ	$h\times10^{-7}$
1	**0.04765**	**4.0788**
2	**0.04300**	**4.3852**
3	**0.01111**	**4.44382**
4	**0.000362**	**4.44747**
5	**0.000005**	**4.44755**

Table 2

Example 3 This is the approximation of $exp(w)$ on the part of the unit circle defined by $\theta \in [0,\pi]$ by a rational function formed by the quotient of two complex polynomials of degree 2: the rational function is normalized by choosing the constant in the denominator to be 1. In this case m = 6 and p = 4. Taking the initial approximation **z** = {(1.0066,0.0003384), (0.51337,0.039861), (0.085256, 0.01995), (–0.48537, 0.037976), (0.073634, -0.015799)}, (with the natural ordering of the coefficients), **t** = { 0, 7α, 22α, 42α, 65α, π }, where $\alpha = \pi/81$, and h=0.0, the algorithm performed as summarized in Table 3.

k	δ	$h\times10^{-7}$
1	**0.02148**	**1.69225**
2	**0.01872**	**1.77062**
3	**0.000452**	**1.77476**
4	**0.000002**	**1.77477**

Table 3

The question remains: how can information about m and p, and also good approximations to the unknowns, be provided? It should be emphasised that extremely good initial approximations may be required: for instance in Example 3, small perturbations of the initial values can lead to divergence. In particular, very good approximations to **t** are often necessary: there is more about this later on. Perhaps the most obvious way of finding initial data is from the solution (or approximate solution) of a discretization of (1), where Γ is replaced by a finite discrete subset, and this gives rise to a class of two-phase methods. Algorithms of this type have been suggested for linear problems (when r is is an affine function of **z**) by, for example, Glashoff and Roleff (1981), Grothkopf and Opfer (1982), Opfer (1982) and Mason and Opfer(1986), based on the formulation of the discrete problem as a semi-infinite linear programming problem. Linear programming methods may be used to approximately solve this problem, and advantage can be taken of the special structure to make this efficient (see, for example, Streit (1985, 1986), Streit and Nuttall (1982,1983)). Notice that this procedure requires a **double** discretization of the original problem (1).

An alternative first phase procedure for the linear problem is suggested by Watson(1988) where the discretization of problem (1) is (directly) solved as a nonlinear programming problem. Advantage is taken of the special form of the problem so that a second order convergent method is obtained, and computational efficiency is achieved by a procedure which restricts attention to small subsets of the original set of discrete points. This method may be adapted to apply to nonlinear problems, although since the convex nature of the problem is lost, the provision of a satisfactory algorithm represents a nontrivial modification and the implementation details have still to be resolved. Some success has been obtained through the use of a Gauss-Newton type of approach (the simple device of ignoring the second term on the right hand side of (4)), and this was used to generate the initial data for Example 3.

Of course both linear and nonlinear discretized problems are examples of nonlinear minimax optimization problems, and standard methods for such problems are available. The main computational effort is expended in identifying the correct index set on which a solution is defined (analogous to the points $w_1, w_2, \ldots, w_m$ of Theorem 1), and most methods are active set methods which iterate towards a correct index set through equality constrained quadratic programming subproblems defined on estimates of this set. Recent developments have been concerned with making better use of the special structure and solution characterization of minimax problems, and for example a method having this goal is given by Yuying Li(1988). A particular feature of the problems considered here is that fine discretizations (and therefore a large number of points) may be necessary, and the relative effectiveness of different approaches has yet to be studied. In particular, the treatment of nonlinear problems of the type (1) and their discretizations has received little attention (with the exception of some rational approximation problems) and so is not something that can be properly dealt with here. For the rest of this paper, therefore, although much of the theory carries over (or may be modified to apply) to the nonlinear case, attention will be confined to the treatment of the linear problem for which

$$r(\mathbf{z}, w) = \sum_{j=1}^{n} z_j \phi_j(w) - F(w), \quad w \in \Gamma, \tag{6}$$

where $F: \Gamma \rightarrow C, \phi_j: \Gamma \rightarrow C, j=1,2,...,n$ are analytic functions. Perhaps the most important consequence of this is that $\nabla^2 f(\mathbf{x}, t)$ is given by the first term on the right hand side of (4) and so is positive semi-definite; therefore so also are H and W provided that the components of $\boldsymbol{\lambda}$ are non-negative. H is of course independent of $\mathbf{x}$.

3. Globalization

The problem (1) may be stated in the form

$$\begin{gathered} \text{minimize } h \\ \text{subject to } f(\mathbf{x}, t) \leq h, \quad t \in [\alpha,\beta], \end{gathered} \tag{7}$$

and a corresponding discrete problem may be written

$$\begin{gathered} \text{minimize } h \\ \text{subject to } f(\mathbf{x}, t_i) \leq h, \quad i=1,2,...,s, \end{gathered} \tag{8}$$

where $t_i \in [\alpha,\beta]$, $i=1,2,...,s$. Clearly if the discrete set contains the points t_i defined by Theorem 1 then the minimum values of both problems will be the same. One approach to solving (7) is therefore to iterate on a sequence of discrete subsets which in the limit contains the appropriate optimal set.

Given $\mathbf{x} \in R^{2n}$ let $t_1, t_2, ..., t_q$ be the set of local maxima (assumed finite) of $f(\mathbf{x}, t)$ in $[\alpha,\beta]$. Then

$$f'(\mathbf{x}, t_j) = 0, \quad t_j \in (\alpha,\beta),$$

and provided that $f''(\mathbf{x}, t_j) \neq 0, \; t_j \in (\alpha,\beta), f'(\mathbf{x}, t_j) \neq 0, \; t_j \in \{\alpha,\beta\}$, the implicit function theorem may be applied to give t_j, $j=1,2,...,q$ as differentiable functions of $\mathbf{x}$. Based on these q local maxima, and given $\boldsymbol{\lambda} \in R^q$ non-negative , define $\mathbf{f} \in R^q, G \in R^{q \times 2n}, H \in R^{2n \times 2n}, \; W \in R^{2n \times 2n} \;, C \in R^{q \times (2n+1)}$ as in Algorithm 1. Let $h = \max_{1 \leq j \leq q} f(\mathbf{x}, t_j)$ and consider the quadratic programming problem

$$\text{minimize } p+\tfrac{1}{2}\mathbf{d}^T W\mathbf{d}$$

$$\text{subject to } G\mathbf{d}-p\mathbf{e} \le h\mathbf{e}-\mathbf{f}. \qquad (9)$$

Theorem 2 For r given by (6), let $p,\mathbf{d}$ solve (9) defined at $\mathbf{x}$. Then

(a) if $p = 0$, $\mathbf{x}$ solves (7),

(b) if $p < 0$, $\mathbf{d}$ is a descent direction for $\|f(\mathbf{x}, .)\|$ at $\mathbf{x}$.

Proof By assumption

$$W = H - A^T \Lambda D^{-1} A$$

is positive semi-definite if it is defined. The Kuhn-Tucker conditions imply the existence of a non-negative vector $\mu \in R^q$ such that

$$\mathbf{e}^T \mu = 1,$$

$$W\mathbf{d} + G^T \mu = 0.$$

Thus

$$\mathbf{d}^T W\mathbf{d} + \mathbf{d}^T G^T \mu = 0,$$

so that $\mathbf{d}^T G^T \mu \le 0$. Also

$$\mu^T (h\mathbf{e}-\mathbf{f}-G\mathbf{d}+p\mathbf{e}) = 0,$$

implies $h+p-\mu^T\mathbf{f} = \mu^T G\mathbf{d} \le 0$, so that $p \le \mu^T\mathbf{f}-h \le 0$.

If $p=0$, $\mathbf{d}^T W\mathbf{d} = 0$, so that if $\mathbf{d} \ne 0$, there is another solution $p = 0$, $\mathbf{d} = 0$, and the Kuhn-Tucker conditions are then equivalent to (2) being satisfied.

Now assume $p < 0$. We may write

$$f_j(\mathbf{x}) = f(\mathbf{x}, t_j(\mathbf{x})) \quad j=1,2,..,q,$$

showing the dependence of t_j on x. Further for $\gamma>0$ small enough

$$\|f(\mathbf{x}+\gamma\mathbf{d}, .)\| = \max_{1\le j\le q} f_j(\mathbf{x}+\gamma\mathbf{d}).$$

Now define J by

$$h = f_j, \quad j \in J.$$

Then $\mathbf{d}^T\nabla f_j \le p < 0$, $j \in J$, so that

$$f_j(\mathbf{x}+\gamma\mathbf{d}) = f_j(\mathbf{x})+\gamma\mathbf{d}^T\nabla f_j+0(\gamma^2)$$

$$<f_j(\mathbf{x})$$

for $\gamma > 0$ small enough. It follows that

$$\|f(\mathbf{x}+\gamma\mathbf{d}, .)\| < \max_{1\le j\le q} f_j(\mathbf{x}) = \|f(\mathbf{x}, .)\|,$$

and the result is proved.

Remark This result goes through for nonlinear problems provided that W is a positive semi-definite matrix (or is replaced by a positive semi-definite matrix).

The new point $\mathbf{x}+\gamma\mathbf{d}$ may be obtained by a line search, and it is usual to choose γ to satisfy the inequality

$$\|f(\mathbf{x}+\gamma\mathbf{d}, .)\| \le \|f(\mathbf{x}, .)\|+\gamma p\tau , \tag{10}$$

where τ is a small positive number, for example 0.0001, and γ is large in [0,1], with the value 1 chosen if possible. Then the following result may be established.

Theorem 3 Let $\{\mathbf{d}^k\},\{p^k\}$ be sequences defined by solutions of (9) at $\mathbf{x}^k$ using (10), where the superscripts k on other quantities imply evaluation at $\mathbf{x}^k$. Then if $\{W^k\},\{\mathbf{d}^k\}$ are bounded, the limit points of $\{\mathbf{x}^k\}$ solve (7).

Proof The result that $\{p^k\}\to 0$ may be established by the application of standard techniques available in optimization theory.

Going to a subsequence if necessary, let $\{\mathbf{x}^k\}\to \mathbf{x}^*$. Now, using boundedness, and going to further subsequences if necessary, $\mu^k\to \mu^*, W^k\to W^*$, $\mathbf{d}^k\to \mathbf{d}^*$ as $k\to\infty$. Thus $\mathbf{d}^*$, p^* (=0) solve (9) at $\mathbf{x} = \mathbf{x}^*, W=W^*$ with objective function value zero. Thus a (possibly different) solution is given by $\mathbf{d} = 0$, $p = 0$ and the conclusion that $\mathbf{x}$ is a solution follows as before.

The following algorithm may be interpreted as being of multiple exchange type.

Algorithm 2

Step 0 Determine the local maxima of f in $[\alpha,\beta]$, say $t_1,t_2,....,t_q$. Set $h = \|f(\mathbf{x}, .)\|$.

Step 1 Solve (9), and unless there is convergence, choose γ so that (10) is satisfied, set $\mathbf{x} = \mathbf{x}+\gamma\mathbf{d}$ and go to Step 0.

If the problem (9) is solved with the constraints fixed as *equalities* , then provided that $\mu \geq 0$ the situation is as before. However the solution may now be obtained in an efficient manner by steps similar to those used in Algorithm 1. An alternative exchange algorithm may therefore be defined as follows.

Algorithm 2a

Step 0 As in Algorithm 2.

Steps 1-5 As in Algorithm 1 (with $\mathbf{f}, G$, etc. defined on the current set of local maxima $t_1,t_2, \cdots t_q$), except that we can set $\mathbf{f}' = 0$ in Step 4, and a full step is not necessarily taken in Step 5 for the new $\mathbf{x}$.

Although the steps of Algorithm 2a are computationally more efficient (and the connection with Algorithm 1 immediately gives a desirable local property), there is no guarantee that all the components of the multiplier vector will be non-negative at solutions to (the modified) problem (9), in particular far from a solution to (1). Therefore an active set strategy (for handling negative μ_j) is required for a practically useful (globally convergent) algorithm. On the other hand, the solution of the inequality constrained problem (9) automatically picks out the correct active set in the event that not all constraints are active. Eventually, the two algorithms should be effectively the same, so that a second order convergent rate is normal with Algorithm 2. An important part of both these algorithms is the calculation of all the local maxima: this is needed not just for Step 0, but also for the implementation of the step length test. The process is not a finite one, and typically is achieved in two stages: firstly a grid search to approximately identify the locations of the maxima; secondly a local procedure (for example Newton's method) to give more accurate values. Incidentally, this is an area in which the availability of a parallel computing facility could lead to great benefits.

An exchange method of either kind can be applied to solve *discretizations* of continuous problems, with fairly obvious modifications to the algorithms (reflecting the absence of (2c) from the conditions to be satisfied): local maxima on the finite set *only* are used, $W = H$ in Step 4 of Algorithm 1 and also in (9) and the first term on the right hand side of the system of equations in Step 4 is absent. However there are potential difficulties, if the number of extrema of discrete and continuous problems do not match. (In this context *extrema* refers to points where the norm is attained.) For example for the approximation of $F(w) = 1/(w-(2+i))$ by a polynomial of degree 3 in the first quadrant of the unit circle, there are 5 extrema of the continuous solution, with parameter values 0, 0.259415, 0.733998, 1.269364, $\pi/2$. However the solution using the algorithm of Watson on $m=101$ equispaced points gives rise to 6 extrema, including the neighbouring pair 0.2513 and 0.2670, and taking $m = 1001$ equispaced points gives 7 extrema, including the triple 0.73042, 0.73356 and 0.73827. For many problems this discrepancy does not occur, and the extent to which it can be attributed to the use of finite precision arithmetic is not clear. Nevertheless, there are two possible dangers associated with this phenomenon: firstly ill-conditioning because of the proximity of active points, and secondly failure of the algorithm through failure to properly locate all of these relevant local maxima as the computation proceeds. If the situation is an inherent property of the problem, it parallels a phenomenon which occurs in real linear Chebyshev approximation: if a continuous real problem (with n real coefficients) is discretized, then in the full rank case there is always a solution with (n+1) extrema, and this is the solution which is obtained if a standard technique is applied. However for non-Chebyshev set problems, the solution to the original continuous problem (assumed unique) may have fewer than (n+1) extrema: such problems have been referred to as *singular* by Osborne and Watson(1969). A difference in the complex case is that it does not seem possible to predict in advance the number of extrema of discretized problems, and indeed this appears to depend on the particular discretization. Nevertheless the same interpretation of *singularity* may be made.

The above phenomenon has no bearing on the direct solution of the continuous complex problem by exchange methods of the type described above, and Algorithm 2 was applied to a number of different problems. Far from the solution, convergence could be slow and two modifications were found to be beneficial : firstly, W was modified to $W+\mu I$, for some $\mu \geq 0$, with μ being allowed to decrease as the algorithm approaches the solution; secondly in the early stages only local maxima on a discrete set were determined (and so W was chosen equal to H), with accurate local maxima obtained only when p became sufficiently small.

Example 4 This is the approximation of w^3 by a polynomial of degree 2 on the first octant of the unit circle. Initially local maxima were sought on a grid of 101 equispaced points. The initial approximation was given by $z_j = (0,0)$, $j=2,3$, $z_1 = (1,0)$ and the performance of the method is summarized in Table 4. The CPU time was 6.63 seconds.

k	q	μ	p	γ	$\|r\|$
1	1	1.0	-5.85	1	1.274
2	2	0.5	-1.60	1	0.647
16	4	0.0001	-1.6E-5	1	0.014709
17	4	0	-8. E-8	1	0.014708
18	4	0	-2. E-8	1	0.0147077
19	4	0	-1. E-8	1	0.0147077

Table 4

An alternative to the above method is to revert to a 2-phase approach, with algorithm 2 (or indeed Algorithm 2a) as a second phase. This makes the 2-phase method more robust (at the expense of extra computation) and permits convergence to a solution of (7) even if incorrect information is supplied at the end of the first phase, or poor approximations only are available. It also means, however, that while global convergence properties are important, the main emphasis can be on local behaviour; if Algorithm 2 is applied with all constraints initially set to equalities, then frequently no change is required to the active set, so little loss in efficiency results in using Algorithm 2 as opposed to Algorithm 2a. In addition W may be used without modification, and $\gamma = 1$ is the expected value satisfying (10). Therefore the next algorithm tested was based on combining the method of Watson with Algorithm 2. In fact a version of the first method was used where solutions were obtained on successive discrete sets defined as follows. Initially $2n+1$ equispaced points were used. Subsequently each set consisted of the current set of extreme points supplemented by all the local maxima of f on the original discrete point set. When a sufficiently small value of p was reached, or a sufficiently small increment vector was obtained, Algorithm 2 was entered.

Algorithm 3

First Phase The algorithm of Watson, modified as described in the previous paragraph.

Second Phase Algorithm 2, with all constraints of (9) initially set to active.

Both phases of this algorithm involve the solution of a sequence of quadratic programming problems, so that in addition to giving a desirable combination of global and rapid local convergence, they also make use of the same software. At each iteration of both phases, a particular subset of discrete points from $[\alpha,\beta]$ is identified and the constraints of the quadratic programming problem are defined by this subset: in the first phase, this remains constant for several iterations, and not all constraints are normally active; in the second phase, the subset changes at each iteration, and all constraints are usually active.

Consider Example 4 solved by Algorithm 3 (from the same initial approximation). Table 5 gives results of the application of the above algorithm applied to the discretized problem. The number k gives the outer iteration count (the number of times a discrete solution is obtained on a *fixed* discrete subset of the original discrete set), q is the number of quadratic programming subproblems solved at each outer iteration, $||\mathbf{r}_k||$ is the value reached of the norm on the current discrete set, and $||\mathbf{r}||$ is the norm evaluated on all 101 points. The number l is the number of local maxima added in at each step, and p denotes the value of p reached on termination. The data input to Algorithm 2 was:

$$\mathbf{z} = \{(0.3679780,0.8883755), (-1.988566,-1.988564), (2.630987,1.089790)\}.$$

After 2 iterations, the maximum modulus component of the increment vector was less than 0.00001 and the solution obtained was

$$\mathbf{z}^* = \{ (0.368115,0.888709), (-1.989059,-1.989059), (2.631323,1.089929)\},$$

$$\mathbf{t}^* = (0,0.198232,0.587165,\tfrac{\pi}{4}),$$

$$\lambda^* = (0.167767,0.332231,0.332236,0.167765),$$

with $||\mathbf{r}^*|| = 0.147077$. The total CPU time was 7.98 seconds.

k	q	$\|\|\mathbf{r}_k\|\|$	$\|\|\mathbf{r}\|\|$	l	p
1	14	0.011666	0.022209	3	-2.0 -7
2	4	0.014607	0.014865	2	-7.1 -10
3	3	0.014706			-4.6 -9

Table 5

For approximation problems defined on the whole of the unit circle, it is not uncommon for there to be a *constant* error curve, in other words for the minimum norm value to be attained at all points on the unit circle. Studies of this phenomenon for polynomial (and rational) approximation have been made by, for example, Trefethen (1981a,1981b). In this case it is clear that Algorithm 2 will eventually break down; on the other hand , for methods of two-phase type, it is easy to identify this situation at the conclusion of the first phase.

Example 5 Let $f(w) = 1/(w-2)$ be approximated by a polynomial of degree 6 on the whole of the unit circle. The performance of the algorithm of Watson(1988) applied on 100 equispaced points from $\mathbf{z} = \mathbf{e}_1$ is summarised in Table 6. At the end of the second iteration $\|\mathbf{r}\| = 0.005208$ attained at all 100 points, and the conclusion that the error curve is a constant on the unit circle follows.

k	q	$\|\mathbf{r}_k\|$	l
1	10	0.002905	7
2	5	0.005208	

Table 6

The possibility of this kind of behaviour points the way to another source of difficulty in accurately computing solutions to continuous complex Chebyshev approximation problems: although the error curve is not constant, it may be *close* to being constant. This means that identification of the positions of the points where the norm is attained may be quite difficult, and the positions of the local maxima as the computation proceeds will be very sensitive to changes in the coefficient values.

Example 6 Consider the approximation of $f(w) = 1/(1+(w+1)^2)^{\frac{1}{2}}$ by $\sum_{j=1}^{n} z_j(1+w)^{1-j}$ on $\{w : w = iy, -20 \le y \le 20\}$. This is an another example of a complex problem which has a solution for which $\mathbf{z}$ is real, and advantage can be taken of this to reduce the size of the computations: in particular symmetry can be used to restrict consideration to non-negative values of y. In Table 7a is shown the result of applying the algorithm as before on 51 equispaced points on $0 \le y \le 20$, taking $n = 4$ and starting from $\mathbf{z} = \mathbf{e}_1$. The minimum value of $\|\mathbf{r}\|$ is 0.016046 attained at the 5 points defined by $\mathbf{t} = (0.4, 0.8, 1.6, 2.0, 20.0)^T$. The points corresponding to local **minima** of $|r_i|$ are 0.0, 1.2 and 3.6 with values 0.015983, 0.016036 and 0.016039 respectively. Table 7b shows the performance of Algorithm 2 applied from the discrete solution; the continuous problem has just 3 extrema and the sensitivity of the local maxima to changes in the coefficients is clearly demonstrated. The total CPU time was 7.81 seconds.

k	q	$\|\mathbf{r}_k\|$	$\|\mathbf{r}\|$	l
1	**16**	**0.000891**	**0.086371**	**1**
2	**8**	**0.014233**	**0.024608**	**1**
3	**4**	**0.015954**	**0.017687**	**2**
4	**3**	**0.016041**	**0.016078**	**2**
5	**2**	**0.016045**	**0.016052**	**1**
6	**2**	**0.016046**		

Table 7a

k	t_1	t_2	t_3	δ	$\|r\|$
1	**0.571362**	**1.777237**	**20.0**	**0.000102**	**0.016059**
2	**0.334464**	**1.650488**	**20.0**	**0.000034**	**0.016051**
3	**0.439621**	**1.610323**	**20.0**	**0.000023**	**0.016049**
4	**0.465623**	**1.639444**	**20.0**	**0.000002**	**0.016049**

Table 7b

Finally the performance of Algorithm 3 applied to some other examples is summarized in Table 8. The different functions used were

$$f_1(w) = (\sin w / w)^{\frac{1}{2}},$$

$$f_2(w) = (w-(2+i))^{-1},$$

$$f_3(w) = (1+(w+1)^2)^{-\frac{1}{2}},$$

and other information defining the problems being solved is given in the table. All initial approximations were $\mathbf{z} = \mathbf{e}_1$. The number δ gives the maximum modulus component of the increment vector on termination, p denotes the final value of p, $\|r\|$ is the final value of the norm, and CPU denotes the CPU time in seconds.

$f(w)$	$\phi_j(w)$	w	α	β	n	$\delta\times10^{-5}$	$p\times10^{-9}$	$\|r\|$	CPU
$f_1(w)$	w^{j-1}	e^{it}	**0**	$\frac{\pi}{2}$	**4**	**.5**	**-2.0**	**0.000281**	**4.02**
$f_2(w)$	w^{j-1}	e^{it}	**0**	$\frac{\pi}{2}$	**4**	**5.0**	**-4.0**	**0.005028**	**15.86**
			0	$\frac{3\pi}{2}$	**6**	**.3**	**-10.0**	**0.003863**	**27.71**
			0	$\frac{3\pi}{2}$	**8**	**2.0**	**-4.0**	**0.000670**	**85.89**
$f_3(w)$	$(1+w)^{1-j}$	it	**0**	**20**	**4**	**.2**	**-3.0**	**0.000281**	**7.81**
			0	**20**	**6**	**.6**	**-0.2**	**0.000140**	**13.70**
	$(w+j)^{-1}$	it	**0**	**20**	**6**	**1.0**	**-7.0**	**0.003809**	**19.63**

Table 8

4. Concluding remarks

Some numerical methods have been presented for solving complex Chebyshev approximation problems defined on a contour of the complex plane. It has been assumed that a one-dimensional parameterization of the contour is available, and this has been exploited to increase efficiency of the methods, although it is not a limiting factor. Numerical results have been presented to demonstrate the effectiveness of different approaches for a variety of linear problems, and examples have been used to illustrate various points. The emphasis has been on the provision of methods which can combine global convergence with satisfactory rapid local convergence and it seems that this is best achieved within the context of a two-phase approach: this is characterized by the use of the solution to a discretization of the original problem (and other information produced) as input to a method with (primarily) good local properties for satisfying the characterization conditions. Conventional methods of this type do not monitor progress in the second phase (and in particular are not descent methods), and rely on the input data being sufficiently good that convergence to a solution of the original problem is achieved without difficulty.

The intention here has been to provide a measure of satisfactory progress in the second phase, so that (at the expense of additional computation) convergence to the solution can be more reasonably guaranteed. While the numerical experience is still somewhat limited, it is hoped that the present work will assist in the quest, and provide a satisfactory framework, for good numerical methods for solving linear complex Chebyshev approximation problems.

Finally, it is clear that nonlinear problems may be tackled by the methods presented here. The second phase (in theory) presents no difficulty, and so the main requirement is for a satisfactory method for solving the discretized problem, and this involves, in particular, deciding how to deal with the Hessian matrix of f. One possibility is to ignore that part of the Hessian matrix which requires second derivatives of r, (the second term on the right hand side of (4)) leading to a method of Gauss-Newton type. While this is simple, it is unlikely to be generally effective, and quasi-Newton or finite difference approximations to the Hessian matrix are more attractive propositions. It may be that there is little to be gained by considering alternatives to standard methods for nonlinear minimax problems, particularly if they are tailored to exploit special structure and properties as is done by Yuying Li(1988). This remains to be investigated and will be the subject of future research.

References

Ben-Tal,A., Teboulle,M. and Zowe, J. (1979), Second order necessary optimality conditions for semi-infinite programming problems, in Semi-Infinite Programming, Hettich, R. (ed.), Springer Verlag, Berlin.

Glashoff,K. and Roleff,K. (1981), A new method for Chebyshev approximation of complex-valued functions, Math. Comp. 36, pp.233-239.

Grothkopf, U. and Opfer, G. (1982), Complex Chebyshev polynomials on circular sectors with degree six or less, Math. Comp. 39, pp. 599-615.

Lorentz, G.G. (1966), Approximation of Functions, Holt, Rinehart and Winston, New York.

Mason, J.C. and Opfer, G. (1986), An algorithm for complex polynomial approximation with nonlinear constraints, in Approximation Theory V, Chui, C.K., Schumaker, L.L. and Ward, J. (eds.), Academic Press, New York.

Opfer, G. (1982), Solving complex approximation problems by semi-infinite-finite optimization techniques: a study on convergence, Num. Math. 39, pp. 411-420.

Osborne, M.R. and Watson, G.A. (1969), A note on singular minimax approximation problems, Jour. Math. Anal. and Appl. 25, pp. 692-700.

Streit, R.L. (1985), An algorithm for the solution of systems of complex linear equations in the l_∞ norm with constraints on the unknowns, ACM Trans. Math. Software 11, pp. 242-249.

Streit, R.L. (1986), Solution of systems of complex linear equations in the L_∞ norm with constraints on the unknowns, SIAM J. Sci. Stat. Comp. 7, pp. 132-149.

Streit, R.L. and Nuttall, A.H. (1982), Linear Chebyshev complex function approximation and an application to beamforming, J. Ass. Soc. Amer. 72, pp. 181-190.

Streit, R.L. and Nuttall, A.H. (1983), A note on the semi-infinite programming approach to complex approximation, Math. Comp. 40, pp. 599-605.

Trefethen, L.N. (1981a), Rational Chebyshev approximation on the unit disc, Num. Math. 37, pp. 297-320.

Trefethen, L.N. (1981b), Near-circularity of the error curve in complex Chebyshev approximation, Jour. Approx. Th. 31, pp. 344-367.

Watson, G.A. (1985), Lagrangian methods for semi-infinite programming problems, in Infinite Programming, Anderson, E.J. and Philpott, A.B. (eds.), Springer Verlag, Berlin.

Watson, G.A. (1988), A method for the Chebyshev solution of an overdetermined system of complex linear equations, IMA Jour. Num. Anal. (to appear).

Yuying Li (1988), An Efficient Algorithm for Nonlinear Minimax Problems, Ph D Thesis, University of Waterloo.

A fast algorithm for linear complex Chebyshev approximation

P. T. P. Tang

Argonne National Laboratory,
Argonne,
Illinois, USA

Abstract A natural generalization of the Remez algorithm from real approximation to complex is presented. The algorithm is the first quadratically convergent of its kind.

Keywords Chebyshev approximation, Remez algorithm, quadratic convergence.

1 Introduction

Given a complex function F analytic on a specified domain in the complex plane, how can one construct the polynomial, of a prescribed degree, that best approximates F in the Chebyshev sense? Applications for such an algorithm can be found in [2], [3], [5], [6], and [8]. In the past, complex Chebyshev polynomial approximation has been far less well understood than its real analogue. In particular, the quadratically convergent Remez algorithm ([7] and [13]) for real approximation has not been satisfactorily generalized to a quadratically convergent algorithm for complex approximation. In this paper, we describe a natural generalization of the real Remez algorithm to complex that converges quadratically when conditions similar to those for the real case are satisfied.

The organization of this paper is as follows. In Section 2 we formulate our approximation problem in a form that facilitates the generalization. In Section 3, we take a short digression on the real Remez algorithm before the complex Remez algorithm is presented. The convergence properties of the algorithm, both global and local, are discussed in Section 4. Although the assumptions in Section 4 for quadratic convergence are not met very often in practice, we are able to relax them by an extension of the algorithm in Section 5. Finally, some concluding remarks are made in Section 6.

2 Formulation

Consider a complex-valued function F analytic in a domain enclosed by a smooth closed curve in the complex plane. Let $\gamma : [0,1] \to \mathbf{C}$, $\gamma(0) = \gamma(1)$, be a smooth parametrization of that curve, which is the boundary of the domain. Define $f, \varphi_1, \varphi_2, \ldots, \varphi_{2n}$, and

p as follows:

$$\begin{aligned} F(t) &:= f(\gamma(t)), \\ \varphi_l(t) &:= \mathrm{Re}\Big(\gamma^{l-1}(t)\Big), \quad l = 1,2,\ldots,n \\ \varphi_{n+l}(t) &:= \imath\mathrm{Im}\Big(\gamma^{l-1}(t)\Big), \quad l = 1,2,\ldots,n \\ p(\boldsymbol{\lambda},t) &:= \sum_{l=1}^{2n} \lambda_l\varphi_l(t) \quad \text{for any } \boldsymbol{\lambda} = [\lambda_1,\lambda_2,\ldots,\lambda_{2n}]^T \in \mathbf{R}^{2n}. \end{aligned}$$

Let $\|\cdot\|$ be the Chebyshev norm over [0,1]. Then the complex Chebyshev approximation problem is to find $h^* \in \mathbf{R}$ and $\boldsymbol{\lambda}^* \in \mathbf{R}^{2n}$ such that

$$h^* = \Big\|f - p(\boldsymbol{\lambda}^*,\cdot)\Big\| \le \Big\|f - p(\boldsymbol{\lambda},\cdot)\Big\| \quad \text{for all } \boldsymbol{\lambda} \in \mathbf{R}^{2n}.$$

3 A Complex Remez Algorithm

In this section, we first revisit the familiar real Remez algorithm and view it as a simplex algorithm that solves the dual of the approximation problem. Although that dual is a semi-infinite linear program, the method of solution needs no discretization or two different phases, as done in [4] and [10]. Once the real Remez is realized as such, simply doubling its dimension yields an algorithm for the complex problem.

3.1 The real Remez algorithm

For the duration of this subsection, assume that f is real and $\gamma = 1$. Thus $F = f$ and $\gamma_l(t) = t^{l-1}, l = 1,2,\ldots,n$. The approximation problem can be viewed as the following minimization problem: Determine $h^* \in \mathbf{R}$ and $\boldsymbol{\lambda}^* \in \mathbf{R}^n$ so as to minimize h, subject to

$$h \ge e^{-\imath\vartheta}\Big(f(t) - p(\boldsymbol{\lambda},t)\Big), \quad \text{for all } (t,\vartheta) \in [0,1] \times \{0,\pi\}.$$

The dual [9] of the minimization problem is as follows: Determine an infinite vector $r : \mathcal{I} \to [0,1]$, where $\mathcal{I} := [0,1] \times \{0,\pi\}$, so as to maximize

$$h = \sum_{(t,\vartheta)\in\mathcal{I}} r(t,\vartheta)e^{-\imath\vartheta}f(t)$$

subject to

$$\sum_{(t,\vartheta)\in\mathcal{I}} r(t,\vartheta) = 1, \quad \text{and}$$

$$\sum_{(t,\vartheta)\in\mathcal{I}} r(t,\vartheta)e^{-\imath\vartheta}\varphi_l(t) = 0, \quad l = 1,2,\ldots,n.$$

The infinite sum can be justified since the optimum is achievable by a vector $\boldsymbol{r}^*$ with at most n+1 nonzero entries.

Given $\boldsymbol{t} = [t_1, t_2, \ldots, t_{n+1}]^T \in [0,1]^{n+1}$ and $\boldsymbol{\vartheta} = [\vartheta_1, \vartheta_2, \ldots, \vartheta_{n+1}]^T \in \{0, \pi\}^{n+1}$, define two $n+1$-vectors and an $(n+1) \times (n+1)$ matrix as follows:

$$\begin{aligned}
\boldsymbol{b} &:= [1, 0, 0, \ldots, 0]^T \\
\boldsymbol{a}(t, \vartheta) &:= [1, e^{-\iota\vartheta}\varphi_1(t), e^{-\iota\vartheta}\varphi_2(t), \ldots, e^{-\iota\vartheta}\varphi_n(t)]^T \\
A(\boldsymbol{t}, \boldsymbol{\vartheta}) &:= [\boldsymbol{a}(t_1, \vartheta_1), \boldsymbol{a}(t_2, \vartheta_2), \ldots, \boldsymbol{a}(t_{n+1}, \vartheta_{n+1})].
\end{aligned}$$

The one-point exchange Remez algorithm for real approximation can be stated as follows:

Algorithm 1

Step 0 Find an initial $\boldsymbol{t} \in [0,1]^{n+1}$ and $\boldsymbol{\vartheta} \in \{0, \pi\}^{n+1}$ such that $A(\boldsymbol{t}, \boldsymbol{\vartheta})$ is nonsingular and $A^{-1}(\boldsymbol{t}, \boldsymbol{\vartheta}) \cdot \boldsymbol{b} > \mathbf{o}$.

Step 1 Define $\boldsymbol{r}(\boldsymbol{t}, \boldsymbol{\vartheta}) := A^{-1}(\boldsymbol{t}, \boldsymbol{\vartheta}) \cdot \boldsymbol{b}$. From $(\boldsymbol{t}, \boldsymbol{\vartheta})$ determine the unique $\boldsymbol{\sigma} := \begin{bmatrix} h \\ \boldsymbol{\lambda} \end{bmatrix}$ such that

$$e^{-\iota\vartheta}\Big(f(t_j) - p(\boldsymbol{\lambda}, t_j)\Big) = h, \qquad j = 1, 2, \ldots, n+1.$$

Step 2 Determine t' such that $|f(t') - p(\boldsymbol{\lambda}, t')| = \|f - p(\boldsymbol{\lambda}, \cdot)\|$. Define

$$\vartheta' := \begin{cases} 0 & \text{if } f(t') - p(\boldsymbol{\lambda}, t') \geq 0; \\ \pi & \text{otherwise.} \end{cases}$$

Since $h \leq \|f - p(\boldsymbol{\lambda}^*, \cdot)\| \leq \|f(t') - p(\boldsymbol{\lambda}, t')\|$, whenever $\|f(t') - p(\boldsymbol{\lambda}, t')\| - h$ is small, h and $\boldsymbol{\lambda}$ are good approximations to the optimal solution and we terminate the algorithm. Otherwise, we move on.

Step 3 Exchange (t', ϑ') with one of the (t_j, ϑ_j)'s, $j = 1, 2, \ldots, n+1$. This exchange can be determined by solving the following small linear programming problem (via the simplex algorithm): Determine

$$r(t_1, \vartheta_1), r(t_2, \vartheta_2), \ldots, r(t_{n+1}, \vartheta_{n+1}), r(t', \vartheta') \in [0,1]$$

so as to maximize the inner product

$$h = \left(\sum_{j=1}^{n+1} r(t_j, \vartheta_j) e^{-\iota\vartheta_j} f(t_j)\right) + r(t', \vartheta') e^{-\iota\vartheta'} f(t')$$

subject to

$$[\,A(\boldsymbol{t}, \boldsymbol{\vartheta}) \quad \boldsymbol{a}(t', \vartheta')\,] \cdot \boldsymbol{r} = \boldsymbol{b}.$$

The optimal basis of this problem is exactly $\{(t_1, \vartheta_1), (t_2, \vartheta_2), \ldots, (t_{n+1}, \vartheta_{n+1})\}$ with one of the elements replaced by (t', ϑ'). Rename the new basis as $(\boldsymbol{t}, \boldsymbol{\vartheta})$. Go back to Step 1.

3.2 The complex Remez algorithm

We are now ready to go back to the complex approximation problem, which can be stated as the following minimization problem: Determine $h^* \in \mathbf{R}$ and $\boldsymbol{\lambda}^* \in \mathbf{R}^{2n}$ so as to minimize h, subject to

$$h \geq \mathrm{Re}\Big(e^{-\iota\vartheta}[f(t) - p(\boldsymbol{\lambda}, t)]\Big), \quad \text{for all } (t, \vartheta) \in [0,1] \times [0, 2\pi].$$

The dual of the minimization problem is as follows: Determine an infinite vector $r : \mathcal{I} \to [0,1]$, where $\mathcal{I} := [0,1] \times [0, 2\pi]$, so as to maximize

$$h = \sum_{(t,\vartheta)\in\mathcal{I}} r(t,\vartheta)\mathrm{Re}\Big(e^{-\iota\vartheta} f(t)\Big)$$

subject to

$$\sum_{(t,\vartheta)\in\mathcal{I}} r(t,\vartheta) = 1, \quad \text{and}$$

$$\sum_{(t,\vartheta)\in\mathcal{I}} r(t,\vartheta)\mathrm{Re}\Big(e^{-\iota\vartheta}\varphi_l(t)\Big) = 0, \quad l = 1, 2, \ldots, 2n.$$

The infinite sum can be justified since the optimum is achievable by a vector $\boldsymbol{r}^*$ with at most $2n+1$ nonzero entries.

Given $\boldsymbol{t} = [t_1, t_2, \ldots, t_{2n+1}]^T \in [0,1]^{n+1}$ and $\boldsymbol{\vartheta} = [\vartheta_1, \vartheta_2, \ldots, \vartheta_{2n+1}]^T \in [0, 2\pi]^{2n+1}$, define two $2n+1$-vectors and an $(2n+1) \times (2n+1)$ matrix as follows:

$$\begin{aligned}
\boldsymbol{b} &:= [1, 0, 0, \ldots, 0]^T \\
\boldsymbol{a}(t,\vartheta) &:= [1, \mathrm{Re}(e^{-\iota\vartheta}\varphi_1(t)), \mathrm{Re}(e^{-\iota\vartheta}\varphi_2(t)), \ldots, \mathrm{Re}(e^{-\iota\vartheta}\varphi_{2n}(t))]^T \\
A(\boldsymbol{t},\boldsymbol{\vartheta}) &:= [\boldsymbol{a}(t_1,\vartheta_1), \boldsymbol{a}(t_2,\vartheta_2), \ldots, \boldsymbol{a}(t_{2n+1},\vartheta_{2n+1})].
\end{aligned}$$

The one-point exchange Remez algorithm for complex approximation is merely a straightforward mimic of Algorithm 1.

Algorithm 2

Step 0 Find an initial $\boldsymbol{t} \in [0,1]^{2n+1}$ and $\boldsymbol{\vartheta} \in [0, 2\pi]^{2n+1}$ such that $A(\boldsymbol{t},\boldsymbol{\vartheta})$ is nonsingular and $A^{-1}(\boldsymbol{t},\boldsymbol{\vartheta}) \cdot \boldsymbol{b} > \mathbf{o}$.

Step 1 Define $\boldsymbol{r}(\boldsymbol{t},\boldsymbol{\vartheta}) := A^{-1}(\boldsymbol{t},\boldsymbol{\vartheta}) \cdot \boldsymbol{b}$. From $(\boldsymbol{t},\boldsymbol{\vartheta})$ determine the unique $\boldsymbol{\sigma} := \begin{bmatrix} h \\ \boldsymbol{\lambda} \end{bmatrix}$ such that

$$\mathrm{Re}\Big(e^{-\iota\vartheta}[f(t_j) - p(\boldsymbol{\lambda}, t_j)]\Big) = h, \qquad j = 1, 2, \ldots, 2n+1.$$

Step 2 Determine t' such that $|f(t') - p(\boldsymbol{\lambda}, t')| = \|f - p(\boldsymbol{\lambda}, \cdot)\|$. Define the angle $\vartheta' := \mathrm{Arg}(f(t') - p(\boldsymbol{\lambda}, t'))$. If $\|f(t') - p(\boldsymbol{\lambda}, t')\| - h$ is small, h and $\boldsymbol{\lambda}$ are good approximations to the optimal solution and we terminate the algorithm. Otherwise, we move on.

Step 3 Exchange (t', ϑ') with one of the (t_j, ϑ_j)'s, $j = 1, 2, \ldots, 2n+1$ as done in the previous algorithm for the real case. Go back to Step 1.

4 Convergence of the Algorithm

The Remez algorithm, whether real or complex, generates a sequence of iterates $\boldsymbol{r}^{(k)}$, $(\boldsymbol{t}^{(k)}, \boldsymbol{\vartheta}^{(k)})$, and $\boldsymbol{\sigma}^{(k)} = \begin{bmatrix} h^{(k)} \\ \boldsymbol{\lambda}^{(k)} \end{bmatrix}$. Questions of convergence are naturally centered on the quantity

$$\eta_k := \left\| f - p(\boldsymbol{\lambda}^{(k)}, \cdot) \right\| - h^{(k)}.$$

Does η_k converge to 0, and, if so, how fast? We answer these questions below in order.

4.1 Global convergence

In the discussion of the real Remez algorithm, [7] shows that $\eta_k \to 0$. The proof is a direct consequence of the fact that $\boldsymbol{r}^{(k)} \geq \Delta > 0$ for some uniform lower bound Δ for all the $\boldsymbol{r}^{(k)}$. Such a bound does not exist in general, however, for two (or higher) dimension real approximation or complex approximation. Nevertheless, similar to the situation in the simplex algorithm, it is reasonable to assume $\boldsymbol{r}^{(k)} > \mathbf{o}$ for all k.

Theorem 1 *If $\boldsymbol{r}^{(k)} > \mathbf{o}$ for $k = 1, 2, 3, \ldots$, then*

$$\liminf_{k \to \infty} \eta_k = 0.$$

Proof A complete proof can be found in [11]. ∎

In practice, $\liminf_{k \to \infty} \eta_k = 0$ means that the algorithm terminates in a finite number of steps for any positive stopping criterion.

4.2 Local convergence

For the real Remez algorithm, it is proved in [7] that the sequence $\{\eta_k\}$ is majorized by a sequence $\{\delta_k\}$, $\delta_k \geq \eta_k$ for all k, that converges to zero quadratically. The standard assumption ([1],[7], and [13]) for real approximation is that the optimal $f - p$ has $n+1$ extrema (alternations) at which the second derivatives are nonzero. The assumption and convergence result are generalized to complex approximation in the next theorem.

Theorem 2 *Let the function $|f(t) - p(\boldsymbol{\lambda}^*, t)|$ have exactly $2n+1$ extrema t_j^*, $1 \leq j \leq 2n+1$, and let $\boldsymbol{r}^* := A^{-1}(\boldsymbol{t}^*, \boldsymbol{\vartheta}^*) \cdot \boldsymbol{b} > \mathbf{o}$, where $\vartheta_j^* := \operatorname{Arg}\big(f(t_j^*) - p(\boldsymbol{\lambda}^*, t_j^*)\big)$, $1 \leq j \leq 2n+1$. Furthermore, let the second derivatives with respect to t of $|\, f(t) - p(\boldsymbol{\lambda}^*, t) \,|$ be nonzero at each of the $2n+1$ extrema. Then there exists a sequence $\{\delta_k\}$,$\delta_k \geq \eta_k$ for all k, and two constants M and K such that*

$$\delta_{2n+1+k} \leq M \delta_k^2 \qquad \text{for } k \geq K.$$

Proof The complete proof is found in [11]. ∎

Unfortunately, The proof in [11] is too long to be included here. That length, however, is mainly due to the fact that the algorithm is a one-point exchange instead of a multiple-exchange one. In the rest of this section, we first explain why quadratic convergence is obtainable. Then, we present a multiple exchange version of the Remez algorithm. Although this version works only when the iterate is close enough to the optimum, the proof of its quadratic convergence is sufficiently short to be presented here.

Consider the nonlinear system of equations

$$(4.1)\qquad \begin{array}{rcll} \mathrm{Re}\Big(e^{-\iota\vartheta_j}[f(t_j)-p(\boldsymbol{\lambda},t_j)]\Big)-h &=& 0 & j=1,2,\ldots,2n+1\\ \frac{d}{dt}|f(t)-p(\boldsymbol{\lambda},t)|\Big|_{t_j} &=& 0 & j=1,2,\ldots,2n+1\\ \mathrm{Im}\Big(e^{-\iota\vartheta_j}[f(t_j)-p(\boldsymbol{\lambda},t_j)]\Big) &=& 0 & j=1,2,\ldots,2n+1 \end{array}$$

Denote the $3(2n+1)$ real parameters by groups of $2n+1$ thus $\boldsymbol{x} := (\boldsymbol{\sigma},\boldsymbol{t},\boldsymbol{\vartheta})$. Moreoever, denote the three groups of $2n+1$ lefthand sides by $\boldsymbol{g}_j(\boldsymbol{x}), j=1,2,3$. Finally, define $\boldsymbol{G}(\boldsymbol{x}) := [\boldsymbol{g}_1^T(\boldsymbol{x}),\boldsymbol{g}_2^T(\boldsymbol{x}),\boldsymbol{g}_3^T(\boldsymbol{x})]$. Suppose the assumptions of Theorem 2 are satisfied; then it is not hard to show that $\boldsymbol{x}^* := (\boldsymbol{\sigma}^*,\boldsymbol{t}^*,\boldsymbol{\vartheta}^*)$ is an isolated zero of Equation 4.1. Moreover, we can prove the next theorem.

Theorem 3 *With the assumptions of Theorem 2, the Jacobian $\boldsymbol{J}(\boldsymbol{x}^*)$ of $\boldsymbol{G}$ at $\boldsymbol{x}^*$ is invertible.*

Proof Partition the Jacobian into 3×3 blocks of dimension $2n+1$ each. Because of the structure of $\boldsymbol{G}$, the derivatives with respect to $\boldsymbol{t}$ and $\boldsymbol{\vartheta}$ are all diagonal matrices. Moreover, because $\boldsymbol{t}^*$ and $\boldsymbol{\vartheta}^*$ are extrema of $\mathrm{Re}(e^{-\iota\vartheta}[f(t)-p(\boldsymbol{\lambda}^*,t)])$ and $|f(t)-p(\boldsymbol{\lambda}^*,t)|$,

$$\begin{array}{rcl} (\boldsymbol{g}_1)_{\boldsymbol{t}}(\boldsymbol{x}) &=& (\boldsymbol{g}_1)_{\boldsymbol{\vartheta}}(\boldsymbol{x}) = \mathbf{o} \quad \text{for all } \boldsymbol{x},\\ (\boldsymbol{g}_2)_{\boldsymbol{t}}(\boldsymbol{x}^*) &=& \mathrm{diag}\Big(\frac{\partial^2}{\partial t^2}|f(t)-p(\boldsymbol{\lambda}^*,t)|\Big|_{t_1^*},\ldots,\frac{\partial^2}{\partial t^2}|f(t)-p(\boldsymbol{\lambda}^*,t)|\Big|_{t_{2n+1}^*}\Big),\\ (\boldsymbol{g}_3)_{\boldsymbol{\vartheta}}(\boldsymbol{x}^*) &=& \mathrm{diag}\Big(-h^*,-h^*,\ldots,-h^*\Big). \end{array}$$

Finally, $(\boldsymbol{g}_1)_{\boldsymbol{\sigma}}(\boldsymbol{x}^*) = -A^T(\boldsymbol{t}^*,\boldsymbol{\vartheta}^*)$. Clearly, the Jacobian at $\boldsymbol{x}$ is block lower triangular with the diagonal blocks invertible. ∎

It is easy to check that the Jacobian is Lipschitz continuous in $\boldsymbol{x}$ and thus Newton iteration would converge quadratically. This approach is not taken in real approximation, however, because of its high cost since the Newton's approach is of dimension $2(n+1)$ instead of $n+1$, the dimension of Remez. The situation is similar in complex approximation. Indeed, we can show that the following multiple exchange complex Remez algorithm, of dimension $2n+1$, converges as fast as Newton iteration does, which is of dimension $3(2n+1)$. The multiple exchange complex Remez algorithm is Algorithm 2 with Steps 2 and 3 replaced as follows:

Step $2'$ Determine $t_1', t_2', \ldots, t_{2n+1}'$ near $t_1, t_2, \ldots, t_{2n+1}$, respectively, such that the first derivatives of $|f(t)-p(\boldsymbol{\lambda},t)|$ with respect to t vanish. Define ϑ_j' as the arguments of $f(t_j')-p(\boldsymbol{\lambda},t_j')$.

Step $3'$ Replace $(\boldsymbol{t},\boldsymbol{\vartheta})$ by $(\boldsymbol{t}',\boldsymbol{\vartheta}')$.

This algorithm works when $(\boldsymbol{t},\boldsymbol{\vartheta})$ is close to the optimum because $A^{-1}(\boldsymbol{t}^*,\boldsymbol{\vartheta}^*)\cdot\boldsymbol{b} > \mathbf{o}$ implies $A^{-1}(\boldsymbol{t},\boldsymbol{\vartheta})\cdot\boldsymbol{b} > \mathbf{o}$ whenever $(\boldsymbol{t},\boldsymbol{\vartheta})$ is close to $(\boldsymbol{t}^*,\boldsymbol{\vartheta}^*)$.

Now, to show the equivalent convergence rate of the Newton and Remez iterations, we consider the updates $\boldsymbol{x}^{(\mathcal{N})}$ and $\boldsymbol{x}^{(\mathcal{R})}$ after one iteration of the Newton and the Remez algorithms, respectively, both starting at a common point $\boldsymbol{x}$ very close to the optimum $\boldsymbol{x}^*$, $\|\boldsymbol{x} - \boldsymbol{x}^*\| \leq \varepsilon$. We will show that $\|\boldsymbol{x}^{(\mathcal{N})} - \boldsymbol{x}^{(\mathcal{R})}\| < M\varepsilon^2$ for some constant M independent of ε.

First, the Newton update $\boldsymbol{x}^{(\mathcal{N})}$ is defined by

$$J(\boldsymbol{x}) \cdot (\boldsymbol{x} - \boldsymbol{x}^{(\mathcal{N})}) = G(\boldsymbol{x}). \tag{4.2}$$

Next, consider $\boldsymbol{x}^{(\mathcal{R})}$. The vector $\boldsymbol{\sigma}^{(\mathcal{R})}$ is defined by

$$(g_1)_{\boldsymbol{\sigma}}(\boldsymbol{x}) \cdot (\boldsymbol{\sigma} - \boldsymbol{\sigma}^{(\mathcal{R})}) = g_1(\boldsymbol{x}).$$

Based on $\boldsymbol{\sigma}^{(\mathcal{R})}$, $\boldsymbol{t}^{(\mathcal{R})}$ and $\boldsymbol{\vartheta}^{(\mathcal{R})}$ are determined by

$$g_2(\boldsymbol{x}^{(\mathcal{R})}) = g_3(\boldsymbol{x}^{(\mathcal{R})}) = \mathbf{o}.$$

Thus,

$$\begin{bmatrix} g_2(\boldsymbol{x}) \\ g_3(\boldsymbol{x}) \end{bmatrix} = \begin{bmatrix} g_2(\boldsymbol{x}) \\ g_3(\boldsymbol{x}) \end{bmatrix} - \begin{bmatrix} g_2(\boldsymbol{x}^{(\mathcal{R})}) \\ g_3(\boldsymbol{x}^{(\mathcal{R})}) \end{bmatrix} = \frac{\partial}{\partial x} \begin{bmatrix} g_2(\boldsymbol{x}) \\ g_3(\boldsymbol{x}) \end{bmatrix} \cdot (\boldsymbol{x} - \boldsymbol{x}^{(\mathcal{R})}) + \boldsymbol{y},$$

where $\|\boldsymbol{y}\| < M_1\varepsilon^2$ for some M_1 indepedent of ε. Thus, $\boldsymbol{x}^{(\mathcal{R})}$ can be characterized as

$$\left(J(\boldsymbol{x}) - \begin{bmatrix} \mathbf{o} & (g_1)_{\boldsymbol{t}}(\boldsymbol{x}) & (g_1)_{\boldsymbol{\vartheta}}(\boldsymbol{x}) \\ \mathbf{o} & \mathbf{o} & \mathbf{o} \\ \mathbf{o} & \mathbf{o} & \mathbf{o} \end{bmatrix} \right) \cdot \left(\boldsymbol{x} - \boldsymbol{x}^{(\mathcal{R})} \right) = G(\boldsymbol{x}) + \boldsymbol{y}. \tag{4.3}$$

Subtracting Equation 4.3 from Equation 4.2 gives

$$J(\boldsymbol{x}) \cdot (\boldsymbol{x}^{(\mathcal{R})} - \boldsymbol{x}^{(\mathcal{N})}) = \boldsymbol{z} \tag{4.4}$$

where $\|\boldsymbol{z}\| < M_2\varepsilon^2$ for some M_2 independent of ε. This is because $\|(g_1)_{\boldsymbol{t}}(\boldsymbol{x})\|$, $\|(g_1)_{\boldsymbol{\vartheta}}(\boldsymbol{x})\|$, and $\|\boldsymbol{x} - \boldsymbol{x}^{(\mathcal{R})}\| < M_3\varepsilon$ for some M_3 independent of ε. Finally, $J(\boldsymbol{x}^*)$ is invertible, and thus $\left\| J^{-1}(\boldsymbol{x}) \right\|$ is uniformly bounded near $\boldsymbol{x}^*$ and the proof is complete.

Numerical examples illustrating the convergence behavior can be found in [12] and [11].

5 An Extension of the Complex Remez Algorithm

In our experience ([11] and [12]), whenever the assumptions in Theorem 2 are not satisfied, the number of extrema is insufficient. Fortunately, in that case we can extend the algorithm slightly to restore quadratic convergence. Details of this extension are presented in [11] and [12]; the idea is roughly as follows. If the number of extrema is d fewer than the ideal number, one can define a function $\boldsymbol{u} : \mathbf{R}^d \to \mathbf{R}^d$ whose zero corresponds to the solution of the original approximation problem. Moreover, to evaluate the function and its Jacobian, we need only to apply the complex Remez algorithm to a problem *that satisfies all the conditions needed for fast convergence.* Consequently, Newton iteration applied on $\boldsymbol{u}$ is efficient. The quadratic convergence of this extended algorithm is fully analyzed in [11].

6 Conclusion

We have shown that linear complex Chebyshev approximation can be naturally viewed as a special case of two-dimensional real approximation for which the Remez exchange algorithm converges quadratically. We believe such a view can help our understanding in other aspects of complex approximation such as Chebyshev approximation by rational functions.

7 Acknowledgment

This work was supported in part by the Strategic Defense Initiative Organization, Office of the Secretary of Defense, under WPD B411.

References

[1] E. W. Cheney. *Introduction to Approximation Theory.* Chelsea, New York, 1986.

[2] H. C. Elman and R. L. Streit. *Polynomial Iteration for Nonsymmetric Indefinite Linear Systems.* Research report YALEU/DCS/RR-380, Department of Computer Science, Yale University, New Haven, Conn., March 1985.

[3] B. Francis, J. W. Helton, and G. Zames. H^∞-optimal feedback controllers for linear multivariable systems. *IEEE Trans. Autom. Control*, 29:888–900, October 1984.

[4] K. Glashoff and K. Roleff. A new method for Chebyshev approximation of complex-valued functions. *Mathematics of Computation*, 36(153):233–239, January 1981.

[5] M. Hartmann and G. Opfer. Uniform approximation as a numerical tool for constructing conformal maps. *Journal of Computational and Applied Mathematics*, 14:193–206, 1986.

[6] J. W. Helton. Worst case analysis in the frequency domain: An H^∞ approach for control. *IEEE Trans. Auto. Control*, 30:1192–1201, December 1985.

[7] M. J. D. Powell. *Approximation Theory and Methods.* Cambridge University Press, Cambridge, 1981.

[8] L. Reichel. On polynomial approximation in the complex plane with application to conformal mapping. *Mathematics of Computation*, 44(170):425–433, April 1985.

[9] T. J. Rivlin and H. S. Shapiro. A unified approach to certain problems of approximation and minimization. *Journal of Soc. Indust. Appl. Math.*, 9(4):670–699, December 1961.

[10] R. L. Streit and A. H. Nuttall. A general Chebyshev complex function approximation procedure and an application to beamforming. *Journal of Acoustical Society of America*, 72:181–190, July 1982.

[11] P. T. P. Tang. *Chebyshev Approximation on the Complex Plane*. PhD thesis, Department of Mathematics, University of California at Berkeley, May 1987.

[12] P. T. P. Tang. A fast algorithm for linear complex Chebyshev approximations. *Mathematics of Computation*, 51(184):721–739, October 1988.

[13] L. Veidinger. On the numerical determination of the best approximation in the Chebyshev sense. *Numerische Mathematik*, 2:99–105, 1960.

PART TWO
Applications

6. Computer Aided Design and Geometric Modelling

Uniform subdivision algorithms for curves and surfaces

N. Dyn, D. Levin

School of Mathematical Sciences, Tel Aviv University, Israel

D. Levin

School of Mathematical Sciences, Tel Aviv University, Israel

J. A. Gregory

Department of Mathematics and Statistics, Brunel University, Uxbridge, Middlesex, UK

Abstract A convergence analysis for studying the continuity and differentiability of limit curves generated by uniform subdivision algorithms is presented. The analysis is based on the study of corresponding difference and divided difference algorithms. The alternative process of "integrating" the algorithms is considered. A specific example of a 4-point interpolatory curve algorithm is described and its generalization to a surface algorithm defined over a subdivision of a regular triangular partition is illustrated.

Key words: Subdivision algorithms, Control polygon, Interpolation, Shape control.

1. Introduction

Subdivision algorithms which generate curves and surfaces play an important role in the subject of computer aided geometric design. The basic idea is that a given initial "control polygon" is successively refined so that, in the limit, it approaches a smooth curve or surface. We will consider uniform binary subdivision algorithms for curves of the following form:

At the k+1'st step of the algorithm, $k = 0,1,2,...$, let f^k denote the control polygon in $\mathbb{R}^N$ with "control point" vertices $f_i^k \in \mathbb{R}^N$, $i \in \mathbb{Z}$. Then the control polygon f^{k+1} has vertices defined by the rule

$$\left.\begin{aligned} f_{2i}^{k+1} &= \sum_{j=\ell}^{m} a_j f_{i+j}^k , \\ f_{2i+1}^{k+1} &= \sum_{j=\ell}^{m} b_j f_{i+j}^k , \end{aligned}\right\} \quad i \in \mathbb{Z} , \tag{1}$$

where $\ell < m$.

Our motivation for studying uniform subdivision schemes of the form (1) is based on the particular example of a 4-point interpolatory rule defined by

$$\left.\begin{aligned} f_{2i}^{k+1} &= f_i^k , \\ f_{2i+1}^{k+1} &= \left(\tfrac{1}{2} + \omega\right)\left(f_i^k + f_{i+1}^k\right) - \omega\left(f_{i-1}^k + f_{i+2}^k\right) , \end{aligned}\right\} \tag{2}$$

see Dyn, Gregory, Levin (1987). Here, ω acts as a shape control parameter. The case $\omega = 0$, namely

$$\left.\begin{aligned} f_{2i}^{k+1} &= f_i^k , \\ f_{2i+1}^{k+1} &= \tfrac{1}{2}\left(f_i^k + f_{i+1}^k\right) , \end{aligned}\right\} \tag{3}$$

has control polygons $f^{k+1} = f^k$ for all k, and hence in this case the limit curve is the initial control polygon f^0. The need to construct a convergence theory for the more general case $\omega \neq 0$ leads us to consider the more general form (1).

Figure 1 illustrates the application of the interpolatory subdivision scheme defined by (2) to a finite open polygon in $\mathbb{R}^2$, where $\omega = 1/16$. (The case $\omega=1/16$ is of significance, since it gives a rule which reproduces cubic polynomials with respect to data defined on a diadic point parameterization.) It should be noted that, since the binary subdivision scheme is local, the scheme is well defined in the case of finite initial data, where control points at each end of the initial polygon act as end conditions on the final limit curve.

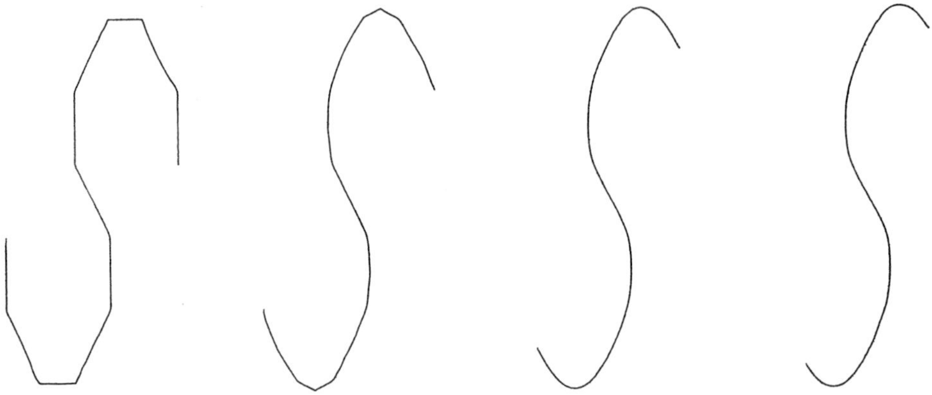

Figure 1. Example of 4-point interpolatory algorithm

Subdivision algorithms have been studied in a general setting by Micchelli and Prautzsch (1987a, 1987b, 1988). Here, however, we review a convergence analysis presented in Dyn, Gregory and Levin (1988) for schemes defined by a rule of the specific form (1). The reader is referred to the 1988 paper for many of the proofs and this allows us to simplify the presentation. Also, our approach here is different in that the analysis is presented as a study of a **fundamental solution** of the scheme. In the final section a new interpolatory subdivision scheme for surfaces is introduced.

For the purposes of the analysis we will consider, without loss of generality, the basic scheme (1) with $\ell = 0$. This scheme with coefficients $\mathbf{a} = [a_0,\ldots,a_m]$ and $\mathbf{b} = [b_0,\ldots,b_m]$ will be denoted symbolically by $S(\mathbf{a},\mathbf{b})$. Thus $S(\mathbf{a},\mathbf{b})$ with initial data $\{f_i^0 \in \mathbb{R}^N / i \in \mathbb{Z}\}$ is defined by the rule

$$\left.\begin{aligned} f_{2i}^{k+1} &= \sum_{j=0}^{m} a_j f_{i+j}^{k} , \\ f_{2i+1}^{k+1} &= \sum_{j=0}^{m} b_j f_{i+j}^{k} , \end{aligned}\right\} \quad i \in \mathbb{Z} \;, \; k = 0,1,2,\ldots \;. \tag{4}$$

2. The fundamental solution

In order to study the limiting behaviour of the sequence of control polygons $\left\{f^k\right\}_{k=0}^{\infty}$ produced by the scheme $S(\mathbf{a},\mathbf{b})$ applied to the initial data $\left\{f_i^0/i \in \mathbb{Z}\right\}$, we will consider the special case of initial data $\left\{\delta_{i,0}/i \in \mathbb{Z}\right\}$. We first, however, describe a parameterization by which the control polygons can be represented in function form.

Since the process is one of binary subdivision, the initial control points f_i^0 are associated with the integer mesh points $t_i^0 := i$, $i \in \mathbb{Z}$, and in general, the control points f_i^k are associated with the diadic mesh points

$$t_i^k = i/2^k \;, \quad i \in \mathbb{Z} \;, \quad k = 0,1,2,\ldots \;. \tag{5}$$

The polygon f^k can now be represented by the piecewise linear interpolant

$$f^k(t) := \left[\frac{t_{i+1}^k - t}{t_{i+1}^k - t_i^k}\right] f_i^k + \left[\frac{t - t_i^k}{t_{i+1}^k - t_i^k}\right] f_{i+1}^k \;, \quad t \in \left[t_i^k, t_{i+1}^k\right) \;, \quad i \in \mathbb{Z} \;. \tag{6}$$

and properties of the limit process can be studied with respect to the parameterization t.

Let $\varphi^k(t)$ be the control polygon with vertices $\left\{\varphi_i^k/i \in \mathbb{Z}\right\}$, for the process $S(\mathbf{a},\mathbf{b})$ applied to the initial data $\left\{\varphi_i^0 = \delta_{i,0}/i \in \mathbb{Z}\right\}$. Thus

$$\left.\begin{aligned} \varphi_{2i}^{k+1} &= \sum_{j=0}^{m} a_j \varphi_{i+j}^{k} , \\ \varphi_{2i+1}^{k+1} &= \sum_{j=0}^{m} b_j \varphi_{i+j}^{k} , \end{aligned}\right\} \quad i \in \mathbb{Z} , \quad k = 0,1,2,\ldots , \tag{7}$$

where

$$\varphi_i^0 = \begin{cases} 1 & \text{for } i = 0 , \\ 0 & \text{otherwise} . \end{cases} \tag{8}$$

Then

$$f^k(t) = \sum_{i \in \mathbb{Z}} f_i^0 \, \varphi^k(t-i) \tag{9}$$

is the control polygon with vertices $\left\{f_i^k / i \in \mathbb{Z}\right\}$ for arbitrary initial data $\left\{f_i^0 / i \in \mathbb{Z}\right\}$. Suppose

$$\lim_{k \to \infty} \varphi^k(t) = \varphi(t) \tag{10}$$

uniformly on $C(-\infty,\infty)$. Then we call $\varphi(t)$ the **fundamental solution** of the binary subdivision process $S(\mathbf{a},\mathbf{b})$ and $\varphi^k(t)$ the k'th **discrete fundamental solution** of the process.

The local nature of the subdivision process is reflected in the fact that φ^k and φ have local support. The support of φ^k is contained in the support of φ which is at most

$$\left.\begin{aligned} &(-2m,1) && \text{for } a_m \neq 0 , \\ &(-2m+1,1) && \text{for } a_m = 0 . \end{aligned}\right\} \tag{11}$$

(In calculating the support we assume $b_m \neq 0$ without loss of generality since otherwise the equivalent process $S(\mathbf{b},\mathbf{a})$ can be considered.) It follows that if $\lim \varphi^k = \varphi$, then

$$\lim f^k(t) = \sum_{i\in\mathbb{Z}} f_i^0 \, \varphi(t-i) \,. \tag{12}$$

Hence the convergence properties of the algorithm for arbitrary initial data are determined by the nature of the convergence of the sequence $\left\{\varphi^k\right\}_{k=0}^{\infty}$ to φ.

3. Conditions for a C^0 limit

We consider conditions under which the sequence of control polygons $\left\{\varphi^k\right\}_{k=0}^{\infty}$ converges uniformly to a continuous limit curve φ. From (7) we immediately obtain:

Lemma 1. A necessary condition for uniform convergence to a continuous limit curve with respect to the diadic point parameterization is that

$$\sum_{j=0}^{m} a_j = \sum_{j=0}^{m} b_j = 1 \,. \tag{13}$$

Assuming the necessary condition (13), we then obtain from (7) the **difference scheme**

$$\left.\begin{aligned} \Delta\varphi_{2i}^{k+1} &= \sum_{j=0}^{m} c_j \, \Delta\varphi_{i+j}^{k} \,, \\ \Delta\varphi_{2i+1}^{k+1} &= \sum_{j=0}^{m} d_j \, \Delta\varphi_{i+j}^{k} \,, \end{aligned}\right\} \quad i \in \mathbb{Z} \,,\ k = 0,1,2,\dots \,, \tag{14}$$

where

$$c_j = \sum_{i=0}^{j} (a_i - b_i) \,, \quad d_j = \sum_{i=0}^{j} (b_i - a_i) + a_j \,. \tag{15}$$

Here

$$\Delta\varphi_i^k := \varphi_{i+1}^k - \varphi_i^k \tag{16}$$

defines a forward difference and we denote the scheme symbolically by $\Delta S(\mathbf{a},\mathbf{b}) := S(\mathbf{c},\mathbf{d})$. We now have the following convergence theorem:

Theorem 2 (convergence) The process $S(\mathbf{a},\mathbf{b})$ has a discrete fundamental solution sequence $\{\varphi^k\}_{k=0}^{\infty}$ which converges uniformly to a continuous fundamental solution φ, if and only if the difference process $\Delta S(\mathbf{a},\mathbf{b})$ has a discrete fundamental sequence $\{\theta^k\}_{k=0}^{\infty}$, say, which converges uniformly to the zero function $\theta(t) \equiv 0$.

Proof. It is sufficient to consider the function sequences defined on the largest possible domain of local support $[-2m,1]$. Observe also that the difference process (14) has control polygons

$$\varphi^k(t+1/2^k) - \varphi^k(t) \equiv \theta^k(t+1) - \theta^k(t)\ , \quad k = 0,1,2,\ldots\ . \tag{17}$$

Suppose $\lim \varphi^k = \varphi$ uniformly in $C[-2m,1]$. Then, from (17), $\theta^k(t+1) - \theta^k(t)$ converges uniformly to zero. Now

$$\theta^k(t) = \theta^k(t) - \theta^k(t+2m+1) = \sum_{i=0}^{2m} \{\theta^k(t+i) - \theta^k(t+i+1)\}\ ,\ t \in [-2m,1]\ ,$$

since $\theta^k(t+2m+1)$ has local support within $(1,2m+2)$. Hence $\theta^k(t)$ converges uniformly to zero. Conversely, suppose $\lim \theta^k = \theta$, $\theta(t) \equiv 0$, uniformly in $C[-2m,1]$. Then the control polygons (17) of the difference process (14) converge uniformly to zero. Consider this difference process from level k to k+L. Then the control polygon at level k+L can be represented as

$$\varphi^{k+L}(t+1/2^{k+L}) - \varphi^{k+L}(t) = \sum_{i\in\mathbb{Z}} \Delta\varphi_i^k\ \theta^L(t2^k-i)$$

(cf. (9)). In particular,

$$\Delta\varphi_j^{k+L} = \sum_{i\in\mathbb{Z}} \Delta\varphi_i^k \, \theta^L(t_j^{k+L}2^k - i)$$

$$= \sum_{i=j_0}^{j_0+2m} \Delta\varphi_i^k \, \theta^L(j/2^L - i) ,$$

where $j_0 = [j/2^L]$ and the summation is restricted to a finite set of 2m+1 integers since θ^L has local support within (-2m,1). Since θ^L converges uniformly to zero it follows that given $0 < \alpha < 1$, there exists L such that

$$\max_j \left|\Delta\varphi_j^{k+L}\right| \le \alpha \max_j \left|\Delta\varphi_j^k\right| . \tag{18}$$

It can now be shown that this contractive property of the differences implies that $\left\{\varphi^k\right\}_{k=0}^{\infty}$ defines a Cauchy sequence in C[-2m,1] (see Dyn, Gregory, Levin (1988)) which completes the proof.

4. A matrix analysis of convergence

Observe from (15) that $c_m = 0$ and $d_m = a_m$ in the difference scheme (14) and define the n_1+1 difference vector

$$\mathbf{\Delta}_{i,k} := \left[\Delta f_i^k, \ldots, \Delta f_{i+n_1}^k\right]^T , \tag{19}$$

where

$$n_1 = \begin{cases} 2m-1 & \text{if} \quad a_m \ne 0 , \\ 2m-2 & \text{if} \quad a_m = 0 . \end{cases} \tag{20}$$

Then the difference scheme (14) gives the two "even" and "odd" matrix transformations

$$\mathbf{\Delta}_{2i,k+1} = \mathbf{C}_0\mathbf{\Delta}_{i,k} , \quad \mathbf{\Delta}_{2i+1,k+1} = \mathbf{C}_1\mathbf{\Delta}_{i,k} , \tag{21}$$

where

$$\mathbf{C}_0=\begin{bmatrix} c_0 & \cdot & c_{m-1} & & & \\ d_0 & \cdot & \cdot & d_m & & \\ & c_0 & \cdot & c_{m-1} & & \\ & d_0 & \cdot & \cdot & d_m & \\ & & \cdot & \cdot & \cdot & \\ & & \cdot & \cdot & \cdot & \cdot \end{bmatrix}, \quad \mathbf{C}_1=\begin{bmatrix} d_0 & \cdot & \cdot & d_m & & \\ & c_0 & \cdot & c_{m-1} & & \\ & d_0 & \cdot & \cdot & d_m & \\ & & c_0 & \cdot & c_{m-1} & \\ & & \cdot & \cdot & \cdot & \cdot \\ & & & \cdot & \cdot & \cdot \end{bmatrix} \tag{22}$$

We call the $(n_1+1) \times (n_1+1)$ matrices $\mathbf{C}_0$ and $\mathbf{C}_1$ the **control point matrices** of the difference scheme $\Delta S(\mathbf{a},\mathbf{b})$. (Here, n_1 has been calculated to give square matrices of lowest possible order.)

From (21) it follows that all transformations between the k'th and k+L'th differences can be accomplished by transformation matrices consisting of all permutations of products of length L of the matrices $\mathbf{C}_0$ and $\mathbf{C}_1$. The contractive property (18) in the proof of Theorem 2 then leads to the following:

Theorem 3 (convergence) The discrete fundamental solutions of $\Delta S(\mathbf{a},\mathbf{b})$ converge uniformly to zero if and only if given $0 < \alpha < 1$, there exists a positive integer L such that

$$\left\| \mathbf{C}_{i_L} \cdots \mathbf{C}_{i_1} \right\|_\infty \le \alpha \quad \forall \;\; i_j \in \{0,1\} \;, \; j = 1,\ldots,L \;. \tag{23}$$

Corollary 4 A necessary condition that the discrete fundamental solutions converge uniformly to zero is that the spectral radii of $\mathbf{C}_0$ and $\mathbf{C}_1$ satisfy

$$\rho(\mathbf{C}_0) < 1 \quad \text{and} \quad \rho(\mathbf{C}_1) < 1 \;. \tag{24}$$

The analysis is, in fact, very rich in matrix theory. For example $\mathbf{C}_0$ and $\mathbf{C}_1$ share nearly all common eigenvalues. They also share common eigenvalues with the $(n_1+2) \times (n_1+2)$ control matrices, $\mathbf{A}_0$ and $\mathbf{A}_1$ say, of the basic scheme $S(\mathbf{a},\mathbf{b})$ (excluding the one eigenvalue unity given by the necessary condition (13)). For details of these results we refer the reader to Dyn et al (1988). (See also

Micchelli and Prautzsch (1987a) for their treatment in terms of invariant subspaces.)

The difference scheme $\Delta S(\mathbf{a},\mathbf{b})$ has control point matrices of one less order than the basic scheme $S(\mathbf{a},\mathbf{b})$. Likewise, if there exist higher order difference schemes these will have control point matrices of lower order. This suggests the application of Theorem 3 to such higher order difference schemes since we have:

<u>Theorem 5</u> Assume the necessary condition (24) and that there exists the difference scheme $\Delta^{\ell}S(\mathbf{a},\mathbf{b})$, $\ell \geq 1$. Then the difference process $\Delta S(\mathbf{a},\mathbf{b})$ has discrete fundamental solutions which converge uniformly to zero if and only if $\Delta^{\ell}S(\mathbf{a},\mathbf{b})$ has discrete fundamental solutions which converge uniformly to zero.

<u>Remarks 6.</u> Given $\Delta^{\ell}S(\mathbf{a},\mathbf{b})$, the existence of $\Delta^{\ell}S(\mathbf{a},\mathbf{b})$ requires that the sum of the coefficient vectors of $\Delta^{\ell-1}S(\mathbf{a},\mathbf{b})$ be identical. For example, if $\Sigma c_j = \Sigma d_j$ for the difference scheme $\Delta S(\mathbf{a},\mathbf{b}) = S(\mathbf{c},\mathbf{d})$, then there exists the scheme $\Delta^2 S(\mathbf{a},\mathbf{b}) = \Delta S(\mathbf{c},\mathbf{d})$. Moreover, these sums will be eigenvalues of the control point matrices $\mathbf{C}_0$ and $\mathbf{C}_1$ and must thus have magnitude less than unity by the necessary condition (24).

5. Conditions for a C^{ℓ} limit

To study differentiability of the limit process we consider the behaviour of the divided differences

$$d_i^k := \left(\varphi_{i+1}^k - \varphi_i^k\right) \Big/ \left(t_{i+1}^k - t_i^k\right) = 2^k \Delta\varphi_i^k \tag{25}$$

of the vertices of the control polygon φ^k. From (14) it follows that the divided differences satisfy the scheme

$$\left.\begin{aligned} d_{2i}^{k+1} &= \sum_{j=0}^{m-1} a_j^{(1)} d_{i+j}^{k} , \\ d_{2i+1}^{k+1} &= \sum_{j=0}^{m} b_j^{(1)} d_{i+j}^{k} , \end{aligned}\right\} \quad i \in \mathbb{Z} , \tag{26}$$

where

$$a_j^{(1)} = 2c_j = 2 \sum_{i=0}^{j} (a_i - b_i), \quad b_j^{(1)} = 2d_j = 2\left\{ \sum_{i=0}^{j} (b_i - a_i) + a_j \right\} . \tag{27}$$

Thus there exists the **divided difference scheme** which we denote by DS(**a**,**b**) := S(**a**,**b**) = 2S(**c**,**d**), where the necessary condition (13) has been assumed. We then have:

Theorem 7. (C^1 convergence) If the divided difference scheme has discrete fundamental solutions which converge uniformly to a C^0 limit, then the basic scheme S(**a**,**b**) has discrete fundamental solutions which converge uniformly to a C^1 limit φ. Moreover, the limit of the divided difference process (26) (i.e. with initial data $\{d_i^0 = \Delta\varphi_i^0, i \in \mathbb{Z}\}$) is φ'.

Theorem 7 suggests that the C^0 convergence theory of sections 3 and 4 can be applied to the divided difference process in order to analyse differentiability. For C^0 convergence of this process, with respect to the diadic point parameterization, it is necessary that

$$\sum_{j=0}^{m-1} a_j^{(1)} = \sum_{j=0}^{m} b_j^{(1)} = 1 . \tag{28}$$

This condition together with (11) is equivalent to:

Proposition 8. A necessary condition for uniform convergence of the divided difference process to a C^0 limit with respect to the diadic point parameterization is that

$$\sum_{j=0}^{m} a_j = \sum_{j=0}^{m} b_j = 1 \quad \text{and} \quad \sum_{j=0}^{m} j(b_j - a_j) = \frac{1}{2} . \tag{29}$$

It can be shown that if (29) holds, then the diadic point

parameterization defined by (5) is an appropriate one for the analysis. If (29) does not hold, there may be some parameterization defined by different points $\{t_i^k\}$ in which the limit curve may be differentiable. (The equivalent condition to (29) is then

$$\sum_{j=0}^{m} a_j = \sum_{j=0}^{m} b_j = 1 \quad \text{and} \quad \sum_{j=0}^{m} t_{i+j}^{k}(b_j - a_j) = t_{2i+1}^{k+1} - t_{2i}^{k+1} ,$$

for parametric points $\{t_i^k\}$ which become dense in the limit.)

An immediate generalization of Theorem 7 is:

Theorem 9. (C^{ℓ} convergence) Suppose there exist the ν'th divided difference schemes $D^{\nu}S(\mathbf{a},\mathbf{b}) = DS(\mathbf{a}^{(\nu-1)},\mathbf{b}^{(\nu-1)}) = S(\mathbf{a}^{(\nu)},\mathbf{b}^{(\nu)})$, $\nu = 0,\dots,\ell$, where

$$\sum a_j^{(\nu)} + \sum b_j^{(\nu)} = 1 , \quad \nu = 0,\dots,\ell . \tag{30}$$

Then if $D^{\ell}S(\mathbf{a},\mathbf{b})$ has discrete fundamental solutions which converge uniformly to a C^0 limit, the basic scheme $S(\mathbf{a},\mathbf{b})$ has discrete fundamental solutions which converge uniformly to a C^{ℓ} limit φ.

Remark 10. Condition (30) implies that each ν'th divided difference scheme has control point matrices with one eigenvalue unity. Since $D^{\nu}S(\mathbf{a},\mathbf{b}) = 2^{\nu}\Delta^{\nu}S(\mathbf{a},\mathbf{b})$, it can then be shown that the control matrices $\mathbf{C}_0$ and $\mathbf{C}_1$ of the difference scheme $\Delta S(\mathbf{a},\mathbf{b})$ (and hence of the basic scheme $S(\mathbf{a},\mathbf{b})$) must have eigenvalues $1/2^{\nu}$, $\nu = 1,\dots,\ell$.

6. A calculus of schemes

Given the **basic scheme**

$$S(\mathbf{a},\mathbf{b}) , \quad \mathbf{a} = [a_0,\dots,a_m] , \quad \mathbf{b} = [b_0,\dots,b_m] \tag{31}$$

where $\Sigma a_i = \Sigma b_i = 1$, we have defined the **divided difference scheme**

$$DS(\mathbf{a},\mathbf{b}) := S(\mathbf{a},\mathbf{b}) , \quad \mathbf{a}^{(1)} = \left[a_0^{(1)},\dots,a_{m-1}^{(1)}\right] , \quad \mathbf{b}^{(1)} = \left[b_0^{(1)},\dots,b_m^{(1)}\right] , \tag{32}$$

with coefficients given by (27). Conversely, there exists an **integral scheme**

$$IS(\mathbf{a},\mathbf{b}) := S(\mathbf{a}^{(-1)},\mathbf{b}^{(-1)}) \ , \ \mathbf{a}^{(-1)} = \left[a_0^{(-1)},\ldots,a_m^{(-1)}\right] \ ,$$

$$\mathbf{b}^{(-1)} = \left[b_0^{(-1)},\ldots,b_{m+1}^{(-1)}\right] \ , \tag{33}$$

whose divided difference scheme is the basic scheme, i.e. $DIS(\mathbf{a},\mathbf{b}) = S(\mathbf{a},\mathbf{b})$. The coefficients of this scheme are given by

$$\left.\begin{array}{l} a_j^{(-1)} = \frac{1}{2}(a_j + b_j) \ , \quad j = 0,\ldots,m \ , \\ b_0^{(-1)} = \frac{1}{2}\, b_0 \ , \ b_j^{(-1)} = \frac{1}{2}(a_{j-1} + b_j) \ , \quad j = 1,\ldots,m+1 \ . \end{array}\right\} \tag{34}$$

Given that $S(\mathbf{a},\mathbf{b})$ has fundamental solution φ, let $DS(\mathbf{a},\mathbf{b})$ and $IS(\mathbf{a},\mathbf{b})$ have fundamental solutions χ and ψ respectively. From Theorems 7 and 9 we have that if $\chi \in C^0(-\infty,\infty)$, then $\varphi \in C^1(-\infty,\infty)$ and $\psi \in C^2(-\infty,\infty)$. More precisely, we can relate the fundamental solutions in the following way:

Consider the divided difference scheme applied to the initial data $\left\{\Delta\varphi_i^0 = \Delta\delta_{i,0}/i \in \mathbb{Z}\right\}$. Then the limit curve is

$$\varphi'(t) \equiv \chi(t+1) - \chi(t) \ . \tag{35}$$

Thus, noting the local support of χ,

$$\varphi(t) = \int_t^{t+1} \chi(s)ds = \int_{-\infty}^{\infty} \chi(t-s)B_1(s)ds = \chi * B_1(t) \ , \tag{36}$$

where

$$B_1(s) = \begin{cases} 1 \ , & -1 \le s \le 0 \ , \\ 0 \ , & \text{otherwise} \ . \end{cases} \tag{37}$$

Applying this convolution result to the integral scheme thus gives:

Theorem 11. Let $S(\mathbf{a},\mathbf{b})$ have discrete fundamental solutions which converge uniformly to $\varphi \in C(-\infty,\infty)$. Then $IS(\mathbf{a},\mathbf{b})$ has discrete fundamental solutions which converge uniformly to

$$\psi = \varphi * B_1 \tag{38}$$

and, in general, $I^{\ell}S(\mathbf{a},\mathbf{b})$, $\ell \geq 1$, has discrete fundamental solutions which converge uniformly to

$$\psi * B_{\ell} := \psi * B_1 * \ldots * B_1 \ . \tag{39}$$

We thus see that ℓ'th order integral schemes (and C^{ℓ} basic schemes) have fundamental solutions which are defined by convolutions with ℓ'th order B-splines, which confirms a conjecture of C.A. Micchelli (1987).

7. An interpolatory subdivision scheme for surfaces

We have so far described a theory for the analysis of convergence of univariate uniform subdivision algorithms defined by a rule of the form (1). The motivation for this work is the specific example of the 4-point interpolatory curve scheme defined by (2). Application of the theory to this specific case (with $L = 2$ in the matrix analysis of section 4) gives $-0.375 < \omega < 0.39$ and $0 < \omega < 0.154$ as sufficient conditions for a C^0 and C^1 limit curve respectively, see Dyn et al (1988). Taking higher values of L and using similarity transformations on the control point matrices suggested by M.J.D. Powell (1988) gives improved ranges for ω. For example, numerical experiments indicate that $|\omega| < 1/2$ is a sufficient condition for a C^0 limit, where for negative ω we have used the result of Micchelli and Prautzsch (1988) using the positivity of the coefficients in (2) ($|\omega| < 1/2$ is also necessary by application of Corollary 5).

We conclude by describing a bivariate interpolatory subdivision scheme for surfaces whose parameterization can be defined on a "type 1" regular triangulation. Clearly, tensor product type surface schemes can be derived immediately from the univariate theory but our interest here is in the development of triangular based schemes. The scheme is defined as follows:

Let f^k denote a control polygon in $\mathbb{R}^3$ with control points $f^k_{i,j} \in \mathbb{R}^3$, $(i,j) \in \mathbb{Z}^2$, and consisting of triangular faces with vertices $\{f_{i,j}, f_{i+1,j}, f_{i,j+1}\}$ and $\{f_{i+1,j}, f_{i+1,j+1}, f_{i,j+1}\}$. Then f^{k+1} has vertices defined by the rule

$$\left.\begin{aligned}
f^{k+1}_{2i,2j} &= f^k_{i,j}\,, \\
f^{k+1}_{2i+1,2j} &= \tfrac{1}{2}\left(f^k_{i,j} + f^k_{i+1,j}\right) + 2\omega\left(f^k_{i,j+1} + f^k_{i+1,j-1}\right) \\
&\quad - \omega\left(f^k_{i-1,j+1} + f^k_{i+1,j+1} + f^k_{i,j-1} + f^k_{i+2,j-1}\right), \\
f^{k+1}_{2i,2j+1} &= \tfrac{1}{2}\left(f^k_{i,j} + f^k_{i,j+1}\right) + 2\omega\left(f^k_{i+1,j} + f^k_{i-1,j+1}\right) \\
&\quad - \omega\left(f^k_{i+1,j-1} + f^k_{i+1,j+1} + f^k_{i-1,j} + f^k_{i-1,j+2}\right), \\
f^k_{2i+1,2j+1} &= \tfrac{1}{2}\left(f^k_{i+1,j} + f^k_{i,j+1}\right) + 2\omega\left(f^k_{i,j} + f^k_{i+1,j+1}\right) \\
&\quad - \omega\left(f^k_{i-1,j+1} + f^k_{i+1,j-1} + f^k_{i,j+2} + f^k_{i+2,j}\right),
\end{aligned}\right\}\ (i,j)\in\mathbb{Z}^2. \qquad (40)$$

As with the univariate rule (2), varying ω in (40) gives some control on the shape of the limit surface. The case $\omega = 0$ gives $f^{k+1} = f^k$ for all k and hence the limit surface is the initial control polygon f^0. The case $\omega = 1/16$ corresponds to a rule which has bivariate cubic polynomial precision with respect to the diadic point parameterization, $f^k_{i,j}$ being defined at

$$t^k_{i,j} := (i/2^k, j/2^k)\,. \qquad (41)$$

The scheme (40) then corresponds to a symmetric rule defined on a uniform subdivision of a "type 1" regular triangulation.

A convergence analysis of this algorithm is currently being developed which suggests that the limit surface will be C^1 for a range of ω which includes $\omega = 1/16$. This indicates the existence of a C^1 interpolant on a regular triangulation whose fundamental solutions (i.e. cardinal basis functions) have local support. The subdivision algorithm is illustrated by the following example:

Figure 2 shows a set of control points defined on the surface of a sphere with two control points pulled away from the spherical surface to give the initial control polygon f^0. Figures 3 show a shaded picture description of the results of the subdivision algorithm through four levels of recursion with $\omega = 1/16$ and where appropriate additional control points have been defined as boundary conditions on the algorithm external to the surface shown. The results indicate a smoothing process suggested by a C^1 limit.

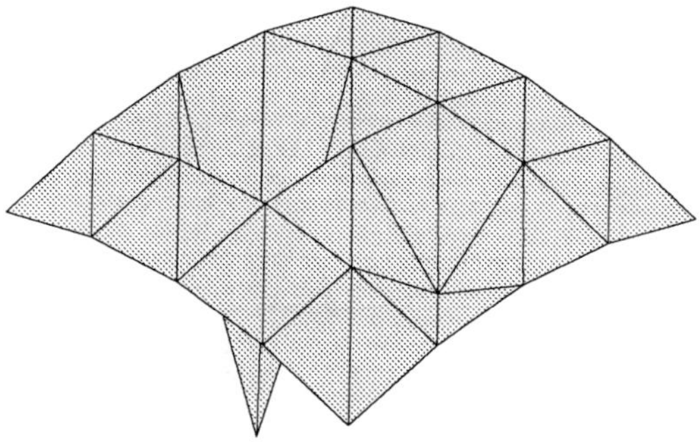

Figure 2. Example data for interpolatory surface algorithm

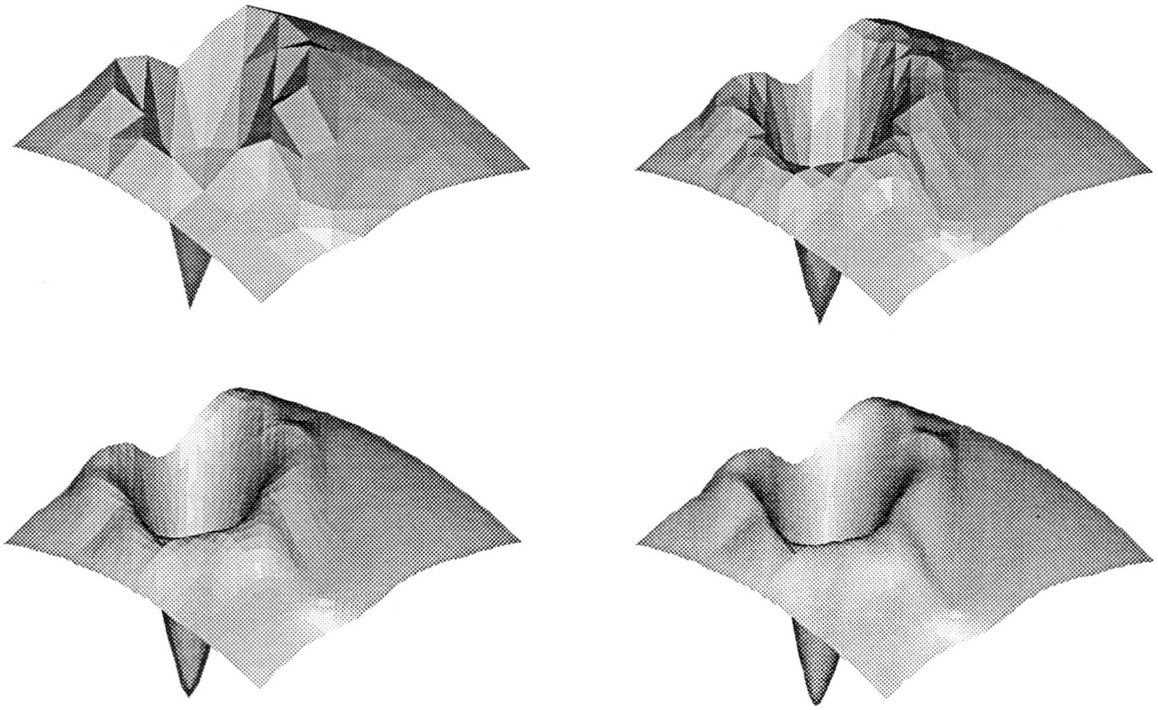

Figures 3. Application of interpolatory surface algorithm

Acknowledgements

This work was supported by the U.K. Science and Engineering Research Council grants GR/E/25139 and 26594; the U.S.A.-Israel Binational Science Foundation grant 86-00243; and the Israel Academy of Science and Humanities grant 548-86. We are also pleased to acknowledge A. Betashvilli in Israel and R. Qu in the U.K. for their contribution to this work.

References

Dyn, N., Gregory, J.A., and Levin, D. (1987), A 4-point interpolatory subdivision scheme for curve design, Computer Aided Geometric Design 4, pp 257-268.

Dyn, N., Gregory, J.A. and Levin, D. (1988), Analysis of uniform binary subdivision schemes for curve design, Brunel University preprint.

Micchelli, C.A. (1987), private communication.

Micchelli, C.A. and Prautzsch, H. (1987a), Uniform refinement of curves, IBM Research Centre preprint.

Micchelli, C.A. and Prautzsch, H. (1988), Refinement and subdivision for spaces of integer translates of a compactly supported function. In Numerical Analysis 1987, Griffiths, D.F. and Watson, G.A. (eds), Pitman Research Notes in Mathematics.

Powell, M.J.D. (1988), private communication.

Prautzsch, H. and Micchelli, C.A. (1987b), Computing curves invariant under halving, Computer Aided Geometric Design 4, pp 133-140.

Approximation by spheres

T. B. Boffey
Statistics and Computational Mathematics Department, University of Liverpool, UK

M. G. Cox
Division of Information Technology and Computing, National Physical Laboratory, Teddington, Middlesex, UK

L. M. Delves, C. J. Pursglove
Statistics and Computational Mathematics Department, University of Liverpool, UK

Abstract It is explained how the method of approximation programming can be applied to find a maximum radius inscribed sphere for a set of data points specified by their Cartesian coordinates in three-dimensional space. A geometric implementation is described and shown to possess computational advantages over a simplex-based algorithm. Methods for other types of approximating spheres are briefly mentioned.
Key words: largest inscribed sphere, linear programming, method of approximation programming, metrological software, sphere fitting.

1. Introduction

An object constructed nominally to be of a certain ideal form will, in practice, have a shape which only approximates the desired one. Thus a "spherical" steel ball, no matter how carefully made, could, if measured accurately enough, be shown not to be spherical. The question therefore arises as to how far it departs from perfect form. The shape obviously depends on the method used to machine the object, and can be assessed by first determining a suitable set of measurements in the surface of the object. It is assumed here that the object is kept at a fixed position and the Cartesian coordinates (x_1, x_2, x_3) of a finite number of points on the surface of the object are determined by a coordinate measuring machine or some other suitable measuring instrument; this leads to a set of points $\Delta = \{\mathbf{u}\}_1^m$, say. Then the "shape" of the object can conveniently be described in terms of the departures of the measured points from a sphere which in some sense provides a "best fit" to the points.

One approximation to the surface of the object is provided by the least-squares sphere (LS-sphere) fit to Δ. The LS-sphere has a radius r and centre $\mathbf{c}$ which minimize $\sum_{i=1}^{m}(|\mathbf{u}_i - \mathbf{c}| - r)^2$. This sphere is often appropriate when the measurement errors are significant compared with the precision to which the object has been manufactured. It can be obtained by standard methods, such as the Gauss-Newton algorithm or a variant thereof [5], and is not considered further here. See [4] for a discussion of least-squares fitting by spheres and other geometric elements.

Other approximating spheres are appropriate when measurement errors are small compared with machining errors [1]. These are the largest sphere containing no points of Δ (the Maximum Inscribed or MI-sphere), and the smallest sphere containing all points of Δ (the Minimum Circumscribing or MC-sphere). All points of Δ lie in the zone which is outside the MI-sphere and inside the MC-sphere, and so these two spheres together give an indication of the variation of the object from being spherical. A possible disadvantage of this description of the variation is that the spheres are not in general concentric. If concentricity is important, as in some aspects of geometric tolerancing in mechanical metrology, it is preferable to determine a pair of concentric spheres of radii R and r $(R \geq r)$ such that $R - r$ is minimized subject to all points of Δ lying between the spheres. These spheres are together called the Minimum Zone or MZ-spheres. The sphere with radius $\frac{1}{2}(R + r)$ that is concentric with these is in fact the minimax (Chebyshev, ℓ_∞) sphere approximant to Δ.

In this paper attention is directed to algorithms for computing MI-spheres. Differences between these algorithms and those for computing MZ- and MC-spheres will briefly be considered. The algorithms described are generalizations to three dimensions of those for computing MI-, MC- and MZ-circles [2].

2. Computing MI-spheres

Let r denote the radius and $\mathbf{c}$ the centre of an MI-sphere for Δ. An MI-sphere can be found by solving the nonlinear programming problem

$$\text{MI-sphere:} \quad \begin{aligned} &\text{maximize } r \\ &\text{subject to} \\ &(\mathbf{u}_i - \mathbf{c})^2 \geq r^2, \quad i = 1, 2, \ldots, m, \\ &\mathbf{c} \in \text{ins}(\Delta), \end{aligned} \tag{1}$$

where $\text{ins}(\Delta)$ denotes the interior of the convex hull of Δ. It will be assumed that the data is sufficiently "good" for the *centre-confining* constraint $\mathbf{c} \in \text{ins}(\Delta)$ to be satisfied automatically and so this constraint will be discarded from the problem formulation; theoretical results on the extent to which this assumption is justified are given in [3].

Suppose now that $\mathbf{c}^{(1)}$ is an initial approximation, with $\mathbf{c}^{(1)} \neq \mathbf{u}_i$ for all i, to the centre of the MI-sphere, and that an improved estimate of the centre is given by $\mathbf{c} = \mathbf{c}^{(1)} + \mathbf{d}$. Expanding $(\mathbf{u} - \mathbf{c}^{(1)} - \mathbf{d})^2$, and neglecting $\mathbf{d}^2$ on the assumption that

$\mathbf{d}$ is small leads to the formulation:

$$\text{LP(1):} \quad \text{maximize } r_L^2 \qquad (2)$$
$$\text{subject to}$$
$$r_L^2 + 2(\mathbf{u}_i - \mathbf{c}^{(1)}).\mathbf{d} + s_i = (\mathbf{u}_i - \mathbf{c}^{(1)})^2, \qquad i = 1, 2, \ldots, m,$$

r being maximized when r_L^2 is. This is a *linear* program with decision variables r_L^2 and the three components (d_1, d_2, d_3) of $\mathbf{d}$, and with slack variables s_i, one for each of the inequality constraints in (1). It can be solved with the simplex algorithm starting with $r_L^2 = 0$ and $\mathbf{d} = \mathbf{0}$.

In the first four simplex iterations in LP(1), r_L^2 and d_1, d_2 and d_3 are successively brought into the basis. Moreover, since r_L^2 increases from its initial value of zero and the components of $\mathbf{d}$ may assume either sign, we may stipulate the following condition.

Basis condition: r_L^2 and d_1, d_2 and d_3 remain basic from the end of iteration 4 onwards, and consequently there will be four nonbasic slack variables.

Let the four nonbasic slack variables be s_e, s_f, s_g and s_h. Then, using (2), for $i \in \{e, f, g, h\}$,

$$s_i = (\mathbf{u}_i - \mathbf{c}^{(1)})^2 - 2(\mathbf{u}_i - \mathbf{c}^{(1)}).\mathbf{d} - r_L^2 = 0,$$

and adding $\mathbf{d}^2$ to both sides of the right-hand equality gives $r^2 - \mathbf{d}^2 = r_L^2$. It follows that the points $\mathbf{u}_e$, $\mathbf{u}_f$, $\mathbf{u}_g$ and $\mathbf{u}_h$ are equidistant from $\mathbf{c}^{(1)}$ and that no other point is nearer. The four data points, which without loss of generality may be assumed non-coplanar [3], define a tetrahedron $T(\mathbf{u}_e, \mathbf{u}_f, \mathbf{u}_g, \mathbf{u}_h)$ called the *basis* tetrahedron. The sphere passing through the vertices of the tetrahedron (which are the four data points) is termed the *emerging sphere*.

There is only a finite number of basis tetraheda and hence only a finite set, E, of possible emerging sphere centres. If LP(1) terminates with a bounded solution, the centre, which will be denoted by $\mathbf{c}^{(2)}$, of the resulting emerging sphere satisfies the following theorem.

Theorem 2.1.

(a) $\mathbf{c}^{(2)} \in E$.

(b) $\mathbf{c}^{(2)}$ is an optimal solution of LP(1) if $\mathbf{c}^{(1)}$ lies inside or on the final basis tetrahedron.

(c) Either $\mathbf{c}^{(2)} = \mathbf{c}^{(1)}$ or $\mathbf{c}^{(2)}$ is strictly further from its nearest data point than $\mathbf{c}^{(1)}$ is from its nearest.

(d) $\mathbf{c}^{(2)}$ is the centre of an MI-sphere if it lies strictly inside the final basis tetrahedron.

Proof. See [3]. □

It frequently happens that $\mathbf{c}^{(2)}$ is the centre of an MI-sphere, but if this is not so a new linear program LP(2) is set up with $\mathbf{c}^{(2)}$ as the starting approximation in place of $\mathbf{c}^{(1)}$. The optimal solution (assumed to be bounded) of LP(2) corresponds to an emerging sphere centre $\mathbf{c}^{(3)}$, say; either this is the centre of an MI-sphere or a

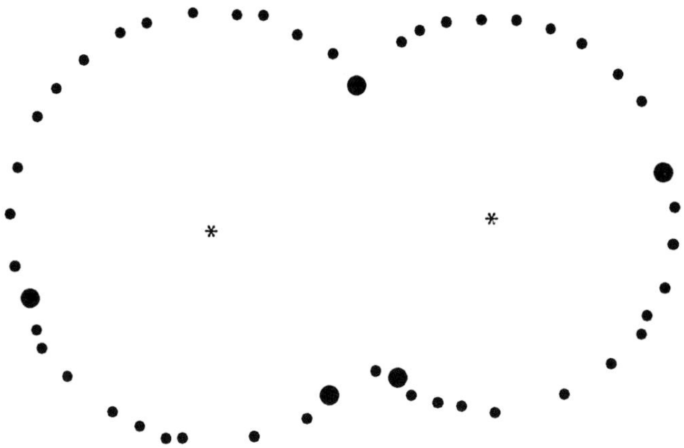

FIG. 1. *An example with two (locally optimal) MI-circles whose centres are indicated by asterisks. The points indicated by large discs lie on one or other (and in one case both) of the MI-circles.*

further linear program is set up, and so on. This is an example of the application of the method of approximation programming - MAP [6, page 64]. Thus a sequence of approximations $\mathbf{c}^{(1)}$, $\mathbf{c}^{(2)}$, $\mathbf{c}^{(3)}$, ..., to the centre of an MI-sphere is formed. That this sequence is finite follows from Theorem 2.1 (applied to LP(k), $k \geq 1$), parts (a) and (c), and the fact that E is finite. If $\mathbf{c}^{(2)} = \mathbf{c}^{(1)}$, define an index $K = 2$. Otherwise, let K be the integer satisfying $\mathbf{c}^{(K)} = \mathbf{c}^{(K-1)} \neq \mathbf{c}^{(K-2)}$. By Theorem 2.1, parts (b) and (d), convergence is to the centre $\mathbf{c}^{(K)}$ of an MI-sphere, except possibly in the case when $\mathbf{c}^{(K)}$ lies *on the surface* of the final basis tetrahedron of LP(K). In this latter, exceptional, case *premature convergence* is said to have taken place; it has not been detected in any of our numerical experiments.

3. Geometric interpretation

It is now shown how the method described above for computing MI-spheres can be implemented directly using geometrical concepts and the advantages of this approach will be mentioned. First, note that the MI-sphere might not be unique. Figure 1 shows, for the analogous 2-dimensional problem of finding an MI-circle [2], how local optima that are inferior to the global optimum can occur. We shall regard any locally optimal solution as giving an MI-sphere. A condition can be specified [3] under which a sphere is a maximum radius MI-sphere.

LP(1) starts with an approximation $\mathbf{c} = \mathbf{c}^{(1)}$ to an MI-sphere centre. A suitable $\mathbf{c}^{(1)}$ is given by the centre of the approximate least-squares sphere obtained by a straightforward extension of the approximate least-squares circle algorithm [2]. The only nonbasic variables are r_L^2, d_1, d_2 and d_3.

In the first iteration r_L^2 becomes basic, increasing to $\min_i(\mathbf{u}_i - \mathbf{c}^{(1)})^2$, and s_e nonbasic. That is, the emerging sphere "expands" from a fixed centre until the first data point, $\mathbf{u}_e$, is reached.

In the second iteration d_1 is made basic and and s_f nonbasic: $\mathbf{c}$ moves parallel to the x_1-axis with the emerging sphere maintaining contact with $\mathbf{u}_e$. Movement continues until a second data point, $\mathbf{u}_f$, is encountered.

In the third iteration d_2 is made basic and s_g nonbasic: the emerging sphere maintains contact with $\mathbf{u}_e$ and $\mathbf{u}_f$, and movement of the centre is away from the midpoint of the line segment joining $\mathbf{u}_e$ and $\mathbf{u}_f$ in the plane $d_3 = 0$ (since d_3 remains nonbasic). Movement continues until a third point, $\mathbf{u}_g$, is encountered.

In the fourth iteration d_3 is made basic and s_h nonbasic. Geometrically, this means that the centre of the emerging sphere moves along the normal to the plane through $\mathbf{u}_e$, $\mathbf{u}_f$ and $\mathbf{u}_g$, away from and through the centre of the circle passing through these points, the sphere maintaining contact with $\mathbf{u}_e$, $\mathbf{u}_f$ and $\mathbf{u}_g$. Movement continues until a fourth data point, $\mathbf{u}_h$, is encountered.

In each subsequent iteration one slack variable is dropped from the basis and another introduced. This corresponds to one of the four data points on the emerging sphere being "discarded" and a transformation of the kind described for iteration 4 performed.

It is shown in [3] that, in the absence of degeneracy, the geometric implementation is equivalent to that based on the simplex tableau in that the same sequence of approximations to the centre of an MI-sphere is generated.

The geometric implementation of the algorithm has several advantages over the linear programming implementation, including the following:

1. Calculations to determine successive emerging sphere centres and basis tetrahedra are straightforward.
2. No simplex tableau is needed.
3. The restarts inherent in simplex-based implementations are eliminated.
4. Less overall computation is required.
5. Calculations are in terms of the original data, and so accuracy is maintained.
6. Possible premature convergence is avoided.
7. The radius (squared) of the emerging sphere, rather than merely an approximation, r_L^2, is always available.

4. Numerical results

Numerical experience with the geometric algorithm for finding an MI-sphere has been very encouraging; some illustrative results are given below.

Two groups of data sets were generated: group A contains five "relatively good" data sets, and group B five "relatively poor" sets. In both cases, points were first generated to lie in the cube with vertices $(\pm1, \pm1, \pm1)$ by sampling from a uniform distribution. Points lying outside the sphere with centre 0 and unit radius were rejected. When sufficient points were obtained they were linearly scaled to lie between

Data group	m	K	Average number of simplex iterations	Min time (seconds)	Max time (seconds)
A	50	0.999	3.2	46	67
B	16	0.8	3.4	31	45

TABLE 1

Results of MI-sphere calculations using a geometric implementation of the method of approximation programming. m denotes the number of data points $\{\mathbf{u}_i\}$ and K is a factor such that $K \leq \|\mathbf{u}_i\| \leq 1$ for $i = 1, 2 \ldots, m$.

concentric spheres centred at the origin and with radii K and 1. Data sets in A each contain 50 data points and K was chosen to be 0.999. Data sets in B each contain 16 points with K set to 0.8.

Results using these data sets are given in Table 1. Note that there is not a great deal of variability in computation times for the data sets in each group. Programs were written in Ada and run on a TORCH microcomputer. It should be noted that the computation times are high for two reasons: floating-point calculations are implemented via software routines, and the code contains extra output used for development purposes. Production versions would run very much faster.

5. MC- and MZ-spheres

MAP algorithms can also be developed for solving the MC-sphere and MZ-spheres problems. There are, however, some differences [3] from the approach described in Sections 2 and 3.

For the MC-sphere problem:

1. There is no need for a centre-confining constraint $\mathbf{c} \in ins(\Delta)$.
2. Premature convergence cannot occur.
3. Local optima cannot occur.
4. The optimum sphere may have only three, or even two, data points on its surface - that is, the optimal solution may not lie at a vertex of the polytope corresponding to the feasible region.

It is straightforward to take account of this last difference in a geometric implementation. If a tableau-based approach is adopted, an appropriate modification to the convex simplex method [7] can be used.

For the MZ-spheres problem:

1. There is no need for a centre-confining constraint.
2. Premature convergence is possible.
3. The optimal solution lies at a vertex.

Although a geometric implementation for the MZ-spheres problem is clearly possible, it is not attractive, being much more complex than that described for MI-spheres.

6. Conclusion

An algorithm has been outlined for computing a largest inscribed sphere for a given set of points in three dimensions. It has been demonstrated that a "geometric implementation" of the algorithm offers advantages over a simplex-based approach. Extensions to minimum circumscribed and minimum zone (minimax) spheres have been indicated.

Acknowledgments
The work described here has been carried out as part of an NPL programme of work on Algorithms and Software for Metrology. We thank G. T. Anthony for his comments on the paper.

References

1. G. T. Anthony and M. G. Cox. The design and validation of software for dimensional metrology. Technical Report DITC 50/84, National Physical Laboratory, 1984.
2. G. T. Anthony and M. G. Cox. Reliable algorithms for roundness assessment according to BS3730. In M. G. Cox and G. N. Peggs, editors, *Software for Co-ordinate Measuring Machines*, pages 30 – 37, Teddington, 1986. National Physical Laboratory.
3. T. B. Boffey, M. G. Cox, L. M. Delves, J. L. Mohamed, and C. J. Pursglove. Fitting spheres to data. Technical report, National Physical Laboratory, 1989. To appear.
4. A. B. Forbes. Least squares best fit geometric elements. In J. C. Mason and M. G. Cox, editors, *Algorithms for Approximation*, London, 1989. Chapman & Hall. To appear.
5. P. E. Gill, W. Murray, and M. H. Wright. *Practical Optimization.* Academic Press, London, 1981.
6. G. R. Walsh. *Methods of Optimization.* John Wiley, 1975.
7. W. I. Zangwill. *Nonlinear Programming.* Prentice-Hall, 1969.

Interpolation of scattered data on a spherical domain

T. A. Foley

Computer Science Department, Arizona State University, Tempe, Arizona, USA

Abstract Given arbitrary points on the surface of a sphere and associated real values, we address the problem of constructing a smooth function defined over the sphere which interpolates the given data. Applications include modeling closed surfaces in CAGD and representing functions which estimate temperature, rainfall, pressure, ozone, etc. at all points on the surface of the earth based on discrete samples. An easily implemented solution is presented which is a C^2 modification of the planar reciprocal multiquadric method.

Key words : Scattered data interpolation, Multivariate interpolation, Surfaces on surfaces, Geometric modeling.

1. Introduction

Given N distinct points p_i on a sphere S and N real values f_i, the problem that we address is the construction of a smooth function $F(p)$ defined on all of S that satisfies $F(p_i) = f_i$ for $i = 1,..., N$. Without loss of generality, assume that S is the unit sphere centered at the origin. Applications include modeling closed surfaces in CAGD and representing functions which estimate temperature, rainfall, pressure, ozone, gravitational forces, etc. at all points on the surface of the earth based on a discrete sample of values taken at arbitrary locations. If the points p_i are sampled in a small region of the sphere, then this region can be mapped to a plane and any number of effective bivariate interpolants can be applied to the transformed data. If the data are sampled over the entire sphere, then since there is no differentiable mapping of the entire sphere to a bounded planar region, there is a need to construct the interpolating function $F(p)$ over the sphere itself.

Very few solutions to this scattered data interpolation problem exist. Recently developed C^1 methods are given in Lawson(1984), Renka(1984), Barnhill et al.(1987) and Nielson and Ramaraj(1987). We present a method which is a C^2 modification of the reciprocal multiquadric (RMQ) of Hardy(1971), generalized to work on a spherical domain. In addition to producing visually smooth plots with small observed errors, this C^2 method is much easier to implement than the more complicated methods referenced earlier. Although the examples presented here involve a modification of the RMQ method, the ideas apply to a much wider class of radial basis distance methods.

Before attacking the problem on a spherical domain, we briefly review some scattered data interpolants defined on a planar domain. Of the many methods tested in Franke(1982) and surveyed further in Franke(1987), three of the more effective methods are the multiquadric (MQ) and reciprocal multiquadric (RMQ) methods in Hardy(1971), and the thin plate spline in Duchon(1977). Recent analytic results on these radial basis methods are given in Micchelli(1986), Powell(1987) and Buhmann(1988), while affine invariant implementation techniques are given in Nielson(1987) and Nielson and Foley(1988). The MQ and RMQ methods can be written in the form

$$F(x,y) = \sum_{i=1}^{N} a_i B[\, d_i(x,y)\,]$$

where $B(t) = (t^2 + R^2)^{1/2}$ for the MQ method, $B(t) = (t^2 + R^2)^{-1/2}$ for the RMQ method, $d_i(x,y)$ is the distance from (x,y) to (x_i, y_i), and R^2 is a positive constant selected by the user. The unknown coefficients a_i are computed by solving the linear system of equations $F(x_j, y_j) = f_j$, for $j = 1,\ldots, N$. The MQ and the RMQ interpolants are C^∞ functions, while the thin plate spline is C^1 at the data points and C^∞ elsewhere.

2. Interpolation on a Spherical Domain

The MQ and RMQ interpolants on a planar domain each depend on the distances from an arbitrary point to each of the data site locations in the plane. For points p and p_i on the unit sphere centered at the origin, the geodesic distance between them can be computed by

$$g_i(p) = arccos(p \cdot p_i)\ . \qquad (1)$$

A straightforward extension of the MQ and the RMQ planar methods is

$$F(p) = \sum_{i=1}^{N} a_i B[\, g_i(p)\,] , \tag{2}$$

where $B(t) = (t^2 + R^2)^{1/2}$ for MQ, $B(t) = (t^2 + R^2)^{-1/2}$ for RMQ, and the coefficients a_i are the computed solutions to the N by N linear system of equations $F(p_j) = f_j$, $j=1,\ldots, N$. Each basis function $B[\, g_i(p)\,]$ is a C^∞ function on the sphere S except at the **antipodal point** p_i^*, which is the farthest point on the sphere from p_i at a distance of π. At the antipodal point p_i^*, the basis function $B[\, g_i(p)\,]$ is not differentiable, thus the interpolant in (2) is only a C^0 function on the sphere.

Two approaches are given which modify the basis functions so that they are C^2 on the sphere. Both methods use piecewise quintic polynomial blending near the antipodal point. The first method modifies the geodesic distance function, while the the second approach operates on the basis function itself. We arbitrarily choose the knot location for the piecewise definitions to be $t = 3$ so that the modified basis functions will equal the MQ and RMQ basis functions on a large portion of $[\, 0, \pi\,]$.

A C^2 modification involving the geodesic distance function in (1) is

$$F(p) = \sum_{i=1}^{N} a_i \; B[\, h_i(p)\,] \tag{3}$$

where $$h_i(p) = \begin{cases} g_i(p) & \text{if } g_i(p) \le 3 , \\ H[\, g_i(p)\,] & \text{otherwise,} \end{cases}$$

and $H(t)$ is the polynomial of degree ≤ 5 that satisfies $H(3) = 3$, $H'(3) = 1$, $H''(3) = 0$, $H(\pi) = c$, $H'(\pi) = 0$ and $H''(\pi) = 0$. We set $c = 3.1$ so that $H(t)$ will be convex and monotone on $[\, 3, \pi\,]$. The modified geodesic distance function $h_i(p)$ is C^2 except at p_i, thus $B[h_i(p)]$ and $F(p)$ in (3) are C^2 on the sphere. This type of modification applies to other planar interpolants that depend on distance. A similar approach, which we use for the following figures, involves directly modifying the basis functions $B[g_i(p)]$ so that they are C^2 on the sphere. The C^2 modified MQ and RMQ methods can be represented by

$$F(p) = \sum_{i=1}^{N} a_i \; D[\, g_i(p)\,] , \tag{4}$$

where $$D(t) = \begin{cases} B(t) & \text{if } t \le 3 , \\ b(t) & \text{if } 3 < t \le \pi , \end{cases}$$

and $b(t)$ is the polynomial of degree ≤ 5 that satisfies $b(3) = B(3)$, $b(\pi) = c$, $b'(3) = B'(3)$, $b'(\pi) = 0$, $b''(3) = B''(3)$ and $b''(\pi) = 0$. The constant c is set to $B(3.1)$ (which depends on

R^2) in our examples so that $D(t)$ is monotone on the interval (3, π). As with the planar interpolant, $B(t) = (t^2 + R^2)^{1/2}$ for the MQ method, $B(t) = (t^2 + R^2)^{-1/2}$ for the RMQ method, and the coefficients a_i are computed by solving the N by N linear system of equations $F(p_j) = f_j$, $j = 1,\ldots, N$. The function $D[g_i(p)]$ is C^2 on the sphere, hence $F(p)$ in (4) is also. Figure 1a) represents the C^0 RMQ basis function $B[g_i(p)]$ in a small neighborhood centered about the antipodal point p_i^*, which is located at π. Figure 1b) is the C^2 modified RMQ basis function $D[g_i(p)]$ in a small neighborhood centered about p_i^*.

In order to construct the C^2 interpolants in (4), the constant R^2 needs to be selected. For the planar MQ and RMQ interpolants, Franke(1982) selected R^2 to be approximately two times the area of the bounding circle divided by N. A similar approach is used in Foley(1987) for a default value, although other options are available. For the examples presented here, we set R^2 to be 2.2 times the surface area of the sphere divided by N, which is approximately $28/N$. This choice is by no means optimal, but it yielded consistently effective results on many test cases involving the C^2 modified RMQ method.

For testing the C^2 interpolants in (4), we constructed the following test functions in the vein of Franke(1982):

$$F_1(x,y,z) = exp[\,-2(\,(x-1)^2 + y^2 + z^2\,)\,] + .5exp[\,-4((\,x^2 + (y-.7)^2 + (z-.7)^2\,)\,] \\ -\ .25exp[\,-4(\,x^2 + (y+.7)^2 + (z+.7)^2\,)\,]\,,$$

$$F_2(x,y,z) = .5 + 4tanh[\,2(\,1-x^2-(y-1)^2-z^2\,)\,] + .5exp[\,-2(\,x^2+(y+.7)^2+(z+.7)^2\,)\,].$$

The plots of these given in Figure 2a) and Figure 3a), represent the test functions evaluated over the entire sphere but plotted over a plane using latitude and longitude for the axes.

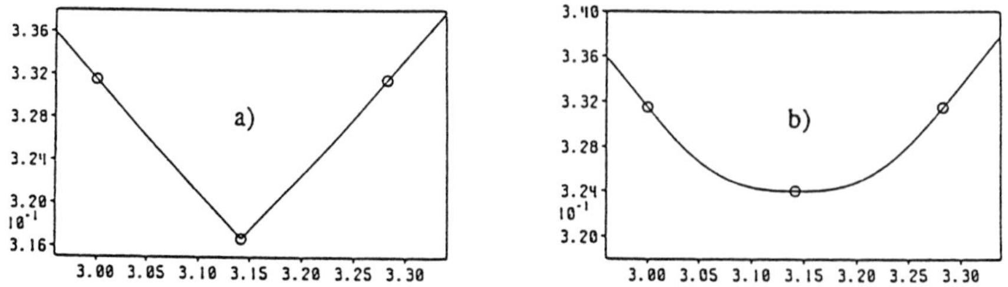

Figure 1 a) The C^0 RMQ basis function $B[g_i(p)]$,
b) the C^2 modified RMQ basis function $D[g_i(p)]$.

The front and back curves are constant and they represent the function evaluated at the north and south poles. The boundary curves on the left and right are equal because they represent the function evaluated along the same meridian. Figure 2b) is a plot of the C^2 modified RMQ method in (4) applied to the $N = 90$ points denoted by the boxes and generated by evaluating $F_1(p)$ at 90 uniformly distributed pseudo-random points on the unit sphere. Figure 2c) is the C^2 modified RMQ method applied to $N = 140$ points, while Figures 3b) and 3c) are the C^2 RMQ method applied to the points generated by evaluating the function $F_2(p)$ at 90 and 140 pseudo-random points on the sphere, respectively. The observed maximum absolute errors are 0.025 in 2b), 0.0029 in 2c), 0.035 in 3b) and 0.020 in 3c). The observed mean absolute errors are 0.0016 in 2b), 0.00024 in 2c), 0.0043 in 3b) and 0.0023 in 3c). Figures 4 and 5 exhibit contour plots of the surfaces shown in Figures 2 and 3, respectively, with the 90 and 140 data sites p_i denoted by marks in parts b) and c).

Figure 6 is another way to visualize the surface shown in Figure 2b), which is the C^2 RMQ method applied to 90 points generated by $F_1(p)$. The transparent shaded surface is the interpolant to the data points projected out radially from the sphere. This figure explains why this problem is often referred to as finding a surface on a surface. Figure 6 also depicts the application of modeling closed surfaces without dealing with periodic end conditions that are generally needed in standard parametric patch methods.

All of the examples presented involve the C^2 RMQ method in (4) because the C^2 MQ method performs poorly on this same data. We were surprised by this behavior because in the planar case, both methods performed well and the MQ method generally yielded slightly better results. A possible explanation for the poor performance of the modified MQ method on the sphere is that the C^2 modification changes the basis function from convex upward to convex downward near the antipodal point. For large values of N, storing and solving the N by N linear system of equations can be costly. However, partitioning or mollifying methods described in Franke(1982), which involve distance, can be used on the spherical domain by simply using geodesic distance. Extensions of this research to non-convex domains have been made and they will be reported on in the future.

Acknowledgements

A portion of this research was performed at Lawrence Livermore National Laboratory and supported by the U.S. Dept. of Energy contract W-7405-ENG-48. This work was also done at Arizona State University and supported by the U.S. Dept. of Energy grant DE-FG-02-87ER25041. Special thanks go to Gregory Nielson of the CAGD group at A.S.U.

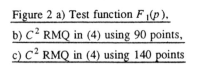

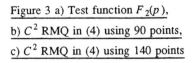

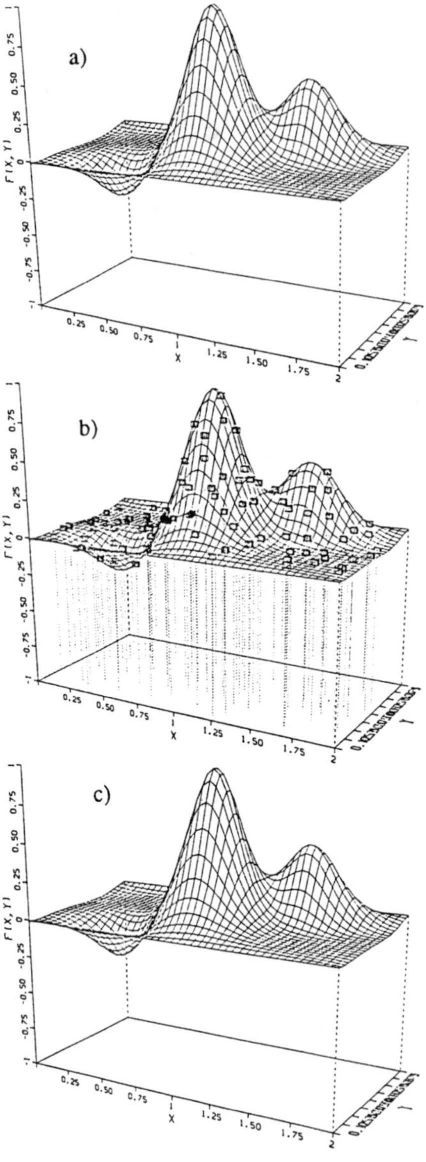

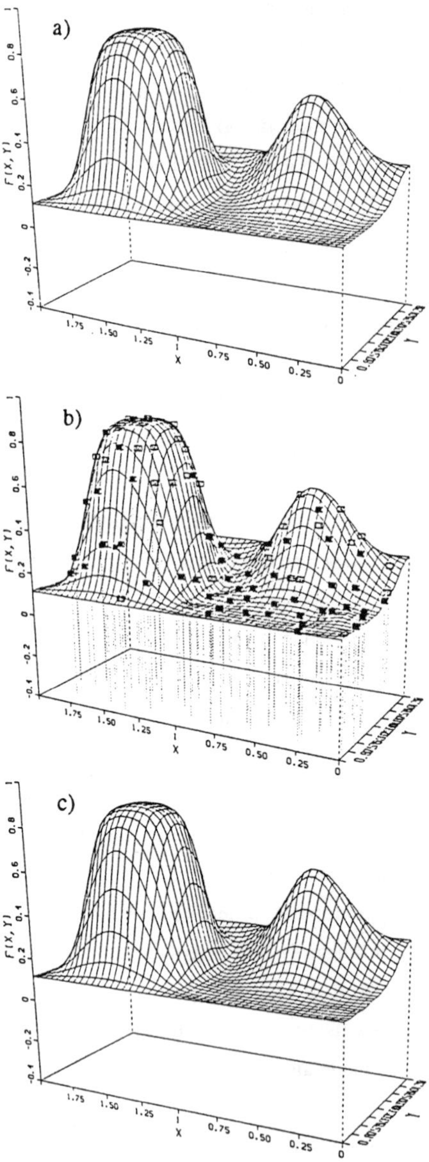

Figure 2 a) Test function $F_1(p)$,
b) C^2 RMQ in (4) using 90 points,
c) C^2 RMQ in (4) using 140 points

Figure 3 a) Test function $F_2(p)$,
b) C^2 RMQ in (4) using 90 points,
c) C^2 RMQ in (4) using 140 points

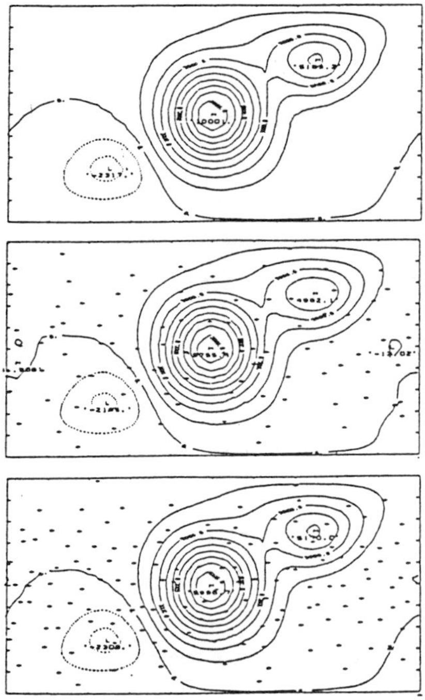

Figure 4 Contours of surfaces in Fig. 2

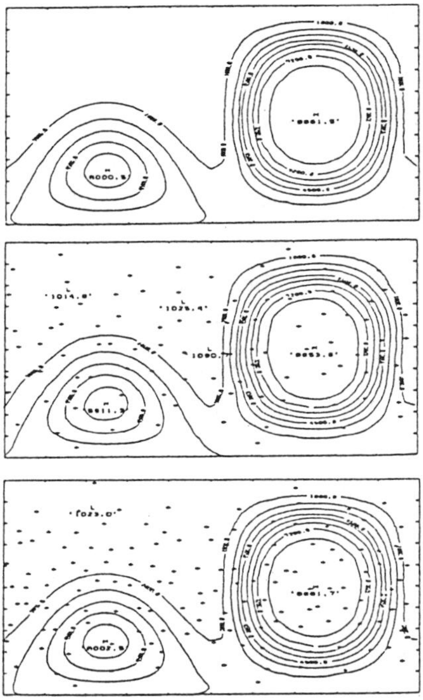

Figure 5 Contours of surfaces in Fig. 3

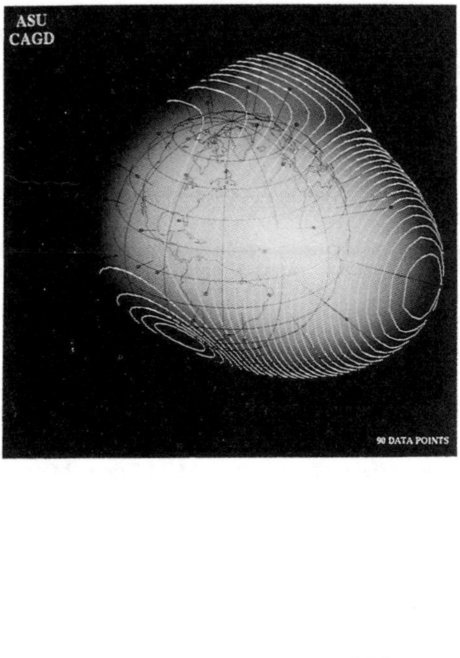

Figure 6 This is the same interpolating function shown in Fig. 2b), but it is rendered using a transparent radial projection through the data points.

References

Barnhill, R.E., Piper, B.R. and Rescorla, K.L. (1987), Interpolation to arbitrary data on a surface. In 'Geometric Modeling', G.E. Farin, ed., SIAM, Philadelphia, pp 281-289.

Buhmann, M.D. (1988), 'Multivariate interpolation in odd dimensional Euclidean spaces using multiquadrics', Tech. Rep. DAMTP 1988/NA6, University of Cambridge.

Duchon, J. (1977), Splines minimizing rotation invariant semi-norms in Sobolev spaces. In 'Constructive Theory of Functions of Several Variables', W. Schempp and K. Zeller, eds., Lecture Notes in Mathematics 571 , Springer-Verlag, New York, pp 85-100.

Foley, T.A. (1987), Interpolation and approximation of 3-D and 4-D scattered data. Comp. and Math. Applic. 13 , 711-740.

Franke, R. (1982), Scattered data interpolation: tests of some methods. Math. Comp. 38 , 181-200.

Franke, R. (1987), Recent advances in the approximation of surfaces from scattered data. In 'Topics in Multivariate Approximation', C.Chui, L.L. Schumaker and F. Utreras, eds., Academic Press, New York, pp 79-98.

Hardy, R.L. (1971), Multiquadric equations of topography and other irregular surfaces. J. Geophys. Res. 76 , 1905-1915.

Lawson, C.L. (1984), C^1 surface interpolation for data on a sphere. Rocky Mountain J. Math. 14 , 177-202.

Micchelli, C.A. (1986), Interpolation of scattered data: distance matrices and conditionally positive definite functions. Const. Approx. 2 , 11-22.

Nielson, G.M. (1987), Coordinate free scattered data interpolation. In 'Topics in Multivariate Approximation', C.Chui, L.L. Schumaker and F. Utreras, eds., Academic Press, New York, pp 175-184.

Nielson, G.M. and Foley, T.A. (1988), A survey of applications of an affine invariant norm. In 'Mathematical Methods in CAGD', T. Lyche and L.L. Schumaker, eds., Academic Press, New York.

Nielson, G.M. and Ramaraj, R. (1987), Interpolation over a sphere. Computer Aided Geometric Design 4 , 41-57.

Powell, M.J.D. (1987), Radial basis functions for multivariate interpolation: a review. In ' Algorithms for the Approximation of Functions and Data', M.G. Cox and J.C. Mason, eds., Oxford University Press.

Renka, R.J. (1984), Interpolation of data on the surface of a sphere. ACM Trans. Math. Soft. 11 , 417-482.

Least squares best fit geometric elements

A. B. Forbes

National Physical Laboratory, Teddington, Middlesex, UK

Abstract We describe algorithms, designed for use by coordinate measuring systems (CMS), for finding the best fit geometric elements (lines, planes, circles, spheres, cylinders, and cones) to metrological data. We first consider the case where the residual errors are calculated normal to the element surface, and concentrate on the development of robust parametrizations and efficient optimization algorithms. In many circumstances it is more appropriate to perform the fitting in a generalized least squares sense, e.g., if the CMS uses cylindrical coordinates, or if the error characteristics have been fully calibrated. We outline corresponding best fit algorithms which can be adapted to fit more general parametrized surfaces in 3-D. The algorithms point to a need for general purpose optimization software which allows the structure of the Jacobian matrix to be fully exploited.

Key words: Geometric element, Least squares, Parametrization

1. Data fitting in dimensional metrology

In this paper, we are concerned with data fitting problems that arise in dimensional metrology: we wish to find best fit curves and surfaces to data supplied by a coordinate measurement system (CMS). Examples of CMSs in common use are i) 3 - coordinate measuring machines (CMMs), ii) photogrammetry, and iii) theodolites. Other systems are being developed, such as iv) 3 - dimensional optical sensing, and v) laser tracking. One common feature of these systems is that the data they produce have measurement errors in all the cartesian coordinates. Most metrological data fitting models assume that the measurement errors in the x, y, and z coordinates are equal and uncorrelated: we call such models *isotropic*. However, this assumption is almost never valid: for example, the architecture of a CMM may result in the z measurement being only half as accurate as the x and y measurements; some CMMs measure in cylindrical or spherical coordinates, and the corresponding cartesian coordinates will consequently have correlated errors. Similar remarks apply to data gathered by other CMSs. We call data fitting models which take into account the more general error characteristics *anisotropic.*

Another aspect of metrological data is that it is usually accurate: the relative error is generally between $1 : 10^3$ and $1 : 10^6$. Also, the data sets can be quite large, with as few as ten or as many as tens of thousands of data points.

The curves and surfaces we are concerned with often have components which are 'geometric elements', i.e., lines, planes, circles, spheres, cylinders or cones. With the advent of computer aided geometric design (CAGD), we are also increasingly involved with more general types of curves and surfaces representing, for example, car body panels, and described in terms of parametric splines, etc.

In modelling the data fitting problem, we have to find parameters to describe the position, orientation, size, and shape of the curve or surface which yield numerically stable optimization problems, the solutions of which will provide the required parameter values. Some parametrizations will be better than others, and we are particularly interested in ones which behave well for near degenerate data, for instance, data representing only a small arc of a circle. We define the residual error at a data point to be a measure of the distance from the point to the curve or surface, and use this to produce a non-linear least squares optimization model. The quality of the data means that a Gauss-Newton type iterative algorithm, [10, page 134], (using line searches or trust regions) is particularly appropriate. For this type of algorithm, we have also to supply starting estimates of the optimization parameters.

The target machine for least squares element fitting algorithms is usually the microcomputer which controls the CMS. There are commercially motivated constraints on how long calculations can take, and how much memory space is available, and it is therefore important that the algorithms exploit any geometrical symmetry or structure in the element fitting model.

2. Isotropic models

In this section, we briefly describe algorithms for finding the best fit least squares geometric elements if we assume that the errors in the x, y and z coordinates are equal and uncorrelated; details can be found in [6]. We have cartesian data points $\mathbf{x}_i$, $i = 1, \ldots m$, and we take for the residual error at the ith data point the (signed) Euclidean distance from $\mathbf{x}_i$ to the curve or surface. (We can also consider weighted distances.)

2.1. Lines and planes

It is straightforward to show that the best fit line or plane passes through the centroid of the data $\bar{\mathbf{x}}$, and the orientation of the line or plane is given by an eigenvector of $J^T J$ where $J(i, *) = (\mathbf{x}_i - \bar{\mathbf{x}})^T$. Thus, the problem reduces to finding the eigenvectors of a 3×3 matrix; we can avoid loss of accuracy in forming $J^T J$ by finding the singular value decomposition of J, [10, pages 40 and 135].

2.2. Circles and spheres

We consider here finding a best fit circle in the plane; the sphere problem is entirely analogous.

A circle is naturally parametrized by the coordinates (x_0, y_0) of its centre and its radius r_0. The residual error ϵ_i at the ith data point is then given by $r_i - r_0$, where

$$r_i = \sqrt{[(x_i - x_0)^2 + (y_i - y_0)^2]}\,, \tag{1}$$

and the resulting non-linear optimization problem can be solved using a Gauss-Newton or Newton algorithm [1]. [1] also discusses a popular approximate circle fitting model which has residual error of the form

$$\epsilon_i = A(x_i^2 + y_i^2) + Bx_i + Cy_i + D\,. \tag{2}$$

If $A = 1$ then $\epsilon_i = r_i^2 - r_0^2$ and we have a linear least squares problem. One important use of this formulation is for furnishing good starting values for the iterative non-linear algorithm.

Both these models essentially use the coordinates of the circle centre as parameters and this gives rise to instability if the data represents only a small arc of the circle, since in this case the circle centre will be far from the data points. If we instead constrain $B^2 + C^2 - 4AD = 1$, [12], then numerical stability is retained at the expense of replacing a linear least squares problem by a generalized eigenvalue problem. The difference in the behaviour of these models is starkly demonstrated when we fit a circle to the points $(\pm\epsilon, 0)$, $(\epsilon, \pm 1)$, for $\epsilon \ll 1$. The least squares best fit circle passes through the points $(0,0)$ and $(\epsilon, \pm 1)$, and this is the solution found by the generalized eigenvalue model. The linear least squares model is nearly rank deficient and produces a solution circle centred at $(\sqrt{\epsilon}, 0)$ with radius $1/\sqrt{2}$, while a Gauss-Newton algorithm again suffers from near rank defficiency, and makes no acceptable progress towards the solution. More details can be found in [7].

2.3. Cylinders

We parametrize a cylinder by specifying i) the point $\mathbf{x}_0$ on the axis nearest the centroid $\bar{\mathbf{x}}$ of the data, ii) a vector $\mathbf{a}$, determining the orientation of the axis, in the form $\mathbf{a} = U(\rho_1, \rho_2)[0, 0, 1]^T$ with U a rotation matrix depending on the two rotation angles ρ_1 and ρ_2, and iii) the radius r_0. The ith residual error is then given by

$$\epsilon_i = \|(\mathbf{x}_i - \mathbf{x}_0) \times \mathbf{a}\| / \|\mathbf{a}\| - r_0\,. \tag{3}$$

To implement a Gauss-Newton algorithm we need to calculate the gradient of ϵ_i and the resulting expressions are a little cumbersome. However, if the cylinder is in *standard position*, i.e., if $\mathbf{a} = [0, 0, 1]^T$ and $\mathbf{x}_0 = [0, 0, 0]^T$, then the residual error ϵ_i and its gradient $\nabla\epsilon_i$ with respect to the radius, position parameters, and rotation parameters, are given by

$$\begin{aligned} \epsilon_i &= r_i - r_0\,, \\ -\nabla\epsilon_i^T &= [r_i,\ x_i,\ y_i,\ x_i z_i,\ y_i z_i]/r_i\,, \end{aligned}$$

with r_i given by Equation 1. We can therefore adopt the following strategy at each iteration: i) translate and rotate the data so that the trial fit cylinder is in standard position, ii) calculate the residuals and gradients as above for this frame of reference, and iii) find the update step and interpret it in the original frame of reference. In this way, we are exploiting more fully the symmetry of the cylinder to produce a cleaner and more efficient algorithm. Using the same technique, second derivatives are also vastly simplified, enabling a Newton algorithm to be implemented. This strategy is also applicable to cone fitting below, and many other situations. However, it does depend on isotropy.

If there are at least 9 data points, starting values for the optimization parameters for cylinders (and cones) can be found by fitting a general quadric to the data, a linear least squares problem (cf. section 2.2). It is not clear what can be done if there are fewer than 9 points.

2.4. Cones

The parametrization of a cone is similar to that for a cylinder. Indeed, we tend to think of a cylinder as a special case of a cone. We need parameters to specify i) a point $\mathbf{x}_0$ on the axis related to the centroid $\bar{\mathbf{x}}$ (see below), ii) an axis orientation vector $\mathbf{a}$ as for the cylinder, iii) the distance t_0 from $\mathbf{x}_0$ to the cone surface, and iv) the angle ϕ the axis makes with a generator. The distance from the ith point to the cone surface is then given by

$$\epsilon_i = \cos\phi \|(\mathbf{x}_i - \mathbf{x}_0) \times \mathbf{a}\|/\|\mathbf{a}\| + \sin\phi(\mathbf{x}_i - \mathbf{x}_0)^T \mathbf{a}/\|\mathbf{a}\| \; - t_0 \, . \tag{4}$$

We do not use the coordinates of the cone vertex for parameters because for a small angled cone the vertex could be very far from the data, giving rise to numerical instability. If the cone is in standard position, i.e., the axis coincides with the z axis and $\mathbf{x}_0 = (0, 0, Z_0)$, then the residual error and its gradient with respect to t_0, ϕ, position and rotation parameters, are given by

$$\epsilon_i = r_i \cos\phi + (z_i - Z_0)\sin\phi - t_0 \, ,$$
$$-\nabla\epsilon_i^T = [r_i, \; -w_i r_i, \; x_i \cos\phi, \; y_i \cos\phi, \; w_i x_i, \; w_i y_i]/r_i \, ,$$

where r_i is as in (1), and $w_i = (z_i - Z_0)\cos\phi - r_i \sin\phi$.1 To optimize the condition of the Jacobian matrix, we want $\sum w_i = 0$, or

$$Z_0 = \bar{z} - \bar{r} \tan\phi \, . \tag{5}$$

($\bar{}$ denotes mean). This equation determines how we choose which point $\mathbf{x}_0$ to use to locate the cone axis. The presence of the $\tan\phi$ term shows that we encounter stability problems when the cone angle approaches $\pi/2$, i.e., the cone opens out to become near planar: $\mathbf{x}_0$ moves further and further away from the cone vertex, reflecting the fact that the position of the vertex is becoming less well-determined by the data. We know of no simple cone parametrization which behaves well for near planar cones.

3. Anisotropic models

We now examine algorithms for finding best fit surface geometric elements to data with correlated errors in the x, y, and z coordinates. (We continue to assume that there are no correlations between the measurement errors for different data points). More details can be found in [8]; see also [4] for the case of circle fitting, and [3] for another example of a generalized least squares fit.

3.1. Mathematical model

We suppose that for each data point $\mathbf{x}_i$, we have a 3×3 variance-covariance matrix V_i describing the measurement error characteristics at that point. We can then use $M_i = V_i^{-1}$ as a weighting matrix for the ith residual error. In practice the variance-covariance matrices can be determined by a full calibration of the CMS which produces a metric $M : B \subset R^3 \longrightarrow M_{3,3}(R)$, associating to each point $\mathbf{x}_i$ in the measurement volume B of the CMS, the corresponding symmetric, strictly positive definite *point metric* matrix M_i.

To find the best fit surface S, parametrized by *surface parameters* $\mathbf{u} = (u_1, \ldots, u_n)$, we define the residual error ϵ_i at $\mathbf{x}_i$ by

$$\epsilon_i^2 = (\mathbf{x}_i^* - \mathbf{x}_i)^T M_i (\mathbf{x}_i^* - \mathbf{x}_i) \; , \tag{6}$$

where $\mathbf{x}_i^*$ is the point on S 'nearest' $\mathbf{x}_i$, i.e., the point on S minimizing ϵ_i^2. Unfortunately, $\mathbf{x}_i^*$ usually cannot be found explicitly, even for relatively simple geometric elements, so for each i, we introduce two further *patch parameters*, α_i and β_i to parametrize $\mathbf{x}_i^*$:

$$\mathbf{x}_i^*(\alpha_i, \beta_i, \mathbf{u}) : D_i \subseteq R^2 \longrightarrow S \subseteq R^3 \; . \tag{7}$$

For example, we can parametrize a point on a cylinder determined by $\mathbf{u} = (r, x_0, y_0, \rho_1, \rho_2)$ by

$$\mathbf{x}_i^* = U(\rho_1, \rho_2) \begin{bmatrix} r_0 \cos \alpha_i & + & x_0 \\ r_0 \sin \alpha_i & + & y_0 \\ \beta_i & + & 0 \end{bmatrix} . \tag{8}$$

Following [5], if we factorize $M_i = L_i^T L_i$, we can model the problem of finding the best fit surface as a (non-linear) least squares optimization

$$\text{minimize} \quad E(\mathbf{A}, \mathbf{B}, \mathbf{u}) = \sum (L_i(\mathbf{x}_i^* - \mathbf{x}_i))^T (L_i(\mathbf{x}_i^* - \mathbf{x}_i)) \tag{9}$$

with respect to the $2m + n$ parameters $\mathbf{A} = (\alpha_1, \ldots, \alpha_m)$, $\mathbf{B} = (\beta_1, \ldots, \beta_m)$, and $\mathbf{u}$. In an anisotropic model, we are no longer able to exploit directly any symmetries of the geometric elements; on the other hand, using patch parameters we can fit more general types of surfaces without much additional effort.

3.2. Structure of the Jacobian

At first sight, it would appear that the introduction of the patch parameters is computationally very expensive: if we have m data points, then the Jacobian matrix J associated with E in (9) is $3m \times (2m+n)$, pointing to an $O(m^3)$ algorithm. However, the fact that only ϵ_i depends on α_i and β_i means that each row of J has only $n + 2$

```
**        ****
**        ****          **        ****
**        ****           *        ****
  **      ****            **      ****
  **      ****             *      ****
  **      ****              **    ****
    **    ****               *    ****
    **    ****                **  ****
    **    ****                 *  ****
      **  ****                  ******
      **  ****                   *****
      **  ****                    ****
        ******                     ***
        ******                      **
        ******                       *
```

FIG. 1. *Structure of the Jacobian matrix and its upper triangular factor*

non-zero elements following a block-angular bordered structure represented schematically in Figure 1. The elements of the ith 3×2 block are the derivatives of the x, y, and z components of $\mathbf{e}_i = L_i(\mathbf{x}_i^* - \mathbf{x}_i)$ with respect to α_i and β_i, while the n border columns contain the derivatives with respect to the surface parameters u_i. The $(2m+n) \times (2m+n)$ upper-triangular orthogonal factor R of J is correspondingly structured as represented in Figure 1.

To construct R from J using Givens plane rotations, for each data point, we have to update an $n \times n$ *border* upper triangle R_n by a 3 row block of J. We first perform row operations to upper triangularize the 3×2 sub-block of the rows. At this stage R_n remains unchanged, and we are left to update R_n by a single n row. With this strategy and using classical Givens plane rotations, it takes approximately $2m(n^2 + 6n + 10)$ operations to produce R. We note that the introduction of the patch parameters costs about $12n + 20$ operations per data point, and so for $n > 10$, say, the bulk of the work is in updating the border triangle R_n. Importantly, any structure that the border may have can usually be exploited. For instance if the surface being fitted is represented by a tensor product of splines, then the border is banded, and we can use library routines for banded matrices (e.g. DASL routine URAB, [2]) to update the border upper triangle.

3.3. Parametrization and derivatives

Our usual parametrization of a point on a surface has the form

$$\mathbf{x}_i^* = U(\mathbf{p}_i + \mathbf{t}) , \tag{10}$$

where $\mathbf{p}_i$ is a point on the surface in a fixed position, for example, a cylinder in standard position, $\mathbf{t}$ is a translation vector depending on 0, 1, 2, or 3 translation parameters, and U is a rotation matrix depending on 0, 1, 2, or 3 rotation angles. It may be that the translation and rotation parameters are the only surface parameters to be determined, for example, when we wish to position a fixed surface or *template* as close as possible to the data. At the other extreme, there may be no rotation or translation parameters in the model, such as in orthogonal distance regression for

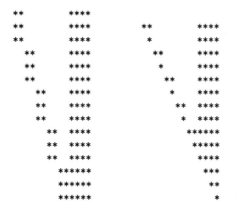

element	$\mathbf{p}_i^T$	$\mathbf{t}^T$
plane	$(\alpha_i, \beta_i, 0)$	$(0, 0, z_0)$
sphere	$(r_0 \cos\alpha_i \cos\beta_i, r_0 \cos\alpha_i \sin\beta_i, r_0 \sin\alpha_i)$ or $(\alpha_i, \beta_i, \sqrt{[r_0^2 - (\alpha_i^2 + \beta_i^2)]}$	(x_0, y_0, z_0)
cylinder	$(r_0 \cos\alpha_i, r_0 \sin\alpha_i, \beta_i)$	$(x_0, y_0, 0)$
cone	$(\tau_i \cos\alpha_i, \tau_i \sin\alpha_i, \beta_i \cos\phi)$, $\tau_i = (t_0 + \beta_i \sin\phi)\cos\phi$	$(x_0, y_0, 0)$

TABLE 1

Parametrization of points on geometric elements

spline surfaces. The parametrizations of points on the surface geometric elements are given in Table 1.

If $\mathbf{x}_i$ is given by (10), then the ith residual 3-vector and its partial derivatives are calculated according to

$$\begin{aligned} \mathbf{e}_i &= L_i(U(\mathbf{p}_i + \mathbf{t}) - \mathbf{x}_i) , \\ \partial\mathbf{e}_i &= (L_i U)(\partial(\mathbf{p}_i + \mathbf{t})) \quad \text{or} \\ \partial\mathbf{e}_i &= L_i \partial U(\mathbf{p}_i + \mathbf{t}) . \end{aligned}$$

The rotation matrix (and its derivatives) will be defined in exactly the same way for many different surfaces, and the only surface dependent information that has to be input into the model is how to calculate $\mathbf{p}_i$, $\mathbf{t}$ and their derivatives. For instance the gradient information required for fitting a cylinder is given by the 3×5 matrix

$$\begin{bmatrix} -r_0 \sin\alpha_i & 0 & \cos\alpha_i & 1 & 0 \\ r_0 \cos\alpha_i & 0 & \sin\alpha_i & 0 & 1 \\ 0 & 1 & 0 & 0 & 0 \end{bmatrix} ,$$

representing the derivatives of $\mathbf{p}_i + \mathbf{t}$ with respect to α_i, β_i, r_0, x_0 and y_0.

3.4. Numerical example - cylinder

We compare the performance of Gauss-Newton isotropic and anisotropic algorithms for the case of cylinders using data representing typical CMS metrological readings. We give both algorithms the same starting estimates for $\mathbf{a}$, $\mathbf{x}_0$ and r_0, and from these we can generate starting estimates for $\mathbf{A}$ and $\mathbf{B}$ for the anisotropic model. For simplicity we use the same point metric

$$L_i = \begin{bmatrix} 1.0 & & \\ 0.0 & 0.9 & \\ 0.02 & 0.01 & 0.5 \end{bmatrix}$$

for each data point for the anisotropic model. The results of the data fitting are given in Table 2. The condition of the Jacobian matrix was similar for both models and the highest correlation coefficient for both models was 0.97 for x_0 with r_0. This is due to the fact that most of the data lay on one half of the cylinder surface. For data points

	a	x_0	r_0	iterations
isotropic	-0.351259	-546.5728	352.1475	4
	0.533806	819.4764		
	0.769200	20.7152		
anisotropic	-0.351256	-546.5734	352.1482	4
	0.533802	819.4758		
	0.769204	20.7150		

TABLE 2

Isotropic and anisotropic cylinder fitting

well distributed over the cylinder surface, there are usually no correlations greater than 0.5 for either model. We note that both algorithms took the same number of iterations.

3.5. Multicomponent surfaces

One further example, an ellipsoidal shell fit, shows how we can fit multicomponent surfaces efficiently. The inner and outer surfaces of the shell are ellipses of revolution, defined by

$$a_k^2(x^2+y^2)+b_k^2z^2=a_k^2b_k^2, \quad k=1,2, \tag{11}$$

and are required to have a common orientation and centre. We parametrize points on the surface as in (10) with $\mathbf{p}_i^T=(b_k\cos\alpha_i\cos\beta_i,\ b_k\cos\alpha_i\sin\beta_i,\ a_k\sin\alpha_i)$ or $\mathbf{p}_i^T=(\alpha_i,\ \beta_i,\ \sqrt{[a_k^2(1-(\alpha_i^2+\beta_i^2)/b_k^2])}$, and $\mathbf{t}^T=(x_0,y_0,z_0)$. The model parameters are $\mathbf{A},\mathbf{B}$, and $\mathbf{u}=(a_1,b_1,a_2,b_2,x_0,y_0,z_0,\rho_1,\rho_2)$. Only the residual errors for points on the inner surface depend on a_1 and b_1, so if we perform a QR factorization first for the inner surface data points we will be working with a border of width 7 rather than 9. We then process the outer surface points in the same way, and finally we update one upper triangle by the other. However, we can rearrange the rows so that in fact we only have to update one *border* triangle by the other *border* triangle. Thus, for multiple component surfaces, we can organise the work so that the computation is $O(mn_{max}^2)$, rather than $O(mn_{tot}^2)$, where n_{max} is the maximal component border width, and n_{tot} is the total number of surface parameters.

3.6. Summary

In summary, we can say that the use of anisotropic models and algorithms as described here has a number of advantages. Firstly, the same element fitting software can be used for different CMSs: the error characteristics of each CMS are encoded in the point metric matrices M_i. Secondly, a wide variety of surfaces and many data fitting problems, including orthogonal regression and data matching, can be tackled using the same basic routines. Thirdly, the algorithms can be made efficient by exploiting the underlying structure. Unfortunately, with most of the standard Library optimization routines, for example those in NAG, [9], it is not easy to take advantage of specific Jacobian structure. However, it is possible to write optimization software which employs reverse communication to allow the user to solve the linear algebra parts of the algorithm independently. At NPL, we have developed linear search routines, adapted from routines from NPL's Numerical Optimization Subroutine Library, [11], which do this.

Acknowledgements

I thank my colleagues G T Anthony, M G Cox, and H M Jones for their many helpful suggestions and comments.

References

1. G. T. Anthony and M. G. Cox. Reliable algorithms for roundness assessment according to BS3730. In M. G. Cox and G. N. Peggs, editors, *Software for Co-ordinate Measuring Machines*, pages 30 – 37, Teddington, 1986. National Physical Laboratory.
2. G. T. Anthony and M. G. Cox. The National Physical Laboratory's Data Approximation Subroutine Library. In J. C. Mason and M. G. Cox, editors, *Algorithms for Approximation*, pages 669 – 687, Oxford, 1987. Clarendon Press.
3. M. G. Cox and H. M. Jones. A nonlinear least squares data fitting problem arising in microwave measurement. In J. C. Mason and M. G. Cox, editors, *Algorithms for Approximation*, London. Chapman & Hall. To appear.
4. M. G. Cox and H. M. Jones. An algorithm for least-squares circle fitting to data with specified uncertainty ellipses. NPL report DITC 118/88, National Physical Laboratory, Teddington, 1988. To appear in *IMA J. Numerical Analysis.*
5. M. G. Cox and P. E. Manneback. Least squares spline regression with block-diagonal variance matrices. *IMA J. Numer. Anal.*, 5:275–286, 1985.
6. A. B. Forbes. Least-squares best-fit geometric elements. NPL report DITC 140/89, National Physical Laboratory, Teddington, 1989.
7. A. B. Forbes. Robust circle and sphere fitting. DITC report, National Physical Laboratory, Teddington, 1989. In preparation.
8. A. B. Forbes. Surface fitting in dimensional metrology. DITC report, National Physical Laboratory, Teddington, 1989. In preparation.
9. B. Ford, J. Bently, J. J. Du Croz, and S. J. Hague. The NAG Library 'machine'. *Software - Practice and Experience*, 9:56–72, 1979.
10. P. E. Gill, W. Murray, and M. H. Wright. *Practical Optimization.* Academic Press, London, 1981.
11. S. M. Hodson, H. M. Jones, and E. M. R. Long. A brief guide to the NPL Numerical Optimization Software Library. Technical report, National Physical Laboratory, Teddington, 1983.
12. V. Pratt. Direct least-squares fitting of algebraic surfaces. *Computer Graphics*, 21(4):145–152, July 1987.

Uniform piecewise approximation on the sphere

W. Freeden

Rheinisch-Westfälische Technische Hochschule Aachen, Aachen, FRG

J. C. Mason

Applied and Computational Mathematics Group, Royal Military College of Science, Shrivenham, Wiltshire, UK

Abstract The purpose of this paper is to develop a piecewise approximation method on the sphere and to show its application in digital terrain modelling. The piecewise basis is defined by the use of trial functions, which may be interpreted as B-splines symmetrical about an axis through a given point. The method of approximation is to form a "Fourier series expansion" in terms of these trial functions and the algorithm simply involves numerical evaluation of the "Fourier coefficients", namely a set of integrals on the sphere. It is shown that uniform convergence can be guaranteed as the element diameter tends to zero, if the approximate integration is based on ideas of equidistribution.

Key words: Piecewise approximation, Finite elements, Sphere, B-splines, Fourier series, Terrain modelling

1. Theoretical Background

As the starting point for our discussion, we define (compare Böhmer (1974)) the following B-splines $B_k:[1,+1]\to \mathbb{R}$, given by

$$B_k(t) = \mu_k \sum_{i=0}^{k} (-1)^{k-i} \binom{k}{i} (t_i-t)_+^{k-1} \Big|_{[-1,+1]} \qquad (1)$$

where $\mu_k(h)$, $k = 2, 3, \ldots$, is a normalization constant

$$\mu_k(h) = \left(\sum_{i=[\frac{k}{2}]+1}^{k} (-1)^{k-i} \binom{k}{i} (t_i - 1)^{k-1} \right)^{-1} \tag{2}$$

chosen so that $B_k(1) = 1$, $k = 2, 3, \ldots$.

and $t_o, \ldots, t_k$ are the equidistant knots

$$t_i = \frac{2i}{k}(1-h)+h, \quad h\varepsilon[0,1). \tag{3}$$

In particular,

$$B_2(t) = \begin{cases} 0 & \text{for } -1 \leqq t \leqq h \\ \frac{t-h}{1-h} & \text{for } h < t \leqq 1 \end{cases},$$

and

$$B_3(t) = \begin{cases} 0 & \text{for } -1 \leqq t \leqq h \\ \frac{3}{2}\left(\frac{h-t}{h-1}\right)^2 & \text{for } h < t \leqq \frac{2+h}{3} \\ 1-3\left(\frac{t-1}{h-1}\right)^2 & \text{for } \frac{2+h}{3} < t \leqq 1. \end{cases}$$

Figure 1 gives a graphical impression of the B-spline for $h=\frac{1}{2}$, $k=3$, and only the bold faced part of the spline is required.

For any two elements ξ, η of the unit sphere Ω in $\mathbb{R}^3$ we set

$$B_k(\xi\eta) = \mu_k(h) \sum_{i=0}^{k} (-1)^{k-i} \binom{k}{i} (t_i - \xi\eta)_+^{k-1} \tag{4}$$

Obviously, $B_k(.\eta):\Omega \rightarrow \mathbb{R}^3$ has symmetry about the axis through the point $\eta\varepsilon\Omega$. Also $B_k(\xi\eta)$ depends only on the spherical distance of the arguments $\xi, \eta\varepsilon\Omega$. Furthermore it is easy to see that

$$\text{supp } B_k(.\eta) = \{\xi\varepsilon\Omega \mid h \leqq \xi\eta \leqq 1\}. \tag{5}$$

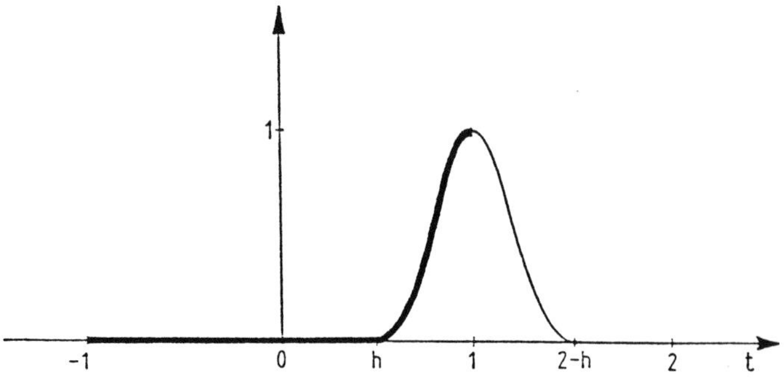

Fig. 1: B-Spline $B_3(t)$ for $h = \frac{1}{2}$

Using the functions B_k, we introduce the kernel K_h: $\Omega \times \Omega \to \mathbb{R}$ defined by

$$K_h(\xi,\zeta) = (\nu_k(h))^{-1} \int_\Omega B_k(\xi\eta)B_k(\zeta\eta)d\omega(\eta) \tag{6}$$

($\xi, \zeta \in \Omega$; $d\omega$: surface element on Ω), where $\nu_k(h)$ is a normalization constant dependent on $h \in [0,1)$ and such that

$$\int_\Omega K_h(\xi,\zeta)d\omega(\zeta) = 1 \tag{7}$$

for all $\xi \in \Omega$. The functions (4) admit an expansion in terms of Legendre polynomials

$$B_k(\xi\eta) = \sum_{n=0}^{\infty} a_n^{(k)}(h)\, P_n(\xi\eta) \,, \tag{8}$$

where $a_n^{(k)}(h)$, $n = 0, 1, \ldots$, are the expansion coefficients

$$a_n^{(k)}(h) = \frac{2n+1}{4\pi} \int_\Omega B_k(\xi\eta)\, P_n(\xi\eta)\, d\omega(\xi) \,. \tag{9}$$

Equivalently we have

$$a_n^{(k)}(h) = \frac{2n+1}{2} \int_h^1 B_k(t)\, P_n(t)dt \,.$$

In particular, for n = 0, we obtain:

$$a_o^{(k)}(h) = \frac{1}{2}\int_h^1 B_k(t)dt \qquad (10)$$

$$= \frac{1}{2}\mu_k(h) \sum_{i=0}^{k} (-1)^{k-i} \binom{k}{i} \int_h^{\min(t_i,1)} (t_i-t)^{k-1}dt \ .$$

Explicitly (compare Böhmer (1974)),

$$a_o^{(k)}(h) = \sigma_k \ \tau_k \ (1-h) \ , \qquad (11)$$

where

$$\sigma_k = \frac{1}{2k}\left(\sum_{i=0}^{[\frac{k}{2}]} (-1)^{k-1} \binom{k}{i} \left(\frac{2i}{k}\right)^k - \sum_{i=[\frac{k}{2}]+1}^{k} (-1)^{k-i} \binom{k}{i} \left[\left(\frac{2i}{k}-1\right)^k - \left(\frac{2i}{k}\right)^k\right]\right)$$

and

$$\tau_k = \left(\sum_{i=[\frac{k}{2}]+1}^{k} (-1)^{k-i} \binom{k}{i} \left(\frac{2i}{k}-1\right)^{k-1}\right)^{-1} .$$

Substituting (8) into (6) we obtain

$$K_h(\xi,\eta) = K_h(\xi\eta) = \nu_k(h)^{-1} \sum_{n=0}^{\infty} \frac{4\pi}{2n+1} (a_n^{(k)})^2 \ P_n(\xi\zeta). \qquad (12)$$

This implies that

$$\nu_k(h) = (4\pi)^2(a_o^{(k)})^2 = (4\pi)^2\sigma_k^2\tau_k^2(1-h)^2 \qquad (13)$$

(observing that $\mu_k(h) = \tau_k(1-h)^{1-k}$). For example,

$$\nu_2(h) = \pi^2 \ (1-h)^2 \ ,$$

$$\nu_3(h) = (\tfrac{8}{9}\pi \ (1-h))^2 \ .$$

Suppose that $u:\Omega \to \mathbb{R}$ is a function of class $C^{(o)}(\Omega)$ or $C^{(\lambda)}(\Omega)$, $\lambda > 0$. For brevity, set

$$\|u\|_{C^{(o)}(\Omega)} = \sup_{\xi\in\Omega} |u(\xi)|$$

for the class $C^{(o)}(\Omega)$, and set

$$\|u\|_{C^{(\lambda)}(\Omega)} = \sup_{\substack{\xi,\eta\in\Omega \\ \xi\neq\eta}} \frac{|u(\xi)-u(\eta)|}{|\xi-\eta|^{\lambda}}.$$

for the class $C^{(\lambda)}(\Omega)$, $\lambda > 0$.

Now let us define by $\hat{u}_h:\Omega \to \mathbb{R}$ the "mean"

$$\hat{u}_h(\xi) = \int_{\Omega} K_h(\xi\zeta)u(\zeta)d\omega(\zeta) \ . \tag{14}$$

The following result then holds.

Lemma 1: Let $u:\Omega \to \mathbb{R}$ be of class $C^{(o)}(\Omega)$.

Then

$$\|\hat{u}_h - u\|_{C^{(o)}(\Omega)} \leq \sup_{-1+2h^2\leq\xi\zeta\leq 1} |u(\xi)-u(\zeta)| \ .$$

If, in addition, u is of class $C^{(\lambda)}(\Omega)$, then

$$\|\hat{u}_h - u\|_{C^{(o)}(\Omega)} \leq 2^{3\lambda/2}(1-h)^{\lambda/2}\|u\|_{C^{(\lambda)}(\Omega)} \ .$$

Proof. For all $\xi\in\Omega$ we find that

$$\hat{u}_h(\xi) - u(\xi) = \int_{-1+2h^2\leq\xi\zeta\leq 1} K_h(\xi\zeta)(u(\zeta)-u(\xi))d\omega(\zeta) \ .$$

The function $u \epsilon C^{(o)}(\Omega)$ is uniformly continuous on the sphere Ω. Hence

$$|\hat{u}_h(\xi) - u(\xi)|$$

$$\leq \int_{-1+2h^2 \leqq \xi\eta \leqq 1} K_h(\xi\zeta)dw(\zeta) \cdot \sup_{-1+2h^2 \leqq \xi\eta \leqq 1} |u(\xi)-u(\zeta)|$$

$$\leq \sup_{-1+2h^2 \leqq \xi\zeta \leqq 1} |u(\xi) - u(\zeta)| .$$

If, in addition, $u \epsilon C^{(\lambda)}(\Omega)$, then

$$|\hat{u}_h(\xi) - u(\xi)| \leq \|u\|_{C^{(\lambda)}(\Omega)} (4-4h^2)^{\lambda/2}$$

$$\leq \|u\|_{C^{(\lambda)}(\Omega)} 2^{3\lambda/2}(1-h)^{\lambda/2} .$$

This completes the proof of Lemma 1. Q.E.D.

In order to formulate adequate discretizations of the mean $\hat{u}_h$, we make the following restrictions on the grid to be used:

Suppose that $\alpha > 0$ and $h \epsilon [0,1)$. A set $\Gamma \subset \Omega$ is said to be an <u>(h,α)-equidistributed grid</u> if there is a partition of Ω into N mutually disjoint parts π_η of equal area, where $N=\#\Gamma$, the number of points in the grid Γ, with the property that each grid point η may be associated with a unique part π_η of the sphere Ω in such a way that

$$\sup_{\xi \epsilon \pi_\eta} |\xi-\eta| \leq \gamma_N(1-h)^\alpha , \tag{15}$$

where γ_N is a positive constant (independent of h).

Remark: Let Γ be an (h,α)-equidistributed grid. Then, for $f \varepsilon C^{(\lambda)}(\Omega)$, it follows that

$$\int_\Omega f(\xi)d\omega(\xi) = \sum_{\eta\varepsilon\Gamma} \int_{\pi_\eta} f(\xi)d\omega(\xi)$$
$$= \frac{4\pi}{N}\sum_{\eta\varepsilon\Gamma} f(\eta) + \sum_{\eta\varepsilon\Gamma} \int_{\pi_\eta} (f(\xi)-f(\eta))d\omega(\xi),$$

where $\sum_{\eta\varepsilon\Gamma}$ means that the summation is extended over all points $\eta\varepsilon\Omega$.

If follows that

$$\left|\int_\Omega f(\xi)d\omega(\xi) - \frac{4\pi}{N}\sum_{\eta\varepsilon\Gamma} f(\eta)\right| \leq 4\pi\, \gamma_N^\lambda\, \|f\|_{C^{(\lambda)}(\Omega)}\, (1-h)^{\alpha\lambda}.$$

In other words, the integral

$$\int_\Omega f(\xi)d\omega(\xi)$$

can be replaced by the arithmetic mean

$$\frac{4\pi}{N}\sum_{\eta\varepsilon\Gamma} f(\eta)$$

for sufficiently small 1-h. The investigations which follow will essentially be based on this observation. The approximation method below can therefore be interpreted formally as a Fourier series approach using "finite elements" on the sphere (compare for example the finite element methods proposed in § 26 of Eskin (1981) to solve elliptic boundary value problems).

2. **Approximation Method**

Given an (h,α)-equidistributed grid, with $N = \#\Gamma$, and given a function $u\varepsilon\, C^{(o)}(\Omega)$, we consider the "first discretization" $\hat{u}_h:\Omega\to R$ of $\hat{u}_h$ defined by

$$\hat{u}_h(\xi) = (\nu_k(h))^{-1} \frac{4\pi}{N} \sum_{\eta \varepsilon \Gamma} \int_{\Omega} u(\zeta) B_k(\eta\zeta) d\omega(\zeta) B_k(\xi\eta), \tag{16}$$

Formula (16) may be rewritten in the form

$$\hat{u}_h(\xi) = \int_{\Omega} K_h^{\Gamma}(\xi,\zeta) u(\zeta) d\omega(\zeta) , \tag{17}$$

where we have used the abbreviation

$$K_h^{\Gamma}(\xi,\zeta) = (\nu_k(h))^{-1} \frac{4\pi}{N} \sum_{\eta \varepsilon \Gamma} B_k(\xi\eta) B_k(\zeta\eta). \tag{18}$$

<u>Lemma 2</u>: Let $\Gamma \subset \Omega$ be an (h,α)-equidistributed grid, $N = \# \Gamma$. Suppose that $u:\Omega \to \mathbb{R}$ is of class $C^{(o)}(\Omega)$. Then there exists a constant $C'_{k,N}$ (independent of h) such that

$$\| \hat{u}_h - \hat{u}_h \|_{C^{(o)}(\Omega)} \leqq C'_{k,N} \|u\|_{C^{(o)}(\Omega)} (1-h)^{\alpha-(k+1)}.$$

<u>Proof</u>. Let ξ be an arbitrary point of Ω. Then it follows that

$$\hat{u}_h(\xi) - \hat{u}_h(\xi) = \int_{\Omega} (K_h(\xi,\zeta) - K_h^{\Gamma}(\xi,\zeta)) u(\zeta) dw(\zeta) .$$

Consequently we find that

$$\hat{u}_h(\xi) - \hat{u}_h(\xi)$$

$$= (\nu_k(h))^{-1} \int_{\Omega} \sum_{\eta' \varepsilon \Gamma} \int_{\pi_{\eta'}} \Big(B_k(\zeta\eta) B_k(\xi\eta) - B_k(\zeta\eta') B_k(\xi\eta') \Big) u(\zeta) d\omega(\eta) d\omega(\zeta).$$

Thus

$$\hat{u}_h(\xi) - \hat{u}_h(\xi) = \int_{\Omega} L_h^{\Gamma}(\xi,\zeta) u(\zeta) d\omega(\zeta) , \tag{19}$$

where

$$L_h(\xi,\zeta) = (\nu_k(h))^{-1} \sum_{\eta' \varepsilon \Gamma} \int_{\pi_{\eta'}} \Big(B_k(\zeta\eta) B_k(\xi\eta) - B_k(\zeta\eta') B_k(\xi\eta') \Big) d\omega(\eta)$$

Consider the integrand on the right hand side in more detail:

$$B_k(\zeta\eta)B_k(\xi\eta) - B_k(\zeta\eta')B_k(\xi\eta')$$

$$= \left(B_k(\zeta\eta) - B_k(\zeta\eta')\ B_k(\xi\eta)\right) + \left(B_k(\xi\eta) - B_k(\xi\eta')\right) B_k(\zeta\eta')$$

Now, from (1), we obtain

$$|B_k(t) - B_k(t')| \leqq (k-1)3^{k-2}|\mu_k(h)|\ |t-t'| \tag{20}$$

for k = 2, 3, ... and any pair t, $t'\varepsilon[-1,1]$.

Therefore,

$$|B_k(\xi\eta)B_k(\zeta\eta) - B_k(\xi\eta')B_k(\zeta\eta')|$$

$$\leqq 2(k-1)3^{k-2}\ |\mu_k(h)|\ |\eta-\eta'| \tag{21}$$

for any pair η, η' $\varepsilon\Omega$ and for all $\xi,\zeta\varepsilon\Omega$. Consequently we are able to show that

$$|L_h(\xi,\zeta)| \leqq 8\pi(k-1)\ 3^{k-2}\ \gamma_N|\mu_k(h)|\ |\nu_k(h)|^{-1}(1-h)^{\alpha} \tag{22}$$

and so we finally obtain:

$$\sup_{\xi\varepsilon\Omega} |\hat{u}_h(\xi) - \hat{u}_h(\xi)|$$

$$\leqq 8(k-1)3^{k-2}\ \sigma_k^{-2}\ \tau_k^{-1}\ \gamma_N(1-h)^{\alpha-(k+1)}\|u\|_{C^{(o)}(\Omega)}.$$

This confirms Lemma 2. Q.E.D.

Suppose now that Λ is an (h,β)-equidistributed grid, $M=\#\Lambda$, and consider the "<u>second discretization</u>" $u_h:\Omega \to \mathbb{R}$ of $\hat{u}_h$ defined by

$$u_h(\xi) = (\nu_k(h))^{-1} \frac{4\pi}{N} \sum_{\eta\in\Gamma} \frac{4\pi}{M} \sum_{\zeta'\in\Lambda} u(\zeta')B_k(\zeta'\eta)B_k(\xi\eta).$$

Lemma 3: Let $\Gamma \subset \Omega$ be an (h,β)-equidistributed grid, $N=\#\Gamma$. Moreover, let $\Lambda\subset\Omega$ be an (h,β)-equidistributed grid, $M=\#\Lambda$. Suppose that $u:\Omega\to\mathbb{R}$ is of class $C^{(\lambda)}(\Omega)$. Then there exists a constant $C''_{K,M}$ (independent of h) such that

$$\|\hat{u}_h - u_h\|_{C^{(o)}(\Omega)} \leq C''_{K,M} \Big(\|u\|_{C^{(o)}(\Omega)} + \|u\|_{C^{(\lambda)}(\Omega)}\Big) (1-h)^{\min(1,\lambda)\beta-(k+1)}$$

Proof. Noting the definitions of u_h, $\hat{u}_h$,

$$\hat{u}_h(\xi) - u_h(\xi) \tag{23}$$

$$= (\nu_k(h))^{-1}\frac{4\pi}{N} \sum_{\eta\in\Gamma} \sum_{\zeta'\in\Lambda} \int_{\pi_{\zeta'}} \Big(u(\zeta)B_k(\eta\zeta) - u(\zeta')B_k(\zeta'\eta)\Big)\, d\omega(\zeta)B_k(\xi\eta)$$

Now,

$$u(\zeta)B_k(\eta\zeta) - u(\zeta')B_k(\eta\zeta')$$

$$= u(\zeta)\Big(B_k(\eta\zeta) - B_k(\eta\zeta')\Big) - B_k(\eta\zeta')\Big(u(\zeta') - u(\zeta)\Big).$$

Thus,

$$|u(\zeta)B_k(\eta\zeta) - u(\zeta')B_k(\eta\zeta')| \tag{24}$$

$$\leq (k-1)3^{k-2}|\mu_k(h)|\,\|u\|_{C^{(o)}(\Omega)}|\zeta-\zeta'| + \|u\|_{C^{(\lambda)}(\Omega)}|\zeta-\zeta'|^{\lambda}.$$

This gives the required result. Q.E.D.

We may summarize our results in the following theorem.

Theorem 1: Let $\Gamma\subset\Omega$ be an (h,α)-equidistributed grid, $N=\#\Gamma$. Let $\Lambda \subset\Omega$ be an (h,β)-equidistributed grid, $M=\#\Lambda$. Suppose that $u:\Omega \to \mathbb{R}$

is of class $C^{(\lambda)}(\Omega)$. Then there exists a constant $C_{k,N,M}$ independent of h, such that

$$\|u-u_h\|_{C^{(o)}(\Omega)} \leq C_{k,N,M}\left(\|u\|_{C^{(o)}(\Omega)} + \|u\|_{C^{(\lambda)}(\Omega)}\right)(1-h)^{\delta},$$

where

$$\delta = \min\left(\lambda/2, \alpha-(k+1), \beta-(k+1), \beta\lambda-(k+1)\right).$$

3. Algorithm

Although Theorem 1 is only a rough estimate for $\|u-u_h\|_{C^{(o)}(\Omega)}$, it shows us that u_h is a uniform approximation to u on Ω for sufficiently small 1-h and appropriately given α, β. This leads to the following algorithm for approximating u:

(i) choose a sufficiently small 1-h

(ii) compute $B_k(\xi\eta)$, $\xi\varepsilon\Lambda$, $\eta\varepsilon\Gamma$

(iii) calculate $\nu_k(h)$

(iv) compute the sums

$$\frac{4\pi}{M} \sum_{\zeta'\varepsilon\Lambda} u(\zeta')B_k(\zeta'\eta) \tag{25}$$

(v) compute the approximating linear combination

$$u_h(\xi) = (\nu_k(h))^{-1} \frac{4\pi}{N} \sum_{\eta\varepsilon\Gamma} \frac{4\pi}{M} \sum_{\zeta'\varepsilon\Lambda} u(\zeta')B_k(\zeta'\eta)B_k(\xi\eta). \tag{26}$$

4. Example

We are interested in applying our finite element method to solve a digital terrain modelling problem.

Suppose that Ω is split by a piecewise smooth curve into two open sets Ω' and Ω''. Let Ω_i' (resp. Ω_e') denote a set with $\bar{\Omega}_i \subset \Omega'$

(resp. $\bar{\Omega}'_e \subset \Omega''$) satisfying dist $(\partial\Omega'_i, \partial\Omega')>\rho$ (resp. dist $(\partial\Omega'_e, \partial\Omega')>\rho$) for some $\rho>0$. The problem is to approximate the restriction $w|_{\bar{\Omega}'}$ of a function $w\varepsilon\ C^{(\lambda)}(\Omega)$ from known discrete data ("heights") $w(\xi)$, $\xi\varepsilon\Lambda', \Lambda' = \Lambda \cap \Omega'$.

For that purpose we consider, instead of $w\varepsilon C^{(\lambda)}(\Omega)$, a function $u\varepsilon C^{(\lambda)}(\Omega)$ with the property that $u(\xi)=w(\xi)$, $\xi\varepsilon\bar{\Omega}'$ and $u(\xi) = 0$, $\xi\varepsilon\bar{\Omega}_e$. The piecewise method now yields an approximating linear combination u_h of the function u; and then the restriction $u_h|_{\bar{\Omega}'}$ is a uniform approximation to $w|_{\bar{\Omega}'}(=u|_{\bar{\Omega}'})$ provided that Γ, Λ are "suitably dense" equidistributed grids.

In practice, however, $(\xi, w(\xi)) \ \varepsilon\ \Lambda'\times \mathbb{R}$ is an a priori given dense (not necessarly equidistributed) data set, while in theory Γ can always be chosen without any practical restraints.

Ideally the region of modelling should be sufficiently large to make the curvature of the earth significant. However, we did not have such physical data available and have therefore had to consider a relatively small region. Figures 2a, 2b, 3a and 3b show an area situated in the north-west of Aachen, called Seffent. The computations (Fig. 3a, b) are based on more than 40,000 data points $(\xi,w(\xi))$, $\xi\varepsilon\Lambda'$, chosen on the grid lines of geographical coordinates. For the computations of the sums (25), all data points have been used. For the determination of Γ every third point of Λ' in each grid of geographical coordinates has been taken. Moreover, the parameters k,h have been chosen as h=0.9, k=3.

Fig. 3a shows the digital terrain model (d.t.m.) for the whole test area Ω'. Because of our construction, of course, the d.t.m. is not acceptable around the boundary $\partial\Omega'$. However, when the erroneous strip around the boundary is cut out (Fig. 3b), the remaining d.t.m. determined by $u_h|_{\bar{\Omega}_i}$ is indeed representative of the resulting area $\bar{\Omega}_i$. More details about the example (Seffent) can be found in the thesis of H.J. Schaffeld (1988), who shows that a very acceptable

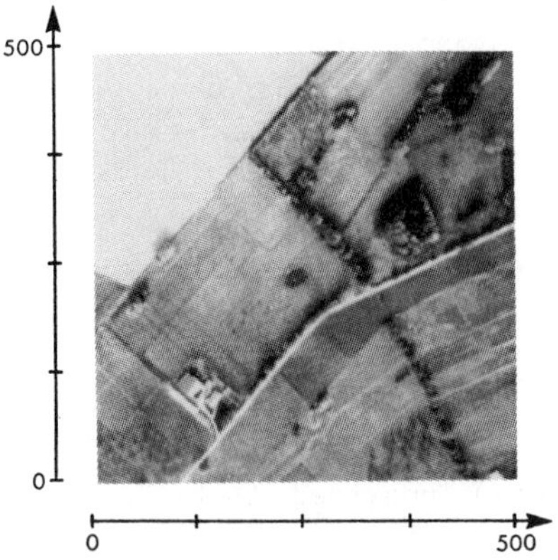

Fig. 2a: Aerial photo of Ω′

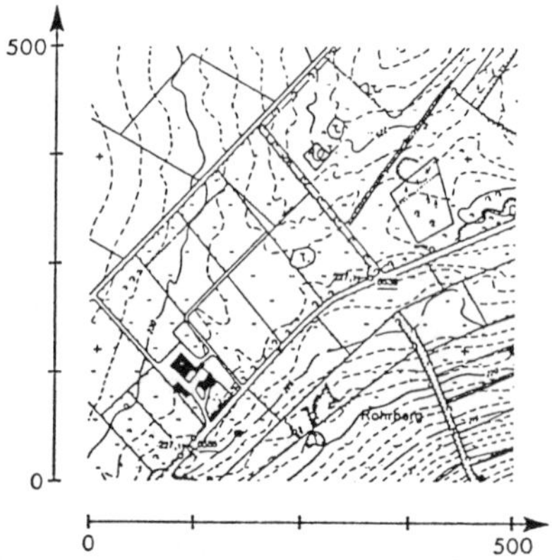

Fig. 2b: Map of Ω′

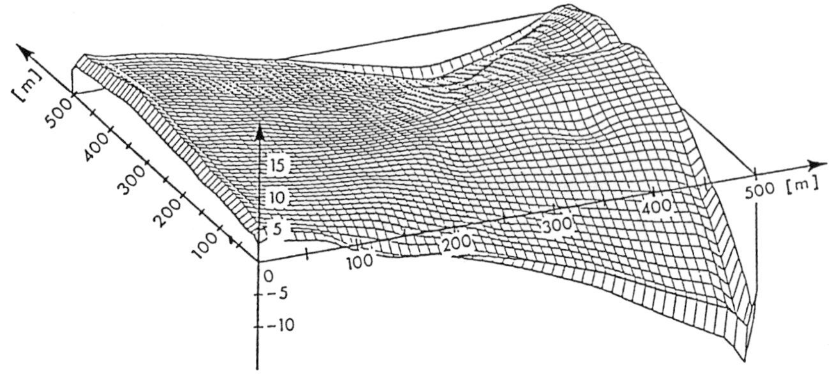

Fig. 3a: Digital terrain model on $\overline{\Omega}'$

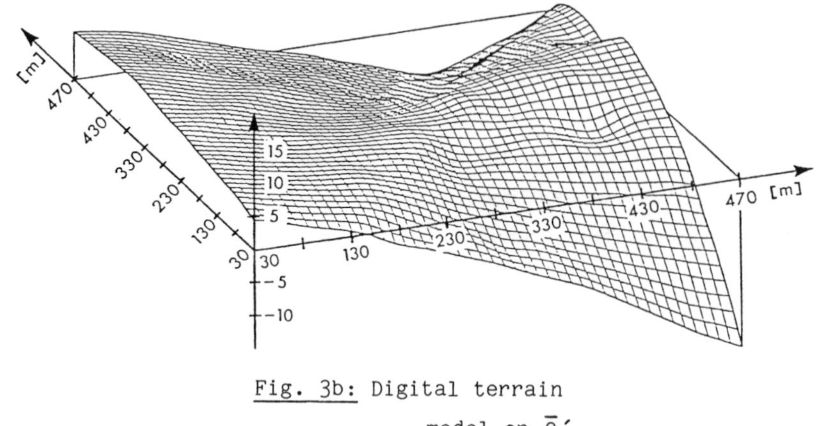

Fig. 3b: Digital terrain model on $\overline{\Omega}_i'$

level of error has been achieved by our method. Schaffeld has also tested the method successfully on a number of synthetic problems, where data were on a truely spherical surface, analogous to a large area of the earth. (Schaffeld's theoretical discussion is not based on uniform norms, as is the present paper).

Thus the approximation method can be very effective, provided that it is adopted over an extended domain which includes the required one.

Acknowledgments

We wish to acknowledge the support of the Science and Engineering Research Council (UK) in providing a Visiting Fellowship for the first author to work with the second. The first author also thanks Prof B. White, Geod. Department, RWTH Aachen, for stimulating discussions.

References

Schaffeld, H.J., (1988), Eine Finite-Elemente-Methode und ihre Anwendung zur Erstellung von Digitalen Geländermodellen. Veröff. Geod. Inst. R.W.T.H. Aachen, Report No. 42.

Böhmer, K., (1974), Spline-Funktionen. Studienbücher Mathematik, Teubuer, Stuttgart.

Eskin, G.I., (1981), Boundary Value Problems for Elliptic Pseudodifferential Equations, Vol 52, Translations of Mathematical Monographs, A.M.S, Providence, Rhode Island.

7. Applications in Numerical Analysis

Approximation theory and numerical linear algebra

L. N. Trefethen

Department of Mathematics, Massachusetts Institute of Technology, Cambridge, Massachusetts, USA

Abstract Many large matrix problems are best solved by iteration, and the convergence of matrix iterations is usually connected with classical questions of polynomial or rational approximation. This paper discusses three examples: (1) the conjugate gradient iteration for symmetric positive definite matrices, (2) Strang's preconditioned conjugate gradient iteration for Toeplitz matrices, and (3) polynomial iterations for nonsymmetric matrices with complex eigenvalues. This last example raises a fundamental problem: what happens to iterative methods in linear algebra, and to the role of approximation theory, when the matrices become highly non-normal? Part of the solution may lie in approximation on a set of "approximate eigenvalues".
Key words: approximation, linear algebra, conjugate gradient iteration, Toeplitz, Hankel, normal, resolvent

1. Introduction

Many large-scale numerical computations require the solution of large systems of equations $Ax = b$, and very often, the best known methods are iterative. This has become especially true since 1970, thanks to the development of multigrid and preconditioned conjugate gradient methods. It is impossible to know what the state of the art may be a generation from now, but for the present, iterative methods of linear algebra occupy a central position in scientific computing.

The purpose of this paper is to explore some connections, mostly already known, between matrix iterative methods and classical problems of polynomial and rational approximation. Since the 1950s, experts in numerical linear algebra have made use of approximation ideas to analyze rates of convergence [16, 27, 57, 58, 60]. The arguments are natural, often elegant, and deserve to be more widely appreciated among approximation theorists. We shall consider three examples:

1. Conjugate gradient iteration. The conjugate gradient iteration converges rapidly if A is a symmetric positive definite matrix with clustered eigenvalues or a low condition number. This phenomenon can be explained by considering an associated problem of polynomial approximation at the eigenvalues of A. The explanation is a beautiful one, one of the gems of modern numerical analysis.

2. Strang's preconditioned Toeplitz iteration. Recently Strang has proposed a method for solving symmetric positive definite Toeplitz matrix problems by a conjugate gradient iteration with a circulant preconditioner. The iteration often converges extremely fast, and the explanation turns out to be a surprisingly deep connection with a certain problem of approximation by rational functions on the complex unit disk. This example is much more specialized than the last one, but is offered for its novelty and nontriviality.

3. Polynomial iterations for nonsymmetric matrices. A less well understood problem is what to do with nonsymmetric matrices with complex eigenvalues. One approach is to construct an iteration based explicitly on a problem of polynomial approximation at these eigenvalues, assuming that something is known about their location. We describe here a particular version of this idea, recently explored by Fischer and Reichel and by Tal-Ezer, in which the approximation problem is solved by interpolation in Fejér points determined by means of conformal mapping.

Approximation theory has many other links with numerical linear algebra besides these. For example:

- The Chebyshev iteration for a symmetric positive definite matrix is connected with polynomial approximation on an interval $[a, b]$ containing the spectrum [21].
- The same is true of the design of "polynomial preconditioners" for accelerating conjugate gradient iterations [30].
- For nonsymmetric matrices, the Chebyshev iteration is connected with polynomial approximation on ellipses in the complex plane [15, 34, 35].
- The iterative solution of symmetric indefinite systems is connected with simultaneous polynomial approximation on two disjoint real intervals $[a, b]$, $[c, d]$ with $b < 0 < c$ [12, 43].
- The alternating direction implicit (ADI) iteration is connected with a problem of rational approximation on a real interval [58].
- Various connections have been described between the Lanczos iteration, continued fractions, and Padé approximation [4, 9, 22]. The history of the LR and QR algorithms is also linked with Padé approximation [22].
- The convergence of multigrid iterations is connected with problems of trigonometric approximation in Fourier space [3, 26].

The list could go on. These citations are by no means comprehensive; references to earlier work can be found therein.

In the final section we discuss an important general problem which is an outgrowth of topic *(3)*: what is the proper way to treat matrices that are not normal — that is, whose eigenvectors are not orthogonal? In such cases there is a gap of size $\kappa(V)$ — the condition number of the matrix of eigenvectors — between approximation at the eigenvalues and convergence of a matrix iteration, and as we show by several examples, in practical problems $\kappa(V)$ can be enormous. We propose that in these cases, approximations should be designed to be accurate on a set of "approximate eigenvalues" rather than just on the set of exact eigenvalues.

2. Conjugate gradient iteration

Let A be a real symmetric positive definite matrix of dimension n, let b be a real n-vector, and let ϕ be the quadratic form

$$\phi(x) = \tfrac{1}{2}x^T A\, x - x^T b, \qquad x \in \mathbb{R}^n. \tag{1}$$

It is readily calculated that the gradient of $\phi(x)$ is $\nabla\phi(x) = Ax - b$, and therefore, $Ax = b$ is satisfied if and only if x is a stationary point of ϕ. Since A is positive definite, the only stationary point is the global minimum, and we are left with the following equivalence:

$$Ax = b \quad \Longleftrightarrow \quad \phi(x) = \inf_{y \in \mathbb{R}^n} \phi(y). \tag{2}$$

We shall denote this solution vector x by x^*.

The conjugate gradient iteration amounts to an iterative minimization of $\phi(x)$ based on a cleverly chosen sequence of search directions [11, 20, 29, 33, 48]. For each k, let K_k denote the *Krylov subspace* of $\mathbb{R}^n$ spanned by the *Krylov vectors* $b, Ab, \ldots, A^{k-1}b$:

$$K_k \;=\; \langle b, Ab, \ldots, A^{k-1}b\rangle \;=\; \langle Ax^*, A^2x^*, \ldots, A^k x^*\rangle, \tag{3}$$

and let x_k be the unique minimizer of ϕ in this subspace:

$$\phi(x_k) = \inf_{y \in K_k} \phi(y) \qquad (x_0 = 0). \tag{4}$$

If $\|\cdot\|_A$ denotes the norm

$$\|y\|_A = \sqrt{y^T A\, y}, \tag{5}$$

and e_k denotes the error in x_k,

$$e_k = x^* - x_k \qquad (e_0 = x^*), \tag{6}$$

then an easy calculation shows that $\|e_k\|_A$ is also minimized at each step:

$$\|e_k\|_A = \inf_{e \in x^* - K_k} \|e\|_A. \tag{7}$$

Since $K_1 \subseteq K_2 \subseteq \cdots \subseteq \mathbb{R}^n$, the values of $\phi(x_k)$ and $\|e_k\|_A$ must decrease monotonically:

$$\phi(x_1) \geq \phi(x_2) \geq \cdots \geq \phi(x^*), \qquad \|e_1\|_A \geq \|e_2\|_A \geq \cdots \geq 0, \tag{8}$$

and the iteration must converge in at most n steps to some limit (in the absence of rounding errors). Since A^{-1} is equal to some polynomial in A (by the Cayley-Hamilton theorem), the limit vector x_k must be x^*.

So far, this idea of optimization in nested subspaces is a general one that may not seem a particularly promising basis for an algorithm. Yet the algorithm turns out to be excellent because of two remarkable properties, both related to the fact that the family of nested subspaces we have chosen is the Krylov sequence. The first is that the optimal vectors $\{x_k\}$ can be computed speedily by the following simple iteration:

Conjugate gradient (CG) iteration

$x_0 := 0,\ r_0 := b,\ \beta_0 := 0,\ p_0 := 0$
For $k := 1, 2, \ldots$

$p_k := r_{k-1} + \beta_{k-1} p_{k-1}$	*(search direction)*
$\alpha_k := r_{k-1}^T r_{k-1} / p_k^T A p_k$	*(distance along search direction)*
$x_k := x_{k-1} + \alpha_k p_k$	*(approximate solution vector)*
$r_k := r_{k-1} - \alpha_k A p_k$	*(residual $b - Ax_k$)*
$\beta_k := r_k^T r_k / r_{k-1}^T r_{k-1}$	

(We shall not reproduce the derivation; see [20].) Since each step involves just $\sim n^2$ floating-point operations (one matrix-vector multiplication $A p_k$), the solution $x_n = x^*$ of $Ax = b$ can in principle be found in $\sim n^3$ operations, or potentially less if A is sparse.

The second remarkable property of the conjugate gradient iteration is that for many matrices A, there is no need to take so many steps: x_k may approximate x^* to machine precision for $k \ll n$. For example, it is not unusual for a matrix problem of dimension 10,000 to be solved to the required precision in 50 or 100 steps. Thus the CG iteration often takes closer to $O(n^2)$ than $O(n^3)$ operations, or even less if A is sparse.[1] It is this phenomenon that we wish to explain by means of polynomial approximation. The underlying ideas go back to the beginnings of the

[1] The mathematical fact that the CG iteration converges rapidly for well-conditioned matrices was known from the beginning, but for twenty years its practical importance was not fully appreci-

subject, and a concise presentation can be found in the book by Luenberger [33]; see also [10, 11, 16, 31].

Let P_k denote the set of polynomials of degree at most k. By (3) and (6), we have

$$x_k = q(A)b, \qquad e_k = p(A)e_0 \tag{9}$$

for some polynomials

$$q \in P_{k-1}, \qquad p \in P_k, \quad p(0) = 1, \tag{10}$$

with $p(z) = 1 - zq(z)$. By (4) and (7), these polynomials are optimal in the following senses:

$$\begin{aligned} \phi(x_k) &= \inf_{q \in P_{k-1}} \phi(q(A)b), \\ \|e_k\|_A &= \inf_{p \in P_k,\ p(0)=1} \|p(A)e_0\|_A. \end{aligned} \tag{11}$$

To put it in words, the polynomial $q(A)$ behaves as much as possible like A^{-1}, as measured by the function ϕ, and $p(A)$ behaves as much as possible like the zero matrix, subject to the constraint $p(0) = 1$, as measured by the A-norm of the error.

With the use of an eigenvector expansion, these statements about matrices become statements about scalars. Let $\{v_j\}$ be an orthonormal set of eigenvectors of A corresponding to eigenvalues $\{\lambda_j\}$, and write

$$e_0 = \sum_{j=1}^{n} a_j v_j, \qquad \|e_0\|_A^2 = \sum_{j=1}^{n} a_j^2 \lambda_j. \tag{12}$$

Then

$$p(A)e_0 = \sum_{j=1}^{n} a_j p(\lambda_j) v_j,$$

which implies

$$\|p(A)e_0\|_A^2 = \sum_{j=1}^{n} a_j^2 \lambda_j p^2(\lambda_j). \tag{13}$$

From (9)–(13) we now obtain the following fundamental theorem on convergence of the conjugate gradient iteration. This bound is independent of the dimension n,

ated. (Credit for the reawakening is usually given to Reid in 1971 [41].) This seems inexplicable to us now, but one must remember that the scale of computations in the 1950s was very small; for the matrices of those days, 50 or 100 steps looked no better than $O(n)$. Also, the idea of preconditioning was not widely known until the 1970s [8].

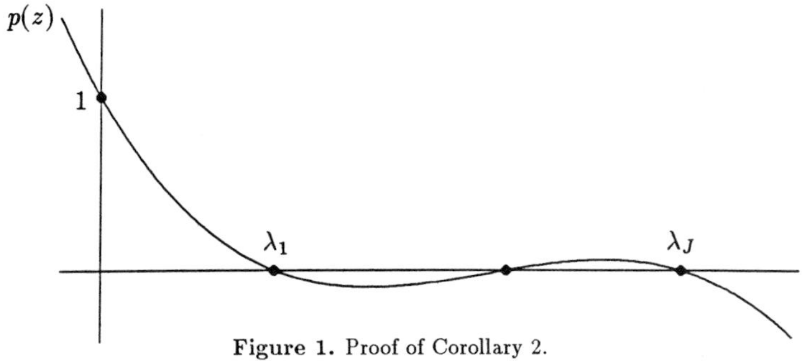

Figure 1. Proof of Corollary 2.

and indeed generalizes to positive definite self-adjoint operators on a Hilbert space [10, 11].

Theorem 1. The CG iteration satisfies

$$\frac{\|e_k\|_A}{\|e_0\|_A} \le \inf_{p\in P_k,\, p(0)=1} \left\{ \sup_{\lambda\in\Sigma} |p(\lambda)| \right\}, \tag{14}$$

where Σ denotes the spectrum of A. □

This theorem associates the CG iteration with the following approximation problem: what is the minimal magnitude that a polynomial $p \in P_k$ can attain on the spectrum of A, subject to the constraint $p(0) = 1$? It is not quite correct to say that the CG iteration implicitly finds the exact solution to this approximation problem, since after all, the exact polynomial $p(A)$ implicit in the iteration depends on e_0, while the approximation problem does not. What is correct is that the error reduction after k steps is guaranteed to be at least this good, and conversely, there are right-hand sides for which it is no better.

Depending on Σ, various consequences can be derived from Theorem 1. One extreme occurs if A happens to have a small number of distinct eigenvalues:

Corollary 2. Suppose A has J distinct eigenvalues $\{\lambda_j\}$. Then the CG iteration converges in at most J steps.

Proof Consider the polynomial $p(z) = \prod_{j=1}^{J}(1 - z/\lambda_j)$ (Figure 1). □

Another well-known corollary is obtained if one assumes merely that the condition number of A, defined by $\kappa = \lambda_{\max}/\lambda_{\min}$, is not too large [10]:[2]

Corollary 3. Suppose A has condition number κ. Then the CG iteration converges at least geometrically, as follows:

$$\frac{\|e_k\|_A}{\|e_0\|_A} \le 2\left(\frac{\sqrt{\kappa}-1}{\sqrt{\kappa}+1}\right)^k \qquad (0 \le k < \infty). \tag{15}$$

[2] Corollary 3 is often cited, but I have been unable to track down its original appearance in print. The factor of 2 is omitted in some texts, but the resulting inequality is invalid except in the limit $\kappa = \infty$.

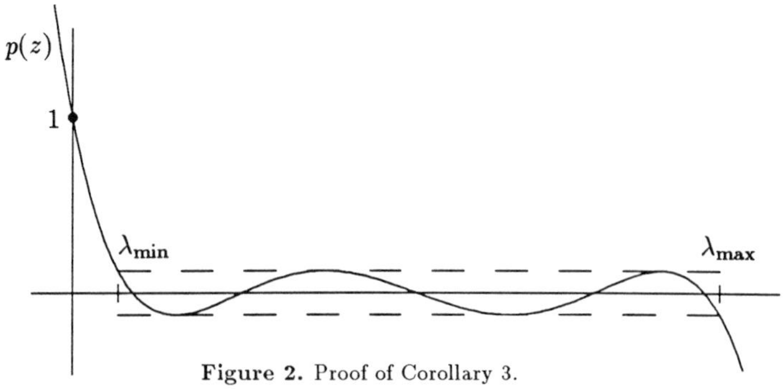

Figure 2. Proof of Corollary 3.

Proof The polynomial $p \in P_k$ that attains the minimum maximal modulus on $[\lambda_{\min}, \lambda_{\max}]$, subject to the constraint $p(0) = 1$, is a shifted and rescaled Chebyshev polynomial:

$$p(x) = \frac{T_k(\gamma - 2x/(\lambda_{\max} - \lambda_{\min}))}{T_k(\gamma)}, \qquad \gamma = \frac{\lambda_{\max} + \lambda_{\min}}{\lambda_{\max} - \lambda_{\min}} = \frac{\kappa + 1}{\kappa - 1}$$

(Figure 2). The supremum of $|p(x)|$ on $[\lambda_{\min}, \lambda_{\max}]$ is $|p(\lambda_{\min})| = |p(\lambda_{\max})| = |T_k(\gamma)|^{-1}$, which tends to be small essentially because Chebyshev polynomials grow faster than any other polynomials, suitably normalized, outside $[-1, 1]$. It can be shown that the right-hand side of (15) is an upper bound for this quantity. □

Corollary 3 implies that the number of iterations required to reduce the error to a specified level ϵ is approximately $\sqrt{\kappa} \log \epsilon$. The corresponding figure for the steepest descent iteration is $\kappa \log \epsilon$, which is much worse unless κ is small.

The significance of Theorem 1 goes far beyond Corollaries 2 and 3: whenever the spectrum of A is favorable for polynomial approximation, the CG iteration will automatically perform well. In particular, the convergence will be rapid if the eigenvalues cluster at one or several points. We turn now to an example of this kind.

3. Strang's preconditioned Toeplitz iteration

We have just seen that if the symmetric positive definite matrix A has clustered eigenvalues, the conjugate gradient iteration will converge quickly. For many problems $Ax = b$, even though the eigenvalues of A are not favorably distributed, a symmetric positive definite matrix C can be found for which the equivalent problem $C^{-1}Ax = C^{-1}b$ does have a favorable eigenvalue distribution. Of course, $C = A$ would be one such matrix, but the point is to pick C so that $C^{-1}y$ ($y \in \mathbb{R}^n$) is easily computable. Such a matrix C is called a *preconditioner,* and the CG iteration applied to $C^{-1}Ax = C^{-1}b$ is known as a *preconditioned conjugate gradient iteration.*

(For implementation details, see [8, 20, 48].) The search for effective preconditioners is a central theme of numerical computation nowadays.

Recently Strang has devised a highly effective preconditioner for problems in which A is Toeplitz — that is, constant along diagonals:

$$A = \begin{bmatrix} a_0 & a_1 & \cdots & & a_{n-1} \\ a_1 & & & & \\ & & \ddots & & \\ & & & & a_1 \\ a_{n-1} & \cdots & & a_1 & a_0 \end{bmatrix}. \qquad (16)$$

Toeplitz matrices occur in a variety of applications, especially in signal processing and control theory, and existing direct techniques for dealing with them include the Levinson-Trench-Zohar $O(n^2)$ algorithms and a variety of $O(n \log^2 n)$ algorithms such as the recent one by Ammar and Gragg [2]. Strang's idea was to treat Toeplitz systems iteratively by a preconditioned conjugate gradient iteration in which C is chosen to be *circulant*:

$$C = \begin{bmatrix} a_0 & a_1 & & \diagdown & a_2 & a_1 \\ a_1 & & & & & a_2 \\ & & & & & \diagdown \\ \diagdown & & \ddots & & & \\ a_2 & & & & & a_1 \\ a_1 & a_2 & \diagdown & & a_1 & a_0 \end{bmatrix}. \qquad (17)$$

A circulant matrix is a Toeplitz matrix in which the diagonals "wrap around," so that the entry a_1 on the first superdiagonal reappears in the lower-left corner, for example. In the present case, as suggested by the dashes, the main diagonal and the first $n/2 - 1$ superdiagonals of C are the same as those of A, but the remaining $n/2$ superdiagonals have been overwritten to achieve the wrap-around.

The idea behind this choice of C is that multiplication by a circulant matrix is equivalent to convolution with a periodic vector, and as a result, circulant systems of equations can be solved in $O(n \log n)$ operations by the Fast Fourier Transform. It follows that each step of a Toeplitz CG iteration with preconditioner C can be executed in $O(n \log n)$ operations, which brings us halfway to an efficient algorithm.

As for the other half, does C precondition this Toeplitz problem effectively? One might expect not, since the corner entries of C and A differ considerably; there is little chance that $C^{-1}A$ will be close to the identity. Yet it turns out that for many Toeplitz matrices A, this preconditioned conjugate gradient iteration converges in 10 or 20 steps. A numerical check reveals that $C^{-1}A$ has a few stray eigenvalues, typically, but that the rest cluster strongly at 1; $C^{-1}A$ is thus close to a low-rank perturbation of the identity. By Theorem 1, the CG iteration is outstanding in such circumstances.

We shall now present the remarkable explanation for this favorable eigenvalue distribution described in a recent paper by Chan and Strang [6]. The essence of the matter is a problem of complex approximation by rational functions on the unit disk. See Chan and Strang for the many details omitted here.

To begin with, assume for simplicity that A has dimension $n = \infty$ (!), which is natural in many applications where A may originate as a finite-rank approximation to an operator. That is, we are going to study the spectrum of the matrices $C_n^{-1}A_n$ in the limit $n \to \infty$, assuming that a fixed sequence of entries $\{a_k\}_{k=0}^{\infty}$ has been prescribed. As usual in the study of Toeplitz matrices, consider the Laurent series

$$f(z) = \sum_{k=-\infty}^{\infty} a_k z^k, \tag{18}$$

whose coefficients $\{a_k\}$ are the entries of A, with $a_k = a_{|k|}$ for $k < 0$. Assume further that these coefficients are absolutely summable,

$$\sum_{k=-\infty}^{\infty} |a_k| < \infty, \tag{19}$$

so that f belongs to the "Wiener algebra" of continuous functions on the complex unit circle $|z| = 1$.

Since A is symmetric and positive definite, it can be shown that $f(z)$ is real and satisfies $f(z) > 0$ for $|z| = 1$. Consider the "spectral factorization" of $f(z)$,

$$f(z) = w(z)w(z^{-1}) \qquad (|z| = 1), \tag{20}$$

where $w(z)$ is a nonzero analytic function in $|z| < 1$. Next, define $v(z)$ by

$$v(z) = \sum_{k=1}^{\infty} v_k z^k = \text{analytic (degree } \geq 1\text{) part of } w(z)/w(z^{-1}), \tag{21}$$

and finally, let H be the infinite Hankel matrix

$$H = \begin{bmatrix} v_1 & v_2 & v_3 & \\ v_2 & v_3 & & \\ v_3 & & \ddots & \\ & & & \end{bmatrix}. \tag{22}$$

(A Hankel matrix is constant along counter-diagonals.) Chan and Strang establish the following connection between the singular values of H (= absolute values of the eigenvalues) and the eigenvalues of $C_n^{-1}A_n$:

Lemma 4. [6] There is a two-to-one correspondence between the singular values of H and the eigenvalues of $C_n^{-1}A_n$: as $n \to \infty$, each singular value σ of H is approached by two eigenvalues of $C_n^{-1}A_n$,

$$\lambda_{\pm} = \frac{1}{1 \pm \sigma}. \quad \Box \tag{23}$$

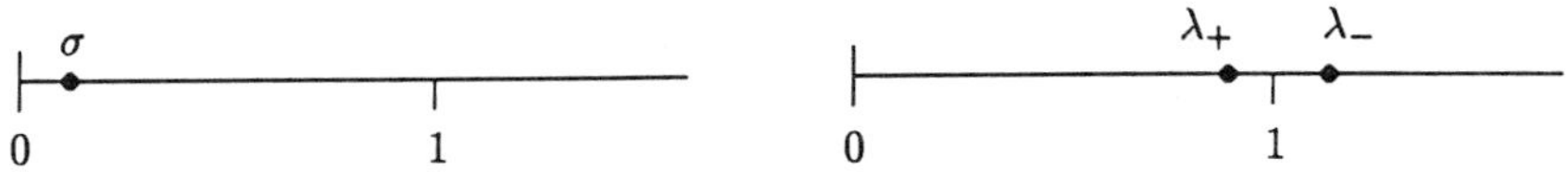

Figure 3. Singular value σ of H and corresponding eigenvalues $\lambda_{\pm}$ of $C_n^{-1}A_n$ (in the limit $n = \infty$).

Figure 3 illustrates this result. We conclude that as the singular values of H decrease to 0, the eigenvalues of $C_n^{-1}A_n$ cluster at $\lambda = 1$. If the decrease is rapid, the clustering will be rapid too, and the CG iteration will converge quickly.

This is where rational approximation unexpectedly enters the picture. According to the theory of "AAK" or "CF" approximation, the singular values $\sigma_0 \geq \sigma_1 \geq \cdots \geq 0$ of H are bounded as follows:

Lemma 5. [1, 53]. The singular values σ_k of H satisfy

$$\sigma_k \leq E_k(v), \tag{24}$$

where $E_k(v)$ denotes the minimal error in sup-norm approximation of $v(z)$ on $|z| = 1$ by rational functions of type (k, k). □

Therefore if $v(z)$ can be efficiently approximated by rational functions, the singular values σ_k must decrease rapidly as $k \to \infty$. In practice, the inequality is often close to an equality, so that the relationship goes approximately in both directions.

Here is the logic in outline:

The entries of A correspond to a smooth function $f(z)$

$\Downarrow$

The function $v(z)$ of (21) is smooth too

$\Downarrow$

$v(z)$ can be well approximated by rational functions on $|z| = 1$

$\Downarrow$ (Lemma 5)

The singular values σ_k of H decrease rapidly to 0

$\Downarrow$ (Lemma 4)

The eigenvalues $\lambda_{\pm k}$ of $C_n^{-1}A_n$ cluster strongly at $\lambda = 1$

$\Downarrow$ (Theorem 1)

The preconditioned CG iteration converges rapidly.

Depending on the precise smoothness assumptions on f, this argument can be fashioned into various theorems. One extreme case is based upon Corollary 2:

Theorem 6. Suppose f is a rational function of type (μ, ν). Then the preconditioned Toeplitz CG iteration converges in at most $1 + 2\max\{\mu, \nu\}$ steps (in the limit $n = \infty$).

Proof If a rational function $f(z)$ is real and positive on $|z| = 1$, then by the Schwarz reflection principle, its poles and zeros lie in pairs symmetric with respect to $|z| = 1$. This implies that the spectral factorization (20) of f is

$$f(z) = w(z)w(z^{-1}) = \frac{p(z)}{q(z)}\frac{p(z^{-1})}{q(z^{-1})},$$

where p and q are polynomials, zero-free in $|z| \leq 1$, of degrees μ and ν respectively. This equation implies

$$\frac{w(z)}{w(z^{-1})} = \frac{p(z)}{q(z)}\frac{q(z^{-1})}{p(z^{-1})},$$

and by (21), the function $v(z)$ is the analytic (degree ≥ 1) part of this expression. By the AAK/CF theory, the number of nonzero singular values of H is equal to the smallest integer N for which $v(z)$ is a rational function of type (N, N) [53] . Careful consideration of the separate cases $\mu \leq \nu$ and $\mu > \nu$ shows that this number is $N = \max\{\mu, \nu\}$. From the two-to-one correspondence of Lemma 4, we conclude that $C_n^{-1}A_n$ has at most $1 + 2\max\{\mu, \nu\}$ distinct eigenvalues (in the limit $n = \infty$), including $\lambda = 1$, and the proof is completed with an application of Corollary 2. □

A weaker smoothness assumption, though still strong, is that f should be analytic on $|z| = 1$:

Theorem 7. [6]. Suppose f is analytic in a neighborhood of $|z| = 1$. Then the errors in the preconditioned Toeplitz CG iteration (in the limit $n = \infty$) satisfy

$$\frac{\|e_k\|_A}{\|e_0\|_A} \leq r^{k^2} \qquad (0 \leq k < \infty) \tag{25}$$

for some $r < 1$.

Proof Let $u(z)$ be the harmonic function in $|z| \leq 1$ which has boundary values $u(z) = \log(\sqrt{f(z)})$ on $|z| = 1$, and let $v(z)$ be the harmonic conjugate of $u(z)$ in $|z| \leq 1$ normalized by $v(0) = 0$. Then the spectral factorization of f is

$$f(z) = w(z)w(z^{-1}) \quad \text{with} \quad w(z) = e^{u(z)+iv(z)},$$

and since $f(z)$ is analytic in a neighborhood of $|z| = 1$, $u(z)$ and $v(z)$ are harmonic in a neighborhood of $|z| \leq 1$, so $w(z)$ is analytic and nonzero in a neighborhood of $|z| \leq 1$. Therefore $w(z)/w(z^{-1})$ is analytic in a neighborhood of $|z| = 1$, which implies that the best rational approximants to its analytic part $v(z)$ converge at least geometrically as $k \to \infty$. By Lemmas 4 and 5, it follows that the eigenvalues $\lambda_{\pm k}$ of $C_n^{-1}A_n$ (in the limit $n = \infty$) approach $\lambda = 1$ geometrically as $k \to \infty$.

We now face a special case of the approximation problem of Theorem 1: how small can a polynomial $p \in P_k$ be, subject to $p(0) = 1$, on a set of points that clusters geometrically at $\lambda = 1$? The upper bound (25) comes from choosing p to have zeros at the outermost eigenvalues $\lambda_{\pm 1}, \ldots, \lambda_{\pm k/2}$. □

Even with a far weaker smoothness assumption, we still obtain super-geometric convergence (cf. Corollary 1 to Theorem 1.4.1 of [10]):

Theorem 8. Suppose f is any function in the Wiener class (i.e. with absolutely convergent Laurent series). Then for *any* $\epsilon > 0$, the errors in the preconditioned Toeplitz CG iteration (in the limit $n = \infty$) satisfy

$$\frac{\|e_k\|_A}{\|e_0\|_A} \le C\epsilon^k \qquad (0 \le k < \infty) \tag{26}$$

for some constant C.

Proof Define $u(z)$, $v(z)$, and $w(z)$ as in the last proof. If f is in the Wiener class, then $u(z) = \log\sqrt{f(z)}$ is certainly Dini-continuous on $|z| = 1$, which is enough to ensure that $v(z)$ and therefore $w(z)/w(z^{-1})$ are continuous on $|z| = 1$ also. By Hartman's Theorem [39], it follows that the infinite Hankel matrix H is compact, which implies that its singular values decrease to 0, however slowly. The same conclusion can be reached in a more pedestrian fashion by combining the Weierstrass approximation theorem with Lemma 5. By Lemma 4, it follows that the eigenvalues of $C_n^{-1}A_n$ cluster at $\lambda = 1$. Given ϵ, the proof is now completed by choosing $p(z)$ to interpolate 0 at a fixed set of outermost eigenvalues of $C_n^{-1}A_n$ chosen so that the remaining eigenvalues lie in a sufficiently small interval about 1; the remaining degrees of freedom provide the geometric convergence asserted in (26). □

The convergence rates of Theorems 6–8 are impressive, but lest it be obscured by our rather casual assumption "$n=\infty$", let us emphasize the conclusion that underlies all of these results: for finite Toeplitz matrices, the number of iterations is bounded independently of the dimension n as $n \to \infty$, so long as the entries $\{a_k\}$ correspond to a continuous function $f(z)$ with a modest degree of regularity. In such cases Strang's Toeplitz iteration is truly an $O(n \log n)$ algorithm.

It would be excellent to absorb these theorems in a general characterization of the rate of convergence of the iteration as a function of the smoothness of f, but this is impossible at present, for not enough is known about rational approximation. In general, rational functions are far more powerful approximators than polynomials, a fact which was discovered by Newman in 1964 [38] and has been widely generalized since then [19]. Unfortunately, Theorems 7 and 8 do not take advantage of this phenomenon. For example, the hypothesis of analyticity in Theorem 7 is necessary and sufficient to guarantee geometric convergence of *polynomial* approximations on $|z| = 1$, which is what we used to obtain the estimate (25); for rational approximations, it is much too strong. Precise conditions in the rational case are not known, although progress on these questions has been made recently.

4. Polynomial iterations for nonsymmetric matrices

So far, our matrices have been symmetric and positive definite. Now we turn to the problem of nonsymmetric matrices A with complex eigenvalues.

Many iterative methods have been devised for nonsymmetric problems, which go by names such as ORTHOMIN, ORTHODIR, GCR, GMRES, and LSQR; see

[14] for a survey. Most of these are related in one way or another to the conjugate gradient iteration, and often they minimize a norm of the error or the residual, exactly or approximately, over a sequence of subspaces. Unfortunately, no method has emerged which has the elegance and power of the conjugate gradient iteration for symmetric positive definite problems.[3] In particular, the good fortune of a 3-term recurrence relation is often lost, so that at the kth step of an iteration, one has to form linear combinations of $O(k)$ vectors. This makes it more important than ever to find methods that converge in a small number of iterations.

This section will describe a particular class of methods which are based on the assumption that the spectrum Σ of A is contained in a known subset K of the complex plane, which may have been determined adaptively:

$$\Sigma \subseteq K \subset \mathcal{C}.$$

(Since A is nonsingular, we assume $0 \notin K$.) Perhaps it is an exceptional problem where an accurate estimate K can be obtained at low cost, but if it can, what use can be made of the information?

This is an old question, to which many researchers have contributed. Varga and his colleagues have described *semi-iterative methods,* in which a simple matrix iteration (e.g. Jacobi or Gauss-Seidel) is accelerated by the formation of linear combinations of the iterates [13, 25, 57]. Others have used the term *polynomial iteration* to describe closely related methods. Omitting details, one ends up in either case, as in Section 2, with a sequence of iterates x_k and errors $e_k = x^* - x_k$,

$$x_k = q(A)b, \qquad e_k = p(A)e_0 \tag{27}$$

for some polynomials

$$q \in P_{k-1}, \qquad p \in P_k, \quad p(0) = 1, \tag{28}$$

with $p(z) = 1 - zq(z)$. (Some methods work with residuals rather than the errors e_k.) However, whereas the conjugate gradient iteration chooses optimal polynomials p and q implicitly and automatically, here we must construct them explicitly.

Thus as in Section 2, we again face two problems of approximation by matrix polynomials: find $q \in P_{k-1}$ such that $q(A) \approx A^{-1}$, or find $p \in P_k$ with $p(0) = 1$ such that $p(A) \approx 0$. To convert these problems from matrices to scalars, it is again natural to consider an eigenvector decomposition. Suppose that A is diagonalizable, and let V be a matrix of normalized eigenvectors and Λ a corresponding diagonal matrix of eigenvalues:

$$A = V\Lambda V^{-1}. \tag{29}$$

[3] Of course one can always apply the conjugate gradient iteration to the normal equations $A^TAx = A^Tb$, implicitly or explicitly, and this is done in LSQR and some other methods. The disadvantage is the possibly large condition number (or unfavorable eigenvalue clustering) of A^TA.

Then $q(A) = Vq(\Lambda)V^{-1}$, which implies

$$A^{-1} - q(A) = V(\Lambda^{-1} - q(\Lambda))V^{-1}, \tag{30}$$

and therefore, making use of the assumption $\Sigma \subseteq K$,

$$\begin{aligned} \|A^{-1} - q(A)\| &\leq \kappa(V)\,\|\Lambda^{-1} - q(\Lambda)\| \\ &= \kappa(V)\,\|z^{-1} - q(z)\|_\Sigma \leq \kappa(V)\,\|z^{-1} - q(z)\|_K \end{aligned} \tag{31}$$

if $\|\cdot\|$ is any p-norm $(1 \leq p \leq \infty)$, where $\kappa(V) = \|V\|\,\|V^{-1}\|$ is the condition number of V. Here $\|f\|_\Sigma$ and $\|f\|_K$ are abbreviations for $\sup_{z\in\Sigma}|f(z)|$ and $\sup_{z\in K}|f(z)|$, respectively. Similarly,

$$\begin{aligned} \|p(A)\| &\leq \kappa(V)\,\|p(\Lambda)\| \\ &= \kappa(V)\,\|p(z)\|_\Sigma \leq \kappa(V)\,\|p(z)\|_K. \end{aligned} \tag{32}$$

For any fixed nondefective matrix A, $\kappa(V)$ is a finite constant, and thus (31) and (32) imply that there is at most a constant gap between our matrix approximation problems and the corresponding scalar approximation problems, which can be formulated as follows:

Problem Q. Find $q \in P_{k-1}$ to minimize $\|z^{-1} - q(z)\|_K$.

Problem P. Find $p \in P_k$, satisfying $p(0) = 1$, to minimize $\|p(z)\|_K$.

(Problems Q and P are not very different; one can be converted to the other by introducing a sup-norm weighted by $|z|$ or $|z|^{-1}$.)

Polynomial approximation in the complex plane is a well understood subject, though not as straightforward as real approximation on the real line [18, 45, 59]. How shall we (approximately) solve Problem Q or P in the context of our linear algebra problem $Ax = b$? In certain special cases, as mentioned in the Introduction, the solution is well known. For symmetric positive definite matrices, K can be taken to be a real interval, and we are led to Chebyshev iteration [20, 21, 57]. For symmetric indefinite matrices K can be taken as a pair of disjoint real intervals [12, 43]. For some nonsymmetric problems good results can be obtained by a Chebyshev iteration based on a choice of K as an ellipse centered on the real axis [15, 34, 35]. For more general problems, Gutknecht has described a method based on Pick-Nevanlinna interpolation [24], and various authors have considered methods based on least-squares approximation and/or orthogonal polynomials [16, 23, 44, 46, 47]. I shall now describe the method of *interpolation in Fejér points*, which has been investigated recently by Reichel and Fischer [17, 40] and Tal-Ezer [50].

For simplicity, assume that K is simply connected (although the method can be generalized via Green's functions to the multiply-connected case [40]), and let $f(z)$ be a conformal map of the exterior of the unit disk $|z| \leq 1$ onto the exterior of K, with $f(\infty) = \infty$. If k interpolation points are needed, let $z_1, \ldots, z_k$ be the

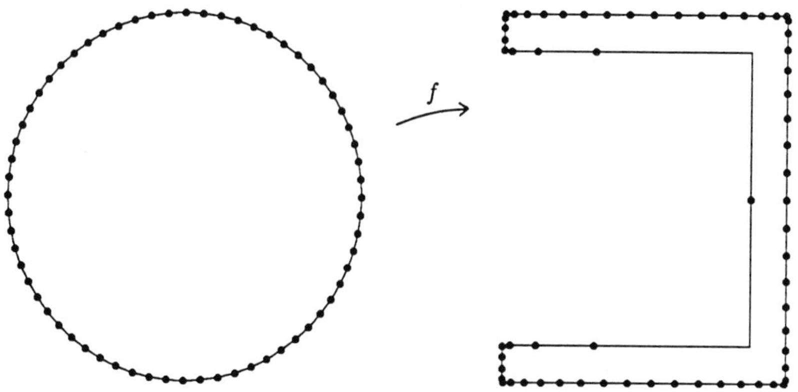

Figure 4. Fejér points for polynomial interpolation ($k = 64$). The density is zero at reentrant corners and infinite at salient corners.

kth roots of unity on $|z| = 1$, and let $w_1, \ldots, w_k$ be their images under f on the boundary of K. (The map f is guaranteed to extend continuously to the boundary if, for example, K is a Jordan region.) These *Fejér points* are shown in Figure 4 for a region K in the shape of a polygonal U.

Interpolation in Fejér points, though not an optimal approximation strategy for any fixed k, is guaranteed to produce the asymptotically optimal order of convergence as $k \to \infty$:

Theorem 9. [18, 59] For each $k \geq 1$, let $q \in P_{k-1}$ be defined by interpolation of z^{-1} in $k-1$ Fejér points on K. Then for any $\epsilon > 0$, there are constants C_1, C_2, and $\rho < 1$ such that

$$\|z^{-1} - q(z)\|_K \leq C_1 (1+\epsilon)^k \inf_{q' \in P_{k-1}} \|z^{-1} - q'(z)\|_K \leq C_2 \rho^k \tag{33}$$

for all sufficiently large k. Similarly, if $p \in P_k$ is defined by interpolation of 0 with $p(0) = 1$, we get the estimate

$$\|p(z)\|_K \leq C_1 (1+\epsilon)^k \inf_{p' \in P_k,\ p'(0)=1} \|p'(z)\|_K \leq C_2 \rho^k. \ \square \tag{34}$$

By combining (31) and (33), or (32) and (34), we obtain bounds on the accuracy of the matrix iteration based on interpolation in Fejér points.[4]

There is still a large piece missing in our numerical algorithm: it depends on the conformal map f, and the computation of conformal maps is not trivial. For

[4] Both Reichel & Fischer and Tal-Ezer point out that in the practical implementation of this iteration, the interpolation points cannot be taken in arbitrary order. For numerical stability, the ordering must correspond to an approximately uniform sampling of the boundary, which can be obtained by numbering the boundary points sequentially in binary and then reversing the bits.

general smooth domains K, effective algorithms for computing f have been devised, but no software is available [28, 54]. If K is a polygon with a reasonably small number of vertices, on the other hand, f can be obtained by a Schwarz-Christoffel transformation computed by the Fortran package SCPACK [52], and that is how Figure 4 was generated. Since f is a map of the exterior of a polygon, its computation would appear to require a modification of SCPACK, which was designed for interior maps. In most cases of practical interest, however, A has real entries, so that Σ and presumably K are symmetric about the real axis. To take advantage of this symmetry one can compute the map of the exterior of the unit disk in the upper half-plane to the exterior of K in the upper half-plane, and then complete the map by the reflection principle. This map of half-planar regions is essentially an interior Schwarz-Christoffel map of the usual sort (with a vertex at infinity), so unmodified SCPACK is applicable after all.

In certain applications by Fischer and Reichel and by Tal-Ezer and his collaborators, the rather complicated iterative method described in this section has performed dramatically well [50, 51]. But its limitations must also be emphasized: accurate estimates of the spectrum Σ of A are not always available, and the conformal map f may be hard to compute. In the final section we will now discuss a more basic and more interesting limitation.

5. Non-normal matrices

The eigenvalues of a nonsymmetric matrix are in general complex, but there is a more fundamental problem: they may be irrelevant! In a sense eigenvalues are never *exactly* the right thing to look at, when A is not normal;[5] the larger the condition number $\kappa(V)$ of the matrix of eigenvectors, the more wrong they may be (see (31) and (32)). In practical applications $\kappa(V)$ is sometimes huge, especially if A is a member of a family of matrices obtained by a process of discretization.

To begin with an extreme example, consider the defective matrix

$$A = \begin{pmatrix} 1 & 1 \\ 0 & 1 \end{pmatrix}, \qquad A^{-1} = \begin{pmatrix} 1 & -1 \\ 0 & 1 \end{pmatrix},$$

with $\kappa(V) = \infty$, and suppose our goal is to approximate A^{-1} by a polynomial $q(A)$. Since the spectrum is the single point $\Sigma = \{1\}$, the polynomial $q(z) = 1$ matches z^{-1} exactly for $z \in \Sigma$, and thus one might be tempted to consider the approximation

$$q(A) = \begin{pmatrix} 1 & 0 \\ 0 & 1 \end{pmatrix} \approx A^{-1}.$$

[5] A normal matrix is one that possesses a complete orthogonal system of eigenvectors. Equivalently, A is normal if and only if $A^HA = AA^H$, where A^H is the conjugate transpose. Hermitian, skew-Hermitian, unitary, and circulant matrices fall in this category.

But of course this matrix is utterly incorrect in the upper-right corner. Approximation at the eigenvalues is not enough.

To a pure mathematician, what went wrong is obvious: since $\lambda = 1$ is a defective eigenvalue of multiplicity 2, $q(z)$ should have been chosen to match both z^{-1} and its derivative. But to a numerical analyst, this remedy is unappealing, for it violates the principle that a good algorithm should be insensitive to small perturbations. Here, an infinitesimal perturbation might separate the eigenvalues and make A non-defective, suggesting approximation again of function values only. Then $q(z)$ would remain a good approximation to z^{-1} on Σ (though no longer exact), but $q(A)$ would still be a bad approximation to A^{-1}. Thus even for nondefective matrices, approximation at the eigenvalues is not enough.

We propose that a better way to treat highly non-normal matrices is to approximate function values only, not derivatives, but replace Σ by a larger region Σ_ϵ of "approximate eigenvalues." Here is the definition:

Definition. Let A be a square matrix of dimension n, and let $\epsilon \geq 0$ be arbitrary. Then Σ_ϵ, the set of *ϵ-approximate eigenvalues* of A, is the set of numbers $z \in C$ that satisfy any of the following equivalent conditions:

(i) z is an eigenvalue of $A + \Delta$ for some matrix Δ with $\|\Delta\| \leq \epsilon$;

(ii) A has an *ϵ-approximate eigenvector* $u \in C^n$ with $\|(A - z)u\| \leq \epsilon$, $\|u\| = 1$;

(iii) $\sigma_n(zI - A) \leq \epsilon$;

(iv) $\|(zI - A)^{-1}\| \geq \epsilon^{-1}$.

In these assertions $\|\cdot\|$ is the 2-norm, σ_n denotes the smallest singular value, and $(zI - A)^{-1}$ is known as the *resolvent.* The proof that the four conditions are equivalent is a routine exercise in the use of the singular value decomposition.

The sets Σ_ϵ form a nested family, and Σ_0 is the same as Σ. For any ϵ, Σ_ϵ contains all the numbers at a distance $\leq \epsilon$ from Σ,

$$\Sigma_\epsilon \supseteq \{z \in C : \operatorname{dist}(z, \Sigma) \leq \epsilon\}, \tag{35}$$

with equality if A is normal. If A is not normal, Σ_ϵ may be much larger. In the 2×2 example above, Σ_ϵ is a disk about $z = 1$ of radius $\sim \epsilon^{1/2}$ as $\epsilon \to 0$, which becomes $\sim \epsilon^{1/J}$ if A is generalized to a Jordan block of dimension J. When J is large, $\epsilon^{1/J}$ is close to 1 even when ϵ is as small as machine precision.

Figure 5 illustrates the idea of approximate eigenvalues for two upper-triangular Toeplitz matrices of the form

$$F = \begin{pmatrix} 0 & 1 & & & & \\ & 0 & 1 & & \text{\Large 0} & \\ & & 0 & 1 & & \\ & & & 0 & 1 & \\ & \text{\Large 0} & & & 0 & 1 \\ & & & & & 0 \end{pmatrix}, \qquad G = \begin{pmatrix} 0 & 1 & 1 & & & \\ & 0 & 1 & 1 & & \text{\Large 0} \\ & & 0 & 1 & 1 & \\ & & & 0 & 1 & 1 \\ & \text{\Large 0} & & & 0 & 1 \\ & & & & & 0 \end{pmatrix},$$

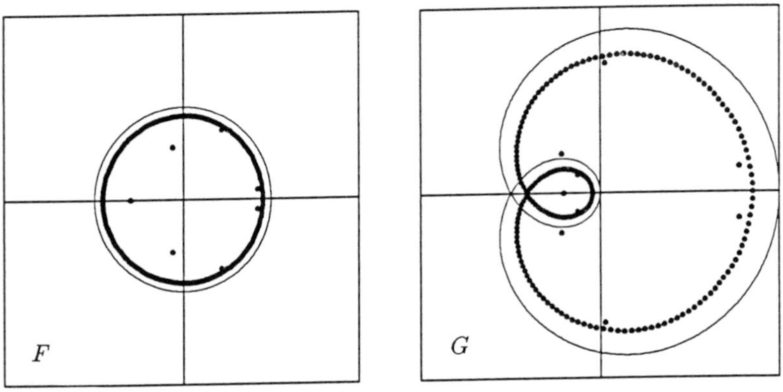

Figure 5. Some ϵ-approximate eigenvalues of two nilpotent matrices F and G of dimension $n = 200$ ($\epsilon = 10^{-8}$). The solid curves are the images of the unit circle under $f(z) = z$ and $g(z) = z + z^2$.

but of dimension $n = 200$; the first is our Jordan block and the second is a kind of generalized Jordan block. (These matrices are again defective, but that could be changed by an arbitrarily small perturbation.) Both F and G are mathematically nilpotent: in each case the spectrum is $\Sigma = \{0\}$. The figure, however, shows eigenvalues of perturbed matrices

$$\tilde{F} = F + \Delta, \quad \tilde{G} = G + \Delta, \qquad \|\Delta\| = 10^{-8},$$

where Δ is a dense random matrix with independent normally distributed elements. The effect of the perturbation on the eigenvalues is enormous and anything but random. If we let f and g denote the "symbols" of these Toeplitz matrices,

$$f(z) = z, \qquad g(z) = z + z^2,$$

and Γ_F and Γ_G the images of the unit circle $|z| = 1$ under f and g, then most of the eigenvalues of $\tilde{F}$ and $\tilde{G}$ evidently lie close to Γ_F and Γ_G.[6] By definition, they are all contained in the corresponding ϵ-approximate eigenvalue regions Σ_ϵ for $\epsilon = 10^{-8}$, and in fact Σ_ϵ becomes exactly Γ_F or Γ_G, together with the regions they enclose, in any limit $\epsilon \to 0$, $n \to \infty$ with $n\epsilon \geq$ const. > 0.

These pictures, coupled with conditions *(i)* and *(ii)* above, give a first indication of why Σ_ϵ may be a more appropriate domain than Σ on which to approximate z^{-1}. If $z \in \Sigma_\epsilon$ for some $\epsilon \ll 1$, then A will behave approximately as if z were an eigenvalue, and if ϵ is as small as machine precision the difference will very likely be undetectable. Under such circumstances, any algorithm which requires a knowledge of Σ in principle had better be supplied with a larger set Σ_ϵ in practice.

[6] If F and G were bi-infinite Toeplitz matrices, their spectra would be Γ_F and Γ_G exactly, and if they were semi-infinite, their spectra would be the regions enclosed by Γ_F and Γ_G.

A more precise statement can be obtained with the aid of condition *(iv)*. It is well known that $A^{-1} - q(A)$ can be represented by the resolvent integral

$$A^{-1} - q(A) = \frac{1}{2\pi i} \oint_{\Gamma} (z^{-1} - q(z))(zI - A)^{-1}\, dz, \tag{36}$$

where Γ is any positively oriented closed contour or union of contours that encloses Σ but not $z=0$ (since z^{-1} has a pole there) [7, 32]. If we define

$$\begin{aligned}
\delta &= \|z^{-1} - q(z)\|_{\Gamma}, \\
L &= \frac{1}{2\pi} \times \text{arc length of } \Gamma, \\
R &= \sup_{z\in\Gamma} \|(zI - A)^{-1}\|, \\
\sigma_{\min} &= R^{-1} = \inf_{z\in\Gamma} \sigma_n(zI - A),
\end{aligned}$$

then (36) leads to the estimate

$$\|A^{-1} - q(A)\| \le LR\,\delta = \frac{L}{\sigma_{\min}}\,\delta. \tag{37}$$

In particular, if Γ is taken as the boundary of Σ_ϵ for some $\epsilon > 0$, then $\sigma_{\min} = \epsilon$ and we get

$$\|A^{-1} - q(A)\| \le \frac{L}{\epsilon}\,\delta. \tag{38}$$

Equations (37) and (38) reflect a basic tradeoff in the approximation of matrix functions by methods of complex analysis: should the domain of approximation be chosen to lie close to the spectrum of A, or not so close? As Γ contracts to Σ, L decreases to 0 but R increases to ∞. If A is normal, the two effects cancel and small contours are best because of the factor δ: z^{-1} can be approximated more accurately on a small region. If A is far from normal, however, then R may be much larger, and the advantage shifts to contours further away.[7] Whether or not A happens to be exactly defective is unimportant.

All of these observations carry over to more general matrix approximation problems in which z^{-1} and A^{-1} are replaced by arbitrary functions $f(z)$ and $f(A)$, so long as Γ lies in a region where $f(z)$ is analytic. An important example is the approximation of matrix exponentials $f(A) = e^{tA}$, which arises in the numerical solution of differential equations [20, 37, 50]. A further generalization would be to permit approximations $q(z)$ other than polynomials. The following theorem restates the results above in this more general context:

[7] This kind of balancing argument underlies the Kreiss Matrix Theorem and its applications in finite-difference computations [42].

Theorem 10. Let A be a square matrix with spectrum Σ, and let $f(z)$ and $q(z)$ be analytic functions defined in a region K containing Σ. Let Γ be a positively oriented closed contour in K that encloses Σ_ϵ; let L, R, and $\sigma_{\min}$ be defined as above; and define $\delta = \|f(z) - q(z)\|_\Gamma$. Then

$$\|f(A) - q(A)\| \;\le\; LR\,\delta \;=\; \frac{L}{\sigma_{\min}}\,\delta. \tag{39}$$

In particular, if Γ is the boundary of Σ_ϵ for some $\epsilon > 0$, then

$$\|f(A) - q(A)\| \;\le\; \frac{L}{\epsilon}\,\delta. \;\Box \tag{40}$$

For comparison, here is the estimate analogous to (31):

$$\|f(A) - q(A)\| \le \kappa(V)\,\delta. \tag{41}$$

This inequality has the advantage that it is independent of Γ, but (39) and (40) have the potentially more important advantage that they are independent of $\kappa(V)$. If A is a member of a family whose eigenvector matrices have unbounded condition numbers, (39) and (40) may possibly provide uniformly valid bounds if Γ is chosen suitably, but (41) cannot.

Although Theorem 10 is stated as an inequality, in practice it gives guidance in both directions. In particular we single out the following rule of thumb:

> *A scalar approximation $q(z) \approx f(z)$ of accuracy δ is of no use for constructing a matrix approximation $q(A) \approx f(A)$ unless it is valid at least on the approximate eigenvalue domain Σ_δ.*

Here is an example to illustrate these ideas. The function

$$h(z) \;=\; \frac{1 + z/4}{1 - z/2} \;=\; 1 + \frac{3}{4}\Big(z + \frac{1}{2}z^2 + \frac{1}{4}z^3 + \cdots\Big)$$

maps the unit disk conformally onto the disk D of radius 1 and center 3/2, with $h(0) = 1$. The corresponding $n \times n$ upper-triangular Toeplitz matrix

$$H \;=\; \begin{pmatrix} 1 & \frac{3}{4} & \frac{3}{8} & & \cdots \\ & 1 & \frac{3}{4} & \frac{3}{8} & \\ & & 1 & \frac{3}{4} & \frac{3}{8} \\ & & & & \\ & 0 & & \ddots & \\ & & & & \end{pmatrix} \tag{42}$$

has spectrum $\Sigma = \{1\}$ but ϵ-approximate spectrum $\Sigma_\epsilon \approx D$ when ϵ is small and n is large. This situation is represented in Figure 6.

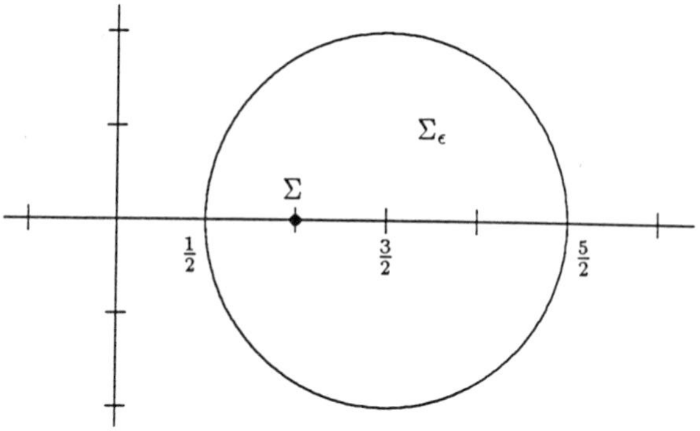

Figure 6. Exact and approximate spectra of the upper-triangular matrix H (ϵ small, n large). The approximate spectrum controls the behavior of a matrix iteration; see Figure 7 below.

Now suppose we solve $Hx = b$ by the one-step Richardson iteration

$$x_{k+1} := x_k + (b - Hx_k)/d, \qquad (x_0 = 0) \tag{43}$$

for some constant d, with corresponding error equation

$$e_{k+1} = (1 - H/d)e_k \qquad (e_k = x^* - x_k). \tag{44}$$

Implicitly we are approximating $f(H) = 0$ by $p_k(H)$ with $p_k(z) = (1 - z/d)^k$. Figure 7 shows the resulting convergence history for dimension $n = 200$ and two values of d.[8] The upper curve represents the "correct" parameter $d = 1$ based on the exact spectrum $\Sigma = \{1\}$, but instead of the instantaneous convergence one would observe if the matrix were normal, the figure shows steady geometric divergence at the rate $(3/2)^k$. The error decreases temporarily around step $k = n$, but rounding errors prevent convergence and soon it is growing again. On the other hand with the "incorrect" parameter $d = 3/2$ in the lower curve, based on the approximate spectrum $\Sigma_\epsilon \approx D$, the iteration converges geometrically at the rate $(2/3)^k$ down to the level of machine precision. Both of these rates are what one would observe for a normal matrix with spectrum D.[9]

This matrix H was manufactured to prove a point, but equally non-normal matrices occur in the wild. One example is the matrix associated with the Gauss-Seidel iteration for the standard discretization of the Poisson equation. For the

[8] The right-hand side was $b = (1, 1, \ldots, 1)^T$, and the matrix H was first transformed by a random orthogonal similarity transformation to ensure the occurrence of rounding errors. This Richardson iteration is the same as a Chebyshev iteration in which the ellipse degenerates to a point.

[9] An interesting question is whether adaptive methods of estimating eigenvalues, such as those employed by Elman, et al. and by Manteuffel [15, 35], tend to come close to exact or approximate eigenvalues in cases where the two are very different. If the latter is true or can be made true by a suitable choice of adaptive scheme, then adaptively estimated eigenvalues may sometimes yield better matrix iterations than exact ones.

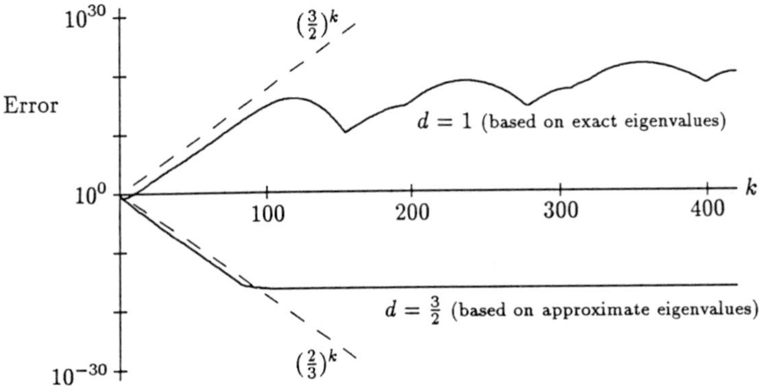

Figure 7. Convergence history of the one-step Richardson iteration (43) for the matrix H ($n = 200$) with two choices of the parameter d.

simplest one-dimensional geometry the eigenvalues are real and positive, but the approximate eigenvalues fill a complex region in the shape of a tennis racket defined by the function $\frac{1}{2}z/(1 - \frac{1}{2}z^{-1})$. As it happens, in this case non-normality does not affect the convergence rate, because the dominant eigenvalue is essentially the same with or without perturbations.

Non-normality has a pronounced effect on convergence, however, in the solution of certain partial differential equations by spectral methods [5]. For example, suppose the model problem

$$u_t = u_x, \qquad x \in [-1, 1], \quad t \geq 0, \quad u(1, t) = 0 \tag{45}$$

is solved by an explicit finite-difference formula in t coupled with a discretization in x consisting of interpolation in N Gauss-Legendre points by a global polynomial $p_N(x)$ satisfying $p_N(1) = 0$ followed by differentiation of $p_N(x)$. This spectral differentiation process is equivalent to multiplication by an $N \times N$ matrix D, which proves to be highly non-normal. The eigenvalues of D are of size $O(N)$, suggesting that the method will be subject to a stability restriction of the form $\Delta t \leq CN^{-1}$ for some constant C [49], but the *approximate* eigenvalues are of size $O(N^2)$ and the actual (Lax-)stability restriction is apparently $\Delta t = O(N^{-2})$ [55, 56]. The stability of spectral methods is poorly understood at present, and perhaps one reason is that the matrices involved tend to be so far from normal.

Although the observations of this section are based on standard mathematics, their application to practical problems has apparently not been explored. It remains to be seen what the implications may be concerning the place of approximation theory in numerical linear algebra. On one hand, the idea of approximating on regions other than exact spectra suggests a new role in which approximation theory may prove useful. On the other hand, the possibility of extreme non-normality should perhaps stand as a warning that beautiful tricks like the conjugate gradient iteration are inevitably tied to special cases — that in general, matrices are truly more complicated than scalars, and approximation theory can provide only some of the answers.

Acknowledgments

Several people have helped me with valuable advice and references; I would like in particular to mention Martin Gutknecht, Tom Manteuffel, Youcef Saad, and Hillel Tal-Ezer. The experiments of the last section were carried out with the help of Satish Reddy.

References

1. V. M. Adamjan, D. Z. Arov, and M. G. Krein. Analytic properties of Schmidt pairs for a Hankel operator and the generalized Schur-Takagi problem. *Math. USSR Sbornik,* 15:31–73, 1971.
2. G. S. Ammar and W. B. Gragg. Superfast solution of real positive definite Toeplitz systems. *SIAM J. Matrix Anal. Applics.*, 9:61–76, 1988.
3. A. Brandt. Multi-level adaptive solutions to boundary value problems. *Math. Comp.*, 31:333–390, 1977.
4. C. Brezinski. *Padé-type Approximation and General Orthogonal Polynomials.* Birkhäuser, 1980.
5. C. Canuto, M. Y. Hussaini, A. Quarteroni, and T. A. Zang. *Spectral Methods in Fluid Dynamics.* Springer, 1988.
6. R. H. Chan and G. Strang. Toeplitz equations by conjugate gradients with circulant preconditioner. *SIAM J. Sci. Stat. Comput.*, to appear.
7. F. Chatelin. *Spectral Approximation of Linear Operators.* Academic Press, 1983.
8. P. Concus, G. H. Golub, and D. P. O'Leary. A generalized conjugate gradient method for the numerical solution of elliptic partial differential equations. In J. R. Bunch and D. J. Rose, eds., *Sparse Matrix Computations,* Academic Press, 1976.
9. G. Cybenko. An explicit formula for Lanczos polynomials. *Lin. Alg. Applics.*, 88:99–115, 1987.
10. J. W. Daniel. The conjugate gradient method for linear and nonlinear operator equations. *SIAM J. Numer. Anal.*, 4:10–26, 1967.
11. J. W. Daniel. *The Approximate Minimization of Functionals.* Prentice-Hall, 1971.
12. C. de Boor and J. R. Rice. Extremal polynomials with application to Richardson iteration for indefinite linear systems. *SIAM J. Sci. Stat. Comp.*, 3:47–57, 1982.
13. M. Eiermann, W. Niethammer, and R. S. Varga. A study of semiiterative methods for nonsymmetric systems of linear equations. *Numer. Math.*, 47:505–533, 1985.
14. H. C. Elman. *Iterative Methods for Large, Sparse, Nonsymmetric Systems of Linear Equations.* PhD thesis, Res. Rep. #229, Dept. of Comp. Sci., Yale U., 1982.
15. H. C. Elman, Y. Saad, and P. E. Saylor. A hybrid Chebyshev Krylov subspace algorithm for solving nonsymmetric systems of linear equations. *SIAM J. Sci. Stat. Comput.*, 7:840–855, 1986.

16. D. K. Faddeev and V. N. Faddeeva. *Computational Methods of Linear Algebra.* W. H. Freeman, 1963.
17. B. Fischer and L. Reichel. A stable Richardson iteration method for complex linear systems. *Numer. Math.*, to appear.
18. D. Gaier. *Lectures on Complex Approximation.* Birkhäuser, 1987.
19. T. Ganelius, W. K. Hayman, and D. J. Newman. *Lectures on Approximation and Value Distribution.* Les Presses de L'Université de Montréal, 1982.
20. G. H. Golub and C. F. Van Loan. *Matrix Computations.* Johns Hopkins U. Press, 1983.
21. G. H. Golub and R. S. Varga. Chebyshev semi-iterative methods, successive over-relaxation iterative methods, and second-order Richardson iterative methods, parts I and II. *Numer. Math.*, 3: 147–56 and 157–168.
22. W. B. Gragg. The Padé table and its relation to certain algorithms of numerical analysis. *SIAM Review*, 14:1–62, 1972.
23. W. B. Gragg and L. Reichel. On the application of orthogonal polynomials to the iterative solution of linear systems of equations with indefinite or non-Hermitian matrices. *Lin. Alg. Applics.*, 88:349–371, 1987.
24. M. H. Gutknecht. An iterative method for solving linear equations based on minimum norm Pick-Nevanlinna interpolation. In C. K. Chui, et al., eds., *Approximation Theory V*, Academic Press, 1986.
25. M. H. Gutknecht. Stationary and almost stationary iterative (k,l)-step methods for linear and nonlinear systems of equations. To appear.
26. W. Hackbusch. *Multi-Grid Methods and Applications.* Springer, 1985.
27. L. A. Hageman and D. M. Young. *Applied Iterative Methods.* Academic Press, 1981.
28. P. Henrici. *Applied and Computational Complex Analysis, v. 3.* Wiley, 1986.
29. M. R. Hestenes and E. Stiefel. Methods of conjugate gradients for solving linear systems. *J. Res. Nat. Bur. Stand.*, 49:409–436, 1952.
30. O. G. Johnson, C. A. Micchelli, and G. Paul. Polynomial preconditioners for conjugate gradient calculations. *SIAM J. Numer. Anal.*, 20:362–376, 1983.
31. S. Kaniel. Estimates for some computational techniques in linear algebra. *Math. Comp.*, 20:369–378, 1966.
32. T. Kato. *Perturbation Theory for Linear Operators.* Springer, 1976.
33. D. Luenberger. *Introduction to Linear and Nonlinear Programming*, 2nd ed. Addison-Wesley, 1984.
34. T. A. Manteuffel. The Tchebychev iteration for nonsymmetric linear systems. *Numer. Math.*, 28:307–327, 1977.
35. T. A. Manteuffel. Adaptive procedure for estimating parameters for the nonsymmetric Tchebychev iteration. *Numer. Math.*, 31:183–208, 1978.
36. G. Meinardus. Über eine Verallgemeinerung einer Ungleichung von L. V. Kantorowitsch. *Numer. Math.*, 5:14–23, 1963.
37. C. Moler and C. Van Loan. Nineteen dubious ways to compute the exponential of a matrix. *SIAM Review*, 20:801–836, 1978.
38. D. J. Newman. Rational approximation to $|x|$. *Michigan Math. J.*, 11:11–14, 1964.
39. S. C. Power. *Hankel Operators on Hilbert Space.* Pitman, 1982.
40. L. Reichel. Polynomials for the Richardson iteration method for complex linear systems. Preprint 84/1, U. Hamburg, 1984.
41. J. K. Reid. On the method of conjugate gradients for the solution of large sparse systems of linear equations. In J. K. Reid, ed., *Large Sparse Sets of Linear Equations*,

Academic Press, 1971.

42. R. D. Richtmyer and K. W. Morton. *Difference Methods for Initial-Value Problems,* 2nd ed. Wiley, 1967.
43. Y. Saad. Iterative solution of indefinite symmetric linear systems by methods using orthogonal polynomials over two disjoint intervals. *SIAM J. Numer. Anal.*, 20:784–811, 1983.
44. Y. Saad. Least squares polynomials in the complex plane and their use for solving nonsymmetric linear systems. *SIAM J. Numer. Anal.*, 24:155–169, 1987.
45. E. B. Saff. Polynomial and rational approximation in the complex domain. *Proc. Symp. Appl. Math.*, 36:21–49, Amer. Math. Soc., 1986.
46. D. C. Smolarski and P. E. Saylor. Optimum parameters for the solution of linear equations by Richardson iteration. Unpublished paper.
47. E. L. Stiefel. Kernel polynomials in linear algebra and their numerical applications. *U.S. Nat. Bur. Stand. Appl. Math. Series*, 49:1–24, 1958.
48. G. Strang. *Introduction to Applied Mathematics.* Wellesley-Cambridge Press, 1986.
49. H. Tal-Ezer. A pseudospectral Legendre method for hyperbolic equations with an improved stability condition. *J. Comp. Phys.*, 67:145–172, 1986.
50. H. Tal-Ezer. Polynomial approximation of functions of matrices and its application to the solution of a general system of linear equations. ICASE Report 87-63, NASA Langley Research Center, 1987.
51. H. Tal-Ezer, J. M. Carcione, and D. Kosloff. An accurate and efficient scheme for wave propagation in linear viscoelastic media. *Geophysics,* to appear.
52. L. N. Trefethen. Numerical computation of the Schwarz-Christoffel transformation. *SIAM J. Sci. Stat. Comput.*, 1:82–102, 1980.
53. L. N. Trefethen. Rational approximation on the unit disk. *Numer. Math.*, 37:297–320, 1981.
54. L. N. Trefethen, ed. *Numerical Conformal Mapping.* North-Holland, 1986.
55. L. N. Trefethen. Lax-stability vs. eigenvalue stability of spectral methods. To appear in Proceedings, 1988 Oxford Conference on Computational Fluid Dynamics.
56. L. N. Trefethen and M. R. Trummer. An instability phenomenon in spectral methods. *SIAM J. Numer. Anal.*, 24:1008–1023, 1987.
57. R. S. Varga. *Matrix Iterative Analysis.* Prentice-Hall, 1962.
58. E. L. Wachspress. *Iterative Solution of Elliptic Systems.* Prentice-Hall, 1966.
59. J. L. Walsh. *Interpolation and Approximation by Rational Functions in the Complex Domain,* 5th ed. Amer. Math. Soc., 1969.
60. D. M. Young. *Iterative Solution of Elliptic Systems.* Academic Press, 1971.

An algorithm for computing minimum norm solutions of finite moment problem

M. Frontini

Dipartimento di Matematica,
Politecnico di Milano, Italy

G. Rodriguez and S. Seatzu

Dipartimento di Matematica,
Universitá di Cagliari, Italy

Abstract A regularization method is proposed for a stable computation of minimum norm solution of Hausdorff's finite moment problem with noisy data. The order of convergence of the approximations is studied and a sufficient condition for the optimal order is also given. Numerical results illustrate both stability and inherent connection between moments number, noise level and accuracy of the method.

Key words: Fredholm equation of first kind, Moment problem, Regularization, Orthogonal polynomials.

1 Introduction

Given an infinite set of moments $\{\mu_i\}$, Hausdorff's problem (Hausdorff 1923; Askey, Schoenberg and Sharma, 1982) consists of the solution of the system

$$\int_0^1 x^{i-1} f(x)dx = \mu_i \qquad (i=1,2,..). \tag{1}$$

Bellman, Kalaba and Lockett (1966) studied, from a numerical point of view, the finite moment problem, namely the system (1) given a finite number of moments. This is a more realistic problem in practice and it is a typical ill-posed problem.

Schoenberg (1973) solved the finite moment problem in the absence of noise.

In this work we study the finite moment problem with noisy data. In §2 we describe a method for the approximation of the minimum norm least squares solution; in §3 we study the convergence rate of the method; in §4 we examine some methods to evaluate the regularization parameter and in §5 the main results of extensive experimentation are presented.

2 The algorithm

For the sake of simplicity we consider the finite moment problem as follows:

$$Af=\underline{b}, \quad \underline{b}\in R^m, \quad (Af)_i = \langle x^{i-1},f\rangle = \int_{-1}^{1} x^{i-1} f(x)dx \quad (i=1,..,m). \tag{2}$$

Let N(A) be the null space of A, Π_n the set of polynomials of degree $\leq$ n and f^+ the minimum norm solution of (2) in $L^2[-1,1]$.
It is easy to show that $f^+\in\Pi_{m-1}$. Therefore if $\{P_i\}_{i=0}^{m-1}$ are the first m Legendre polynomials, orthonormal in $L^2[-1,1]$, we can write

$$f^+ = \sum_{j=1}^{m} \alpha_j^+ P_{j-1},$$

where $\hat{H}\underline{\alpha}^+=\underline{b}$, $(\hat{H})_{ij} = \langle x^{i-2},P_{j-1}\rangle$ (i, j=1,..,m).
Because of the orthogonality of $\{P_i\}$ the matrix $\hat{H}$ is nonsingular; moreover, $\hat{H}$ is lower triangular.

Alternatively, f^+ could be expanded by means of the powers $\{x^{i-1}\}_{i=1}^m$, but the power expansion is not recommended because:
(a) the conditioning of the power expansion is worse than the orthogonal expansion (Gautschi, 1972);
(b) if H is the matrix related to the power expansion then $\text{cond}(H)=(\text{cond}(\hat{H}))^2$.
The proof follows easily from the two mentioned expansions by means of some straightforward algebraic manipulations. We therefore prefer the use of the $\{P_i\}$.

The method proposed is based on the evaluation in Π_{m-1} of the minimizer $f_\lambda^{[k]}$ of the regularization functional

$$J_\lambda^{[k]}(f) = \lambda|f|_k^2 + \tfrac{1}{m}\|Af-\underline{b}\|^2,$$

where $|f|_k^2 = \|f^{(k)}\|^2_{L^2[-1,1]}$, $\|\cdot\| = \|\cdot\|_2$.
$\lambda\in\mathbb{R}^+$ is the regularization parameter and k=0,1,2 is an index identifying the functional.
Π_{m-1} is natural as a regularization space, since we want to approximate f^+, which belongs to Π_{m-1}, by means of $f_\lambda^{[k]}$.
The choice of k depends upon the type of control we want to perform on the regularized function $f_\lambda^{[k]}$ (k=0 for value; k=1 for gradient; k=2 for second derivative).

Since $Af=A\left(\sum_1^m \alpha_j P_{j-1}\right)=\hat{H}\underline{\alpha}$, $J_\lambda^{[k]}$ is characterized by the quadratic form

$$Q_\lambda^{[k]}(\underline{\alpha}) = \lambda\underline{\alpha}^T\Omega^{[k]}\underline{\alpha} + \tfrac{1}{m}\|\hat{H}\underline{\alpha}-\underline{b}\|^2, \quad \Omega^{[k]}=\langle P_{i-1}^{(k)},P_{j-1}^{(k)}\rangle \quad (i,j=1,..,m).$$

$\Omega^{[k]}$ is a symmetric semidefinite positive matrix with $\text{rank}(\Omega^{[k]})=m-k$.

Since $N(\Omega^{[k]}) \cap N(\hat{H})=\{0\}$, for every vector $\underline{b} \in R^m$ there is only one minimizer $f_\lambda^{[k]}$ of $J_\lambda^{[k]}$.

Our numerical examples show that the choice of k is important and that the best one is k=2, followed by k=1. For this reason we will present the case k=2 and for simplicity remove the index k. Moreover, given a vector $\underline{a} \in R^m$, we will generally decompose it as follows:

$$\underline{a}=\begin{bmatrix} \underline{a}' \\ \underline{a}'' \end{bmatrix}, \quad \underline{a}'=[a_1,a_2]^T, \quad \underline{a}''=[a_3,a_4,\dots,a_m]^T.$$

Consequently, denoting by $\hat{\Omega}$ the matrix obtained from Ω by deleting the first two rows and columns, Q_λ becomes

$$Q_\lambda(\underline{\alpha}) = \lambda \underline{\alpha}''^T \hat{\Omega} \underline{\alpha}'' + \frac{1}{m}\|\hat{H}\underline{\alpha}-\underline{b}\|^2.$$

The regularization term depends only on $\underline{\alpha}''$ and the residual term only on $\underline{\alpha}$, so it is useful to decompose the matrix $\hat{H}$ into the product of an orthogonal Householder matrix Q and a matrix K arranged as follows:

$$\hat{H}=QK, \qquad Q^TQ=I, \qquad K=\begin{bmatrix} K_{11} & K_{12} \\ 0 & K_{22} \end{bmatrix},$$

where K_{11} is a nonsingular upper triangular matrix of order 2. This decomposition is numerically stable and inexpensive, because it needs only the first two steps of the Householder triangularization method.

$\hat{\Omega}$ is symmetric and positive definite, so by Cholesky decomposition it may be written as LL^T, where L is a nonsingular lower triangular matrix.

Defining $\underline{c}=Q^T\underline{b}$, Q_λ becomes:

$$Q_\lambda(\underline{\alpha}) = P_\lambda(\underline{\alpha}'') + \frac{1}{m}\|K_{11}\underline{\alpha}'+K_{12}\underline{\alpha}''-\underline{c}'\|^2,$$

where

$$P_\lambda(\underline{\alpha}'') = \lambda\|L\underline{\alpha}''\|^2 + \frac{1}{m}\|K_{22}\underline{\alpha}''-\underline{c}''\|^2.$$

Then, if $\underline{\alpha}''_\lambda$ is the minimizer of P_λ, the minimizer $\underline{\alpha}_\lambda$ of Q_λ is:

$$\underline{\alpha}_\lambda = \begin{bmatrix} \underline{\alpha}'_\lambda \\ \underline{\alpha}''_\lambda \end{bmatrix}, \qquad \underline{\alpha}'_\lambda = K_{11}^{-1}\left(\underline{c}'-K_{12}\underline{\alpha}''_\lambda\right). \tag{3}$$

Equation (3) immediately follows from these properties:

(1) $\min_{\mathbb{R}^{m-2}} P_\lambda \le \min_{\mathbb{R}^m} Q_\lambda$,

(2) given $\underline{\alpha}''_\lambda$ there is only one $\underline{\alpha}'_\lambda$ so that $\|K_{11}\underline{\alpha}'_\lambda + K_{12}\underline{\alpha}''_\lambda - \underline{c}'\|=0$.

Setting $\underline{\beta}=L\underline{\alpha}''$ we obtain the standard form

$$\hat{P}_\lambda(\underline{\beta}) = \lambda\|\underline{\beta}\|^2 + \frac{1}{m}\|K_{22}L^{-1}\underline{\beta}-\underline{c}\|^2.$$

Finally, denoting by UDV^T the singular value decomposition of $K_{22}L^{-1}$, $\hat{P}_\lambda$ can be written in the diagonal form:

$$\hat{P}_\lambda(\underline{\tilde{\beta}}) = \lambda\|\underline{\tilde{\beta}}\|^2 + \frac{1}{m}\|D\underline{\tilde{\beta}}-\underline{\tilde{c}}\|^2, \text{ where } \underline{\tilde{\beta}}=V^T\underline{\beta}, \quad \underline{\tilde{c}}=U^T\underline{c}''.$$

The characteristic equation of the minimizer of $\tilde{P}_\lambda$ is

$$\left(m\lambda I+D^2\right)\underline{\tilde{\beta}}_\lambda = D\underline{\tilde{c}}. \tag{4}$$

With the same notation, the vector $\underline{\alpha}^+$ of the normal solution is defined by the system

$$\begin{cases} (\underline{\alpha}^+)'' = L^{-1}\underline{\beta}^+, \quad \underline{\beta}^+=V\underline{\tilde{\beta}}^+, \quad \underline{\tilde{\beta}}^+=D^{-1}\underline{\tilde{c}}, \\ (\underline{\alpha}^+)' = K_{11}^{-1}(\underline{c}'-K_{12}(\underline{\alpha}^+)''). \end{cases} \tag{5}$$

From (4) and (5) the vector $\underline{\tilde{\beta}}_\lambda$ of the regularized solution may be written in these two forms:

$$\underline{\tilde{\beta}}_\lambda = (m\lambda I+D^2)^{-1}D\underline{\tilde{c}} = (m\lambda I+D^2)^{-1}D^2\underline{\tilde{\beta}}^+. \tag{6}$$

Equation (6) says that $\|\underline{\tilde{\beta}}_\lambda\|$ monotonically decreases from $\|\underline{\tilde{\beta}}^+\|$ to zero as λ increases from zero to ∞.

Remark 1 The described algorithm can easily be generalized to solve the weighted moment problem. We consider this problem with Chebyshev weight, that is

$$Af=\underline{b}, \quad \underline{b}\in\mathbb{R}^m, \quad (Af)_i = \langle x^{i-1},f\rangle = \int_{-1}^{1}(1-x^2)^{-1/2}x^{i-1}f(x)dx. \tag{7}$$

Since $f^+\in\Pi_{m-1}$ and the inner product is characterized by Chebyshev weight, it is natural to expand f^+ by means of Chebyshev polynomials $\{T_i\}_{i=0}^{m-1}$, orthonormal on $L^2[-1,1]$ with respect to the weight $(1-x^2)^{-1/2}$. Consequently, if $f^+=\sum_1^m\delta_j^+T_{j-1}$, then $\underline{\delta}^+$ can be obtained by the same transformations used to obtain $\underline{\alpha}^+$.

The previous considerations are valid with or without noise. However, in practice it is necessary to solve the problem with noisy data. Therefore, to verify the stability of the method, given a model function f and the related moment vector $\underline{b}=Af$, we consider the noisy vector $\underline{b}^\varepsilon=\underline{b}+\sigma\underline{\varepsilon}$, where ε_i is a standard normal random variable with $E[\varepsilon_i]=0$ and $E[\varepsilon_i\varepsilon_j]=\delta_{ij}$, σ is the standard deviation and δ_{ij} is the Kronecker symbol.

3 Convergence of the method

In this section, we consider the moment problem both with exact and noisy data. We will use the index ε to indicate the presence of noise.

Lemma 1 With the foregoing notation we have:

$$\|\tilde{\underline{\beta}}_\lambda^\varepsilon-\tilde{\underline{\beta}}^+\| \le \|D^{-1}\| \ \|(I-D(\lambda))\tilde{\underline{c}}-D(\lambda)\tilde{\underline{\varepsilon}}\|, \tag{8}$$

where

$$D(\lambda) = (m\lambda I+D^2)^{-1}D^2, \qquad \tilde{\underline{\varepsilon}} = \tilde{\underline{c}}^\varepsilon-\tilde{\underline{c}}. \tag{9}$$

Proof Let $\tilde{\underline{\beta}}_\lambda^\varepsilon$ be the approximate solution of $\tilde{\underline{\beta}}^+$ given by the method. Then the error may be written:

$$\tilde{\underline{\beta}}_\lambda^\varepsilon-\tilde{\underline{\beta}}^+ = [\tilde{\underline{\beta}}_\lambda^\varepsilon-\tilde{\underline{\beta}}_\lambda] + [\tilde{\underline{\beta}}_\lambda-\tilde{\underline{\beta}}^+]. \tag{10}$$

From (6) and (9)

$$\tilde{\underline{\beta}}_\lambda^\varepsilon-\tilde{\underline{\beta}}_\lambda = (m\lambda I+D^2)^{-1}D(\tilde{\underline{c}}^\varepsilon-\tilde{\underline{c}}) = D^{-1}D(\lambda)\tilde{\underline{\varepsilon}}, \tag{11}$$

and from (6), since $\tilde{\underline{\beta}}^+=D^{-1}\tilde{\underline{c}}$,

$$\tilde{\underline{\beta}}_\lambda-\tilde{\underline{\beta}}^+ = D^{-1}(D(\lambda)-I)\tilde{\underline{c}}. \tag{12}$$

(8) then follows immediately from (10), (11) and (12).

Lemma 2 If $\lambda=O(\sigma^\alpha)$ and $\alpha>0$, then

$$E\left[\|\tilde{\underline{\beta}}_\lambda^\varepsilon-\tilde{\underline{\beta}}^+\|\right] = O(\sigma^\beta), \qquad \beta = \min\left\{1,\alpha\right\}. \tag{13}$$

Proof From (11)

$$\|\tilde{\underline{\beta}}_\lambda^\varepsilon-\tilde{\underline{\beta}}_\lambda\|^2 = \tilde{\underline{\varepsilon}}^TD(\lambda)(m\lambda I+D^2)^{-1}\tilde{\underline{\varepsilon}} < \tilde{\underline{\varepsilon}}^T(m\lambda I+D^2)^{-1}\tilde{\underline{\varepsilon}},$$

since $D(\lambda)$ is a diagonal matrix and $\|D(\lambda)\|_\infty<1$, $\forall\lambda\in R^+$.

Moreover, taking into account the transformations produced by the method,

$$\tilde{\underline{\varepsilon}} = U^TQ^T[(\underline{b}^\varepsilon)''-\underline{b}''] = \sigma U^TQ^T\underline{\varepsilon}''.$$

Besides, since $\lambda=O(\sigma^\alpha)$, we have

$$E\left[\|\tilde{\underline{\beta}}_\lambda^\varepsilon-\tilde{\underline{\beta}}_\lambda\|\right] \le \sigma\sqrt{Tr(D^{-2})} = O(\sigma).$$

From (12)

$$\tilde{\underline{\beta}}_\lambda-\tilde{\underline{\beta}}^+ = -m\lambda(m\lambda I+D^2)^{-1}\tilde{\underline{\beta}}^+$$

and also, setting $\tilde{\underline{\beta}}^+=D^2\underline{\gamma}$,

$$\|\tilde{\underline{\beta}}_\lambda-\tilde{\underline{\beta}}^+\| = m\lambda\|D(\lambda)\underline{\gamma}\| < m\lambda\|\underline{\gamma}\| = O(\sigma^\alpha).$$

Finally, since

$$\tilde{\underline{\beta}}_\lambda^\varepsilon-\tilde{\underline{\beta}}^+ = [\tilde{\underline{\beta}}_\lambda^\varepsilon-\tilde{\underline{\beta}}_\lambda] + [\tilde{\underline{\beta}}_\lambda-\tilde{\underline{\beta}}^+],$$

(13) follows.

Theorem 1 If $\lambda=O(\sigma^\alpha)$ and $\alpha>0$, then

$$E\left[\|f_\lambda^\varepsilon-f^+\|\right] = O(\sigma^\beta), \qquad \beta = \min\left\{1,\alpha\right\}. \tag{14}$$

Proof From the orthogonality of Legendre polynomials,

$$\|f_\lambda^\varepsilon-f^+\| = \|\underline{\alpha}_\lambda^\varepsilon-\underline{\alpha}^+\|. \tag{15}$$

Moreover, because of the foregoing transformations,

$$(\underline{\alpha}_\lambda^\varepsilon)''-(\underline{\alpha}^+)'' = L^{-1}(\underline{\beta}_\lambda^\varepsilon-\underline{\beta}^+) = L^{-1}V(\tilde{\underline{\beta}}_\lambda^\varepsilon-\tilde{\underline{\beta}}^+),$$

$$(\underline{\alpha}^{\varepsilon}_{\lambda})' - (\underline{\alpha}^{+})' = K_{11}^{-1}K_{12}[(\underline{\alpha}^{\varepsilon}_{\lambda})'' - (\underline{\alpha}^{+})''].$$

Taking into account the decompositions of $\underline{\alpha}^{\varepsilon}_{\lambda}$ and $\underline{\alpha}^{+}$, from the orthogonality of V and from the properties of the euclidean norm:

$$\|\underline{\alpha}^{\varepsilon}_{\lambda} - \underline{\alpha}^{+}\| \le \left(1 + \|K_{11}^{-1}K_{12}\|\right) \|L^{-1}\| \; \|\underline{\tilde{\beta}}^{\varepsilon}_{\lambda} - \underline{\tilde{\beta}}^{+}\|. \tag{16}$$

Finally, because of (15), (16) and Lemma 2, (14) follows.

Theorem 1 implies that the maximum possible order of convergence is 1 and it is obtained when $\alpha=1$. Such order is not obtainable for the infinite moment problem, since the maximum order is 2/3 for the regularization in infinite dimensional spaces (Groetsch, 1984).

4 Evaluation of the regularization parameter

It is well known that in regularization methods, as in smoothing, the value of the regularization parameter is crucial (Craven and Wahba, 1979; Groetsch, 1984; Morozov, 1984).

From (15), (16) and (8) we obtain

$$\|f^{\varepsilon}_{\lambda} - f^{+}\| \le \left(1 + \|K_{11}^{-1}K_{12}\|\right) \|L^{-1}\| \; \|D^{-1}\| \; [R(\lambda)]^{1/2},$$

where $R(\lambda) = \|(I-D(\lambda))\underline{\tilde{c}} - D(\lambda)\underline{\tilde{\varepsilon}}\|^2$.

Unfortunately we cannot minimize $R(\lambda)$ because $\underline{b}$ and $\underline{\varepsilon}$, and consequently $\underline{\tilde{c}}$ and $\underline{\tilde{\varepsilon}}$, are unknown. However, when the variance σ^2 on the data vector $\underline{b}^{\varepsilon}$ is specified, an unbiased estimate of $E[R(\lambda)]$ is given by

$$\hat{R}(\lambda) = \|(I-D(\lambda))\underline{\tilde{c}}^{\varepsilon}\|^2 - \sigma^2 \mathrm{Tr}(I-D(\lambda))^2 + \sigma^2 \mathrm{Tr}(D^2(\lambda)).$$

Therefore, if σ^2 is known, the minimizer of $\hat{R}$ can be taken as an effective approximation of optimal regularization parameter $\tilde{\lambda}$. This estimate is the same as that introduced by Mallows (1973) in the context of ridge regression. Morozov's discrepancy principle (1984) is also applicable.

As is well known, when there is a lack of reliable information on σ^2, GCV is very effective in smoothing (Craven and Wahba, 1979; Utreras, 1981), in ridge regression (Golub, Heath and Wahba, 1979) and in the regularization of Fredholm equations of the first kind (Wahba, 1977; Wahba, 1982). Our experiments prove that when they are both applicable, GCV and Mallows' principle are equivalent, but Morozov's criterion proves to be less effective. Since GCV does not require the variance to be known, the recovery of the model functions refer to the regularization parameter obtained by GCV.

5 Numerical results

We present numerical results for the following model functions:

$$f_1(x) = \frac{1}{5\pi} \sin \frac{5}{2}\pi(x+1) - \frac{1}{10\pi} \sin 5\pi(x+1),$$

$$f_2(x) = (1+125x^2)^{-1}.$$

We chose, as an index of quality of our approximations, the relative error

$$e(f) = \|\tilde{f}-f\|/\|f\|,$$

where $\tilde{f}$ is the recovered function.

For an immediate comparison of the results, we adopt as a measure of the noise level, the relative standard deviation

$$\tilde{\sigma}=\sigma/\mu, \qquad \mu=\sum_1^m |b_i^{\varepsilon}|/m.$$

Numerical results, obtained for $20 \le m \le 40$, prove that, given the noise level, there is an optimal number of moments. Increasing the number of moments, without an accompanying reduction of noise level, gives poorer recoveries. This property appears to be intrinsic to the problem, and it has been verified by the application of the three different techniques used to evaluate the optimal regularization parameter. This result confirms a similar analytical conclusion obtained by Talenti (1987) in a more limited context.

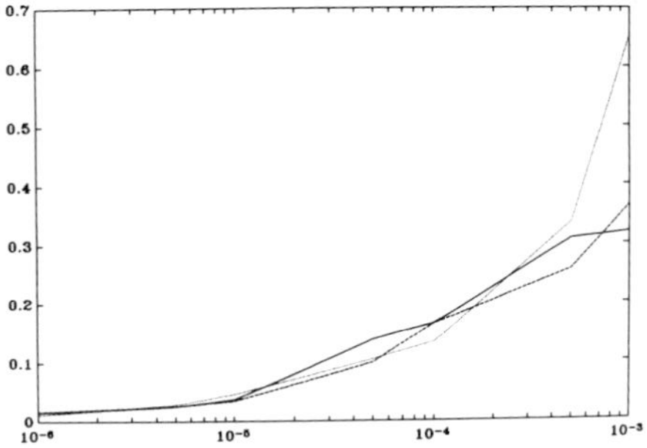

Figure 1

To exemplify, we illustrate this property for model function f_1. Figure 1 shows $e(f_1)$ as a function of $\tilde{\sigma}$ for m=20(—),30(--),40(··).

We feel the existence of an optimal number of moments is a consequence of the convergence to zero of the moments.

m=30 appears to be optimal when $10^{-6} \le \tilde{\sigma} \le 10^{-4}$. Therefore we report in figures 2 and 3 respectively, for m=30, the functions $f_1, \tilde{f}_1$ and $f_2, \tilde{f}_2$.

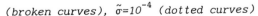

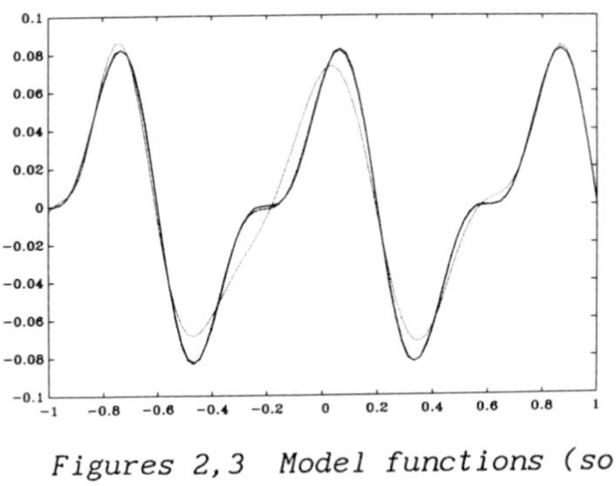

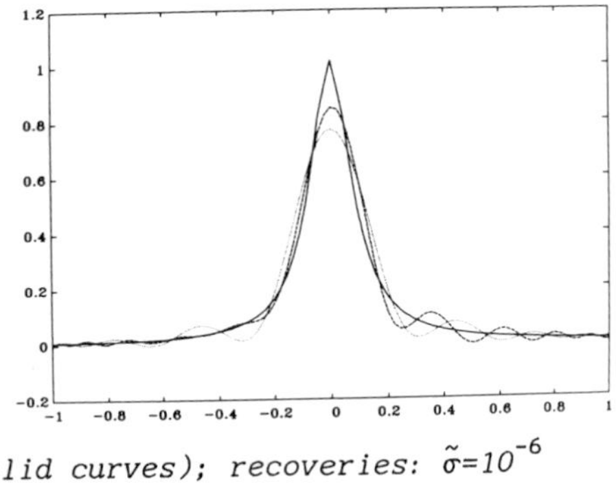

Figures 2,3 Model functions (solid curves); recoveries: $\tilde{\sigma}=10^{-6}$ (broken curves), $\tilde{\sigma}=10^{-4}$ (dotted curves)

Remark 2 Numerical results show no significant difference in the recoveries of bounded functions by means of both the moment problem (2) and the weighted moment problem (7).
However, if the function to be recovered has the form $f(x)=g(x)(1-x^2)^p$, where $-1/2\le p<0$ and g is bounded on $[-1,1]$, accurate recoveries can only be obtained by solving the moment problem with Chebyshev weight.

Acknowledgements

This work was carried out with the support of Italian MPI.

References

Askey, R., Schoenberg, I. J. and Sharma, A. (1982), Hausdorff's moment problem and expansions in Legendre polynomials, J. Math. Anal. Appl. 86, 237-245
Bellman, R. E., Kalaba, R. E. and Lockett, J. A. (1966), 'Numerical Inversion of the Laplace Transform', Amer. Elsevier, New York
Craven, P. and Wahba, G. (1979), Smoothing noisy data with spline functions: estimating the correct degree of smoothing by the method of Generalized Cross Validation, Numer. Math. 31, 377-403
Gautschi, W. (1972), The condition of orthogonal polynomials, Math. Comp. 26, 923-924
Golub, G., Heath, M. and Wahba, G. (1979), Generalized Cross Validation as a method for choosing a good ridge parameter, Technometrics 31, 315-324
Groetsch, Ch. W. (1984), 'The theory of Tikhonov Regularization for Fredholm Equations of the First Kind', Research Notes in Mathematics 10, Pitman, London
Hausdorff, F. (1923), Moment probleme für ein endliches intervall, Math. Z. 16, 220-248
Mallows, C. L. (1973), Some comments on C_p, Technometrics 15, 661-675
Morozov, V. A. (1984), 'Methods for Solving Incorrectly Posed Problems', Springer, Berlin
Schoenberg, I. J. (1973), Remarks concerning a numerical inversion of the Laplace transform due to Bellman, Kalaba and Lockett, J. Math. Anal. Appl. 43, 823-838
Talenti, G. (1987), Recovering a function from a finite number of moments, Inverse Problems 3, 501-517
Utreras, F. I. (1981), Optimal smoothing of noisy data using spline functions, SIAM J. Sci. Stat. Comput. 2, 249-362
Wahba, G. (1977), Practical approximation solutions to linear operator equations when the data are noisy, SIAM J. Numer. Anal. 14, 651-667
Wahba, G. (1982), Constrained regularization for ill-posed linear operator equations, with applications in Meteorology and Medicine, In Statistical Decision Theory and Related Topics III, vol. 2, S. S. Gupta and J. O. Bergen eds., Academic Press

Numerical solution of the biharmonic equation using different types of bivariate spline functions

R. H. J. Gmelig Meyling

Department of Applied Mathematics, University of Twente, Enschede, The Netherlands

Abstract In this paper, we discuss the numerical solution of a boundary value problem by the method of finite elements. The partial differential equation we consider is the so-called biharmonic equation, which is elliptic and of fourth order. This equation describes the deformation of a thin elastic plate subject to load. As approximate solutions we choose piecewise polynomial functions of minimal degree with respect to a triangular mesh. In order to obtain conforming finite elements, suitable spline functions will be of class C^1 at least. The aim of our research is to compare accuracy as well as computational cost of three different (conforming) finite elements. The elements we study are either polynomial or piecewise polynomial per triangle. A preconditioned conjugate gradient method is used to solve the resulting large (but sparse) systems of linear equations. A comparison is made between the different elements applied to certain model problems described in the literature.

Key words: Biharmonic equation, Plate bending problem, Thin plate functional, Piecewise polynomials, Conforming finite elements, Bézier-Bernstein representation, Split-triangle elements, Conjugate gradient method.

1. Plate bending problem

Consider the deformation of a thin elastic plate subject to perpendicular forces. The vertical displacement s(x,y) of the plate at a point (x,y) is given by the biharmonic equation

$$\Delta^2 s(x,y) = \frac{1}{D} f(x,y), \quad \text{for} \quad (x,y) \in \Omega . \tag{1}$$

Here, $f \in L_2(\Omega)$ is a load function defined over a polygonal domain Ω in R^2 and D is the plate stiffness. In practical applications, the deformation of the plate will be constrained by additional boundary conditions.

Due to the high-order derivatives, the partial differential equation (1) is in this form unsuitable for direct numerical calculation. However, the plate bending problem can be stated in variational form. Namely, the solution of the elliptic equation (1) also minimizes the potential energy of the plate, which can be described by the thin plate functional

$$\Pi = \frac{1}{2} D \iint_{\Omega} [s_{xx}^2 + 2\nu s_{xx} s_{yy} + s_{yy}^2 + 2(1-\nu) s_{xy}^2] dxdy - \iint_{\Omega} fs dxdy , \tag{2}$$

where ν denotes Poisson's ratio.

In the finite element method, the domain Ω is partitioned into subdomains. As approximants we choose piecewise polynomial functions with respect to triangles, which are the basic elements in two dimensions. If $\underline{q}$ denotes a parameter vector defining a spline s, then the functional Π turns into a quadratic function

$$\Pi(\underline{q}) = \frac{1}{2} \underline{q}^T S \underline{q} - \underline{q}^T \underline{F} , \tag{3}$$

with S the symmetric and positive definite stiffness matrix. For approximants of class C^1 (i.e. conforming elements) convergence of the approximant towards the exact solution is guaranteed when the triangular partition is refined.

2. Smooth piecewise polynomials

During the last few years, considerable progress has been achieved within the field of approximation by piecewise polynomials, especially splines. Splines in two variables with respect to triangles can be conveniently expressed in Bézier-Bernstein (BB)-form (cf. Dahmen (1986) and Farin (1986)). The BB-form greatly facilitates the construction of smooth surfaces composed of many different polynomial pieces. Moreover, computationally attractive schemes exist for evaluating and differentiating piecewise polynomials in this form.

Let θ be a triangulation of the planar domain Ω. Suppose that θ consists of T (nondegenerate) triangles, V vertices and E edges. We assume that V_0 vertices and E_0 edges of θ are located in the interior of Ω. In addition, let Σ denote the number of singular vertices in θ. A singular vertex is an interior vertex, where precisely four edges meet which have only two different slopes (see Figure 1c).

Let $[\underline{x}^1,\underline{x}^2,\underline{x}^3]$ be an arbitrary triangle in θ having $\underline{x}^1,\underline{x}^2,\underline{x}^3 \in R^2$ as its vertices. If $\lambda_1, \lambda_2, \lambda_3$ are the (nonnegative) barycentric coordinates of a point $\underline{x}$ in the triangle, then any bivariate polynomial p of degree $\le$ d over the triangle can be expressed in BB-form

$$p(\underline{x}) = \sum_{|\underline{\alpha}|=d} b_{\underline{\alpha}} \frac{d!}{\alpha_1!\,\alpha_2!\,\alpha_3!} \lambda_1^{\alpha_1} \lambda_2^{\alpha_2} \lambda_3^{\alpha_3} . \tag{4}$$

In formula (4), the symbol $\underline{\alpha} = (\alpha_1,\alpha_2,\alpha_3) \in Z_+^3$ denotes a multi-index with $|\underline{\alpha}| = \alpha_1+\alpha_2+\alpha_3$. The Bézier ordinates $b_{\underline{\alpha}}$ are generally associated with domain points $\underline{v}_{\underline{\alpha}} = (\alpha_1\underline{x}^1+\alpha_2\underline{x}^2+\alpha_3\underline{x}^3)/d$, for $|\underline{\alpha}|=d$. The domain points are arranged in a regular mesh (the Bézier network) over the triangle. In turn, the $\binom{d+2}{2}$ control points $(\underline{v}_{\underline{\alpha}},b_{\underline{\alpha}}) \in R^3$, with $|\underline{\alpha}|=d$, determine uniquely the shape of the d-th degree polynomial p over the triangle.

It is a straightforward task to enforce a C^1-smooth transition between two adjacent polynomials in BB-form across an interior edge. The procedure is illustrated by Figure 1a for the cubic case (d=3), where all four control points belonging to each of the d shaded quadrilaterals are coplanar. We point out that (in Figures 1 and 2) thick lines denote triangle boundaries, whereas thin lines indicate Bézier networks. Obviously, over the entire triangulation θ there are redundancies within the collection of these smoothness constraints. Observe in Figure 1b that two conditions on the control points at the inner disk around any interior vertex are redundant. In the quadratic and cubic case, an additional redundancy exists for every singular vertex (see Figure 1c and compare Gmelig Meyling and Pfluger (1985)).

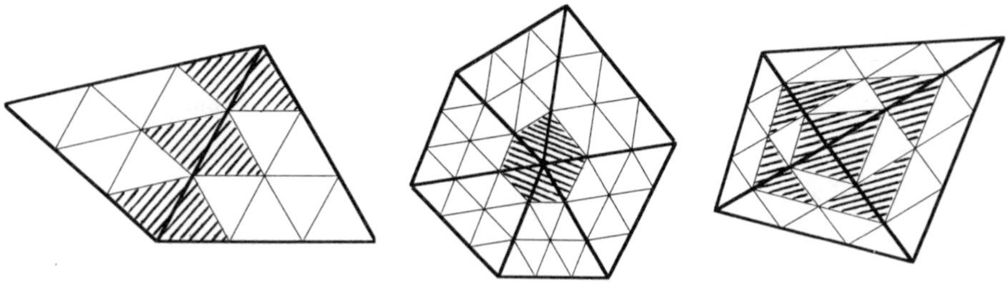

(a) (b) (c)

Figure 1 C^1-smoothness conditions in terms of control points

3. Conforming finite elements

In this section, we consider two different techniques for the construction of smooth piecewise polynomials of minimal degree as conforming finite elements. In the first (global) technique, the approximating spline is a polynomial of a suitable degree on every triangle in θ. C^1-continuity is enforced by imposing smoothness constraints across the triangle boundaries (as explained before). Schumaker (1984) and Gmelig Meyling and Pfluger (1985) have shown that piecewise quadratics on arbitrary triangulations do not provide sufficient degrees of freedom for approximation. Hence, as finite element I we consider a cubic polynomial over a triangle (see the Bézier network in Figure 2a and compare Gmelig Meyling (1987)). In this case, the approximating spline over Ω is defined by V+2E+T control points and constrained by $3E_0-2V_0-\Sigma$ smoothness conditions.

As a second (local) approach, one first divides the (macro-)

triangles in θ into a number of subtriangles and then chooses a polynomial on every subtriangle. Conforming elements obtained by this method are often referred to as split-triangle elements. Finite element II is based on C^1-piecewise cubics with respect to the Clough&Tocher-split (1965) of a macro-triangle into three subtriangles (Figure 2b). It is possible to lower the degree of the polynomials by choosing a more complicated partition. In Figure 2c, the twelve subtriangles are shown which support the quadratic C^1-element III by Heindl (1979).

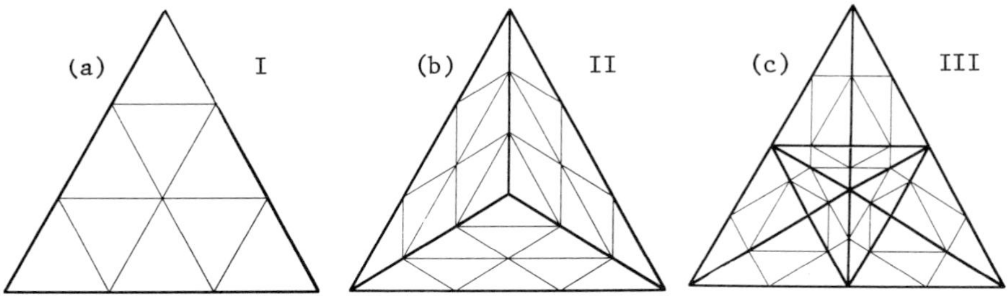

Figure 2 Bézier networks for three conforming finite elements

Elements II and III have the advantage that C^1-smoothness can be guaranteed locally by choosing a suitable parametrisation, namely function and first derivative values at the vertices and first normal derivatives at the edge midpoints of the macro-triangles. Hence, adjacent elements match C^1-smoothly along common boundaries (cf. Gmelig Meyling (1988)). A spline based on elements II or III over θ is defined by 3V+E parameters. An alternative quadratic C^1-element has been derived by Powell and Sabin (1977) and an application of bivariate B-splines to plate bending problems has been described by Traas (1988).

Assembling the stiffness matrix S is most efficiently accomplished by exploiting the special second derivative formulas for polynomials in BB-form by Farin (1986). Minimization of the quadratic function $\Pi(\underline{q})$, subject to boundary conditions and smoothness constraints of the form $C\underline{q}=\underline{0}$, is equivalent to the solution of the linear system $\begin{pmatrix} S & C^T \\ C & 0 \end{pmatrix}\begin{pmatrix} \underline{q} \\ \underline{\lambda} \end{pmatrix}=\begin{pmatrix} \underline{F} \\ \underline{0} \end{pmatrix}$. Preconditioned conjugate gradients by Golub and van Loan (1983) are used to solve this system for the spline parameters $\underline{q}$ and the Lagrange multipliers $\underline{\lambda}$, while maintaining the sparseness of the matrices S and C.

4. Numerical experiments

In this section, we study two model problems, where we set $\nu = 1/6$, $D = 1$ and we assume the plate being uniformly loaded (i.e. $f(x,y) \equiv 1$). The objective is to investigate how closely the three C^1-elements approximate the exact solutions of these plate bending problems.

In Schwarz (1980), a 2x2-square plate is examined, with the left side of the plate clamped ($s=s_n=0$) and the opposite side simply supported ($s=0$). If one introduces a symmetry condition ($s_n=0$), then it is sufficient to consider only the upper half of the plate (namely $\Omega = [-1,1] \times [0,1]$). The domain Ω is triangulated uniformly by a $2m \times m$ -unidiagonal grid. Schwarz provides solutions (correct in four digits) for the vertical displacement s at four test points A=(-1/2,0), B=(-1/2,1), C=(1/2,0) and D=(1/2,1).

Table 1 Absolute errors (x100) in displacement for three elements

Element	m	A	B	C	D
I	1	0.2507	0.2726	0.2113	0.4983
I	2	0.0074	0.0246	0.0064	0.0337
I	3	0.0060	0.0036	0.0082	0.0156
I	4	0.0020	0.0033	0.0032	0.0067
I	5	0.0016	0.0011	0.0017	0.0030
II	1	0.2926	0.1272	0.2403	0.2688
II	2	0.0130	0.0295	0.0068	0.0013
II	3	0.0071	0.0053	0.0061	0.0036
II	4	0.0019	0.0048	0.0016	0.0002
II	5	0.0014	0.0018	0.0013	0.0006
III	1	0.9704	0.8981	0.8042	0.9335
III	2	0.2432	0.1974	0.1993	0.2281
III	3	0.0970	0.0875	0.0864	0.1011
III	4	0.0540	0.0474	0.0487	0.0565
III	5	0.0340	0.0304	0.0310	0.0362

It follows from Table 1 that both cubic elements (I,II) provide good estimates for the solution at the test points, whereas the estimates obtained by the quadratic element III are less accurate. For all elements convergence towards the exact solution occurs if the triangular grid is refined.

In Gmelig Meyling (1988), a pentagonal plate is considered having sides of unit length. The plate is clamped along one of its sides, simply supported along two sides, and free at the remaining boundaries. See Figure 3, where three non-uniform triangulations of the plate are displayed.

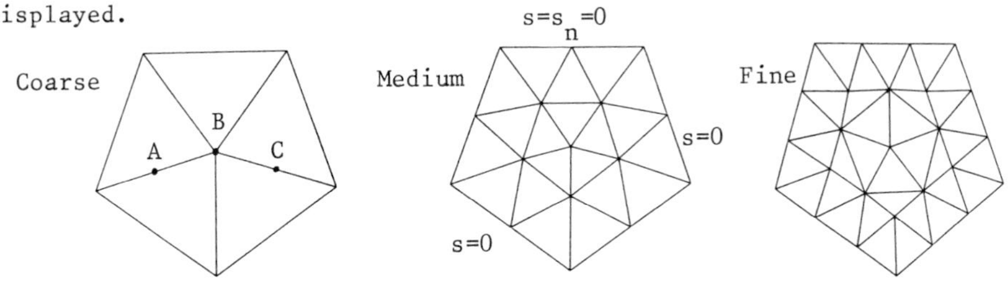

Figure 3 Three triangular partitions of a pentagonal plate

Table 2 lists the vertical displacement at three test points A,B and C using the three elements on triangulations of increased grid fineness. Note that also in this experiment there is considerable agreement between numerical results obtained by different elements. It is difficult to establish which element to use in practical computations. Generally, element II produces the most accurate results. On the other hand, for element I (a single triangle) assembling the stiffness matrix and evaluating the spline can be done very efficiently. Finally, the quadratic element III is most suitable for contour plotting and for numerical integration of (piecewise constant) second order derivatives.

Table 2 Estimated displacement (x100) for three elements

	Coarse mesh			Medium mesh			Fine mesh		
Element	A	B	C	A	B	C	A	B	C
I	1.6646	2.2655	2.1349	1.9532	2.9039	2.3018	2.0117	3.0097	2.4277
II	1.8421	2.7553	2.3076	2.0153	3.0001	2.4310	2.0344	3.0227	2.4687
III	1.6900	2.5962	2.1618	1.9388	2.9063	2.3580	1.9792	2.9575	2.4189

References

Clough, R.W. and Tocher, J.L. (1965), Finite element stiffness matrices for analysis of plates in bending, In 'Conference on Matrix Methods in Structural Mechanics', Wright Patterson, Ohio

Dahmen, W. (1986), Bernstein-Bézier representation of polynomial surfaces, ACM SIGGRAPH86, Dallas

Farin, G. (1986), Triangular Bernstein-Bézier patches, Computer Aided Geometric Design 3, 83-127

Gmelig Meyling, R.H.J. and Pfluger, P.R. (1985), On the dimension of the spline space $S_2^1(\Delta)$ in special cases, In 'Multivariate Approximation Theory III', W. Schempp, eds., Birkhäuser, Basel, 180-190

Gmelig Meyling, R.H.J. (1987), Approximation by cubic C^1-splines on arbitrary triangulations, Numer. Math. 51, 65-85

Gmelig Meyling, R.H.J. (1988), On the construction of conforming finite elements based on smooth piecewise polynomials over triangles, In preparation

Golub, G.H. and van Loan, C.F. (1983), 'Matrix Computations', North Oxford Academic, Oxford

Heindl, G. (1979), Interpolation and approximation by piecewise quadratic C^1-functions of two variables, In 'Multivariate Approximation Theory', W. Schempp and K. Zeller, eds., Birkhäuser, Basel, 146-161

Powell, M.J.D. and Sabin, M.A. (1977), Piecewise quadratic approximation on triangles, ACM Trans. Math. Software 3, 316-325

Schumaker, L.L. (1984), Bounds on the dimension of spaces of multivariate piecewise polynomials, Rocky Mountain J. Math. 14, 251-264

Schwarz, H.R. (1980), 'Methode der finiten Elemente', Teubner, Stuttgart

Traas, C.R. (1988), An application of bivariate B-splines to plate bending problems, In '7th Symposium on Trends in Applications of Mathematics to Mechanics', J.F. Besseling and W. Eckhaus, eds., Wassenaar

Quadrature solution of integral equations: a uniform treatment of Fredholm and Volterra equations

G. Olaofe

Department of Mathematics,
University of Ibadan,
Oyo State, Nigeria

Abstract A differential quadrature method has been presented by Ọlaọfẹ and Mason (1988) for the solution of ordinary and partial differential equations. The method is noteworthy not only for its simplicity and efficiency, but also for its uniformity in the treatment of a wide selection of boundary and initial conditions. In this paper a new integral quadrature method, analoguous to the differential quadrature, is presented for the solution of integral equations. The integral quadrature method, is, similarly, uniformly applicable to a broad range of linear, nonlinear singular, and non-singular Fredholm or Volterra integral or integro-differential equations. Both the differential and integral quadrature methods exploit formulae which link definite or indefinite integrals with an orthogonal polynomial series, with unknown parameters equal to function values at the polynomial zeros. And by using the same nodes for both the differential and integral quadrature methods, extension of the method to solution of integro-differential equations is immediate. Results presented here for simple linear examples demonstrate the remarkably superior accuracy and efficiency of the method over, for example, the finite difference method. For although integral quadrature requires the solution of an $N_Q \times N_Q$ dense linear system, while the finite difference method involves only a triangular $N_F \times N_F$ system, comparable accurary forces N_F to be much greater than N_Q. Thus the finite difference method involves not only several orders more operations but also, in general, several orders more function evaluations of the kernel . Some nonlinear and singular cases will be presented elsewhere.

Keywords : Numerical Quadrature; Differential Quadrature; Integral Quadrature; Quadrature Coefficients; Field or Boundary Nodes; Superfluous Nodes

1. The Differential Quadrature

1.1 Differential Quadrature Equations

The differential quadrature method has been described by Ọlaọfẹ and Mason(1988) for solving a first order initial value ordinary differential equation:

$$dy/dx \quad = \quad f(x,y); \quad x \in [a,b]; \quad y(a) = y_a. \tag{1}$$

by approximating the desired function $y(x)$ globally in the interval $[a,b]$ by the Lagrange interpolating polynomial $p_{N-1}(x)$, of degree $N-1$:

$$y(x) = p_{N-1}(x) + \varepsilon(x) = \sum_{\nu=1}^{N} L_\nu(x)y(x_\nu) + \varepsilon(x). \tag{2}$$

Here

$$\varepsilon(x) = T_N(x)y^{(N)}(x)/N! \tag{3}$$

is the global error of the approximation and is given by

$$\|\varepsilon(x)\| \leq \frac{1}{N!}\|y^{(N)}(x)\| \; \|T_N(x)\|, \tag{4}$$

and $L_\nu(x)$ are the Lagrangian interpolants defined, for $\nu = 1, \ldots, N$, by

$$L_\nu(x) = \prod_{\substack{k=1 \\ k\neq\nu}}^{N} \frac{x - x_k}{x_\nu - x_k} = T_N(x)/[(x - x_\nu)T_N'(x_\nu)], \tag{5}$$

where T_N is defined as the 'product' or associated polynomial,

$$T_N(x) = (x - x_1)(x - x_2)\ldots(x - x_N), \tag{6}$$

which in this paper shall be chosen to be the Chebyshev polynomial of the first kind. Since the Lagrangian interpolants satisfy exactly the equation

$$dL_\nu(x)/dx = \sum_{k=1}^{N} a_{k\nu}^{(1)} L_k(x), \tag{7}$$

and are orthogonal for every orthogonal product polynomial, the differential quadrature equivalent of the initial value problem is a set of N 'field' quadrature equations, one at each 'field' or internal node, $\{x_\nu = \cos(N - \nu + \frac{1}{2})\frac{\pi}{N}\}$, namely,

$$\left.\frac{dy}{dx}\right|_{x\,=\,x_\nu} = \sum_{k=1}^{N} a_{\nu k}^{(1)} Y_k = f(x_\nu, Y_\nu), \quad \nu = 1, 2, \ldots, N, \tag{8}$$

together with the differential quadrature representation of the initial condition

$$y(a) = y_a = \sum_{k=1}^{N} L_k(a)Y_k \tag{9}$$

where Y_k is the computed value of $y(x_k)$. These equations for $\{Y_k\}$ are in general nonlinear and standard Newton-type iteration methods are adopted; for linear problems Eqn.(8) is a linear algebraic system. The procedure for solution is to choose any $N-1$ field quadrature equations (8) together with the one boundary quadrature equation (9). The residual obtained from the remaining 'superfluous' field equation serves as an error check for the quadrature solution.

1.2 The Differential Quadrature Coefficients

From the exact relation:

$$d^\mu L_k(x_\nu)/dx^\mu = a_{\nu k}^{(\mu)}, \tag{10}$$

differential quadrature coefficients $a_{\nu k}^{(\mu)}$ may be computed. (Note that at the nodes, the Lagrangian interpolants reduce to the Kronecker Delta, i.e., $L_k(x_\nu) = a_{\nu k}^{(0)} = \delta_{\nu k}$.) Hence the first order ($\mu = 1$) differential quadrature coefficients are defined by

$$\begin{aligned} a_{i\nu}^{(1)} &= T_N'(x_i)/[(x_i - x_\nu)T_N'(x_\nu)]; & i \neq \nu \\ &= T_N''(x_i)/[2T_N'(x_\nu)]; & i = \nu. \end{aligned} \tag{11}$$

However for $T_N(x) = \cos(N \arccos x)$, and x in the real interval $[-1, +1]$, the relation

$$\frac{T_N(x)}{x - x_j} = \frac{-2}{T_{N+1}(x_j)} \times \sum_{k=0}^{N-1}{}'' T_k(x)T_k(x_j), \tag{12}$$

where the dashes denote that the first term is to be halved, (see Rivlin (1974)), may be exploited to obtain the simplification

$$a_{i\nu}^{(1)} = \frac{2}{N}\sum_{k=0}^{N-1}{}'' T_k'(x_i)T_k(x_\nu). \tag{13}$$

It may also be observed that $\varepsilon(x) = 0$ for polynomials of degree $N-1$ or less and hence by substituting the members of the Chebyshev set $\{y(x) = [1 = x^0, x, x^2, ..., x^{N-1}]\}$ in turn, in equation (2), an alternative representation may be derived in matrix form

$$AV = W \qquad , i.e. \qquad A = WV^{-1} \tag{14}$$

where the matrix of quadrature coefficients A, the Vandermonde matrix V and the matrix of function values W are defined by

$$A = [a_{ij}^{(1)}], \quad V = [v_{ij}] = [x_i^{j-1}], \quad W = [w_{ij}] = [(j-1)x_i^{j-2}]. \tag{15}$$

It is a matter of interest to note that the coefficients $c_{\nu j}$ of the inverse V^{-1} of the Vandermonde matrix V, may be written explicitly in the form

$$c_{\nu j} = \frac{(-1)^{N-j}\sum_{i_p \neq j}^{N} \prod x_{i_1}x_{i_2}x_{i_3}\ldots x_{i_{N-\nu}}}{\prod_{k\neq j}^{N}(x_j - x_k)}. \tag{16}$$

Although this appears to be an entirely new result, this representation is of little practical use except for $N < 10$.

2. The Integro-Differential Equations

2.1 The Integral Quadrature Equations

Quadrature method for solving Fredholm integral equations has been given in the literature (see for example Delves and Walsh(1977), Baker(1977) and, for more general references, Brunner(1986)). In this paper a novel general 'integral' quadrature method is provided for handling both Fredholm and Volterra equations. Consider for simplicity the linear integro-differential equation:

$$\gamma(x)\frac{df}{dx} = f(x) + \int_a^{\rho(x)} K(x,y)f(y)dy + g(x); \quad a \leq x \leq b, \tag{17}$$

where $g(x)$, $\gamma(x)$, the Kernel $K(x,y)$, and $f(a) = f_a$ say, are prescribed, and $f(x)$ is to be determined. For Fredholm type integral equations $\rho(x) = b$ and for Volterra type equations, $\rho(x) = x$. Without loss of generality we set $a = -b = -1$. It is assumed that $\gamma(x), g(x)$ and the kernel $K(x,y)$ are sufficiently differentiable in their respective arguments so that $f(x)$ is also sufficiently differentiable. The differential part of the integro-differential equation has been taken care of by the last section: hence we turn attention to the integral part. If both the integral and the differential terms of (17) can be represented in terms of function values at the same set of polynomial zeros, then a complete quadrature representation is achieved. There is therefore little else to do than to proceed to derive the analogous integral quadrature representation of the integral term of (17) through the use of the following integral quadrature coefficents.

2.2 The Integral Quadrature Coefficients

In a manner analogous to the differential quadrature system described above we may write

$$\int_{-1}^{x} f(\zeta)d\zeta = \sum_{i=1}^{N} L_i(x) \int_{-1}^{x_i} f(\zeta)d\zeta. \tag{18}$$

The Volterra integral quadrature coefficients $\{g_{ij}\}$ are therefore defined by

$$\sum_{j=1}^{N} f_j \int_{-1}^{x_i} L_j(\zeta)d\zeta = \sum_{j=1}^{N} g_{ij} f_j. \tag{19}$$

By appropriate integration and using Eqn. (12), it is easy to show that

$$\begin{aligned} g_{ij} = {} & N^{-1}[1 + x_i + x_j({x_i}^2 - 1) \\ & + \sum_{k=2}^{N-1} T_k(x_j)\{\frac{T_{k+1}(x_i) - (-1)^{k+1}}{k+1} - \frac{T_{k-1}(x_i) - (-1)^{k-1}}{k-1}\}]. \end{aligned} \tag{20}$$

Clearly by setting $x_i = 1$ in the above, the corresponding Fredholm integral quadrature coefficients , $\{g_j^*\}$, are derived from this simplification, namely

$$g_j^* = \frac{2}{N}[1 - \sum_{k=2}^{N-1} T_k(x_j)\frac{1+(-1)^k}{k^2-1}]. \tag{21}$$

We believe that the result (20) is completely new, while (21) is slightly different from the well-known quadrature weights since (21) is exact for polynomials of degree $N-1$, or less. It is of interest to note that the corresponding Volterra integral quadrature coefficients, in the case when the product polynomial is the Legendre Polynomial $P_N(x)$, are given by:

$$g_{ij} = \frac{1}{(1-x_j^2)P'^2_N(x_j)} \sum_{k=0}^{N-1} P_k(x_j)[P_{k+1}(x_i) - P_{k-1}(x_i)]. \tag{22}$$

Again, analogous to the differential quadrature case, an alternative representation of the matrix of integral quadrature coefficients $G = [g_{ij}]$ is given by the symbolic form

$$GV = \Gamma \qquad or \qquad G = \Gamma V^{-1}, \tag{23}$$

where V has been defined in (14) and (15) above and

$$\Gamma = [\Gamma_{ij}] = [x_i^j - (-1)^j]/j. \tag{24}$$

2.3 Error Estimate

In a manner similar to that used for estimating the differential quadrature error, the integral quadrature error may be estimated by the Lanczos tau-method. Consider for example equation (17) for $\gamma(x) = 0$. Since the quadrature solution is exact for the perturbed equation

$$F(x) + \int_a^{\rho(x)} K(x,y)F(y)dy + g(x) = \tau T_N^*(x), \tag{25}$$

where $F(x)$ is the quadrature approximation to $f(x)$, an estimate of the error $|\varepsilon(x)|$ may be derived from

$$\varepsilon(x) + \int_a^{\rho(x)} K(x,y)\varepsilon(y)dy = \tau T_N^*(x), \tag{26}$$

from which we observe that, if $F_\nu = F(x_\nu)$, then

$$|\tau| = |\varepsilon(-1)| = |g(a) - \sum_{\nu=1}^{N} L_\nu(-1)F_\nu|. \tag{27}$$

Clearly, if $K(x,y)$ and its derivatives are bounded, it can be shown that $|\varepsilon(-1)| = O(\tau)$.

3. Reciprocal Operators A and G

Next the relationship between the differential quadrature operator A and the integral quadrature operator G is examined. By writing the differential quadrature equations (8) and (9) in matrix form :

$$A\mathbf{y} = \mathbf{f}; \qquad \Lambda\mathbf{y} = \mathbf{z} \tag{28}$$

and the integral quadrature equations in the form :

$$\mathbf{y} = \mathbf{z} + G\mathbf{f}. \tag{29}$$

equation (28) may be substituted into (29) to derive

$$GA = I - \Lambda, \tag{30}$$

where I is the identity operator and Λ is the singular matrix operator with all rows equal to $\{L_1(-1), L_2(-1), ..., L_N(-1)\}$. Similarly upon substituting (29) into (28) we derive

$$AG = I. \tag{31}$$

Alternatively, these relations may be proved by noting that

$$[GA]_{ik} = \sum_{j=1}^{N} g_{ij} a_{jk}^{(1)}, \tag{32}$$

from which it follows that

$$[GA]_{ik} = \sum_{j=1}^{N} \int_{-1}^{x_j} L_j(\eta) L'_k(\eta_j) d\eta, \tag{33}$$

the right hand side of which sums by (2) to

$$[GA]_{ik} = \int_{-1}^{x_i} L'_k(\eta) d\eta = \delta_{ik} - L_k(-1). \tag{34}$$

Similarly

$$[AG]_{ik} = \sum_{j=1}^{N} a_{ij}^{(1)} g_{jk} \tag{35}$$

$$[AG]_{ik} = \frac{d}{dx} \sum_{j=1}^{N} L_j(x) \int_{-1}^{x_j} L_k(\eta) d\eta \,\Big|_{x = x_i} \tag{36}$$

Hence the result follows from

$$[AG]_{ik} = \frac{d}{dx} \int_{-1}^{x} L_k(\eta) d\eta \,\Big|_{x = x_i}. \tag{37}$$

4. Numerical Results

Table 1 shows, for the following three linear Fredholm equations, or Problems 1 to 3 (Equations (38) - (40)) , the maximum errors of the quadrature solutions in comparison with the exact or analytical solutions which are given in the adjoining square brackets, namely

$$1 : \quad f(x) - \int_0^1 (x+y)f(y)dy = e^x \quad : [f(x) = e^x - 6x(1+e) - (4e+2)]\,; \quad (38)$$

$$2 : \quad f(x) - \int_0^1 (1+xy)f(y)dy = e^x \quad : \left[f(x) = e^x + \frac{2}{3} - \frac{8e}{3} + 2x(1-e)\right];(39)$$

$$3 : \quad f(x) - \int_0^1 |(x-y)|f(y)dy = -(2x^3+2-9x)/6 \quad : [f(x) = x]\,. \quad (40)$$

Table 1: Errors of Quadrature Solution of Fredholm Problems

Problem No.	N=4	N=8	N=12	N=16
1	0.33649E-02	0.32629E-08	0.12275E-05	0.10755E-13
2	0.50414E-03	0.43358E-09	0.56050E-15	0.10319E-14
3	0.16283E-16	0.15425E-16	0.13965E-16	0.14260E-16

(For this and other tables, numerical calculations have been performed on the VAX 8700 at RMCS using the NAG subroutine F04ATF.) Comparison is made for values of $N = 4, 8, 12$ and 16 in all cases : while Problem Nos. 2 and 3 show increasing accuracy with increasing N, the first produces, curiously, a better result for $N = 8$ than for $N = 12$. This needs to be further investigated. Table 2 also shows the corresponding maximum errors of the quadrature solutions of nine linear Volterra equations , Problems 4 to 12 (Equations (41) to (49)) :

$$4: \quad f(x) - 17\int_0^x [\tfrac{1}{1+9y}f(y)]dy = 1 \quad : \left[f(x) = e^{\{\frac{17}{9}\ln(1+9x)\}}\right] \quad (41)$$

$$5: \quad f(x) + \int_0^x (x-y)\cos(x-y)f(y)dy = \cos(x) : \left[f(x) = \tfrac{1}{3}(2\cos x\sqrt{3}+1)\right] (42)$$

$$6: \quad f(x) - \int_0^x \sin(x-y)f(y)dy = x \quad : \left[f(x) = x + \tfrac{1}{6}x^3\right] \quad (43)$$

$$7: \quad f(x) + \int_0^x (x-y)f(y)dy = \sin(x) \quad : \left[f(x) = \tfrac{1}{2}\{\sin(x)+\sinh(x)\}\right] \quad (44)$$

$$8: \quad f(x) - 2\int_0^x \cos(x-y)f(y)dy = e^x \quad : [f(x) = e^x(1+x)^2] \quad (45)$$

$$9: \quad f(x) - \int_0^x f(y)dy = \cos(x) \quad : \left[f(x) = \tfrac{1}{2}\{\sin(x)+\cos(x)+e^x\}\right] \quad (46)$$

$$10: \quad f(x) + \int_0^x \cosh(x-y)f(y)dy = \sinh(x) \quad : \left[f(x) = \tfrac{2}{\sqrt{5}}\sinh(\tfrac{\sqrt{5}}{2}e^{-\frac{1}{2}x})\right] \quad (47)$$

$$11: \quad f(x) - \lambda\int_0^x f(y)dy = 1; \ \lambda = -12 \quad : [f(x) = e^{-12x}] \quad (48)$$

$$12: \quad f(x) - \tfrac{1}{5}\int_0^x xyf(y)dy = x \quad : \left[f(x) = xe^{x^3/15}\right]. \quad (49)$$

Table 2: Errors of Quadrature Solution of Volterra Problems

Problem No.	N=4	N=8	N=12	N=16
4	0.48023E-01	0.29644E-02	0.47616E-04	0.99003E-06
5	0.92562E-03	0.13809E-06	0.21008E-11	0.22819E-16
6	0.39074E-04	0.26523E-08	0.23518E-14	0.18341E-17
7	0.30562E-04	0.57248E-10	0.44173E-16	0.37960E-16
8	0.14571E-01	0.66781E-07	0.76776E-13	0.22008E-15
9	0.11458E-03	0.96418E-10	0.90924E-16	0.98424E-16
10	0.24740E-03	0.14549E-07	0.14883E-12	0.33577E-16
11	0.14366E-00	0.18994E-02	0.58227E-05	0.67077E-08
12	0.41118E-04	0.12541E-08	0.19909E-08	0.18362E-16

Since Problem 4 converges slowly with increasing N, it is of interest to note that the maximum error is 0.15568E-11 when $N = 32$. Equations (42) to (47) are those given on page 785 of Baker(1977) whilst Equation (48) can be found on page 781. Using Table 6.5 of Baker, the efficiency and accuracy of our quadrature solutions are compared in Table 3 with the finite difference method. Here N_F denotes the number of nodes used in the finite difference solution while N_Q denotes the corresponding number for the integral quadrature solution. Although the finite difference algorithm is explicit, and requires, apart from function evaluations, only $O\{\frac{1}{2}N_F^2\}$ operations for each problem, the integral quadrature method correspondingly involves solving a dense matrix equation requiring $O\{\frac{1}{3}N_Q^3\}$ operations. However, as Table 3 shows, for the same tolerance or accuracy for both methods, $N_F >> N_Q$ with the result that $\frac{1}{2}N_F^2$ is much bigger than $\frac{1}{3}N_Q^3$, and indeed, for the general kernel, such as $\cos(x-y)$ in Eqn. (45), the finite difference method would involve far more function evaluations ($N_F^2/2$) than would the integral quadrature method (N_Q^2).

Table 3: Comparison of N_F with N_Q

Problem No.	N_Q=4	N_Q=8	N_Q=12	N_Q=16
5	N_F= 4	14	147	> 200
6	N_F= 7	90	> 200	
7	N_F= 7	> 200	> 200	
8	N_F= 6	120	> 200	
9	N_F= 4	110	> 200	
10	N_F= 6	48	> 200	
11	N_F= 7	35	120	> 200

Table 4 shows the estimate of the maximum error for Problems (5) to (11) (Equations (42)-(49) mentioned above). Since $||K||_1 \leq 1$ for Problems (5), (7) and (8), $||\varepsilon||$ is easily determined from equation (26) and is $O(\tau)$. For the rest, iteration by Picard's method shows that $||\varepsilon||$ differs from $O(\tau)$ by a factor $\frac{1}{N}$.

Table 4: A Posteriori Error Estmates :$||\varepsilon||$

Problem No.	$\|\|\varepsilon\|\|$	True Max. Error	$\|\tau\|$	$\|f(0) - F(0)\|$
5	0.4757E-11	0.2110E-11	0.2941E-11	0.2034E-11
6	0.5020E-14	0.2352E-14	0.2510E-14	0.1789E-14
7	0.4000E-16	0.4417E-16	0.2000E-16	0.1261E-16
8	0.7191E-12	0.7678E-13	0.1140E-12	0.7447E-13
9	0.1069E-15	0.9092E-16	0.2000E-16	0.1290E-16
10	0.2109E-12	0.1488E-12	0.1137E-12	0.8312E-13
11	0.7476E-05	0.5823E-05	0.3738E-05	0.4695E-05

In conclusion we claim that the integral quadrature method, much like the differential quadrature method, is extremely efficient and accurate and constitutes a considerable saving of time and effort.

5. Acknowledgement

The author is greatly indebted to the Royal Military College of Science at Shrivenham for providing financial assistance, including extensive computational facilities while on leave at Shrivenham, and to the Nigerian Federal Ministry of Science and Technology for giving a travel grant to attend the Conference. The author is also particularly grateful to Professor John Mason at RMCS for technical advice and other support during the preparation of this work.

6. References

Baker, C. T. H. (1977)

The Numerical Treatment of Integral Equations.
Oxford University Press, Oxford

Brunner, H. (1986)

On the History of Numerical Methods for Volterra Integral Equations
CWI Newsletter, Number 11, Centre for Mathematical and Computer Sciences, Kruislaan 413, 1098 SJ Amsterdam, The Netherlands.

Delves, L. M., and Walsh, J. (1974)

Numerical Solution of Integral Equations.
Clarendon Press, Oxford.

Ọlaọfẹ, G. Olurẹmi, and Mason, John C. (1988)

Quadrature Solution of Ordinary and Partial Differential Equations,
Mathematical and Computer Modelling, Vol II pp 652-655
(Proceedings of the 6th International Conference on Mathematical Modelling, held in St. Louis, USA, 3-7 August, 1987).
Pergamon Press, Oxford.

Rivlin, T. J.(1974)

The Chebyshev Polynomials
John Wiley and Sons

Increasing the convergence modulus of an asymptotic expansion: an algorithm for numerical differentiation

G. Walz
Fakultät Mathematik und Informatik,
Universität Mannheim, FRG

Abstract: Very often in numerical analysis one is concerned with sequences of numbers or functions $\{\sigma_n\}_{n\in IN}$, which converge to a given value (or function) F and possess an asymptotic expansion with respect to the index n. In this case one can apply several well-known extrapolation methods in order to obtain faster converging sequences und thus good approximations for the value F. The aim of this paper is the presentation of a method for the construction of a.e.'s with arbitrarily large modulus. We will apply the method to the numerical differentiation of functions, or, in other words, the approximation of the function $\frac{d^p}{dz^p}(f(z))$ for given $f(z)$ and $p \geq 1$.
Key words: Asymptotic expansion, Extrapolation, Numerical Differentiation.

1. Definitions and introductory examples

We start with a precise definition of the type of asymptotic expansion we shall make use of.

Definition 1: Let I denote a real interval or a complex domain, $\eta > 0$ a real number and $N \geq 1$ an integer; furthermore, let $\{\sigma(n, z)\}_{n\in IN}$ denote a sequence of functions defined on I.

This sequence possesses an *asymptotic expansion (a.e.)* of *order* N with *(convergence) modulus* η, if there exist functions $F(z)$ and $c_\nu(z)$, $\nu = 1, \ldots, N$, all defined on I and independent of n, such that for all $z \in I$ the following relation holds:

$$\sigma(n,z) = F(z) + \sum_{\nu=1}^{N} \frac{c_\nu(z)}{n^{\eta\cdot\nu}} + O(n^{-\eta(N+1)}) \qquad \text{for } n \to \infty . \tag{1}$$

We of course always suppose η to be maximal, which means that there is no $\eta_1 > \eta$ with the property (1).

Before introducing the extrapolation procedure, we give two simple, but instructive examples of such a.e.'s:

Examples: 1. Let z be an arbitrary complex number, and define

$$\sigma(n,z) := (1 + z/n)^n , \quad n \in \mathbb{N} .$$

Then

$$\sigma(n,z) = \exp(z) + \sum_{\nu=1} \frac{c_\nu(z)}{n^\nu} ,$$

thus σ possesses an a.e. with modulus 1. We omitted the upper summation limit as well as the "O"-term to indicate that this is an expansion of arbitrary order. For a proof, see Walz (1987).

2. (Cp. Rutishauser (1963)). Let f be a function, which is analytic in a neighbourhood of the point $z \in \mathbb{C}$, and define for all $n \in \mathbb{N}$ the difference operators

$$\delta^1(n) := \delta^1(n,f,z) := n \cdot (f(z + \frac{1}{n}) - f(z)) .$$

Using the series expansion of f in $z + 1/n$, it is easily seen that

$$\delta^1(n,f,z) = f'(z) + \sum_{\nu=1} \frac{c_\nu(z)}{n^\nu} ,$$

which means that $\delta^1(n)$ possesses an a.e. with modulus 1. In section 3 we will come back to this example in some detail.

Given an a.e. of type (1), we can define the following extrapolation procedure ($K \in \mathbb{N}$, $K < N$):

$$\left.\begin{aligned} y_i^{(0)}(z) &= \sigma(2^i, z) , \qquad i = 0(1)K \\ y_i^{(k)}(z) &= y_{i+1}^{(k-1)}(z) + \frac{1}{2^{\eta\cdot k} - 1} \cdot (y_{i+1}^{(k-1)}(z) - y_i^{(k-1)}(z)) \; \begin{cases} k = 1(1)K \\ i = 0(1)K - k, \end{cases} \end{aligned}\right\} \tag{2}$$

and it is well-known (cp. Joyce (1971), Meinardus (1979)) that the resulting sequences $\{y_i^{(k)}(z)\}$ possess the a.e.

$$y_i^{(k)}(z) = F(z) + \sum_{\nu=k+1}^{N} \frac{c_\nu^{(k)}(z)}{2^{i\cdot\eta\cdot\nu}} + O(2^{-i\cdot\eta(N+1)}) \qquad \text{for } i \to \infty .$$

Asymptotic expansions and the related extrapolation methods appear in many fields of numerical analysis, and their properties were investigated over the years; we refer to the papers of Joyce (1971), Meinardus (1979) and Walz (1987).

The aim of this paper is to present a method for the construction of a.e.'s of type (1) with arbitrarily large modulus. We will apply the method to the numerical differentiation of functions, or, in other words, the approximation of the function $\frac{d^p}{dz^p}(f(z))$ for given $f(z)$ and $p \geq 1$.

2. The E-method

Consider a given sequence of functions $\sigma(n,z)$ of type (1) with $\eta = 1$ and *arbitrary* order N, thus

$$\sigma(n,z) = F(z) + \sum_{\nu=1} \frac{c_\nu(z)}{n^\nu} , \quad n \in \mathbb{N} \tag{3}$$

(this type occurs in most applications). In many cases (cp. Rutishauser (1963)) the functions

$$\tilde{\sigma}(n,z) := \frac{1}{2}(\sigma(n,z) + \sigma(-n,z)) \tag{4}$$

are well-defined and possess an a.e. with modulus 2, see also Walz (1987), Bem. 3.11. This construction can be generalized to the following mean-value method:
We denote, for fixed $m \in \mathbb{N}$, by ς_m the primitive $m-$th root of unity:

$$\varsigma_m := \exp(\frac{2\pi i}{m}) \qquad (i := \sqrt{-1}) .$$

With the function σ from (3) we define for $\mu = 0(1)m-1$:

$$\sigma^{[\mu]} := \sigma^{[\mu]}(n,z) := \sigma(\varsigma_m^\mu \cdot n, z) , \tag{5}$$

where it is assumed that the values on the right-hand side are well-defined; then these $\sigma^{[\mu]}$ also possess an a.e. with modulus 1:

$$\sigma^{[\mu]} = F(z) + \sum_{\nu=1} \frac{c_\nu(z)}{\varsigma_m^{\mu\nu}} \cdot \frac{1}{n^\nu}$$

$$= F(z) + \sum_{\nu=1} \frac{c_\nu^{[\mu]}(z)}{n^\nu} \tag{6}$$

with $|c_\nu^{[\mu]}(z)| = |c_\nu(z)|$ for all $z \in I$ and $\nu = 1(1)\infty$, $\mu = 0(1)m-1$ (examples will be given later). Using these auxiliary functions, we now construct an expansion with modulus m :

Theorem 2: Let $\{\sigma(n,z)\}$ be a sequence of functions of type (3) such that $\sigma^{[\mu]}(n,z)$, $\mu = 0(1)m-1$, is well-defined by (5) and possesses an a.e. of type (6) $(m \in \mathbb{N})$; then for all $n \in \mathbb{N}$ and $z \in I$ the functions

$$\Sigma^{[m]}(n,z) := \frac{1}{m} \cdot \sum_{\mu=0}^{m-1} \sigma^{[\mu]}(n,z) \tag{7}$$

possess an a.e. of the form (1) with modulus m.

Proof: This is just an application of properties of ς_m , in particular that

$$\sum_{\mu=0}^{m-1} \frac{1}{\varsigma_m^{\mu\nu}} = \begin{cases} m & , \quad \text{if } m|\nu \text{ , hence } \varsigma_m^\nu = 1 \\ 0 & \quad \text{otherwise .} \end{cases}$$

We have

$$\begin{aligned} \Sigma^{[m]}(n,z) &= \frac{1}{m} \cdot \sum_{\mu=0}^{m-1} \sigma^{[\mu]}(n,z) \\ &= F(z) + \frac{1}{m} \sum_{\mu=0}^{m-1} \sum_{\nu=1} \frac{c_\nu(z)}{\varsigma_m^{\mu\nu}} \cdot \frac{1}{n^\nu} \\ &= F(z) + \frac{1}{m} \sum_{\nu=1} \frac{c_\nu(z)}{n^\nu} \cdot \sum_{\mu=0}^{m-1} \frac{1}{\varsigma_m^{\mu\nu}} \\ &= F(z) + \sum_{\lambda=1} \frac{c_{m\lambda}(z)}{n^{m\lambda}} . \end{aligned}$$

Obviously, there are no restrictions concerning the integer m , so (7) defines a method for the construction of a.e.'s with *arbitrarily large modulus*. Due to the strong connection of this method with the roots of unity (= "Einheitswurzeln" in German), it is called *E-method*.

3. An Application: Numerical Differentiation

In order to illustrate the efficiency of the E–method, we will now apply it in some detail to the numerical differentiation operator δ^1 of example 2. It will turn out that this yields – in combination with the procedure (2) – quite powerful algorithms.

Obviously, the sequence $\{\delta^1(n)\}$ approximates $f'(z)$ by evaluating the behavior of $f(w)$, as w approaches z parallel to the real axis. But f is an analytic function, so we could as well let w approach z from other sides, and, using the resulting values of $f(w)$, we can be quite sure that we end up with a better approximation of $f'(z)$. This is precisely what the E–method does:

Lemma 3: Let ς_m again denote the primitive m– th root of unity and define, according to (7), the difference operator

$$\Delta^{1,m}(n,f,z) := \frac{1}{m} \cdot \sum_{\mu=0}^{m-1} \delta^1(\varsigma_m^\mu \cdot n, f, z) \quad ; \tag{8}$$

then $\Delta^{1,m}$ has, according to theorem 2, an a.e. with modulus m :

$$\Delta^{1,m}(n,f,z) = f'(z) + \sum_{\nu=1} \frac{c_{m\nu}(z)}{n^{m\nu}} \quad . \tag{9}$$

Remark 4: If we replace in (8) δ^1 by the term $n(f(z+1/n)-f(z))$, we get

$$\Delta^{1,m}(n,f,z) = \frac{1}{m} \cdot \sum_{\mu=0}^{m-1} \varsigma_m^\mu \cdot n \cdot (f(z+\frac{1}{\varsigma_m^\mu \cdot n}) - f(z)) \ . \tag{10}$$

This illustrates the assertions from above: each $\Delta^{1,m}(n,f,z)$ evaluates the function f on m equidistant points on a circle around z with radius $\frac{1}{n}$. In the above notation : it evaluates the behavior of $f(w)$, as w approaches z from m different sides.

Remark 5: Using the integration path $\gamma_n(t) := z + \frac{1}{n}e^{-it}$, $n \in I\!N$, $0 \le t \le 2\pi$, the CAUCHY integral formula for the first derivative of f reads

$$f'(z) = -\frac{1}{2\pi} \int_0^{2\pi} e^{it}\, n \cdot (f(z+\frac{1}{e^{it} n}) - f(z))\, dt \ . \tag{11}$$

Thus (10) is precisely the discrete version of (11); $\Delta^{1,m}$ could therefore be denoted as *discrete CAUCHY integral formula.*

There is no need to restrict ourselves to the approximation of the *first* derivative:

Lemma 6: Let $p \geq 1$ be an integer and define for all $n \in \mathbb{N}$ the operator

$$\delta^p(n,f,z) := n^p \cdot \sum_{\nu=0}^{p} \binom{p}{\nu} (-1)^\nu \cdot f(z + \frac{p-\nu}{n}) \tag{12}$$

Then we have for all $n \in \mathbb{N}$

$$\delta^p(n,f,z) = f^{(p)}(z) + \sum_{\nu=1} \frac{c_\nu(p,f,z)}{n^\nu} . \tag{13}$$

Here $f^{(p)}$ denotes the $p-$th derivative of f, and the $c_\nu(p,f,z)$ are certain functions, which are not depending on n.

Proof: Formula (12) results from a $p-$fold application of the operator δ^1; therefore (13) is evident.

Now again the E-method can be applied:

Theorem 7: Let f be an function which is analytic in a neighbourhood of the point $z \in \mathbb{C}$, and p and m integers ≥ 1; furthermore, let ζ_m again denote the primitive $m-$th root of unity. Define for all $n \in \mathbb{N}$

$$\Delta^{p,m}(n,f,z) := \frac{n^p}{m} \cdot \sum_{\nu=0}^{p} \binom{p}{\nu} (-1)^\nu \sum_{\mu=0}^{m-1} \zeta_m^{p\mu} f(z + \frac{p-\nu}{\zeta_m^\mu n}) \tag{14a}$$

or, in a more symmetric form,

$$\tilde{\Delta}^{p,m}(n,f,z) := \frac{1}{m} \left(\frac{n}{2}\right)^p \cdot \sum_{\nu=0}^{p} \binom{p}{\nu} (-1)^\nu \sum_{\mu=0}^{m-1} \zeta_m^{p\mu} f(z + \frac{p-2\nu}{\zeta_m^\mu n}) . \tag{14b}$$

Then both $\Delta^{p,m}$ and $\tilde{\Delta}^{p,m}$ possess an a.e. of the form

$$\Delta^{p,m}(n,f,z) = f^{(p)}(z) + \sum_{\nu=1} \frac{c_\nu(p,f,z)}{n^{m\nu}} .$$

In connection with extrapolation procedures like (2), the operators given by (14) are powerful instruments for numerical differentiation. We illustrate this by the following

<u>Example:</u> We want to compute the second derivative of a function f in a point $z \in \mathbb{C}$. An approximate solution of this problem is given by

$$\begin{aligned}
&\tilde{\Delta}^{2,4}(n,f,z) \\
&= \frac{1}{4}\left(\frac{n}{2}\right)^2 \cdot \sum_{\nu=0}^{2} \binom{2}{\nu} (-1)^{\nu} \sum_{\mu=0}^{3} (-1)^{\mu} f\left(z + \frac{2-2\nu}{i^{\mu}\, n}\right) \\
&\underset{(n \to 2n)}{=} \cdots = \frac{n^2}{2} \cdot \left[f\left(z+\frac{1}{n}\right) + f\left(z-\frac{1}{n}\right) - f\left(z-\frac{i}{n}\right) - f\left(z+\frac{i}{n}\right) \right] \qquad (15) \\
&= f^{(2)}(z) + \sum_{\lambda=1} \frac{\gamma_{\lambda}(z)}{n^{4\lambda}} \quad . \qquad (16)
\end{aligned}$$

The computation of (15) essentially requires 4 evaluations of the function f; with this input only, it yields an approximating sequence whose error tends to 0 with n^{-4} and, what is even more important in view of the application of an extrapolation procedure, has an a.e. with modulus 4.

For $f(z) = \sin(z)$ and $z = (1+i)\cdot\frac{\pi}{2}$, tables 1 and 2 show the *deviations* of the $y_i^{(k)}(z)$, computed using (15) and (2), to the true value $f^{(2)}(z) = -2.50917847\ldots$ for $k = 0(1)3$, $i = 0(1)4-k$, (1), and the *quotients* of two subsequent errors (2). Obviously, these quotients tend to $2^{(k+1)4}$, as predicted by (16).

<u>Table 1: Errors of the extrapolated values $y_i^{(k)}$</u>

i \ k	0	1	2	3
0	.6971323E-02	.8643661E-07	.1405399E-13	.4672005E-22
1	.4356267E-03	.3376290E-09	.3431103E-17	.7115464E-27
2	.2722635E-04	.1318860E-11	.8376708E-21	
3	.1701646E-05	.5151796E-14		
4	.1063528E-06			

Table 2: Quotients of two subsequent errors from above

i \ k	0	1	2	3
0	.1600298E 02			
		.2560106E 03		
1	.1600019E 02		.4096056E 04	
		.2560007E 03		.6565988E 05
2	.1600001E 02		.4096003E 04	
		.2560000E 03		
3	.1600000E 02			

References

Joyce, D.C. (1971). *Survey of Extrapolation Processes in Numerical Analysis.* SIAM Review 13, 435 – 488

Meinardus, G. (1979), *On the Asymptotic Behavior of Iteration Sequences.* Proceedings of the Fifth South African Symposium on Numerical Analysis, Durban

Rutishauser, H. (1963) *Ausdehnung des Rombergschen Prinzips.* Numerische Mathematik 5, 48 – 54

Walz, G. (1987), *Approximation von Funktionen durch asymptotische Entwicklungen und Eliminationsprozeduren.* Dissertation, Universität Mannheim

Approximation and parameter estimation in ordinary differential equations

J. Williams
*Department of Mathematics,
University of Manchester, UK*

Abstract The important problem of determining unknown parameters in systems of ordinary differential equations will be discussed. Many such problems arise naturally in the mathematical modelling of processes. From observed values of the dependent variables the problem leads to nonlinear least squares fitting. Here particular emphasis is given to the conditioning of the underlying problem with respect to the unknown parameters. Aspects of the mathematical software are also discussed and illustrative examples are described.
Key words: Parameter estimation, Mathematical modelling, Ordinary differential equations, Least squares, Conditioning.

1. The parameter estimation problem

1.1 Description of the problem

The mathematical model to be treated is in the form of a system of ordinary differential equations,

$$y' = f(y,p,t), \qquad t \in [0,T] \tag{1}$$

where the dependent variables consist of $y = (y_1, y_2, \ldots, y_m)^T$ and

$p = (p_1, p_2, \ldots, p_n)^T \in R$ are the unknown parameters. Here $R \subset \mathbb{R}^n$ consists of the set of feasible parameters and may incorporate requirements such as positivity in some components p_i. The initial conditions may be known but it is more likely that they are unknown and take the form $y(0) = g(p)$. The data consists of the results of N experiments at times $0 \leqslant t_1 < t_2 < \ldots < t_N \leqslant T$ and yield the values $Y_1, Y_2, \ldots, Y_N$, which we may regard as being approximations to the "true", but unknown function $Y(t)$ which describes the process exactly for $t \in [0, T]$. From this data where $mN > n$, the idea is to either determine if the model is correct (model discrimination) or, for a known correct model $(y(p,t) \equiv Y(t)$ for some $p)$ estimate p for use in subsequent prediction. In either case the objective here is to investigate if a model is reasonably well suited to parameter estimation and we follow closely the approach of Varah (1987).

1.2 The fitting criterion, maximum likelihood

The most widely used fitting criterion is by way of maximum likelihood analysis, Bard (1974). This leads to the minimisation with respect to $p \in R$ of an objective function whose form depends upon the assumptions regarding the structure of the error in the data $\{t_i, Y_i\}_1^N$. Firstly, under the assumption of a correct model the experimental error satisfies $Y_i = y(p,t_i) + \varepsilon_i$, $i = 1,2,\ldots,N$ (t_i always assumed exact). Secondly we assume the following:

i) experimental errors at different times are uncorrelated,

ii) the errors at each t_i are normally distributed with zero mean and covariance matrix V_i.

The maximum likelihood estimate of p is now given by p^*, where $S_N(p^*) \leqslant S_N(p)$ for all $p \in R$, and

$$S_N(p) = \sum_{i=1}^{N} (Y_i - y(p,t_i))^T V_i^{-1} (Y_i - y(p,t_i)) \qquad (2)$$

In practice a further common assumption is that each V_i is diagonal with different variances in which case the objective function $S_N(p)$ reduces to the familiar weighted sum of squares. If $V_i = \text{diag}\left[\sigma_{i1}^2, \sigma_{i2}^2, \ldots, \sigma_{im}^2\right]$, then in the absence of any other information a practical estimate of V_i may take the form, $\sigma_{ik} = c_k Y_{ik}$, where Y_{ik} is the kth component of the observed value Y_i and c_k may be estimated from the data. This form reflects relative errors in the experiments whereas $\sigma_{ik} = c_k$, could be used in the case of absolute errors.

1.3 Description of the exact problem

For each $k = 1,2,\ldots,m$ let $\{w_{ik} > 0 : i = 1,2,\ldots,N\}$ denote the weights of an N-point quadrature rule with abscissae $0 \leqslant t_1 < t_2 < \ldots < t_N \leqslant T$, so that for $g \in C[0, T]$

$$\int_0^T g(t)dt - \sum_{i=1}^{N} w_{ik}\, g(t_i) \longrightarrow 0 \quad \text{as} \quad N \longrightarrow \infty$$

For each k the rate of convergence will depend on the degree of precision of the quadrature rule and on the precise smoothness properties of the integrand. Let $V(t) = \text{diag}(\sigma_1{}^2(t), \sigma_2{}^2(t), \ldots, \sigma_m{}^2(t))$ where $\sigma_k \in C[0,T]$ and define for $p \in R$, the model decrepancy with respect to the unknown "true" $Y(t)$, by

$$I(p) = \int_0^T (Y(t) - y(p,t))^T\, V(t)^{-1}(Y(t) - y(p,t))dt$$

$$= \sum_{k=1}^{m} \int_0^T \left[\frac{Y_k(t) - y_k(p,t)}{\sigma_k(t)}\right]^2 dt \qquad (3)$$

It is now assumed that the model is correct so that $Y(t) \equiv y(p^*,t)$, where $\min_{p \in R} I(p) = I(p^*) = 0$. In the case of relative errors in the experiment the variances will be obtained from $\sigma_k(t) = c_k y_k(p^*,t)$. This is the idealised exact curve fitting problem which may be approximated using the above quadrature rules to give $I(p) \simeq I_N(p)$,

$$I_N(p) = \sum_{k=1}^{m} \sum_{i=1}^{N} w_{ik} \left[\frac{Y_k(t_i) - y_k(p,t_i)}{\sigma_k(t_i)}\right]^2$$

$$= \sum_{i=1}^{N} (Y(t_i) - y(p,t_i))^T W_i (Y(t_i) - y(p,t_i)) \qquad (4)$$

Now $Y(t_i)$ are exact values (assuming no experimental errors) and $W_i = \mathrm{diag}(w_{i1}, w_{i2}, \ldots, w_{im}) V(t_i)^{-1}$. It follows that for $p \epsilon R$, $I_N(p) \longrightarrow I(p)$ as $N \longrightarrow \infty$, and so the true problem may be approximated arbitrarily closely by choosing N sufficiently large.

In (2) the diagonal weighting matrices V_i^{-1} are associated with variances in the ith experiment. In general a given set of quadrature rules may be interpreted as introducing by way of W_i fictitious variances but subsequent statistical analysis would be invalid. One case, however, which overcomes this difficulty is when the weights are all equal, the obvious rules being the simple Riemann sums, with $h = T/N$,

$$R_N(g) = h \sum_{i=1}^{N} g(ih), \quad \bar{R}_N(g) = h \sum_{i=1}^{N} g((i-1)h), \quad M_N(g) = h \sum_{i=1}^{N} g((i-\tfrac{1}{2})h)$$

The first two rules converge like $O(N^{-1})$ for $g \epsilon C^1[0, T]$ and the midpoint rule like $O(N^{-2})$ for $g \epsilon C^2[0,T]$. For each of these rules (4) takes the form

$$I_N(p) = h \sum_{i=1}^{N} (Y(t_i) - y(p,t_i))^T V(t_i)^{-1} (Y(t_i) - y(p,t_i)) \qquad (6)$$

with appropriate t_i. Apart from the factor h, this compares directly with the original problem (2) and may therefore be used to examine the conditioning with respect to p of the exact problem (3).

2. Model sensitivity

2.1 Sensitivity coefficients

Of considerable importance in the formulation of the model (1) is the sensitivity of the solution $y(p,t)$ with respect to changes in the parameters p. Assuming appropriate differentiability,

$$y_i(p+\delta p,t) = y_i(p,t) + \sum_{j=1}^{n} \frac{\partial y_i}{\partial p_j} \delta p_j + O(||\delta p||^2)$$

The first order sensitivity coefficients are defined by

$$s_i = \left[\frac{\partial y_1}{\partial p_i}, \frac{\partial y_2}{\partial p_i}, \ldots, \frac{\partial y_m}{\partial p_i}\right]^T, \quad i = 1,2,\ldots,n$$

and may be regarded as a measure of the <u>local</u> sensitivity of the solution at (p,t) with respect to <u>small</u> changes in p. Differentiating the model equations yields the variational equations from which the s_i may be computed, $i = 1,2,\ldots,n$,

$$s_i' = \left[\frac{\partial f_i}{\partial y_j}(y,p,t)\right] s_i + b_i, \quad b_i = \left[\frac{\partial f_1}{\partial p_i}, \frac{\partial f_2}{\partial p_i}, \ldots, \frac{\partial f_m}{\partial p_i}\right]^T \tag{7}$$

The associated initial conditions in the standard case where a component of $y(0)$ is either a parameter value or a known value, gives $s_i'(0)$ whose elements are either 1 or 0. Here the s_i, will be needed in carrying out the minimization of $S_N(p)$. Problems with small m and n will be considered here for which it is practicable to combine (1) and (7) and solve as one augmented system.

3. Conditioning with respect to the parameters

3.1 Conditioning of the nonlinear least squares problem

The expression (6) may be writen as a sum of squares

$$I_N(p) = \sum_{i=1}^{N} ||Y(t_i) - y(p,t_i)||_{V_i}^2 = \sum_{j=1}^{mN} r_j^2(p) = ||r(p)||_2^2, \quad r \in \mathbb{R}^{mN} \tag{8}$$

for the appropriate inner product norms defined with respect to $hV_i^{-1} = D_i^2$. The considerations relating to the conditioning of this problem are well known, see Björck (1987) for both methods and software. The Jacobian J of $r(p)$ consists of the blocks $J_i = D_i[s_1\ s_2\ \ldots\ s_n]$, $i = 1,2,\ldots,N$ where J_i is evaluated at

(t_i,p) and the vectors s_k are defined via (7). Problem (8) can be regarded as arising from an overdetermined system of nonlinear equations $r(p) = 0$, which in the neighbourhood of a local minimum p^* can be linearised to yield the overdetermined linear system,

$$r(p^*) + J(p^*)(p-p^*) = 0 \qquad (9)$$

The local conditioning of problem (8) in a neighbourhood of p^* can now be interpreted from the linear problem (9). If $\text{rank}(J(p^*)) = n$, then the relative sensitivity of p^* with respect to relative changes in the data is measured by $\kappa = \kappa(J(p^*)) = \sigma_1/\sigma_n$, where $\sigma_1 \geqslant \sigma_2 \geqslant \ldots \geqslant \sigma_n > 0$ are the singular values of $J(p^*)$. The size of κ will be regarded here as an overall measure of the conditioning of problem (8); it is stressed however that this is a local condition number, which depends on p^* and the points $\{t_i\}_1^N$. A desirable property of the model (1) is that κ should not be too large, so that p^* may be reasonably well determined.

3.2 Conditioning of the exact problem

As explained in Section 1.3 the exact problem (3) may be approximated in practice by the discretized form (6) in which an exact model is assumed. Following Varah (1987) the condition of (3) may be investigated by the following procedure:

(i) select N and a Riemann sum quadrature rule

(ii) choose a parameter vector p in the range of interest and generate exact data $\{Y(t_i)\}_1^N$ by solving the model equations

(iii) solve the parameter estimation problem (6) and compute the local condition number κ. For sufficiently large N, κ will be an adequate estimate of the condition number of the exact problem.

Since each problem in (iii) has a known global minimum with a zero residual, we may regard the problem as one of unconstrained minimisation, since excellent starting values are always available.

The above procedure may reveal that the model is very ill-conditioned, in which case there seems little point in solving the actual problem (2) for the experimental data. The model then needs to be reconsidered and possibly rejected completely. Otherwise, for an acceptably conditioned exact problem it is meaningful to investigate fully, to see if the conditioning has not deteriorated too much. Finally full statistical analysis involving confidence intervals for the parameters can be carried out.

4. Software

4.1 The ODE solver and minimization routine

The procedure of section 3.2 can be implemented in Fortran using the public domain software available via electronic mail from the NETLIB system, Dongarra and Gross (1987).

The ODE solver used was LSODE (1987 version) which contains, along with sophisticated features, stiff and nonstiff options. This is contained in the package ODE PACK. LSODE is sufficiently general in its variety of options to treat a wide range of parameter estimation problems. From (8) it is required to carry out the unconstrained minimisation of a sum of squares. From the MINPACK package this can be achieved using the implementation of the modified Levenberg-Marquardt method. Two routines were used. Firstly the routine LMDER, which requires the user to provide the Jacobian matrix of the function $r(p)$. This was available from the sensitivity coefficients s_i, $i = 1,2,\ldots,n$; here the s_i were computed from the augmented system (1) + (7). Secondly, the routine LMDIF, which computes automatically approximations to the jacobian via forward difference approximations; only $r(p)$ need be provided. Overall, LMDIF provides the simplest way of forming the complete software for the investigation of conditioning. As expected, for the problems treated, it was also much more efficient than employing directly the variational equations in LMDER.

All the associated linear algebra routines were obtained from the LINPACK package. In addition, the singular values of $J(p^*)$ used in the condition number $\kappa = \sigma_1/\sigma_n$ were computed using the routine DSVDC from LINPACK. This part of the calculation was simplified by taking advantage of the output from LMDER/LMDIF which provides the final computed jacobian in terms of the $n \times n$ upper triangular matrix R, where in terms of the permutation matix P, $P^T J^T J P = R^T R$. Hence only R need be passed to DSVDC. Clearly for models with a small or moderate number of parameters, this is a negligible part of the overall computational cost.

5. Numerical Examples

Using the N point mid-point rule, the procedure of Section 3.2 was applied to the following problems with variances obtained from $\sigma_k(t) = c y_k(p^*, t)$; both LMDER and LMDIF were used.

5.1 Problem A. Predator prey model, Varah (1987)

$$y_1' = p_1 y_1 - p_2 y_1 y_2, \quad y_2' = p_2 y_1 y_2 - p_3 y_2, \quad t \in [0,5]$$

$$y_1(0) = p_4, \quad y_2(0) = p_5,$$

where the "true" parameter values p^* are given by $p^* = (0.85, 2.13, 1.91, 1.02, 0.25)^T$ and $c = 0.001$. The estimates of the condition number κ_A of the exact problem are given in the table; the problem may be regarded as being well suited to parameter estimation.

5.2 Problems B and C. Chemical kinetics

$$y_1' = \frac{p_1 y_1 + p_2}{p_4 + p_3 y_1 + y_1^2}, \quad y_1(0) = p_5, \quad t \in [0,60]$$

This is Problem B, the original unscaled problem with $p^* = (0.031, -4.5E-6, -14, -0.048, 0.08)^T$, $c = 0.05$. The condition number estimates κ_B clearly indicate that the problem should be recast. Scaling $y_1(0)$ and the parameters to unity, so that $p^* = (1,1,1,1,1)^T$ gives for appropriate k_i, Problem C,

$$y_1' = \frac{p_1k_1y_1+p_2k_2}{1+p_3k_3y_1+p_4k_4y_1^2}, \quad y_1(0) = p_5, \quad t \in [0,60]$$

The corresponding estimates κ_C show the considerable improvement but also show the inherent rather poor conditioning of the problem.

Table 1

N	6	10	20	50	100
κ_A	17.26	16.63	16.38	16.31	16.30
κ_B	5.44E9	4.05E9	3.73E9	3.74E9	3.74E9
κ_C	4.55E4	2.98E4	2.69E4	2.68E4	2.68E4

Acknowledgement

The author is grateful to Dr. W.G. Bardsley of the Department of Obstetrics and Gynaecology, University of Manchester for making available the rational function model of Section 5.

References

Bard, Y. (1974), 'Nonlinear Parameter Estimation', Academic Press, New York.

Björck, A. (1987), Least Squares Methods. In 'Handbook of Numerical Analysis', Vol. 1, eds. Ciarlet, P.G. and Lions, J.L., Elsevier/North Holland.

Dongarra, J. and Grosse, E. (1987), Distribution of Mathematical software via electronic mail. CACM, 30, 403-407.

Varah, J.M. (1987), On the conditioning of parameter estimation problems. In proceedings of NPL conference, England, 1987; to appear.

8. Applications in Other Disciplines

Applications of discrete L_1 methods in science and engineering

C. Zala

Barrodale Computing Services Ltd, Victoria, British Columbia, Canada

I. Barrodale

Barrodale Computing Services Ltd, and Department of Computer Science, University of Victoria, Canada

Abstract Applications are reviewed for which the use of the L_1 (least absolute values) norm has been found to have advantages over the L_2 (least squares) norm. Desirable properties of L_1 norm solutions to both overdetermined and underdetermined linear systems of equations are outlined. Applications involving overdetermined systems are described which make use of the properties of the L_1 norm in outlier rejection, incorporation of linear constraints and retention of phase information. These applications span the fields of deconvolution, state estimation, inversion, and parameter estimation. The use of L_1 solutions to underdetermined linear systems of equations to achieve high resolution is discussed and some applications of the L_1 norm involving nonlinear systems are also described.

Key words: L_1, Least absolute values, Applications, Deconvolution, Approximation, Parameter estimation, Sparse solutions, Robust.

1 Introduction

The use of the L_1 (least absolute values) norm can often provide a useful alternative to the L_2 (least squares) norm in the analysis of data (Barrodale and Zala (1986); Branham (1982); Claerbout and Muir (1973); Dielman (1984); Narula and Wellington (1982); Narula (1987)). The far greater popularity of the L_2 norm stems largely from two reasons: the long-time existence of the calculus-based normal equations algorithm for computing L_2 solutions and the development of statistical theory based on the Gaussian error distribution, for which the L_2 norm is optimal. The effective use of L_1 methods had to await the development of linear programming and, while certain error distributions exist for which the L_1 norm is optimal, the statistical basis for L_1 methods is currently very much underdeveloped.

The development of fast special-purpose algorithms for the solution of the overdetermined L_1 problem (Barrodale and Roberts (1973, 1974)) and the demonstration that L_1 methods can offer real advantages over L_2 in some applications (Claerbout and Muir (1973)) have fueled interest in L_1 methods in recent years. It is the aim of this article to review those applications where L_1 methods have been shown to possess significant advantages over L_2 methods. Investigations in which L_1 does not appear to offer any advantages compared to L_2 have therefore not been included. Also, no attempt has been made to review and evaluate the various algorithms for computing L_1 solutions. Fortran subroutines for solving the overdetermined L_1 problem are contained in the widely used NAG and IMSL software libraries.

2 Overdetermined linear systems of equations

2.1 Statement of the problem

The linear overdetermined constrained L_1 problem may be defined as follows. Given a system of linear equations

$$\mathbf{Ax} = \mathbf{b} + \mathbf{r} \tag{1}$$

subject to the linear constraints

$$\left.\begin{array}{rcl} \mathbf{Cx} & = & \mathbf{d} \\ \mathbf{Ex} & \leq & \mathbf{f} \end{array}\right\} \tag{2}$$

we wish to find $\mathbf{x}$ to minimize $\|\mathbf{r}\|_1 = \sum_{i=1}^{m} |r_i|$. Here, $\mathbf{A}$ is an m by n design matrix, with $m > n$, m is the number of rows of $\mathbf{A}$ (number of data points), n is the number of columns of $\mathbf{A}$ (number of parameters), $\mathbf{x}$ is the vector of parameters whose solution is sought, $\mathbf{b}$ is the right hand side vector (function values or data points), $\mathbf{r}$ is a vector of residuals, $\mathbf{C}$ and $\mathbf{E}$ are constraint matrices, and $\mathbf{d}$ and $\mathbf{f}$ are vectors containing the constraint values.

If either the function to be minimized or the constraints cannot be expressed in this form, the problem is then nonlinear. If no constraints exist, then the problem becomes one of unconstrained minimization.

2.2 Properties

The L_1 norm shares certain properties with the L_2 norm, but there are important differences, which can yield vastly different results in certain cases. Many of the properties of the L_1 norm compared to L_2 which are described below may be simply illustrated by comparing the L_1 and L_2 solutions to the problem of fitting a number of data points by a constant. Here the L_1 estimate is the median, while the L_2 estimate is the mean.

2.2.1 Existence and uniqueness

An L_1 solution to an unconstrained overdetermined system of equations always exists; for the constrained problem, the constraints must also be consistent. The L_1 solution is not necessarily unique, in contrast to the L_2 solution, where the solution is always unique when $\mathbf{A}$ is of full rank. The example of the median demonstrates the possibility of nonuniqueness. For an odd number of points the median is the single centre point, while for an even number, any value between (and including) the two centre points gives a minimum L_1 error.

The nonuniqueness of some L_1 solutions, usually illustrated by simple integer examples, is sometimes cited as a drawback. However, in our experience the solutions obtained are always unique in practice.

2.2.2 Interpolation of data points

L_1 solutions interpolate some subset of the data points. A minimum L_1 norm solution to the overdetermined $m \times n$ system always exists which passes through at least n of the m data points, assuming $\mathbf{A}$ is of full rank. In contrast, the L_2 solution does not in general interpolate any of the points. This is obvious in the median-mean example. This interpolatory property yields a simple algorithm for very small m: simply form the C_n^m possible solutions, compute the residuals and select that solution with the smallest L_1 norm.

A further consequence of the interpolatory property is that the residuals of L_1 solutions to slightly overdetermined systems are sparse. In contrast, the residuals of L_2 solutions are in general all nonzero, irrespective of the degree of overdetermination.

2.2.3 Robustness

L_1 solutions are robust, meaning that the solution is resistant to some large changes in the data. In the median-mean example, the median remains unchanged if any of the other points are changed by an arbitrarily large amount, provided the sign of the residuals for those points remains unchanged. The mean will in general be affected by any change in any of the points.

L_1 solutions stay close to most of the data, while ignoring any points (outliers) which deviate strongly from the fit. This outlier rejection property arises solely from the properties of the L_1 norm and does not involve any assumptions or thresholds. It is an extremely useful property when the data are known to be contaminated with occasional wild points or spiky noise. The L_1 estimate is also resistant to the case where all the outliers lie on the same side of the data, (i.e., all the outlier residuals have the same sign), while the L_2 norm is strongly affected. The L_1 norm can thus be used in some cases where the error distribution is asymmetric. For data containing Gaussian noise, the L_1 solution is usually observed to be very close to the L_2 solution.

It is often the case that data are nonlinearly transformed to allow application of a linear regression model. What was a Gaussian distribution of errors in the original data may be highly non-Gaussian in the domain where the regression is performed. In such situations the L_1 norm may be capable of producing more accurate estimates of the model parameters than the L_2 norm because of its superior performance with some non-Gaussian error distributions.

2.2.4 Relation to linear programming

L_1 problems are equivalent to linear programming (LP) problems, and algorithms for solving L_1 problems are based on the techniques of linear programming. The converse is also true: linear programming problems may also be formulated as L_1 problems. Special-purpose L_1 algorithms (Barrodale and Roberts (1973, 1974)) make use of time-saving techniques such as multiple pivoting, which is not possible in the general linear programming technique. Thus it may be advantageous to formulate some linear programming applications as L_1 problems and use these algorithms, at a significant saving in time.

2.2.5 Incorporation of linear constraints

The incorporation of linear constraints in L_1 solutions is straightforward. Because L_1 algorithms use the methods of linear programming, which are designed to incorporate linear constraints in the minimization of an objective function, such constraints may be conveniently included in L_1 algorithms (Barrodale and Roberts (1978, 1980); Bartels, Conn and Sinclair (1978)).

2.2.6 Complex L_1 approximation

The application of L_1 methods to complex-valued data is difficult. If $r_j = a_j + ib_j$ is complex, the L_1 problem becomes nonlinear, since the quantity to be minimized is $\sum_{j=1}^{m} \sqrt{a_j^2 + b_j^2}$. The methods of linear programming cannot be applied to this problem. (In contrast, L_2 methods may easily be applied to complex data.) It is still possible, however, to modify the definition of the L_1 norm to allow its application to complex data (Barrodale (1978)), by defining the modified norm $\|\mathbf{r}\|_+ = \sum_{j=1}^{m} |a_j| + |b_j|$. The real and imaginary parts of the equations are thus treated independently. This modified norm will always be within $\sqrt{2}$ of the true L_1 norm and will yield results which have the properties described above. With complex data containing outliers, for example, the use of this modified norm may be desirable even though the result obtained is not the true L_1 solution.

2.2.7 Summary

In summary, L_1 solutions have certain appealing properties not shared by L_2 solutions. The feature of producing interpolatory fits which approximate most of the data well while ignoring outliers can be of considerable benefit in many practical situations. In this connection, Claerbout and Muir (1973) made the analogy that "When a traveller reaches a fork in the road, the L_1 norm tells him to take either one way or the other, but the L_2 norm instructs him to head off into the bushes."

2.3 Applications

2.3.1 Curve-fitting with non-Gaussian error distributions

The single most useful feature of L_1 solutions is their ability to yield robust solutions when the error distribution in the data is non-Gaussian, and especially when the data are contaminated with outlier values. We summarize here a number of applications for which various investigators have found a substantial benefit in the use of L_1 curve-fitting techniques.

Astrophysical parameter estimation

The question of how to identify and treat outlier values in astrophysical data has been a subject of discussion for more than a century (see reviews by Branham (1982) and Beckman and Cook (1983)). Since the development of digital computers and efficient L_1 algorithms based on linear programming, increasing use has been made of L_1 methods, especially for the estimation of orbital data.

Mudrov et al (1968) recommended the use of L_1 rather than L_2 in this application, finding that the L_1 orbital estimates displayed greater stability. They also pointed out that no rejection parameter was required by the L_1 technique, unlike the L_2 procedure they employed.

In a series of papers, Branham (1982, 1985, 1986a, 1986b) reported substantial advantages in the use of L_1 methods for parameter estimation from astronomical data. In a particularly extensive study involving more than 21,000 observations of the minor planets, the non-Gaussian nature of the distribution of residuals was clearly shown (Branham (1986b)). This distribution was asymmetric, more peaked than the Gaussian distribution and more long-tailed as well (indicating the presence of outliers).

Geophysical parameter estimation

In the course of exploratory geophysical surveys, large amounts of data are often obtained, which are subject to various sources of error. The existence of outlier values can substantially degrade results, even after smoothing of the data. In order to prevent these effects from seriously affecting the accuracy of the results, Dougherty and Smith (1966) proposed a technique for locally smoothing the raw data based on the L_1 norm.

In a landmark paper, Claerbout and Muir (1973) suggested that L_1 methods would be particularly appropriate for many types of geophysical data processing. Numerous examples and suggestions for the effective use of the L_1 norm were presented. These included the alignment of seismograms, the use of ratios computed from noisy data, the identification of arrival times for earthquakes (which requires that the norm be adaptable to asymmetric error distributions), and the use of running medians to remove outliers from noisy data. In addition, they anticipated some advantages of the L_1 norm in deconvolution filter design, as well as the usefulness of the sparseness property of L_1 in underdetermined problems. Some of these ideas were subsequently developed further by other workers and are discussed below.

Weibull parameter estimation

The L_1 norm has been successfully applied to improved estimation of Weibull parameters for failure time data (Lawrence and Shier (1981)). The Weibull probability density function $f(t)$ is given by

$$f(t) = (a/b)(t/b)^{a-1} \exp[-(t/b)^a] \quad (3)$$

and the cumulative distribution $F(t)$ by

$$F(t) = 1 - \exp[-(t/b)^a], \quad (4)$$

where $t \geq 0, a > 0$ and $b > 0$.

The Weibull parameter a represents the shape of the distribution while b represents its scale parameter, or characteristic life. The parameters a and b may be estimated by linear regression after logarithmic transformation of the data into either of the linear forms:

$$\ln [\ln \frac{1}{1 - F(t)}] = a \ln t - a \ln b \quad (5)$$

or

$$\ln t = (1/a) \ln [\ln \frac{1}{1 - F(t)}] + \ln b. \quad (6)$$

In an extensive set of studies with synthetic data, Lawrence and Shier (1981) found that the estimates of a obtained using the L_1 norm in regressions were consistently more accurate than those using L_2. The mean squared error of the estimated relative to the true value for a was usually 10 to 30 percent higher for L_2 estimates. In these studies, b was accurately estimated by both L_1 and L_2 procedures and the mean squared error was usually very small. In further work, Shier and Lawrence (1984) recommended the use of L_1 methods to provide starting parameter values for iterative techniques which were directed at further improvement of the estimation of the Weibull parameters. Although the reasons for the better performance of L_1 were not addressed, a likely explanation is that the logarithmic transform resulted in a non-Gaussian distribution of errors for which the L_1 norm was better suited than the L_2 norm.

2.3.2 Incorporation of linear constraints in inversion

The ease with which linear constraints may be specified has led to the use of the L_1 norm in several types of geophysical inverse problems. Vigneresse (1977) examined the L_1, L_2 and L_∞ norms applied to gravity profile inversion. The observation that unconstrained L_1 solutions were more stable than solutions in the other norms was attributed to the stability of L_1 to outliers in the data. The L_2 solution was stabilized considerably when nonnegativity constraints on the parameters were introduced and the suggestion was made (but not implemented) to combine the advantages of using the L_1 norm and nonnegativity constraints.

Jurkevics et al (1980) adopted the approach of expressing the inverse problem in body-wave inversion exclusively in terms of five different types of constraints on the solution. This led to an overdetermined system of inequality constraints, which was solved in L_1 directly as though it were an overdetermined system of equations, but incorporating different weights for positive and negative deviations. In this way, certain constraints were satisfied exactly and the sum of absolute deviations of the others was minimized. Selection of suitable weights effectively guaranteed that certain constraints would be respected.

The L_1 solution in the presence of linear inequality constraints was applied to the problem of moment tensor inversion by Tanimoto and Kanamori (1986). The method was found to give stable solutions even for shallow events and could also be used to determine automatically the compatibility of two data sets.

2.3.3 Deconvolution

Convolutional models are employed in many areas of scientific analysis and the inverse problem of deconvolution is consequently of considerable interest. The convolution of two discrete series w (k elements) and s (n elements) to yield an m ($= n + k - 1$) element trace t along with additive noise r may be expressed as $t = w * s + r$, (where '$*$' denotes convolution) or in matrix notation as

$$\mathbf{t} = \mathbf{W}\mathbf{s} + \mathbf{r}, \tag{7}$$

where $\mathbf{W}$ is an $m \times n$ banded matrix formed from the elements of w as follows:

$$W_{ij} = \begin{cases} w_{i-j+1} & \text{if } 1 \leq i - j + 1 \leq k, \\ 0 & \text{otherwise.} \end{cases} \tag{8}$$

In many situations w may be interpreted as a known wavelet or impulse response function which is of short duration, while s is an unknown sparse spike train to be estimated. The above system of equations is only slightly overdetermined and it is natural to examine ways in which the properties of L_1 solutions may be applied to the estimation of sparse spike trains. We first consider applications of L_1 methods to the problem of estimating s, given t and w. Then we discuss the general problem of estimating both w and s simultaneously, given only t.

Estimation of s for known t and w

A common problem in deconvolution is that w is often band-limited, so that $\mathbf{W}$ is ill-conditioned. Consequently attempts to solve (7) directly using any norm, including L_1, will yield unstable and unacceptable solutions. Taylor et al (1979) and Taylor (1980) addressed this problem by formulating a deconvolution scheme to minimize the expression

$$\sum_{i=1}^{m} |r_i| + \lambda \sum_{j=1}^{n} |s_j|, \tag{9}$$

where λ is an adjustable parameter. For large values of λ, the output spike train is zero, and as λ is decreased the number of nonzero spikes in the solution increases. In practice, the choice of an 'optimal' value for λ involves considerable computation and experimentation.

Another approach to estimation of a sparse spike train was based on a partial stepwise L_1 solution of (7) by extracting spikes one at a time (Chapman and Barrodale (1983), Barrodale et al (1984)). At each stage, the position and amplitude of the new spike was determined as that which would yield the greatest decrease in the L_1 norm of the difference between the trace and the fit. Spike extraction was terminated when a sufficiently close fit of the convolution of the spike train and the wavelet to the trace had been achieved. In a comprehensive series of experiments with numerous model wavelets and spike trains, Barrodale et al (1984) found that the L_1 norm generally yielded superior results to those obtained with the corresponding algorithm based on the L_2 norm, even in the presence of Gaussian noise.

Estimation of both s and w (deconvolution filter synthesis)

In the above application, it was assumed that the series w was known. The more general deconvolution problem is the simultaneous estimation of both w and s, given only t. This problem is, of course, ill-posed, but may be solved in some cases when w or s are known to be constrained in some way. We consider here the constraint that s be sparse, and formulate the problem in such a way that the spikes correspond to residuals. In this case the tendency of the L_1 norm to yield residuals with a few large outliers may be exploited. An added feature of the L_1 solution is the ability to recover the correct phase of the wavelet.

Assume that w is invertible, i.e., that some inverse filter f exists for which

$$w * f = f * w = (0, \ldots, 0, 1, 0, \ldots, 0). \tag{10}$$

Then, convolving $t = w * s$ with the filter f, we obtain $f * t = f * w * s = s$ or, in matrix notation, $\mathbf{Tf} = \mathbf{s}$. Thus, if an inverse filter f can be estimated for the trace t, the spike train may be obtained by convolution of f with the trace, and the wavelet may be recovered by taking $w = f^{-1}$, the inverse of f.

If a normalization criterion is applied to f, this system of equations may be expressed in terms of the trace t. Normalizing f so that its value at zero delay is unity (i.e., $f_0 = 1$) and assuming j forward (positive subscripts) and k backward (negative subscripts) coefficients in the filter, with $p = j + k + 1$ elements overall, we have

$$\begin{bmatrix} t_1 & \ldots & t_{j+1} & \ldots & t_p \\ t_2 & \ldots & t_{j+2} & \ldots & t_{p+1} \\ t_3 & \ldots & t_{j+3} & \ldots & t_{p+2} \\ \vdots & \ldots & \vdots & \ldots & \vdots \\ \vdots & \ldots & \vdots & \ldots & \vdots \\ t_{m-p+1} & \ldots & t_{m-k} & \ldots & t_m \end{bmatrix} \begin{bmatrix} f_j \\ \vdots \\ f_0(=1) \\ \vdots \\ f_{-k} \end{bmatrix} = \begin{bmatrix} s_{j+1} \\ s_{j+2} \\ s_{j+3} \\ \vdots \\ \vdots \\ s_{m-k} \end{bmatrix}. \tag{11}$$

Moving the column of the matrix which corresponds to the zero delay of the filter (i.e., that column starting with t_{j+1}) to the right hand side, we obtain a system of forwards and backwards prediction equations based only on the trace, with the (unknown) spikes representing the residuals:

$$
\begin{bmatrix}
t_1 & \cdots & t_j & t_{j+2} & \cdots & t_p \\
t_2 & \cdots & t_{j+1} & t_{j+3} & \cdots & t_{p+1} \\
t_3 & \cdots & t_{j+2} & t_{j+4} & \cdots & t_{p+2} \\
\vdots & \cdots & \vdots & \vdots & \cdots & \vdots \\
\vdots & \cdots & \vdots & \vdots & \cdots & \vdots \\
t_{m-p+1} & \cdots & t_{m-k-1} & t_{m-k+1} & \cdots & t_m
\end{bmatrix}
\begin{bmatrix}
f_j \\ \vdots \\ f_1 \\ f_{-1} \\ \vdots \\ f_{-k}
\end{bmatrix}
=
\begin{bmatrix}
-t_{j+1} + s_{j+1} \\ -t_{j+2} + s_{j+2} \\ -t_{j+3} + s_{j+3} \\ \vdots \\ \vdots \\ -t_{m-k} + s_{m-k}
\end{bmatrix} . \quad (12)
$$

In this overdetermined system of prediction equations, a desirable prediction filter would give a close fit to most of the right hand side values where residuals (spikes) are near zero, except where a spike is present; at these points the residual of the prediction could be large. Thus if the spike train s is sparse, it is highly advantageous to solve (12) using the L_1 norm, as the spikes arise in the form of outliers. This application of the L_1 norm is quite distinct from the situation where outliers occur as invalid points in the data; here it is *the model itself* which is formulated in such a way as to allow the advantageous application of the L_1 norm.

Least squares methods are not at all suited to this application because of their inability to accommodate a few large residuals. A further disadvantage of least squares in this context is the destruction of phase information by L_2 methods, due to the symmetric nature of the autocorrelation matrix $\mathbf{A}^{\mathrm{T}}\mathbf{A}$ which characterizes L_2 solutions. No such restrictions apply to the L_1 solution, and *the phase of the wavelet may be correctly recovered.* By a suitable choice of the overall and relative numbers of forwards and backwards prediction coefficients, it is possible to estimate wavelets of mixed phase, as well as minimum and maximum phase. Thus the frequent assumption of minimum phase in wavelet estimation may be overcome by using this L_1-based procedure.

This application of L_1 to the problem of wavelet estimation was first described by Scargle (1977, 1981) and independently found by ourselves (Barrodale and Zala (1986)). However, there are certain limitations in its implementation. First, the spike train must be sparse. Second, a criterion must be developed for determining the optimal numbers of forwards and backwards coefficients; it is not sufficient simply to compare the reduction in L_1 norm given by trial solutions. Finally, the method cannot yet be applied to band-limited data, for which an inverse filter does not exist. Solution of the equations in this case yields filters which give almost perfect prediction but which are not invertible, so that neither the spike series nor the wavelet can be recovered. Since most data are band-limited, it would be desirable to develop procedures or constraints which would then allow the application of this potentially very useful L_1 technique to such data.

2.3.4 Autoregression

The system of equations (12) is similar to that which arises in autoregression, and the above technique may be viewed as a general autoregression, where the forwards and backwards coefficients may differ. The use of L_1 methods in autoregression has been examined by Gross and Steiger (1979) and Bloomfield and Steiger

(1983), who found that for the forwards-only autoregressive model, the L_1 norm estimate was strongly consistent and produced comparable or better estimates of the coefficients than the L_2 norm for various types of innovational series.

The possibility of developing an L_1 algorithm similar to the autoregressive algorithm of Burg (1967) was suggested by Claerbout and Muir (1973) and realized by Denoel and Solvay (1985) in their analysis of speech signals. The algorithm was comparable in computational complexity to Burg's L_2 algorithm, and the stability of the resulting filters was also guaranteed. The performance of L_1 in forwards-only prediction was also examined by Yarlagadda et al (1985), Bednar et al (1986) and Scales and Treitel (1987); it was pointed out that the forwards-only filter is not guaranteed to be stable (unlike the L_2 case) and that the use of the L_1 norm admits the possibility of handling non-minimum phase wavelets.

Additional work in the application of L_1 methods to autoregressive problems has been done by Ward (1984), who considered the on-line estimation problem and by Ruzinsky and Olsen (1989), who applied the Karmarker algorithm to L_1 and L_∞ minimization. Interestingly, in their implementations, the Karmarker algorithm was about five times slower than a simplex-type scheme, even for large (1000×100) problems.

2.3.5 Median filtering

Median filtering provides a robust alternative to the more common practice of filtering by convolution with an averaging filter (e.g., running mean). It is an L_1 norm-based technique which, unlike averaging methods, has the often desirable property of retaining sharp transitions between regions of the data while still reducing noise. For one-dimensional data, repeated application of a median filter eventually yields an invariant series known as a root signal, which is locally monotone. The root signal may also have regions between plateaus where the slope is steep. In certain applications, such as acoustic impedance profile estimation for layered media, this blockiness of the output is a highly desirable feature. Application of median filtering to two-dimensional problems has the advantage that noisy signals may be smoothed while maintaining the edges of the features. There is a large and growing literature on median filters and their effective implementation, which we shall not attempt to review. Some examples of median filtering techniques in geophysics and image processing may be found in Tukey (1977); Evans (1982); Bednar (1983); Bednar and Watt (1984); Fitch et al (1984,1985); Lee and Kassam (1985); and Liao et al (1985).

3 Underdetermined linear systems of equations

3.1 Statement of the problem

Given a system of linear equations

$$\mathbf{Ax} = \mathbf{b} \tag{13}$$

we wish to find $\mathbf{x}$ to minimize $\|\mathbf{x}\|_1 = \sum_{i=1}^{n} |x_i|$, where $\mathbf{A}$ is an m by n matrix, with $m < n$, m is the number of rows of $\mathbf{A}$ (number of constraints), n is the number of columns of $\mathbf{A}$ (number of parameters), $\mathbf{x}$ is the vector of parameters whose solution is sought, and $\mathbf{b}$ is the right hand side vector.

3.2 Properties

An L_1 solution to an underdetermined system of equations always exists if the constraints are not inconsistent, but may not be unique.

A unique L_1 solution to an m by n system of equations ($m < n$) will have at most m nonzero elements in $\mathbf{x}$. Thus the number of nonzero elements in the L_1 solution will be at most equal to the number of equations, no matter what the number of parameters. For many more parameters than equations, this property results in a sparse solution with only a relatively few nonzero parameters in the solution. In the minimum norm L_2 solution to the underdetermined problem, all the parameters are nonzero in general.

There is no corresponding property of robustness in the underdetermined L_1 solution, since even small changes in one constraint can give rise to substantial changes in the solution, with previously zero elements becoming nonzero and vice versa.

The underdetermined L_1 problem, like the overdetermined case, is a linear programming problem. Linear constraints may thus be incorporated in a natural way.

3.3 Applications

The underdetermined L_1 technique has been employed to yield high resolution estimates of a real-valued signal when a measured function of this signal is band-limited but the signal itself is known to be sparse. The procedure, which is a method for achieving "superresolution," may be formulated as follows. Assume that the Fourier transform X of an unknown real-valued series x is available for a particular band of frequencies. The Fourier transform within this band may be defined as follows:

$$\left.\begin{array}{rcl} \frac{1}{N}\sum_{k=0}^{N-1} x_k \cos(k\omega_j) & = & \Re(X_j) \\ -\frac{1}{N}\sum_{k=0}^{N-1} x_k \sin(k\omega_j) & = & \Im(X_j), \end{array}\right\} j_1 \le j \le j_2 \tag{14}$$

where j_1 and j_2 define the band of frequencies for which X is defined, N is the number of points in x and defines the resolution, x is the unknown series for which a sparse solution is sought, ω_j is the Fourier frequency defined by $\omega_j = 2\pi j/N$, and X_j is the known Fourier coefficient at frequency ω_j.

For $N > 2(j_2 - j_1 + 1)$, (14) is an underdetermined system of equations with $m = 2(j_2 - j_1 + 1)$ rows and $n = N$ columns. A solution may be obtained by specifying that x be minimized in some norm. If the L_1 norm is chosen, the number of nonzero elements of x will normally be at most $2(j_2 - j_1 + 1)$ and a sparse, high resolution result will be obtained, no matter how many points N are present.

In contrast to the high resolution result given by L_1, the L_2 solution may be shown to be equal simply to the inverse Fourier transform of X padded with zeros. This estimate of x is therefore band-limited and consequently smooth and continuous, in contrast to the high resolution result obtained using the L_1 norm.

An application of this method to deconvolution was described by Levy and Fullagar (1981), where X was the Fourier transform estimate of a spike train obtained by frequency domain division of the trace by the known wavelet in the frequency band where the signal-to-noise ratio of the wavelet was high. In using the L_1 norm, these workers also recommended the formulation of the problem in terms of inequality constraints, by introducing frequency-dependent tolerances in the above equations (and hence doubling the number of constraints). The procedure was found to give excellent estimates of sparse spike trains when only a portion of the amplitude spectrum was reliable.

Similar applications have been discussed in optical image restoration (Mammone and Eichmann (1982); Mammone (1983); Abdelmalek and Otsu (1985)), spectral estimation (Figueiras-Vidal et al (1985); Martinelli el al (1986); Levy et al (1982)) and and beamforming (Zala et al (1985, 1987)). In the beamforming application, X was the set of complex spatial correlations between the elements of a linear array and x was the spatial signal intensity. Using the underdetermined L_1 technique, it was found that both discrete signals and isotropic noise fields could be jointly estimated by making minor modifications in the formulation of the equations.

4 Nonlinear systems of equations

Several applications of L_1 methods to nonlinear overdetermined problems have appeared in recent years. The robustness and interpolatory properties of overdetermined L_1 solutions also apply to the nonlinear problem, and the studies described here have exploited this property in applications involving outliers or non-Gaussian distributions of residuals.

4.1 State estimation

The use of L_1 methods in static state estimation of power systems was investigated by Kotiuga and Vidyasagar (1982) and extended by Kotiuga (1985) to allow on-line tracking of the changes in system state by processing measurements as they become available. In this state estimation application, misassignments in the status of lines in the model can give rise to outliers. The use of weighted L_1 methods may then be expected to yield more reliable estimates of the system state than is possible with the L_2 norm.

Static state estimation is a nonlinear problem in which a set of nonlinear equations relate the $n = 2(N-1)$ state variables x_j (consisting of $N-1$ voltage magnitudes and $N-1$ voltage phase angles, where N is the number of nodes) to the m measurements z_i (real and reactive power injection and flow). These equations may be represented by

$$\mathbf{z} = \mathbf{f}(\mathbf{x}) + \mathbf{r}, \tag{15}$$

where $\mathbf{r}$ is a vector of residuals. An estimate of the state may be obtained by using a weighted L_1 estimator to minimize

$$\sum_{i=1}^{m} w_i|z_i - f_i(\mathbf{x})|, \tag{16}$$

where w_i is a weight associated with a particular measurement. Estimation of the L_1 solution for $\mathbf{x}$ was performed by first linearizing the equations by approximation with a first order Taylor series and then employing a Newton iteration procedure.

The conclusion reached in these studies was that the L_1 estimate of the state in the presence of bad data was as good or better than the L_2 estimate in the absence of bad data. Accurate estimates of the state were thus obtained without the need of any bad data detectors because the L_1 norm simultaneously detected and rejected these outliers.

4.2 Network analysis

In a similar application to that above, Bandler et al (1987) used L_1 techniques in network analysis of electrical circuits. Several different applications of the L_1 norm were developed. For fault isolation, a nonlinear model was fitted to the measured data, which were obtained by applying excitations to the network. Deviations in the network parameters from their nominal values (faults) were identified by inspection of the residuals of the L_1 fit. A one-sided L_1 norm minimization technique was also applied to the problem of contiguous-band multiplexer design. This modified L_1 method was used to identify "bad channels" and provide starting values for a subsequent minimax optimization. An example involving more than sixty nonlinear design variables was presented. Finally, the L_1 norm was applied to a nonlinear parameter estimation problem in the presence of large isolated measurement errors, and its performance was illustrated with a sixth order filter example. A special-purpose algorithm based on the work of Hald and Madsen (1985) was employed for these studies.

4.3 Estimation of pharmokinetic parameters

A common problem in pharmacology and other areas of biology is fitting compartmental models to concentration time-curve data. In these experiments a known amount of drug is administered orally and its concentration in the blood is followed. This will be a function of the rate of absorption and the rate of elimination, for which the parameters to be estimated are nonlinear. For the simple one compartment open model, the expected concentration is of the form

$$f(x_i, \mathbf{a}) = \frac{a_3 a_1}{a_1 - a_2}[\exp(-a_2(x_i - a_4)) - \exp(-a_1(x_i - a_4))], \tag{17}$$

where x_i is the ith time point, a_1 and a_2 are rate constants for absorption and elimination, a_3 is related to the fraction of drug ultimately absorbed, and a_4 is a delay time. All parameters a_i must be positive.

Frome and Yakatan (1980) found advantages in the use of the L_1 norm in fitting the model, as occasional stray points may appear in the data. This was accomplished using an iterative reweighted least squares procedure.

5 Conclusions

There are often substantial advantages to be gained by the use of the L_1 norm rather than the L_2 norm. For curve-fitting applications, L_1 methods give more robust estimation of parameters when the data contain outliers or "wild points". Thus data which contain certain non-Gaussian error distributions or which have been transformed prior to fitting a model are usually more reliably analyzed using L_1 rather than L_2 methods. Algorithms for solving the linear overdetermined L_1 problem are provided in widely used numerical subroutine libraries such as NAG and IMSL, and are comparable in computational time to L_2 methods. Algorithms for the solution of nonlinear L_1 problems are becoming increasingly powerful. Given the availability and potential advantages of overdetermined L_1 techniques, it is recommended that initial analysis of data routinely include the use of both L_1 and L_2 norms.

Underdetermined L_1 methods have been successfully applied to achieve high resolution reconstruction of band-limited data; such increases in resolution are not possible using L_2 methods.

The L_1 norm has found numerous applications in a variety of scientific and engineering fields, including astronomy, geophysics, electrical engineering and inverse problems. With the availability of algorithms and the increasing awareness of the unique and useful properties of the L_1 norm, the number of beneficial applications of L_1 will certainly continue to grow.

References

[1] Abdelmalek, N.N. and Otsu, N. (1985) Restoration of images with missing high-frequency components by minimizing the L_1 norm of the solution vector. *Appl. Opt.* 24:1415-1420.

[2] Bandler, J.W., Kellermann, W., and Madsen, K. (1987) A nonlinear L_1 optimization algorithm for design, modeling, and diagnosis of networks. *IEEE. Tr. Circ. Syst.* CAS-34:174-181.

[3] Barrodale, I. (1978) Best approximation of complex-valued data. In *"Numerical Analysis" (G.A. Watson, ed.), Proceedings, Biennial Conference, Dundee 1977*, pp. 14-22, Springer-Verlag.

[4] Barrodale, I. and Roberts, F.D.K. (1973) An improved algorithm for discrete L_1 linear approximation. *SIAM J. Numer. Anal.* 10:839-848.

[5] Barrodale, I. and Roberts, F.D.K. (1974) Algorithm 478: Solution of an overdetermined system of equations in the L_1 norm. *Comm. ACM.* 17:319-320.

[6] Barrodale, I. and Roberts, F.D.K. (1978) An efficient algorithm for discrete L_1 linear approximation with linear constraints. *SIAM J. Numer. Anal.* 15:603-611.

[7] Barrodale, I. and Roberts, F.D.K. (1980) Algorithm 552 - Solution of the constrained L_1 linear approximation [F4]. *ACM TOMS* 6:231-235.

[8] Barrodale, I., Zala, C.A. and Chapman, N.R. (1984) Comparison of the L_1 and L_2 norms applied to one-at-a-time spike extraction from seismic traces. *Geophysics* 49:2048-2052.

[9] Barrodale, I. and Zala, C.A. (1986) L_1 and L_∞ curve fitting and linear programming: algorithms and applications. In *"Numerical Algorithms" (J.L. Mohamed and J.E. Walsh, eds.)*, Oxford University Press, Oxford, Chapter 11, pp. 220-238.

[10] Bartels, R.H., Conn, A.R. and Sinclair, J.W. (1978) Minimization techniques for piecewise differentiable functions: The L_1 solution to an overdetermined linear system. *SIAM J. Numer. Anal.* 15:224-241.

[11] Beckman, R.J. and Cook, R.D. (1983) Outlier..........s. *Technometrics* 25:119-149.

[12] Bednar, J.B. (1983) Applications of median filtering to deconvolution, pulse estimation, and statistical editing of seismic data. *Geophysics* 48:1598-1610.

[13] Bednar, J.B. and Watt, T.L. (1984) Alpha-trimmed means and their relationship median filters. *IEEE Tr. Acoust. Sp. Sig. Proc.* ASSP-32: 145-153.

[14] Bednar, J.B., Yarlagadda, R. and Watt, T. (1986) L_1 deconvolution and its application to seismic signal processing. *IEEE Tr. Acoust. Sp. Sig. Proc.* ASSP-34:1655-1658.

[15] Bloomfield, P. and Steiger, W.L. (1983) *"Least absolute deviations: Theory, applications and algorithms" (Progress in probability and statistics; Vol. 6)*, Birkhauser, Boston.

[16] Branham, R.L., Jr. (1982) Alternatives to least squares. *Astronom. J.* 87:928-937.

[17] Branham, R.L., Jr. (1985) A new orbit of comet 1961 V (Wilson-Hubbard). *Celes. Mech.* 36:365-373.

[18] Branham, R.L., Jr. (1986a) Error estimates with L_1 solutions. *Celes. Mech.* 39:239-247.

[19] Branham, R.L., Jr. (1986b) Is robust estimation useful for astronomical data reduction? *Q. J. R. astr. Soc.* 27:182-193.

[20] Burg, J.P. (1967) Maximum entropy spectral analysis. Paper presented at the 37th Annual Meeting of the Society of Exploration Geophysicists, Oklahoma City, Oklahoma.

[21] Chapman, N.R. and Barrodale, I. (1983) Deconvolution of marine seismic data using the L_1 norm. *Geophys. J. R. astr. Soc.* 72:93-100.

[22] Claerbout, J.F. and Muir, F. (1973) Robust modeling with erratic data. *Geophysics* 38:826-844.

[23] Denoel, E. and Solvay, J.-P. (1985) Linear prediction of speech with a least absolute values criterion. *IEEE Tr. Acoust. Sp. Sig. Proc.* ASSP-33:1397-1403.

[24] Dielman, T.E. (1984) Least absolute value estimation in regression models: an annotated bibliography. *Commun. Stat. Theor. Meth.* A13:513-541.

[25] Dougherty, E.L. and Smith, S.T. (1966) The use of linear programming to filter digitized map data. *Geophysics* 31:253-259.

[26] Evans, J.R. (1982) Running median filters and a general despiker. *Bull. Seis. Soc. Am.* 72:331-338.

[27] Figueiras-Vidal, J.R., Casar-Corredera, J.R., Garcia-Gomez, R. and Paez-Borrallo, J.M. (1985) L_1-norm versus L_2-norm minimization in parametric spectral analysis: A general discussion. In *Proc. Internat. Conf. Acoust. Sp. Sig. Proc. (Tampa, Fla.)*, pp. 304-307.

[28] Fitch, J.P., Coyle, E.J. and Gallagher, N.C.,Jr. (1984) Median filtering by threshold decomposition. *IEEE Tr. Acoust. Sp. Sig. Proc.* ASSP-32:1183-1188.

[29] Fitch, J.P., Coyle, E.J. and Gallagher, N.C.,Jr. (1985) Root properties and convergence rates of median filters. *IEEE Tr. Acoust. Sp. Sig. Proc.* ASSP-33:230-240.

[30] Frome, E.L. and Yakatan, G.J. (1980) Statistical estimation of the pharmokinetic parameters in the one compartment open model. *Commun. Stat. Sim. Comput.* B9:201-222.

[31] Gross, S. and Steiger, W.L. (1979) Least absolute deviation estimates in autoregression with infinite variance. *J. Appl. Prob.* 16:104-116.

[32] Hald, J. and Madsen, K. (1985) Combined LP and quasi-Newton methods for nonlinear L_1 optimization. *SIAM J. Numer. Anal.* 22:68-80.

[33] Jurkevics, A., Wiggins, R. and Canales, L. (1980) Body-wave inversion using travel time and amplitude data. *Geophys. J. R. astr. Soc.* 63:75-93.

[34] Kotiuga, W.W. (1985) Development of a least absolute value power system tracking state estimator. *IEEE. Tr. Power. Appar. Sys.* PAS-104:1160-1166.

[35] Kotiuga, W.W. and Vidyasagar, M. (1982) Bad data rejection properties of weighted least absolute value techniques applied to static state estimation. *IEEE Tr. Power. Appar. Sys.* PAS-101:844-851.

[36] Lawrence, K.D. and Shier, D.R. (1981) A comparison of least squares and least absolute deviation regression models for estimating Weibull parameters. *Commun. Stat. Sim. Comput.* B10:315-326.

[37] Lee, Y.H. and Kassam, S.A. (1985) Generalized median filtering and related nonlinear filtering techniques. *IEEE. Tr. Acoust. Sp. Sig. Proc.* ASSP-33:672-683.

[38] Levy. S. and Fullagar, P.K. (1981) Reconstruction of a sparse spike train from a portion of its amplitude spectrum and application to high resolution deconvolution. *Geophysics* 46:1235-1243.

[39] Levy, S., Walker, C., Ulrych, T.J. and Fullagar, P.K. (1982) A linear programming approach to the estimation of the power spectra of harmonic processes. *IEEE Tr. Acoust. Sp. Sig. Proc.* ASSP-30:675-679.

[40] Liao. G.-Y., Nodes, T.A., and Gallagher, N.C.,Jr. (1985) Output distributions of two-dimensional median filters. *IEEE. Tr. Acoust. Sp. Sig. Proc.* ASSP-33:1280-1295.

[41] Mammone, R. and Eichmann, G. (1982) Superresolving image restoration using linear programming. *Appl. Opt.* 21:496-501.

[42] Mammone, R.J. (1983) Spectral extrapolation of constrained signals. *J. Opt. Soc. Am.* 73:1476-1480.

[43] Martinelli, G., Orlando, G. and Burrascano, P. (1986) Spectral estimation by repetitive L_1-norm minimization. *Proc. IEEE.* 74:523-524.

[44] Mudrov, V.I., Kushko, V.L., Mikhailov, V.I. and Osovitskii, E.M. (1968) Some experiments on the use of the least-moduli method in processing orbital data. *Cosmic Res.* 6:421-431.

[45] Narula, S.C. (1987) The minimum sum of absolute errors regression. *J. Qual. Tech.* 19:37-45.

[46] Narula, S.C. and Wellington, J.F. (1982) The minimum sum of absolute errors regression: a state of the art survey. *Internat. Stat. Rev.* 50:317-326.

[47] Ruzinsky, S. A. and Olsen, E. T. (1989) L_1 and L_∞ minimization via a variant of Karmarkar's algorithm. *IEEE Tr. Acoust. Sp. Sig. Proc.* ASSP-37:245-253.

[48] Scales, J.A. and Treitel, S. (1987) On the connection between IRLS and Gauss' method for L_1 inversion: Comments on "Fast algorithms for L_p deconvolution". *IEEE Tr. Acoust. Sp. Sig. Proc.* ASSP-35:581-582.

[49] Scargle, J.D. (1977) Absolute value optimization to estimate phase properties of stochastic time series. *IEEE Tr. Info. Theory,* January:140-143.

[50] Scargle, J.D. (1981) Studies in astronomical time series analysis. 1. Modelling random processes in the time domain. *Astrophys. J. Suppl.* 45:1-71.

[51] Shier, D.R. and Lawrence, K.D. (1984) A comparison of robust regression techniques for the estimation of Weibull parameters. *Commun. Stat. Sim. Comput.* B13:743-750.

[52] Tanimoto, T. and Kanamori, H. (1986) Linear programming approach to moment tensor inversion of earthquake sources and some tests on the three- dimensional structure of the upper mantle. *Geophys. J. R. astr. Soc.* 84:413-430.

[53] Taylor, H.L., Banks, S.C. and McCoy, J.F. (1979) Deconvolution with the L_1 norm. *Geophysics* 44:39-52.

[54] Taylor, H.L. (1980) The L_1 norm in seismic data processing. In *"Developments in Geophysical Exploration Methods - 2" (A.A. Fitch, ed.)*, Applied Science Publishers, Barking, U.K.

[55] Tukey, J.W. (1977) *"Exploratory Data Analysis"*, Addison-Wesley, Reading, Mass.

[56] Vigneresse, J.L. (1977) Linear inverse problem in gravity profile interpretation. *J. Geophys.* 43:193-213.

[57] Ward, R. K. (1984) An on-line adaptation for discrete L_1 linear estimation. *IEEE Tr. Automat. Contr.* AC-29:67-71.

[58] Yarlagadda, R., Bednar, J.B. and Watt, T.L. (1985) Fast algorithms for L_p deconvolution. *IEEE Tr. Acoust. Sp. Sig. Proc.* ASSP-33:174-182.

[59] Zala, C.A., Barrodale, I. and Kennedy, J.S. (1985) Comparison of algorithms for high resolution deconvolution of array beamforming output. In *"Acoustical Imaging", Proceedings, Vol. 14 (eds. A.J. Berkhout, J. Ridder and L.F. van der Wal)*, Plenum Press, New York, pp. 699-702.

[60] Zala, C.A., Barrodale, I. and Kennedy, J.S. (1987) High-resolution signal and noise field estimation using the L_1 (least absolute values) norm. *IEEE J. Ocean Eng.* OE-12:253-264.

Constrained complex approximation algorithms in communication engineering

J. C. Mason
Applied and Computational Mathematics Group, Royal Military College of Science, Shrivenham, Wiltshire, UK

S. J. Wilde
Applied and Computational Mathematics Group, Royal Military College of Science, Shrivenham, Wiltshire, UK

A. E. Trefethen
Thinking Machine Corp., Cambridge, Massachusettes, USA

Abstract The design and adaptation of antenna arrays, for the transmission and reception of signals and the suppression of interference, is a vital problem in communication engineering, which may be viewed as a complex polynomial approximation problem subject to equality or inequality constraints. Various parameter choices are described for obtaining initial designs, and links are noted with the Chebyshev polynomials of first and second kinds. A number of novel algorithms are then discussed for suppressing broad ranges of interference, based on both least squares and minimax criteria. For practical reasons the emphasis is on phase-only and amplitude-only parameter changes, and on planar and polygonal antenna arrays.
Key words Antennas, quiescent patterns, least squares, minimax, Chebyshev polynomials, phased arrays, rectangular arrays, octagonal arrays.

1. Introduction

Complex approximation problems occur frequently in communication engineering, where the physical system is represented by a series of elements, each of which is a complex function. Changes in the amplitudes of these functions represent changes in power, and changes in the phases (i.e. arguments) represent time delays. In some applications it is physically realistic to vary only amplitudes of elements, and considerable thought has been given to such "amplitude-only" algorithms by Steyskal (1982) and many others. There are also, however, "phase-only" applications in which it is cheaper to change phases of elements and fix amplitudes. For the purpose of illustration we restrict attention here to the important problem of designing antenna arrays for radar and communications.

A linear array of antenna elements, equally spaced by d, is shown in Figure 1, with a general field point located at angle θ to the normal. The "complex response" of such an array to a received signal of wavelength λ takes the form

$$p(u) = \sum_{j=1}^{N} x_j \, e^{id_j u}, \qquad (-1 \leq u \leq 1) \tag{1}$$

where x_j is the complex excitation of the j th element,

$$u = \frac{2d}{\lambda} \sin \theta , \qquad (-\pi/2 \leq \theta \leq \pi/2)$$

and

$$d_j = \pi\left[\frac{n-1}{2} - (j-1)\right] = -d_{N-j+1} \quad (j=1,\ldots,N). \tag{2}$$

The amplitude $|p|$ of p represents the receiving power of the array. On the other hand, if the array is a transmitter, then p (given by (1)) represents the "field pattern" and $|p|$ the transmitting power. Assuming that $d = \frac{1}{2}\lambda$ (half wavelength spacing), (1) gives

$$p(u) = e^{i(N-1)\pi u/2} \sum_{j=1}^{N} x_j \, z^{j-1} \tag{3}$$

where

$$z = e^{-i\pi u} \quad \text{and } u = \sin \theta . \tag{4}$$

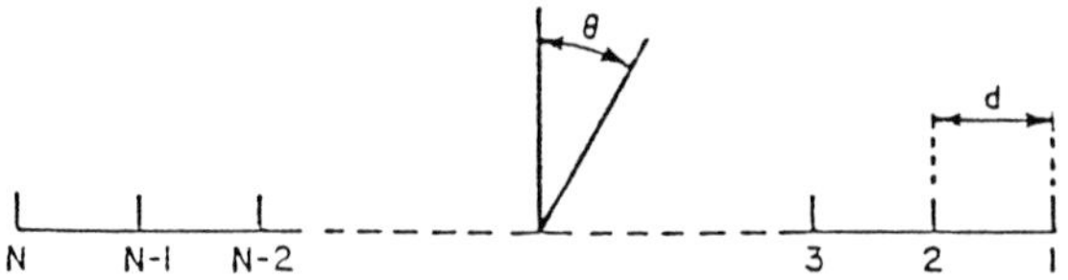

Figure 1. Antenna Array Geometry (ID)

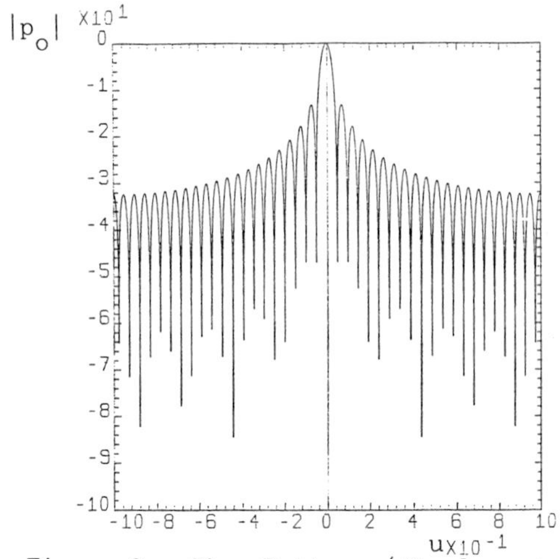

Figure 2. Sinc Pattern (41 elements)

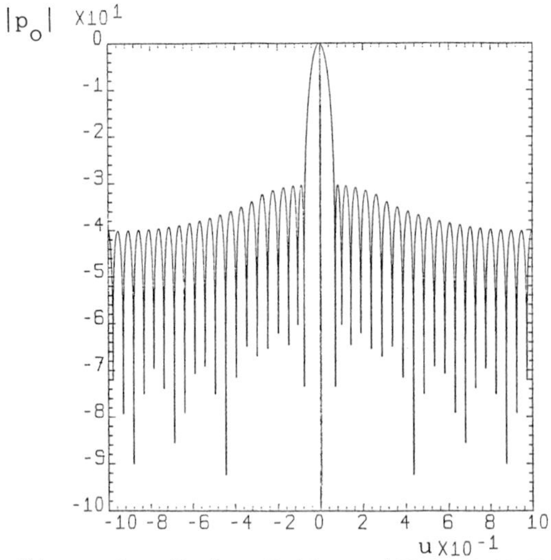

Figure 3. Taylor Pattern (30dB,N=41,N'=6)

Apart from a phase change of $(N-1)\pi u/2$, $p(u)$ is a complex polynomial in z, where z moves on the unit circle. In practice we are mainly concerned with the power $|p(u)|$ of the antenna, and so phase changes to p are of little consequence. Typically a "quiescent" pattern

$$p_o(u) = \sum_{j=1}^{N} W_j \, e^{id_j u} \qquad (-1 \le u \le 1) \qquad (5)$$

is a simple initial pattern chosen to provide a broadly acceptable response.

Given (5), $\{x_j\}$ are subsequently chosen (in (1)) so that the response $p(u)$ approximates $p_o(u)$ in a chosen norm, subject to certain physical constraints imposed to provide suitable modifications. For example, we typically require that $|p(u)|$ should be small or zero (null) over a set of given θ values, so that interference in that region is suppressed, and we may also impose zero values on any of the x_j which are known to be faulty or which we wish to omit.

Two obvious choices of norm are the L_2 norm

$$||p-p_o||_2 = \left[\int_{-1}^{1} \left[\sum_{j=1}^{N} (x_j - W_j) \, e^{id_j u} \right]^2 du \right]^{½}, \qquad (6)$$

and the L_∞ norm

$$||p-p_o||_\infty = \max_{-1 \le u \le 1} \left| \Sigma (x_j - W_j) \, e^{id_j u} \right| . \qquad (7)$$

(Here we measure the error over a continuum, but in practice it may be reasonable to replace [-1,1] by a discrete set of u values.) The majority of researchers have adopted the L_2 norm, partly because it leads to fast linear algorithms in simple cases, and partly because it relates to energy considerations (See Hudson (1981) for example). However, there are also advantages in adopting L_∞. Firstly, data are often relatively accurate, so that L_∞ is a realistic and ideal measure and secondly speed of computation may not be paramount, so that the relative slowness of L_∞ methods may not be crucial. Moreover, by adopting linear or nonlinear programming techniques, for L_∞, a wide variety of problems can readily be tackled, and we can reap unexpected rewards when, for example, we find that few of our antenna parameters need to be varied.

2. Quiescent Patterns - Initial Designs

There are three quiescent patterns $p_o(u)$ which we consider here, namely the so called "sinc", "Chebyshev", and "Taylor" patterns.

The sinc pattern is based on the very simple choice of equal unit weights ($W_j = N^{-1}$), for which (5) becomes

$$p_o(u) = N^{-1} \sum_{j=1}^{N} e^{id_j u} = N^{-1} e^{i(N-1)\pi u/2} \sum_{j=1}^{N} e^{-i\pi(j-1)u}.$$

On summing the geometric series, we obtain

$$p_o(u) = N^{-1} \sin(N\pi u/2) / \sin(\pi u/2) . \qquad (8)$$

The power of the pattern $|p_o(u)|$ has the form shown in Figure 2, with a "main beam" and N-1 "sidelobes" decreasing monotonically away from the centre. We count the pair of extrema at $u=\pm 1$ as a single extremum. Note that $p_o(u)$ in (8) is in fact a multiple of a Chebyshev polynomial of the second kind in a related variable:

$$p_o(u) = N^{-1} U_{N-1}(t), \text{ where } t=\cos(\pi u/2). \qquad (9)$$

In some problems we require the set of "sidelobes" to be as uniform as possible. Indeed Dolph (1946) showed that it was relatively simple to generate a "Chebyshev pattern" with a main beam of unity and N-1 equal sidelobes, by forming a minimax approximation to zero over an interior interval. Indeed, from (5), if we shift the range of u to [0,2] for convenience (by making phase shifts to the weights W_j) and take W_j to be real, we obtain

$$|p_o(u)|^2 = \sum_{j=1}^{N} W_j e^{-i(j-1)\pi u} \sum_{j=1}^{N} W_j e^{i(j-1)\pi u} \qquad (0 \le u \le 2)$$

$$= b_1 + 2 \sum_{j=2}^{N} b_j \cos (j-1)\pi u$$

where $$b_j = W_j W_1 + W_{j+1} W_2 + \ldots + W_N W_{N-j+1} .$$

Since cos $(j-1)\pi u$ is a polynomial of degree $2(j-1)$ in $t = \cos(\pi u/2)$, it follows that $|p_o(u)|^2$ is a polynomial $\phi_{2N-2}(t)$ of degree 2N-2 in t. By an appropriate choice of W_j, this polynomial may be taken to have the form

$$\phi_{2N-2}(t) = \tfrac{1}{2} A \, [1+T_{2N-2}(t/c)] \qquad (10)$$

where $A = 2\,[1+T_{2N-2}(1/c)]^{-1}$ and c (≤ 1) is a parameter. It then follows that $|p_o(u)| = [\phi_{2N-2}(t)]^{\frac{1}{2}}$ has N-1 equal sidelobes of "level" $A^{\frac{1}{2}} (\le 1)$ on the range $[-c,c]$ of t, and a main beam of level unity (at $t=\pm 1$). This provides the so-called Chebyshev pattern, and the ratio of main beam level to sidelobe level, namely $A^{-\frac{1}{2}}$, can be adjusted by choice of c. (The ratio approaches 1 as c approaches 1, and it approaches ∞ as c approaches 0).

A third choice of quiescent pattern, introduced by Taylor (1955), is a compromise in which $p_o(u)$ has N'(say) of the zeros of a Chebyshev pattern followed by (N-1)-N' zeros of a sinc pattern. This leads to a useful pattern in which the "near in" sidelobes are roughly constant and the "far out" sidelobes decay. The computational details are well described by Villeneuve (1984) and an example is shown in Figure 3. We remark that this pattern suggests the potential for a theoretical study of polynomials which have as their zeros a mixture of the zeros of the Chebyshev polynomials of first and second kinds. We are not aware of any significant work in this area.

3. Least Squares Methods

Steyskal (1982) developed a very efficient algorithm, based on orthogonal functions, for fitting a pattern to a quiescent pattern subject to zero constraints, in the case where x_j takes <u>any</u> real or complex value. Shore (1983) later described an analogous algorithm for phase-only fitting, where x_j and W_j are related in (6) by

$$x_j = W_j e^{i\phi_j} \quad (\phi_j \text{ real}) , \qquad (11)$$

which was based on the solution of a nonlinear optimisation problem. In fact the Steyskal/Shore algorithms are rather easily derived by adopting slightly different approaches, based on Lagrange multipliers and a modified choice of objective function, as we now show. Note first that, in L_2, the fitting problem with x_j complex reduces to a real problem, if real and imaginary parts of the fit are considered separately. It suffices therefore to consider the cases where x_j is real or of the form (11).

3.1 Real problems

In the case where x_j is real,

$$(\|p-p_o\|_2)^2 = \int_{-1}^{1} \left| \sum_{j=1}^{N} (x_j - W_j)\, e^{-ij\pi u} \right|^2 du = \sum_{j=1}^{N} |x_j - W_j|^2 ,$$

and we minimise this quantity subject to

$$L(u_r) = \sum_{j=1}^{N} x_j e^{-ij\pi u_r} = 0 \qquad (r=1,\ldots,R) \quad ,$$

where u_r are R "nulls" of the pattern. Using Lagrange multipliers λ_r, we solve

$$\frac{\partial}{\partial x_j} \sum_{n=1}^{N} (x_j - W_j)^2 + \frac{\partial}{\partial x_j} \sum_{r=1}^{R} \lambda_r L(u_r) = 0$$

subject to $L(u_r)=0$. Thus we obtain

$$2(x_j - W_j) + \sum_{r=1}^{R} \lambda_r\, e^{-ij\pi u_r} = 0 \qquad (j=1,\ldots,N) \qquad (12)$$

$$\text{and} \quad \sum_{j=1}^{N} x_j\, e^{-ij\pi u_r} = 0 \qquad (r=1,\ldots,R) \qquad (13)$$

Substituting for x_j in terms of $\{\lambda_r\}$ from (12) into (13) gives a system of R simultaneous linear equations for $\{\lambda_r\}$. Once $\{\lambda_r\}$ have been determined, each x_j is given explicitly by (12). Thus the size of the problem reduces to the number of nulls imposed, and the algorithm is extremely efficient.

3.2 Phase-only problems

For phase-only weight changes, we may set $x_j = W_j\, e^{i\phi_j}$ and obtain

$$F = (\|p-p_o\|_2)^2 = \int_{-1}^{1} \left| \sum_{j=1}^{N} W_j (e^{i\phi_j}-1) e^{-ij\pi u} \right|^2 du$$

$$= \sum_{j=1}^{N} |W_j|^2 \; |e^{i\phi_j}-1|^2 \quad . \tag{14}$$

We minimise this quantity subject to

$$L_r = L(u_r) = \sum_{j=1}^{N} W_j\, e^{i\phi_j} e^{id_j u_r} = 0 \quad (r=1,\ldots,R) \; . \tag{15}$$

Suppose for simplicity that p_o is real and hence that the element weights are conjugate symmetric:

$$W_{N-j+1} = \bar{W}_j \qquad (j=1,\ldots,N) \; . \tag{16}$$

Then it has been shown (Shore (1984)) that the resulting phase-perturbations and coefficients are odd-symmetric:

$$\phi_{N-j+1} = -\phi_j \;\; , \;\; d_{N-j+1} = -d_j \; . \tag{17}$$

Now, if we write W_j in the form

$$W_j = |W_j|\; e^{id_j u_s} \quad ,$$

where u_s is a suitable fixed u, then we deduce from (15) that

$$L_r = \sum_{j=1}^{N} |W_j| \cos [\phi_j + d_j(u_r - u_s)] = 0 \ (r=1,\ldots,R) \qquad (18)$$

(since $\sum_{j=1}^{N} |W_j| \sin [\phi_j + d_j(u_r - u_s)] = 0$ by (17)).

The objective function (14) takes the form

$$F = \sum_{j=1}^{N} |W_j|^2 \, 2(1-\cos\phi_j) \qquad (19)$$

and the problem is to solve

$$\frac{\partial F}{\partial \phi_j} + \sum_{r=1}^{R} \lambda_r \frac{\partial L_r}{\partial \phi_j} = 0 \quad (j=1,\ldots,N).$$

Differentiating, we obtain the system of equations

$$2\,|W_j|^2 \sin \phi_j - \sum_{r=1}^{R} \lambda_r |W_j| \sin [\phi_j + d_j(u_r - u_s)] = 0 \quad (20)$$

together with equations (18). A little manipulation of (20) gives

$$\tan \phi_j = \sum_{r=1}^{R} \lambda_r \sin[d_j(u_r - u_s)] \Big/ \left[2|W_j| - \sum_{r=1}^{R} \lambda_r \cos[d_j(u_r - u_s)]\right]$$

(21)

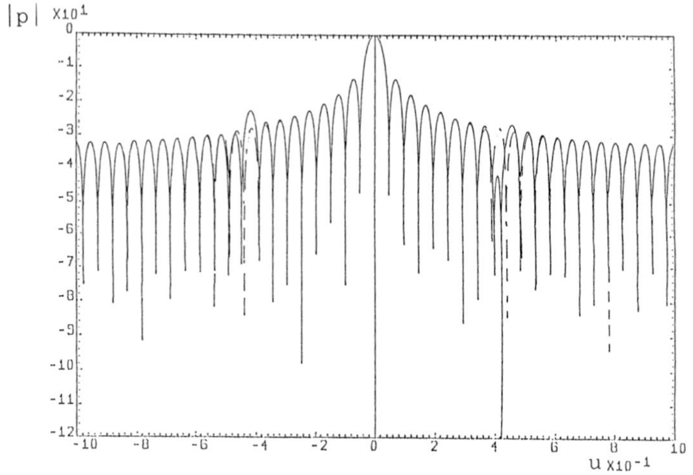

Figure 4. Single Null - Symmetric Enhancement

Null direction (rad)	Optimal Beam Coefficients	Average Phase Pertubation (rad)	Null Depth (DBA)	Loss at Main Beam
0.43633	0.13468	4.32306E-02	-334.54672	3.902157

Table 1 Single Null

Null direction (rad)	Optimal Beam Coefficients	Average Phase Pertubation (rad).	Null Depth (DBA)	Loss at Main Beam
0.59556	-0.12590	4.01644E-02	-300.86929	3.3491
0.59556 0.65575	-0.12430 -0.11685	4.96558E-02	-320.59109 -323.25893	6.2658
0.59556 0.65575 0.71887	-0.12161 -0.11331 -0.10434	5.28214E-02	-282.87559 -280.68316 -275.63945	8.7066

Table 2 Multiple Nulls

Thus ϕ_j has been expressed in terms of $\{\lambda_r\}$ for each j, and hence (18) comprise N nonlinear equations for λ_r. Once $\{\lambda_r\}$ have been determined, $\{\lambda_j\}$ are given explicitly by (21). Again the algorithm appears to be extremely efficient, effectively reducing an N parameter problem to an R parameter one.

We note that the form (21) was also obtained by Shore (1983), who minimised a different objective function while solving an equivalent problem. The parameters which Shore terms the "beam coefficients" are in fact the negative Lagrange multipliers divided by 2. At present the algorithm has been tested by the (equivalent) method of minimising $F + \lambda_r L_r$ subject to $L_r=0$, using the NAG routine E04VDF, which employs a sequential quadratic programming method [Gill et al (1981)]. We give numerical results in Table 1, which agree with those of Shore (1983), for a problem with N=41, $W_j = 1$, $u_s=0$, with a single null located at $\theta = .43633$. The algorithm has proved to be remarkably robust in providing useful antenna patterns.

3.3 Small phase perturbations

If phase perturbations are assumed to be small, then

$$\tan \phi_j \simeq \phi_j \quad \text{and} \quad e^{i\phi_j} \simeq 1+i\phi_j\ , \tag{22}$$

and we may rewrite the "cancellation pattern" $p-p_o$ as the sum of two beams as follows:

$$\Delta p_o(u) = p(u)-p_o(u) = \tfrac{1}{2} \sum_{r=1}^{R} \lambda_r \sum_{j=1}^{N} c_j^{-1} \left[e^{id_j(u-2u_s+u_r)} - e^{id_j(u-u_r)} \right]$$

$$\text{where} \quad c_j = 2 - |W_j|^{-1} \sum_{r=1}^{R} \lambda_r \cos\left[d_j(u_r-u_s)\right].$$

One beam is in the direction of the null u_r and cancels the original pattern, while the second beam is in a symmetrical location causing an enhancement to the pattern. This is illustrated in Figure 4 for the numerical example of Table 1, where a null at $u_r=.43633$ is reflected by pattern enhancement at $u=-.43633$. (Here the broken curve is the original pattern, and the continuous curve is the fitted pattern.) One consequence of such symmetrical enhancement is that it is not possible to synthesise nulls at symmetrical locations using the linear approximations (22). Indeed this problem leads to an inconsistent system of linear algebraic equations.

Note that, for the linear approximations (22), if coefficients λ_r are small (relative to 2) and $|W_j|=1$, then, neglecting contributions from other cancellation beams at $u=u_r$,

$$\Delta p_o(u_r) = -N\lambda_r/4 \qquad (r=1,\ldots,R)$$

Since $p(u_r)=o$, $\Delta p_o(u_r)=-p_o(u_r)$ and hence

$$\lambda_r = 4\, p_o(u_r)\,/N \tag{23}$$

Thus we have very simple approximate expressions for λ_r. Table 2 shows results obtained when nulls are placed at $\phi=0.59556$, 0.65575, 0.71887. Here the true beam coefficients λ_r should be -0.12646, -0.11923, and -0.11349, and the calculated (approximate) values are clearly of the same order. Note that the "null depth" is the value of the pattern at the prescribed null; this is measured in dBA (i.e. 20 $\log_{10}$ of the pattern), so that, for example, a value of -280dBA corresponds to 10^{-14} in traditional units. We see that very deep nulls have been obtained in this example. However, the linearising algorithm is far less successful when locations of distinct nulls approach each other, since the coefficients λ_r then cease to be objectively independent. This is also true if we attempt to impose nulls close to locations symmetrical to other nulls.

3.4 Higher order nulls

Since there are difficulties in imposing nulls at locations which are close together, it is useful to impose higher order nulls at u_r

(r=1,...,R), by setting to zero derivatives up to some order m:

$$\frac{d^{\nu}}{du^{\nu}} \quad p(u_r)=0 \qquad (r=1,\ldots,R) \; (\nu=0,\ldots,m) \qquad (24)$$

Such an idea was used for amplitude-only problems by Steyskal (1982). In the case of phase-only problems, the T=R(m+1) constraints (24) are given by

$$C_t = \sum_{j=1}^{N} (d_j)^{\nu} \; |W_j| \cos\left[\phi_j + d_j(u_r - u_s)\right] = 0 \;\; (\nu \text{ even})$$

$$C_t = \sum_{j=1}^{N} (d_j)^{\nu} \; |W_j| \sin\left[\phi_j + d_j(u_r - u_s)\right] = 0 \;\; (\nu \text{ odd})$$

where each t=0,1,...T corresponds to a pair of indices r, ν.

The Lagrangian now takes the form

$$L = F - \sum_{t=1}^{T} \lambda_t C_t$$

and the condition for a minimum is that

$$\frac{\partial F}{\partial \phi_j} - \sum_{t=1}^{T} \lambda_t \frac{\partial C_t}{\partial \phi_j} = 0 \qquad (j=1,\ldots,N) \qquad (25)$$

and

$$C_t = 0 \qquad (t=1,\ldots,T). \qquad (26)$$

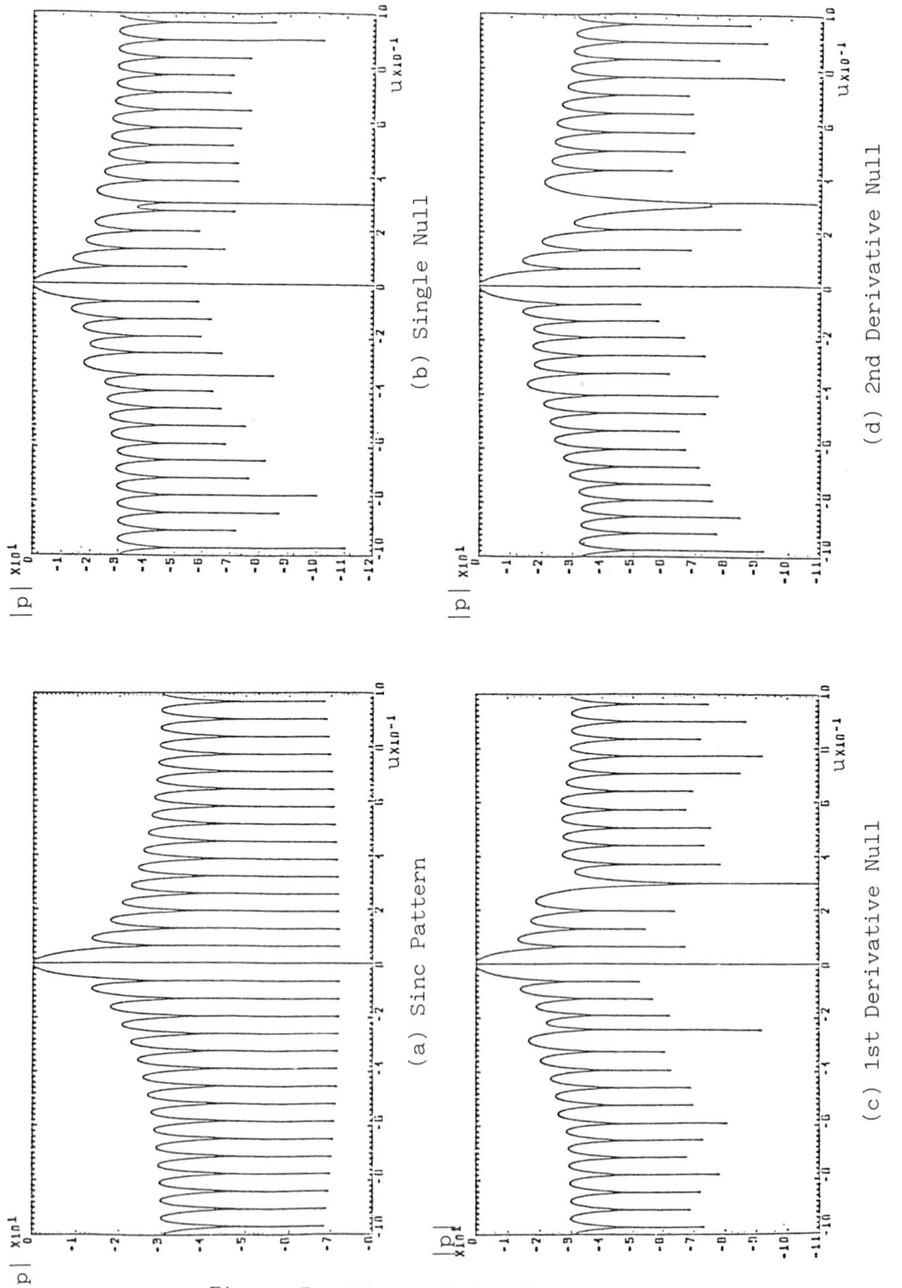

Figure 5. Higher Order Constraints

After some manipulations, we deduce that

$$\tan \phi_j = \frac{\sum_{r=1}^{R} \sum_{\nu=0}^{m} \lambda_t (d_j)^{\nu} f_{\nu}(d_j(u_r - u_s))}{2|W_j| + \sum_{r=1}^{R} \sum_{\nu=0}^{m} \lambda_t (d_j)^{\nu} g_{\nu}(d_j(u_r - u_s))} \tag{27}$$

where $f_{\nu}(.) = \begin{cases} -\sin(.) & \text{for } \nu \text{ even} \\ +\cos(.) & \text{for } \nu \text{ odd}, \end{cases}$

$g_{\nu}(.) = \begin{cases} \cos(.) & \text{for } \nu \text{ even} \\ -\sin(.) & \text{for } \nu \text{ odd} \end{cases}$

and $t = t(r,\nu) = (r-1)(m+1) + (\nu+1)$

Now ϕ_j has been expressed in terms of $\{\lambda_t\}$, and, by substitution in (26), a system of T equations is obtained for the multiplier λ_t. Thus the problem has been reduced to a nonlinear algebraic system, whose order is equal to the number of constraints applied (allowing for multiplicities).

The effect of higher order constraints is shown in Figure 5. Figure 5(a) is a quiescent sinc pattern with 31 elements (u_s=0 and W_j=1, all j). Figure 5(b) shows a fitted pattern with a single null at u_r=0.3 (indicated by the vertical line). Figures 5(c) and 5(d) show fitted patterns with first and second order nulls (ν=1 and ν=2) at u_s=0.3 and illustrate the broadening of the null. Note that there is some pattern enhancement at the symmetrical position in all cases.

3.5 Two-dimensional problems

In practice antenna arrays are often 2- or 3- dimensional, since they are placed in a 3-D environment. A two-dimensional planar array, of N_1 elements with spacing $d^{(1)}$ in the x-direction and N_2 elements with spacing $d^{(2)}$ in the y-direction, has the complex response

$$p = \sum_{j=1}^{N_1} \sum_{k=1}^{N_2} x_{jk} \, e^{i\theta_{jk}}$$

where x_{jk} is the complex excitation of the j,k element and where

$$\theta_{jk} = \frac{2}{\lambda} \left[d_j^{(1)} \; u + d_j^{(2)} \; v \right]$$

$$d_j^{(1)} = \pi[(j-1)-\tfrac{1}{2}(N_1-1)] \; , \qquad u=\sin\psi \cos\phi,$$

$$d_j^{(2)} = \pi[(j-1)-\tfrac{1}{2}(N_2-1)] \; , \qquad v=\sin\psi \sin\phi.$$

The angles ϕ and ψ are illustrated in Figure 6.

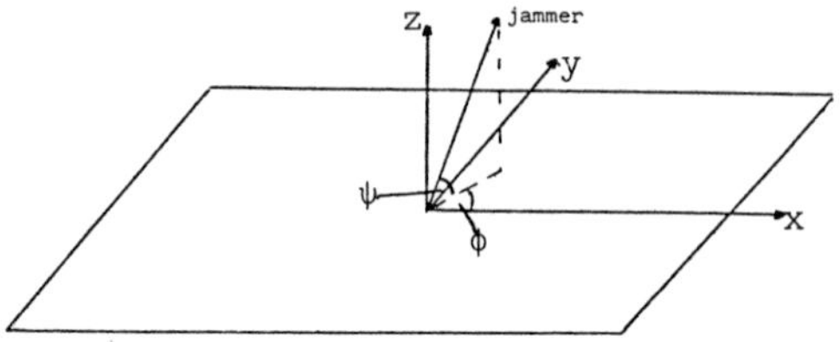

Figure 6. Coordinate system for 2-D array

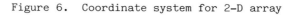

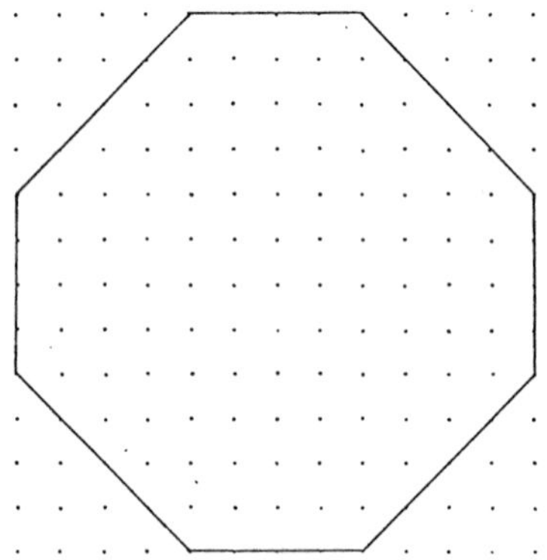

Figure 7. Embedded octagon in 2-D array

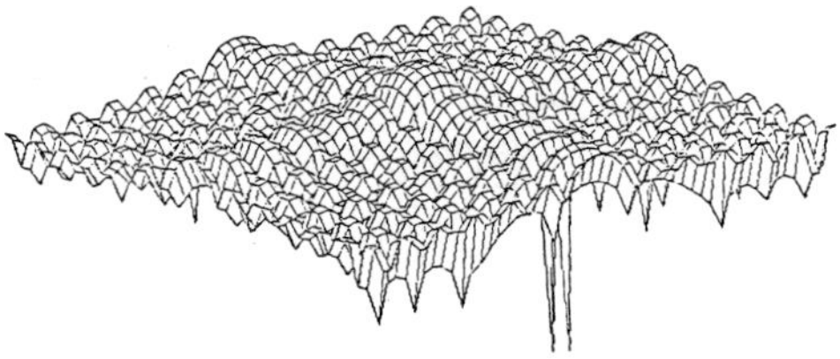

Figure 8(a) Rectangular Array - 2 Nulls

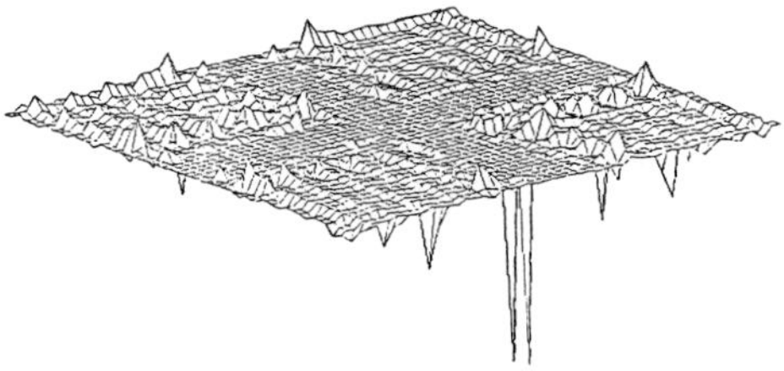

Figure 8(b) Rectangular Perturbation

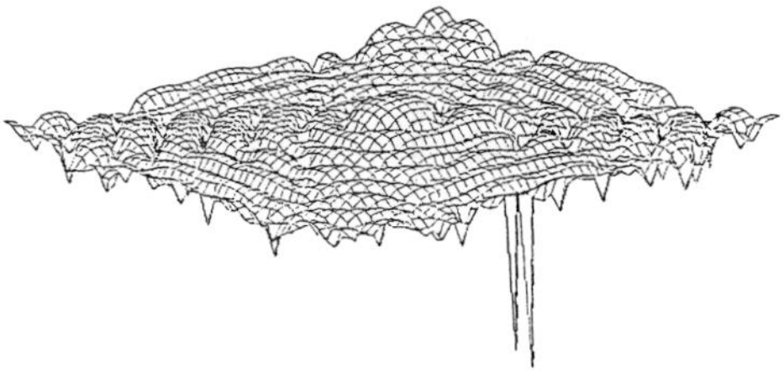

Figure 9(a) Octagonal Array - 2 Nulls

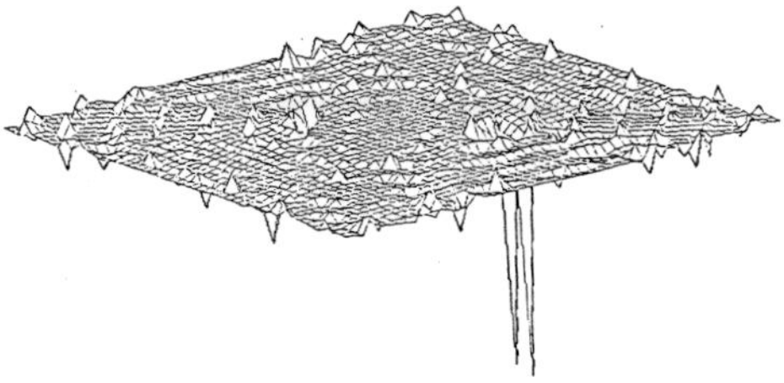

Figure 9(b) Octagonal Perturbation

Table 3. Rectangular Phased Array - 2 Nulls

Antenna Arrays : Phase Only Nulling

There are 169 elements and 2 contraints.

Constraint	1 is at position	.28000	.32000
Constraint	2 is at position	.32000	.36000

Initial value for constraint 1 is .10000
Initial value for constraint 2 is .10000

Required accuracy is .10000E-07
Optimal Beam Coefficients
.86483E-01
-.89406E-01

Coefficients for Perturbed pattern

1	.53802E-01	58	-.15846E-01
2	-.11934E-02	59	-.20047E-02
3	-.46955E-01	60	.48291E-02
4	-.38105E-01	61	-.36409E-03
5	.37794E-02	62	-.27727E-02
6	.33931E-01	63	.69819E-02
7	.25908E-01	64	.16175E-01
8	-.49172E-02	65	.56081E-02
9	-.20995E-01	66	.18598E-01
10	-.11646E-01	67	.32407E-01
11	.34537E-02	68	.12171E-01
12	.57008E-02	69	-.13147E-01
13	-.10019E-02	70	-.16315E-01
14	.62571E-02	71	-.33453E-02
15	-.43776E-01	72	.44103E-02
16	-.41216E-01	73	-.18350E-03
17	-.13255E-02	74	-.33739E-02
18	.31713E-01	75	.58532E-02
19	.28181E-01	76	.16483E-01
20	-.14293E-02	77	.82735E-02
21	-.20031E-01	78	-.17961E-01
22	-.13013E-01	79	.32191E-01
23	.21166E-02	80	.15194E-01
24	.56119E-02	81	-.10800E-01
25	-.86034E-03	82	-.16529E-01
26	-.75122E-03	83	-.46347E-02
27	-.39847E-01	84	.39226E-02
28	-.43719E-01	85	.00000
29	-.64120E-02		
30	.29024E-01		
31	.29987E-01		
32	.20832E-02		
33	-.18744E-01		
34	-.14180E-01		
35	.74719E-03		
36	.54348E-02		
37	-.70521E-03		
38	-.14511E-02		
39	.89014E-02		
40	-.45568E-01		
41	-.11409E-01		
42	.25907E-01		
43	.31302E-01		
44	.55600E-02		
45	-.17150E-01		
46	-.15130E-01		
47	-.63361E-03		
48	.51725E-02		
49	-.53892E-03		
50	-.21284E-02		
51	.80033E-02		
52	.15610E-01		
53	-.16247E-01		
54	.22414E-01		
55	.32112E-01		
56	.89418E-02		
57	-.15275E-01		

Null depth = -196.59 DBA
at Constraint 1 .

Null depth = -196.11 DBA
at Constraint 2 .

Table 4. Octagonal Phased Array - 2 Nulls

Antenna Arrays : Phase Only Nulling

There are 129 elements and 2 contraints.

Constraint 1 is at position .28000 .32000
Constraint 2 is at position .32000 .36000

Initial value for constraint 1 is .10000
Initial value for constraint 2 is .10000

Required accuracy is .10000E-07
Optimal Beam Coefficients
-.59158
.37090

Coefficients for Perturbed pattern

5	.10705
6	-.17203
7	-.19683
8	-.73508E-01
9	.62827E-01
17	.14239
18	-.14722
19	-.20278
20	-.91101E-01
21	.46339E-01
22	.12795
23	.11257
29	.17337
30	-.11836
31	-.20582
32	-.10779
33	.29386E-01
34	.11997
35	.11885
36	.37580E-01
37	-.58779E-01
41	.19944
42	-.86054E-01
43	-.20569
44	-.12337
45	.12141E-01
46	.11046
47	.12300
48	.52316E-01
49	-.45232E-01
50	-.10985
51	-.10702
53	.22033
54	-.51102E-01
55	-.20213
56	-.13764
57	-.52235E-02
58	.99605E-01
59	.12505
60	.65926E-01
61	-.30714E-01
62	-.10472
63	-.11400
64	-.46184E-01
65	.72443E-01
66	-.14530E-01
67	-.19490
68	-.15037
69	-.22537E-01
70	.87578E-01
71	.12508
72	.78154E-01
73	-.15528E-01
74	-.97679E-01
75	-.11943
76	-.60697E-01
77	.56321E-01
78	.17034
79	-.18386
80	-.16135
81	-.39627E-01
82	.74551E-01
83	.12316
84	.88790E-01
85	.00000

Null depth = -268.88 DBA
at Constraint 1

Null depth = -286.87 DBA
at Constraint 2

If the elements and corresponding coefficients are ordered by rows, then the coefficients may be represented by a vector $x_{jk}=\underline{x}(\ell)$, $\ell = k N_1 + j$.

The field pattern then takes the vector form

$$p = \sum_{j=1}^{N} x_j e^{i\theta_j} ,$$

where θ_j is a vector form of θ_{jk}, and we have a (large) one-dimensional problem. Suppose that nulls are placed at $(u,v)=(u_r,v_r)$ $(r=1,\ldots,R)$; then the computational detail follows in precisely the same way as in §3.2, and indeed the formula (27) is again obtained if we assume that field coefficients x_j take the form

$$x_j = W_j e^{i\phi_j} .$$

Thus an algorithm is obtained which requires the solution of only R equations, where R is the total number of null constraints.

It is not difficult to consider certain polygonal arrays by embedding these in a rectangular array and setting to zero the antenna weights which correspond to nodes outside the polygon. For example an octagon of 129 elements, with "sides" of lengths alternately proportional to 1 and 2, may readily be obtained by "chopping" 10 elements off each corner of a 13x13 rectangular array, as illustrated in Figure 7. The resulting octagon has five elements along each side.

The phase-only algorithm of §3.2 was used to adapt a sinc pattern for a 13x13 rectangular array as well as for a 129 element octagonal array (embedded in the rectangular array). The quiescent (sinc) pattern is nearly circular for the octagonal array, but has two prominent ridges for the rectangular array. In figures 8a, 8b we show the perturbed rectangular pattern after imposing a pair of nulls at (0.28,0.32) and (0.32, 0.36), as well as the discrepancy between this and the quiescent pattern. In Figures 9a, 9b we show corresponding graphs for the octagonal array. The phase perturbations ϕ_j for the respective (rectangular, octagonal) perturbed patterns are given in Tables 3 and 4. The perturbations are conjugate symmetric about the central element (for the pattern to be real) and so only the first half of the coefficients needs to be listed. Note that the octagonal pattern yields a deeper null and a larger average perturbation.

The efficiency of this algorithm deserves to be emphasised. For example, the pair of nulls have been imposed on the 169 element array in Figure 8 by solving just 2 simultaneous equations. Moreover the algorithm has proved remarkably robust for a wide range of examples.

4. Minimax Methods

A minimax method is discussed in detail for phase-only problems by Mason and Wilde (1989). If nulls, or near-nulls, are to be imposed at a number of points, then the minimax method is very much less efficient than the least squares method of §3.2, for the size of the problem in the minimax method is dictated by the number N of weights rather than by the number of nulls. However, there is a great gain in that, by constraining phase changes to $[0,2\pi]$, many of the weights are unchanged in phase from their quiescent value. Thus there is a great deal of external computation but very little physical adaptation to be imposed. Clearly then the relative advantages of least squares and minimax methods will depend on the nature of the practical problem. If real-time computation has to be very fast, then the least squares method is essential. If changes to antenna elements need to be kept to a minimum, then the minimax method is ideal.

To illustrate the effectiveness of minimax methods we determine an amplitude-only adaptation of a sinc pattern. Here the perturbed and quiescent patterns are, respectively,

$$p(u) = \sum_{j=1}^{N} x_j \; e^{-ij\pi u} \qquad (x_j \text{ real})$$

and

$$p_o(u) = \sum_{j=1}^{N} e^{-ij\pi u}$$

Choosing a discretization $\{U_\ell\}$ of $[-1,1]$, and "near-nulls" $\{u_r\}$, we follow Mason and Owen (1987) and solve

$$\underset{x_j}{\text{Min}} \; \underset{U_\ell}{\text{Max}} \left\{ \text{Max} \quad (\text{Re } |p-p_o| \;,\; \text{Im } |p-po|) \right\} \tag{28}$$

subject to $-1 \leq x_j \leq 1$, (29)

$$\pm \text{Re}(p_o(u_r)) \leq \varepsilon \quad , \quad \pm \text{Im}(p_o(u_r)) \leq \varepsilon \tag{30}$$

where ε is a chosen depth of null. This constrained linear programming problem may be solved by an algorithm of Roberts and

$x_i=1$ (i=1 to 41)
except
$x_1=x_{41}=0.278353$
$x_2=x_{40}=0.659333$
$x_3=x_{39}=0.963233$

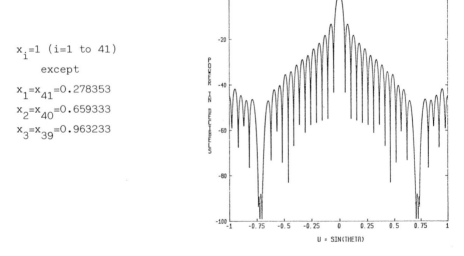

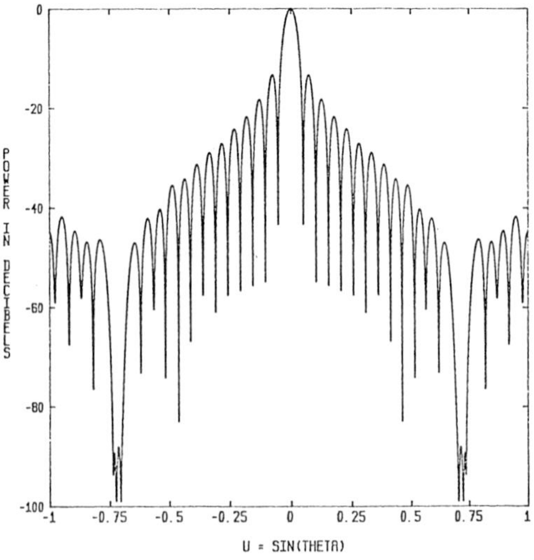

Figure 10. Perturbed sinc pattern: 4 nulls (.7, .71, .72, .73)

Barrodale (1979). In Figure 10 we show the fitted pattern for N=41 with 4 nulls (0.7,0.71,0.72,0.73) , and a wide and deep null area is achieved. Note that the 41 coefficients are all equal to 1 except for 3 pairs of smaller values. The adaptation is thus very simple. We remark that we have not yet failed to solve the nonlinear programming problem for sensible choices of $\{u_r\}$.

5. Continuum methods

In §3 we were able to fit the perturbed pattern to the quiescent pattern by least squares over the continuum [-1,1] of u. In §4, however, we apparently needed to discretise u in order to apply a minimax method.

In fact it is possible to consider a minimax fit over the continuum [-1,1], by taking $\{U_\ell\}$ to be all points in [-1,1] in (22), and this leads to a semi-infinite programming problem. A method of solution, using iteration from a discretised solution and based on a dual LP problem, has been described by Mason and Opfer (1986).

Acknowledgements

The financial support of the US Office of Aerospace Research and Development (Grant AFOSR-87-0206 DEF), the Science and Engineering Research Council, and the Admiralty Research Establishment is gratefully acknowledged.

References

Dolph, C.L. (1946), A current distribution for broadside arrays which optimises the relationship between beam width and side-lobe level. Proc. IRE and W. and E. June 1946, pp 335-348.

Gill, P.E., Murray, W. and Wright, M.H. (1981), "Practical Optimisation", Academic Press, New York.

Hudson, J.E. (1981), 'Adaptive Array Principles', Peter Peregrinus, London.

Mason, J.C. and Opfer, G. (1986), An algorithm for complex polynomial approximation with nonlinear constraints, In "Approximation Theory V", C.K. Chui, L. Schumaker, and J. Ward (Eds.), Academic Press, New York, pp 471-4

Mason, J.C. and Owen, P. (1987), Some simple algorithms for constrained complex and rational approximation. In: "Algorithms for Approximation", J.C. Mason and M.G. Cox (Eds.), Oxford U. Press, pp 357-372.

Mason, J.C. and Wilde, S.J. (1989), Phase only nulling in isotropic antenna arrays - a new minimax algorithm'. This volume.

Roberts, F.D.K. and Barrodale, I. (1979) "Solution of the constrained linear approximation problem". Report DM-132-IR, Department of Computer Science, University of Victoria, Victoria BC, Canada.

Shore, R.A. (1983), 'The use of beam space representation and nonlinear programming in phase-only nulling'. RADC-TR-83-124, Rome Air Development Corporation.

Shore. R.A., (1984), A proof of odd-symmetry of the phases for minimum weight perturbation phase-only null synthesis. IEEE Trans. Antennas Prop. AP-32 (1984), 528-530.

Steyskal, H. (1982), Synthesis of antenna patterns with prescribed nulls. IEEE Trans on Antennas and Prop. AP-30(1982), 273-279.

Taylor, T.T. (1955), Design of line-source antennas for narrow beam-width and low sidelobes. IRE Trans. Antennas Prop. AP-7, 16-28.

Villeneuve, A.T. (1984), 'Taylor patterns for discrete arrays. IEEE Trans on Antennas and Prop. AP-32(1984), 1089-1093.

Integration of absolute amplitude from a decibel B-spline fit

R. W. Allen and J. G. Metcalfe

Plessey Research (Caswell) Ltd, Towcester, Northamptonshire, UK

Abstract In a signal processing application, a signal passes through several devices each of which acts as a frequency dependent linear filter. The signal and transfer function for each device are each represented as a cubic B-spline, decibels against wavelength. The logarithmic representation means that the splines can be added. The final waveform needs to be integrated as absolute amplitude however. A general method of evaluating the function at many points before integrating would not be as fast or as accurate as the method described in this paper. The decibel function in each knot range is expanded in terms of a rapidly converging power series in absolute amplitude, integrated analytically and then summed over the ranges.

Key words: B-spline, Coefficients, Decibels, Integration.

1 Background to problem

In a signal processing application a signal passes through a number of devices. It is desired to find the total transmitted power at the end of the chain. Each device has a wavelength response given in decibels, and the final response is the product of the transfer functions for each of the devices in the chain.

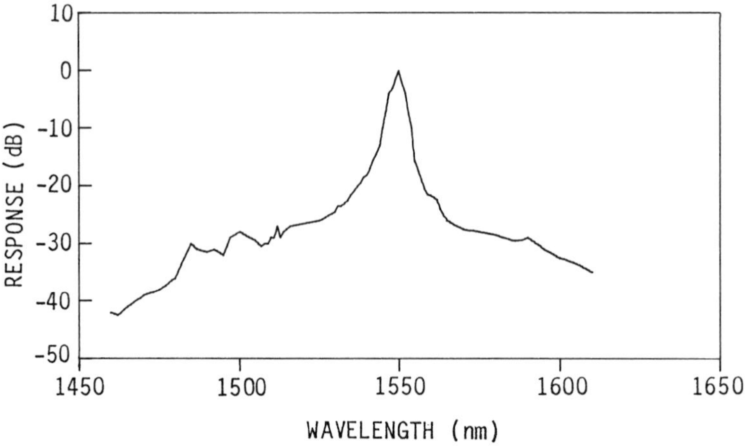

Figure 1   Typical transfer function

The response of each device is represented as a cubic spline in decibels against wavelength which means, if the splines have a common knot set, that the spline coefficients may just be added to yield the overall response. In practice, because the devices are defined over widely varying ranges it is found easier to fit these transfer functions with cubic splines using knots at multiples of a power of 2 in order to get as many common knots as possible.

The problem then remains of integrating the final response, both accurately and efficiently. If the cubic spline was itself to be integrated then a straightforward manipulation of the coefficients would suffice, but the objective is to integrate its anti-log.

The cubic spline can be represented as a linear combination of B-splines. We show how the integration process can be carried out in terms of these B-splines and their coefficients. The first stage is to express each B-spline as a set of polynominal pieces. Knowledge of the explicit form of these pieces facilitates the required integration.

2 Expansion of B-spline

A cubic spline, with N interior knots $k_1 < k_2 < \ldots < k_N$ is represented in terms of B-spline as:-

$$S(x)=\sum_{J=1}^{N+4} C_J . N_{4,J}(x), \text{ where } N_{n,J}(x) \text{ is the B-spline of}$$

order n defined on the knots $k_{J-n}, k_{J-n+1}, \ldots k_J$, and k_J for $J<1$ and $J>N$ are suitably chosen "exterior" knots such that $k_J \leqslant k_{J+1}$ for all J(see, e.g., Cox,1982). The B-splines satisfy the properties.

$N_{n,J}(x) > 0$ if $k_{J-n} < x < k_J$ and

$N_{n,J}(x) = 0$ if $x < k_{J-n}$ or $x > k_J$

They may be generated stably (Cox, 1972) by the three-term recurrence relation

$$N_{n,J}(x)= \frac{(x-k_{J-n})\, N_{n-1,J-1}(x)}{k_{J-1} - k_{J-n}} + \frac{(k_J-x)\, N_{n-1,J}(x)}{k_J - k_{J-n+1}}$$

with initial conditions

$$N_{1,J+1}(x) = \begin{cases} 1, & \text{if } k_J \leqslant x < k_{J+1} \\ 0, & \text{otherwise.} \end{cases}$$

If $k_J \leqslant x < k_{J+1}$, where J = 0, 1,...,N, then we can express $N_{4,M}(x)$, for M=J+1,J+2,J+3,J+4,in terms of x and the surrounding knots k_{J-2} to k_{J+3}.

$$S(x)= C_{J+1}.N_{4,J+1}(x)+C_{J+2}.N_{4,J+2}(x)+C_{J+3}.N_{4,J+3}(x)+C_{J+4}.N_{4,J+4}(x).$$

Let (cf Cox and Hayes, 1973), $C1=C_{J+1}$, $C2=C_{J+2}$, $C3=C_{J+3}$, $C4=C_{J+3}$, $K1=k_{J-2}$, $K2=k_{J-1}$, $K3=k_J$, $K4=k_{J+1}$, $K5=k_{J+2}$, $K6=k_{J+3}$ where $K3 \leqslant = x < K4$. Then the algebraic use of the recurrence relation yields.

$$S(x) = C1.\frac{(K4-x).(K4-x).(K4-x)}{(K4-K1).(K4-K2).(K4-K3)} + C2.\frac{(x-K1).(K4-x).(K4-x)}{(K4-K1).(K4-K2).(K4-K3)}$$

$$+C2.\frac{(K5-x).(x-K2).(K4-x)}{(K5-K2).(K4-K2).(K4-K3)} + C2.\frac{(K5-x).(K5-x).(x-k3)}{(K5-K2).(K5-K3).(K4-K)}$$

$$+C3.\frac{(x-K2).(x-K2).(K4-x)}{(K5-K2).(K4-K2).(K4-K3)} + C3.\frac{(x-K2).(K5-x).(x-K3)}{(K5-K2).(K5-K3).(K4-K}$$

$$+C3.\frac{(K6-x).(x-K3).(x-K3)}{(K6-K3).(K5-K3).(K4-K3)} + C4.\frac{(x-K3).(x-K3).(x-K3)}{(K6-K3).(K5-K3).(K4-K3)}$$

We now introduce a shifted variable z, with origin at the centre of the interval: $x = z + \frac{(K4+K3)}{2}$ and use the algebraic computing package MACSYMA to simplify $S(z) \equiv S(x)$ to the form of the explicit cubic polynomial:

$s(z) = S3.z^3 + S2.z^2 + S1.z + S0$, where S0,S1,S2,S3 are provided by the following Fortran subroutine:-

```
 SUBROUTINE EXPANDBSPLINE(L,KNOT,COEFF,S0,S1,S2,S3)
  INTEGER*4 L
  REAL*8 K1,K2,K3,K4,K5,K6,C1,C2,C3,C4,G1,G2,G3,G4,G5,G6,G7,G8,G9,
  &  S0,S1,S2,S3,KNOT(*),COEFF(*)
   K1=KNOT(L+1)
   K2=KNOT(L+2)-K1
   K3=KNOT(L+3)-K1
   K4=KNOT(L+4)-K1
   K5=KNOT(L+5)-K1
   K6=KNOT(L+6)-K1
   C1=COEFF(L)
   C2=COEFF(L+1)
   C3=COEFF(L+2)
   C4=COEFF(L+3)
   G2=K4-K2
   G3=K4-K3
   G4=K5-K2
   G5=K5-K3
   G6=K6-K3
   G7=K4*G2*G3*G4*G5*G6
   G8=G5*G6
   G9=C2-C3
   S0=(K4*(4*K4*((C2*K5-C3*K2)*(G2*G6*K5-G8*K2)-C3*K3*G4*G2*K6)+
  & K3*K3*(K4*(G8*G9+G2*((3*C4+C3)*G4-G9*G6))-(C2-3*C1)*G4*G8)+K4*
  & (2*K4*(G2*(C3*G4*K6+G6*(C3*K2-(C2+G9)*K5))+G8*(C2*(K5+K2)-2*C3*
  & K2))+(K4*(G2*(G9*G6+(C3-3*C4)*G4)-G8*G9)-G4*G8*(C2+3*C1))*K3))
  & +2*K4*K3*(K3*(G2*(C3*G4*K6+G6*((C2+G9)*K5-C3*K2))+G8*(-C2*(K5+
  & K2)+2*C3*K2))+2*(C3*K2-C2*K5)*(G2*G6*K5-K2*G8))-(G9*K4*G8-K4*
  & G2*(G9*G6+(C4-C3)*G4)-(C2+C1)*G4*G8)*K4*K4*K4+(K4*G8*G9-K4*G2*
  & (G9*G6+(C4+C3)*G4)+(C2-C1)*G4*G8)*K3*K3*K3)/G7/8
```

```
S1=(K4*((4*K4*G2*(C3*G4*K6+G6*(C3*K2-(C2+G9)*K5))+2*(K4*G8*
& G9+K4*G2*(G9*G6+(C3-3*C4)*G4)-(C2-3*C1)*G4*G8)*K3)+4*K3*(-C3*
& G2*G4*K6+G8*(-C2*(K5+K2)+2*C3*K2))+4*(C2*K5-C3*K2)*(G2*G6*K5+
& G8*K2))+(-K4*G8*G9+3*K4*G2*(G9*G6+(C4-C3)*G4)-(C2+3*C1)*G4*G8)*
& K4*K4+(3*K4*G8*G9+K4*G2*((3*C4+C3)*G4-G9*G6)+3*(C2-C1)*G4*G8)*
& K3*K3)/G7/4
S2=(K4*(2*(G2*(C3*G4*K6+G6*(K5*(C3-2*C2)+C3*K2))+G8*(K2*(2*C3-C2)
& C2*K5))+(G8*(K4*G9-(C2-3*C1)*G4)+3*K4*G2*(G9*G6+(C4-C3)*G4)))+-
& (3*G8*(K4*G9+G4*(C2-C1))+K4*G2*(G9*G6+G4*(C3-3*C4)))*K3)/G7/2
S3=(K4*G6*G9*(G5+G2)+G4*(K4*G2*(C4-C3)+G8*(C2-C1)))/G7
RETURN
END
```

3 Integration of absolute amplitude

For the interval (k_r, k_{r+1}), we wish to determine:-

$$I_n = \int_{k_r}^{kr+1} 10^{(e_r x^3 + f_r x^2 + g_r x + h_r)/10} dx,$$

where $e_r x^3 + f_r x^2 + g_r x + h_r$ represents the cubic polynomial in x to which the spline reduces for $k_r < x < k_{r+1}$.

Putting $z = x - \frac{k_{r+1}+k_r}{2}$ and $DX = \frac{k_{r+1}-k_r}{2}$,

$$I_n = \int_{-DX}^{DX} 10^{(S3.z^3 + S2.z^2 + S1.z + S0)/10} dz$$

where S0,S1,S2,S3 are the coefficients given by the routine:- EXPANDBSPLINE for the spline between the knots k_r, k_{r+1}.

This integral may be rewritten:-

$$I_n = \exp\frac{(S0.\ln 10)}{10} \int_{-DX}^{DX} \exp\left(\frac{S3.\ln 10.x^3}{10} + \frac{S2.\ln 10.x^2}{10} + \frac{S1.\ln 10.x}{10}\right).dx$$

or

$$I_n = d_n \int_{-DX}^{DX} e^{a_n.x^3} .e^{b_n.x^2} .e^{c_n.x} .dx,$$

where $a_n = (S3.\ln 10)/10$

$b_n = (S2.\ln 10)/10$

$c_n = (S1.\ln 10)/10$

$d_n = e^{((S0.\ln 10)/10)}$.

Expand these exponentials as power series and multiply out to give

$$I_n = d_n \int_{-DX}^{DX} \left(1+\left(b_n+\frac{c_n^2}{2!}\right).x^2+\left(a_nc_n+\frac{b_n^2}{2!}+\frac{b_nc_n^2}{2!}+\frac{c_n^4}{4!}\right).x^4.+\ldots\right)dx$$

omitting terms of odd order, since the integral over the stated range is zero. We then integrate I_n analytically and substitute the limits to give

$$I_n = 2.d_n.\left(DX+\left(b_n+\frac{c_n^2}{2!}\right)\frac{(DX)^3}{3}+\left(a_nc_n+\frac{b_n^2}{2!}+\frac{b_nc_n^2}{2!}+\frac{c_n^4}{4!}\right)\frac{(DX)^5}{5!}+\ldots\right.$$

Let $A(N) = \frac{(a_n(DX)^3)^{N-1}}{(N-1)!}$, $B(N) = \frac{(b_n(DX)^2)^{N-1}}{(N-1)!}$, $C(N) = \frac{(c_n\ DX)^{N-1}}{(N-1)!}$

$$I_n = 2.d_n.DX.\left(1 + \frac{B(2)+C(3)}{3} + \frac{C(2)A(2)+B(3)+B(2)C(3)+C(5)+\ldots}{5}\right)$$

$$= 2.d_n.DX.\sum_{N=1}^{N=N_F} \frac{F(N)}{2N-1},$$

where

F(1)=1,

F(2)=C(3)+B(2),

F(3)=C(5)+B(2)C(3)+B(3)+C(2)A(2),

etc.

Let $E(N) = \sum C(J).B(K)$

J = N, N-2, N-4,. . 2 or 1 (steps of -2)
K = 1, 2, 3, . . .

and $F(N) = \sum E(J).A(K)$

J = 2N-1, 2N-4, 2N-7,. . 3 or 2 or 1 (steps of -3)
K = 1, 2, 3, . . .

The subroutine FORMPOWERSERIES works as follows.

Let e = 10^{-15}.

The terms A(N), B(N), C(N) are evaluated until they are less than the product of e and the maximum term.

Let N_E= maximum of $2N_B$ and N_C, N_F= (maximum of $3N_A$ and N_E+1)/2.

Over 1000 integrations the average values used were:

N_A = 7, N_B = 7, N_C = 11, N_E = 11, N_F = 15.

The numbers quoted are for VAX FORTRAN DOUBLE PRECISION. Data involving 1089 integrals took only 6.6 seconds total cpu time to complete on a VAX8700 computer.

We believe that the method compares favourably with approaches based on numerical integration both in respect of obtainable accuracy and computation time.

```
C-------------------------------------------------------------
      SUBROUTINE INTEGRATELOSS(LOSS,SOURCENAME,DET_NAME)
      INTEGER*4 I,J,K,L,NSIGMA,COEFNO,ADDRECS,N,TCOUNT,NCOEFF
      REAL*8 DX,X,S0,S1,S2,S3,KNOT(2008),COEFF(2004),LOSS,SIGMA(300)
      CHARACTER*10 SOURCENAME,DETNAME
      CALL GETSPLINE(KNOT,COEFF,NCOEFF)
          IF(NCOEFF.EQ.0) THEN
          LOSS=-9.99D2
          ELSE
          DO 10 I=NCOEFF+1,4,-1
          KNOT(I)=KNOT(I-3)
 10       CONTINUE
          KNOT(1)=KNOT(4)
          KNOT(2)=KNOT(4)
          KNOT(3)=KNOT(4)
          KNOT(NCOEFF+2)=KNOT(NCOEFF+1)
          KNOT(NCOEFF+3)=KNOT(NCOEFF+1)
          KNOT(NCOEFF+4)=KNOT(NCOEFF+1)
          LOSS=0.0D0
          DO 30 L=1,NCOEFF-3
          CALL EXPANDBSPLINE(L,KNOT,COEFF,S0,S1,S2,S3)
          DX = (KNOT(L+4)-KNOT(L+3))/2.0D0
          CALL FORMPOWERSERIES(DX,S0,S1,S2,S3,SIGMA,NSIGMA)
          X=0.0D0
          DO 20 I=NSIGMA,1,-1
          X=X+SIGMA(I)/(2*I-1)
  20      CONTINUE
          X=2.0D0*X*DX*DEXP(S0)
          LOSS=LOSS+X
  30      CONTINUE
          LOSS=1.0D1*DLOG10(LOSS)
          ENDIF
          RETURN
          END
```

```
C-----------------------------------------------------------------
      SUBROUTINE FORMPOWERSERIES(DX,SO,S1,S2,S3,F,NF)
      INTEGER*4 NF,I,J,K,NC,NB,NA,NE
      REAL*8 DX,S,SO,S1,S2,S3,CL,SMAX,EPSILON,
      &   A(100),B(100),C(100),F(300),E(300)
      L=DLOG(1.0D1)/1.0D1
      SO=CL*SO
      S1=CL*S1*DX
      S2=CL*S2*DX*DX
      S3=CL*S3*DX**3
      EPSILON=1.0D-15
      S=1.0D0
      SMAX=S
      I=0
      DO 10 WHILE (DABS(S).GT.EPSILON*SMAX)
      I=I+1
      C(I)=S
      S=S*S1/I
      SMAX=DMAX1(SMAX,DABS(S))
   10 CONTINUE
      NC=I
      S=1.0D0
      SMAX=S
      I=0
      DO 20 WHILE (DABS(S).GT.EPSILON*SMAX)
      I=I+1
      B(I)=S
      S=S*S2/I
      SMAX=DMAX1(SMAX,DABS(S))
   20 CONTINUE
      NB=I
      S=1.0D0
      SMAX=S
      I=0
      DO 30 WHILE (DABS(S).GT.EPSILON*SMAX)
      I=I+1
      A(I)=S
      S=S*S3/I
      SMAX=DMAX1(SMAX,DABS(S))
   30  CONTINUE
       NA=I
       NE=MAX0(2*NB,NC)
       NF=(MAX0(3*NA,NE)+1)/2
       E(1)=C(1)
```

```
      DO 50 I=1,NE
      J=I
      K=1
      E(I)=0.0D0
      DO 40 WHILE (J.GT.0 .AND. I.LE.NC .AND. K.LE.NB)
      E(I)=E(I)+C(J)*B(K)
      K=K+1
      J=J-2
   40 CONTINUE
   50 CONTINUE
      F(1)=E(1)
      DO 70 I=1,NF
      J=2*I-1
      K=1
      F(I)=0.0D0
      DO 60 WHILE (J.GT.0 .AND. K.LE.NA)
      F(I)=F(I)+E(J)*A(K)
      K=K+1
      J=J-3
   60 CONTINUE
   70 CONTINUE
      RETURN
      END
C-----------------------------------------------------
```

Acknowledgement

This work is sponsored by the Rome Air Development Center (RADC) of the U.S. Air Force System Command.

References

Cox, M.G. 1972 The numerical evaluation of B-splines. J. Inst. Math. Appl. 10, 134-149. Also National Physical Laboratory NAC Report No. 4, 1971.

Cox, M.G. and Hayes, J.G. 1973 Curve fitting: a guide and suite of algorithms for the non-specialist user. National Physical Laboratory NAC Report No. 26.

Cox M.G. 1982 Practical spline approximation. Lecture Notes in Mathematics 965: Topics in Numerical Analysis, P.R. Turner, Ed., Berlin, SpringerVerlag, 79-112. Also National Physical Laboratory DITC Report No. 1/82, 1982.

A nonlinear least squares data fitting problem arising in microwave measurement

M. G. Cox and H. M. Jones

National Physical Laboratory,
Teddington, Middlesex, UK

Abstract An automatic network analyser can be used to measure the complex transmission coefficient, T_d, of a microwave device. The Q-factor which characterises the device is determined by fitting a nonlinear model to data which is collected by measuring T_d at a number of distinct known frequencies. Both real and imaginary parts of the measurements of T_d contain errors. Also there is correlation between the errors in the real and imaginary components of any particular measurement of T_d, but the errors in measurements at different frequencies are independent. Therefore, a block-diagonally weighted nonlinear least squares problem is formulated and solved. The solution method used is iterative and requires starting values which are obtained automatically from a pair of linear least squares problems.

Key words: Automatic starting values, Block-diagonal weighting, Correlated errors, Data fitting, Linear least squares, Nonlinear least squares, Microwave measurement.

1. Introduction

The problem considered here is that of estimating values of parameters which characterise a microwave device. This involves fitting a nonlinear model to data obtained using an automatic network analyser (ANA). The analyser measures values of the transmission coefficient, T_d, of the device for a number of signals at different (known) frequencies. T_d is represented by the real and imaginary parts of a complex number. For a given frequency, f,

$$T_d = \frac{T_0 e^{i\phi}}{1 + iQ\left(\frac{f}{f_0} - \frac{f_0}{f}\right)} + E, \tag{1}$$

where T_0 is the transmission coefficient at the resonant frequency, f_0, Q is the Q-factor and ϕ and E are physical parameters relating to the phase error and leakage in the measurement system; see Grant and Phillips (1975, pp209-214). In particular, we wish to determine Q and its statistical uncertainty. The measured values of T_d contain errors which are correlated in the manner described in section 2. In section 3 we derive the mathematical model and formulate the least squares fitting problem. This formulation incorporates a block-diagonal weighting matrix which takes account of the error structure. The problem is analysed in section 4 and recast in a form suitable for solution by standard library software. The method used for solving the least squares problem is iterative and so requires starting values. These are obtained by solving a pair of linear least squares problems; see section 5. We present the results obtained for a typical data set in section 6. Section 7 contains some concluding remarks.

2. Data

The data collected by the ANA is denoted here by f_i, x_i, y_i, $i = 1, 2, \ldots, m$, where x_i and y_i are the measured real and imaginary parts of the complex transmission coefficient at the known frequency f_i. The number of distinct frequencies is usually quite small; typically $m = 8$. The ANA takes many (eg 60) readings of the transmission coefficient for each f_i, so the values x_i and y_i provided are the result of averaging these readings; estimates of $\text{var}(x_i)$, $\text{var}(y_i)$ and $\text{cov}(x_i, y_i)$ are also provided by the ANA. In general $\text{cov}(x_i, y_i)$ is nonzero and, since measurements at *different* frequencies are statistically independent,

$$\text{cov}(x_i, x_j) = \text{cov}(y_i, y_j) = \text{cov}(x_i, y_j) = 0, \; i \neq j.$$

The frequency of the input signal can be controlled to an accuracy which is very much greater than that of the measurements x_i and y_i, so it is reasonable to regard the f_i as exact.

Data with this error structure can arise in other measurement problems, for instance, in the use of coordinate measuring machines to measure mechanical components (Cox, 1985) and in photogrammetry to measure a wide range of objects (Granshaw, 1980). See also Forbes (these proceedings). The correlation may not be in pairs but could be in groups of 3, 4, or more.

3. Model and problem formulation

In (1), T_d is regarded as the complex-valued dependent variable, and f the independent variable. The unknown parameters T_0, f_0, Q, and ϕ are all real and $E = E_x + iE_y$ is complex. Setting $T_d = x + iy$ and $\tilde{Q} = Q/f_0$, and separating real and imaginary parts we obtain

$$x = \frac{T_0 \cos\phi + T_0 \sin\phi \, \tilde{Q} \left(\frac{f^2 - f_0^2}{f}\right)}{1 + \tilde{Q}^2 \left(\frac{f^2 - f_0^2}{f}\right)^2} + E_x$$

and

$$y = \frac{T_0 \sin\phi - T_0 \cos\phi \, \tilde{Q} \left(\frac{f^2 - f_0^2}{f}\right)}{1 + \tilde{Q}^2 \left(\frac{f^2 - f_0^2}{f}\right)^2} + E_y.$$

The change of variable from Q to $\tilde{Q}$ reduces the correlation between the parameters and thereby makes the fitting process more robust.

If we set $\mathbf{u} = (T_0, \phi, \tilde{Q}, f_0, E_x, E_y)^T$, these equations can be written as:

$$x = X(\mathbf{u}, f)$$

and

$$y = Y(\mathbf{u}, f),$$

where both X and Y are nonlinear functions of the model parameters $\mathbf{u}$ and the independent variable f. Since the measurement errors occur in the dependent variables x and y, this pair of equations has right hand sides which only involve data which can be regarded as exact.

Assuming the model is correct we have, for $i = 1, 2, \ldots, m$,

$$x_i = X(\mathbf{u}, f_i) + e_{2i-1}$$

and

$$y_i = Y(\mathbf{u}, f_i) + e_{2i},$$

where $\mathbf{e} = (e_1, e_2, \ldots, e_{2m})^T$ is the error vector. Components of $\mathbf{e}$ with odd subscripts relate to measurements of x and the remainder to measurements of y. The appropriate problem to solve is

$$\min_{\mathbf{u}} = \mathbf{e}^T W \mathbf{e},$$

where the weighting matrix $W = V^{-1}$, the inverse of the variance matrix associated with $\mathbf{e}$. This formulation takes full account of the nonuniformity of the data errors; see for example, Mardia, Kent and Bibby (1979, pp172-173). In the next section we show that although V (and hence W) is nondiagonal it has a block structure which can be exploited.

4. Analysis

The diagonal elements of the variance matrix V are the variances of the measured data and the off-diagonal terms their covariances. In our case, because the measurements are only correlated pairwise, V takes the block-diagonal form

$$V = \begin{pmatrix} V_1 & & & \\ & V_2 & & 0 \\ 0 & & \ddots & \\ & & & V_m \end{pmatrix}$$

with

$$V_i = \begin{pmatrix} \mathrm{var}(x_i) & \mathrm{cov}(x_i, y_i) \\ \mathrm{cov}(x_i, y_i) & \mathrm{var}(y_i) \end{pmatrix}.$$

If the data were correlated in bigger groups the structure would be identical but with larger blocks on the diagonal.

V^{-1} has a similar structure and can be factorised to give

$$V^{-1} = L^T L,$$

where L has the block lower triangular structure

$$L = \begin{pmatrix} L_1 & & & \\ & L_2 & & 0 \\ 0 & & \ddots & \\ & & & L_m \end{pmatrix}$$

with
$$L_i = \begin{pmatrix} \alpha_i & 0 \\ \beta_i & \gamma_i \end{pmatrix},$$

where
$$\alpha_i = \frac{1}{\sqrt{\text{var}(x_i)}},$$

$$\beta_i = \frac{-\text{cov}(x_i, y_i)}{\text{var}(x_i)\sqrt{\text{var}(y_i)(1-\rho_i^2)}}$$

and
$$\gamma_i = \frac{1}{\sqrt{\text{var}(y_i)(1-\rho_i^2)}}.$$

In the above, ρ_i is the coefficient of correlation between x_i and y_i and is given by $\rho_i = \text{cov}(x_i, y_i)/\sqrt{\text{var}(x_i)\text{var}(y_i)}$. A derivation of these formulae is given in Appendix A of Cox and Jones (1988).

The problem becomes

$$\min_{\mathbf{u}} \mathbf{e}^T L^T L \mathbf{e} = \min_{\mathbf{u}} \tilde{\mathbf{e}}^T \tilde{\mathbf{e}},$$

where
$$\tilde{e}_{2i-1} = \alpha_i e_{2i-1}$$

and
$$\tilde{e}_{2i} = \beta_i e_{2i-1} + \gamma_i e_{2i}.$$

This is now a nonlinear least squares data fitting problem with a modified model. There are a number of recognised techniques for solving such problems, including Gauss-Newton, Levenberg-Marquardt, Quasi-Newton and modified Newton (see, eg, Gill, Murray and Wright, 1980, pp133-140). These and any other method for solving nonlinear problems will be iterative, so good starting values are required in order to ensure satisfactory convergence and to reduce the risk of convergence to possibly inferior local solutions.

5. Starting values

In some cases the experimentalist may be able to provide estimates of the parameters $\mathbf{u}$ which can be used as starting values for the iteration. However it is often difficult to estimate some of the parameters accurately, particularly if the data is analysed remote from the experiment. So an automatic method of providing the starting values is preferred.

For experimental reasons the data is collected at frequencies close to the resonant frequency f_0, so an estimate of this value should always be available. If the experiment did not record this estimate then the mean of the frequencies at which the readings have been made should provide an adequate starting value.

The remaining starting values are obtained from the model. First define

$$\hat{f} = \frac{f}{f_0} - \frac{f_0}{f},$$

and express (1) in the form

$$x + iy = \frac{T_0(\cos\phi + i\sin\phi)}{1 + iQ\hat{f}} + E_x + iE_y.$$

Multiplying throughout by $1 + iQ\hat{f}$ and taking real and imaginary parts gives two equations:

$$x - Qy\hat{f} = T_0\cos\phi + E_x - E_yQ\hat{f}$$

and

$$y + Qx\hat{f} = T_0\sin\phi + E_y + E_xQ\hat{f}.$$

These equations, unlike those used to solve the full problem, give x as a function of both y and f and y as a function of x and f. It is difficult to take account of the error structure in the data when solving equations in this form. However, this is not crucial since we are using them solely to generate starting values which will then be improved by solving a problem with the correct model.

Each of these equations can be expressed as a linear model:

$$x = b_1 + b_2y\hat{f} - b_3\hat{f},$$
$$y = b_4 - b_5x\hat{f} + b_6\hat{f},$$

where

$$b_1 = T_0\cos\phi + E_x,$$
$$b_2 = Q,$$
$$b_3 = E_yQ,$$
$$b_4 = T_0\sin\phi + E_y,$$
$$b_5 = Q$$

and

$$b_6 = E_xQ.$$

Estimates of b_j, $j = 1, 2, \ldots, 6$, are obtained by solving the two resulting linear regression problems independently. Starting values are then given by

$$Q = (b_2 + b_5)/2,$$
$$E_y = b_3/Q,$$
$$E_x = b_6/Q,$$
$$\phi = \tan^{-1}\left(\frac{b_4 - E_y}{b_1 - E_x}\right)$$

and

$$T_0 = \sqrt{(b_1 - E_x)^2 + (b_4 - E_y)^2}.$$

It should be noted that no approximation, other than ignoring the measurement error, is made in obtaining these starting values. The method therefore has the property that it will provide the *exact* solution for exact data and can be expected to provide good estimates for data with small errors.

Frequency f	Number of readings N	Real part x	Imaginary part y	$N\times$ var(x)	var(y)	cov(x, y)
8989.6500	61	-0.125313	-0.667130	0.070666	0.072623	0.006347
8989.6786	63	-0.220381	-0.739569	0.082335	0.072512	0.005690
8989.7071	60	-0.359474	-0.837055	0.071147	0.074034	0.009399
8989.7357	63	-0.542974	-0.790293	0.070657	0.070277	0.010966
8989.7643	63	-0.706873	-0.614188	0.076543	0.077347	0.030132
8989.7929	59	-0.726177	-0.443217	0.069627	0.064567	0.008923
8989.8214	62	-0.675334	-0.271518	0.073091	0.072928	0.024311
8989.8500	62	-0.582775	-0.194680	0.067468	0.062324	0.009345

TABLE 1
Data provided by the ANA

Iteration number	Number of function evaluations	Sum of squares of residuals S	Norm of gradient vector
0	1	1.73×10^{-2}	8.56×10^{-1}
1	2	1.69×10^{-3}	5.61×10^{-6}
2	3	1.69×10^{-3}	1.08×10^{-11}
3	6	1.69×10^{-3}	1.47×10^{-13}
4	13	1.69×10^{-3}	6.27×10^{-24}
Q-Factor = 54595.		Standard error = 2104.	

TABLE 2
Results obtained by optimization

6. Results

The method described above has been applied successfully to a number of ANA data sets. We present here the results corresponding to one of these sets in which approximately 60 measurements of the complex transmission coefficient were made for each of 8 nominally uniformly spaced frequencies. Table 1 shows the data provided by the ANA. The first (numerical) row of Table 2 corresponds to the initial estimates of the six parameters as provided by the automatic start method of section 5. Subsequent rows show the progress of the routine LSQFDN - a nonlinear least squares algorithm based on a modified Newton method from NOSL - the NPL Numerical Optimization Software Library (Hodson, Jones and Long, 1982). We note that convergence to the solution was very rapid with almost all of the reduction in the residual sum of squares, S, occuring in the first iteration. The value of S corresponding to the starting values exceeded that at the solution by a factor of only 10, thus demonstrating that the initial estimation procedure is quite effective. Figure 1 shows both the real and imaginary parts of the data and the fitted function.

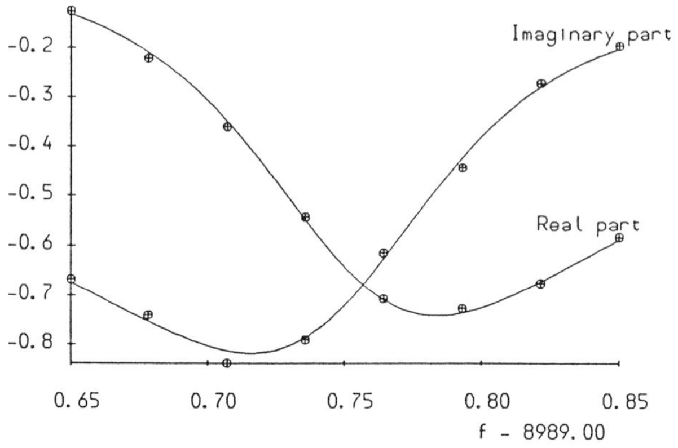

FIG. 1. *Input data and fitted function*

7. Concluding remarks

In this paper we have described a method based on the use of optimization software for solving a generalised least squares data fitting problem in which the model is nonlinear in its parameters and the errors in the data have a particular structure. We have shown how starting values for the optimization can be obtained by solving a related linear least squares problem. This linear problem was obtained by ignoring the measurement errors but without making any mathematical approximation, and thus has the desirable property that it will provide exact values of the parameters for exact data.

Data with a similar error structure can arise in mathematical models of other physical processes. Although many of these models can also be solved using nonlinear optimization, it may only be possible to obtain a linear starting problem by using an *approximation* to the real problem; an example of this is given in Cox and Jones (1988).

A class of *linear* data fitting problems with block-diagonal variance matrices, in which the mathematical model is a polynomial spline, is considered in Cox and Manneback (1985).

8. Acknowledgements

The authors are grateful to Bob Clarke of the Division of Electrical Science at NPL for bringing this problem to their attention and for supplying the ANA data sets. We also thank Gerald Anthony and Alistair Forbes for discussions on this problem, and Chris Donovan and Tina Eagling who developed pilot code for solving it.

References

1. M G Cox. Software for computer-aided design. *Laboratory Practice*, 34 (6):59–63, 1985.
2. M G Cox and H M Jones. An algorithm for least squares circle fitting to data with specified uncertainty ellipses. Technical Report DITC 118/88, National Physical Laboratory, 1988.
3. M G Cox and P E Manneback. Least squares spline regression with block-diagonal variance matrices. *IMA J Numer Anal*, pages 275–286, 1985.
4. A B Forbes. Least squares best fit geomtric elements. These proceedings.
5. P E Gill, W Murray, and M H Wright. *Practical Optimization*. London, Academic Press, 1981.
6. S I Granshaw. Bundle adjustment methods in engineering photogrammetry. *Photogrammetry Record*, 10:181–207, 1980.
7. I S Grant and W R Phillips. *Electromagnetism*. Chichester, Wiley, 1975.
8. S M Hodson, H M Jones, and E M R Long. A brief guide to the NPL Numerical Optimization Software Library. Technical report, National Physical Laboratory, 1982.
9. K V Mardia, J T Kent, and J M Bibby. *Multivariate Analysis*. London, Academic Press, 1979.

A complex minimax algorithm for phase-only adaptation in antenna arrays

J. C. Mason and S. J. Wilde

Applied and Computational Mathematics Group,
Royal Military College of Science,
Shrivenham, Wiltshire, UK

Abstract Suppression of electromagnetic radiation in antenna array patterns may be achieved by imposing constraints on the directional power of the antenna. Current interest is in applying time (phase) delays at each element of the array to meet the required constraints. The problem is essentially a complex polynomial approximation problem with inequality constraints on the polynomial and equality constraints on the magnitude of each coefficient. We present an algorithm for computing these phase delays, based on an l_∞ measure of fit.

Keywords: Antenna, Constrained, Fitting, Minimax, Nonlinear, Phase-only.

1 Introduction

A phased array antenna (an example of a linear array is given in Figure 1) is a configuration of a multitude of individual antennas. In practice these antennas are equispaced and isotropic (i.e non-directional) and may receive or transmit signals. The radiation pattern of the array is determined by the relative phases and amplitudes of the currents at each individual antenna. An important property is to be able to suppress strong directional interference (jammers) when receiving signals and to be able to constrain transmission to specific locations. This may be achieved by altering the current weighting of each antenna element in phase (i.e. argument) or amplitude or both, so as to produce constrained regions in the directional power distribution (known as the far field pattern).

The power (amplitude) of each element is usually preset by the hardware design of the array and so amplitude adjustment is an undesirable option both physically and commercially. The amplitude weightings are chosen so as to give the desired far field pattern in the quiescent (no interference) environment. Phase-only adaptation is of particular importance since in a phased array antenna the phase shifters are already present, saving additional cost.

The objective is to minimize the perturbation between the quiescent and updated patterns whilst satisfying the directional power constraints. We develop an algorithm which gives a minimax fit to the desired quiescent pattern, subject to these constraints. The algorithm establishes a bound for the maximum perturbation error and can easily support the failure of elements in the array.

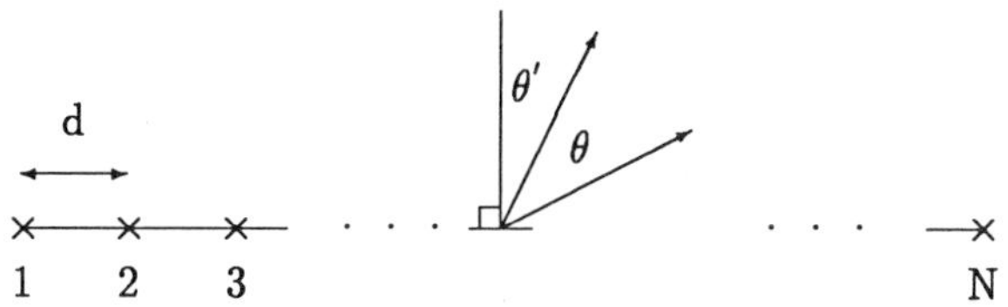

Figure 1: N isotropic antenna elements - interelement spacing d.

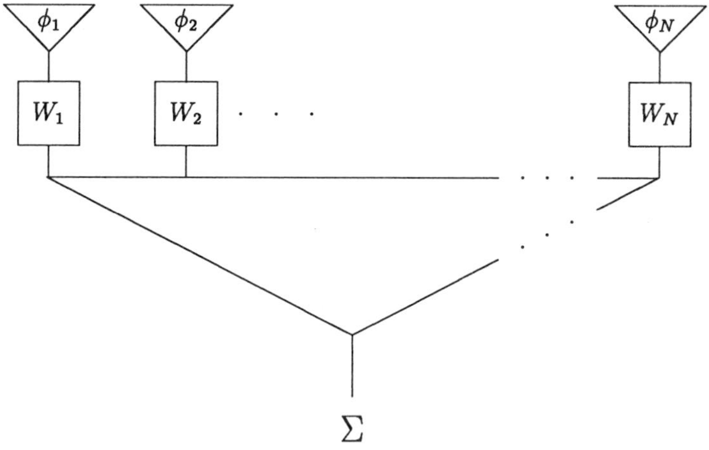

Figure 2: Basic structure of a phased array.

2 The Quiescent Pattern

The transmission and reception patterns of an antenna are identical and so for convenience we shall restrict the discussion to signal reception. The directive field pattern (Elliot (1981)) of a linear antenna array of N equispaced isotropic elements (Figures 1 & 2) is proportional to

$$P(u) = \sum_{j=1}^{N} x_j e^{i[d_j u + \phi_j]} \tag{1}$$

where $\{x_j\}$ are the real preset current amplitudes and

$$d_j = \left[\frac{N-1}{2}\right] - (j-1) \qquad j = 1, \ldots, N,$$

$$u = \frac{2\pi}{\lambda}\, d\,(\sin\theta - \sin\theta'),$$

where

θ	=	angle from boresight[1],
θ'	=	direction of boresight,
d	=	inter-element spacing,
λ	=	wavelength,
ϕ_j	=	element phase shifters.

In the quiescent (i.e. no interference) environment $\{\phi_j\} = 0$. A typical Taylor weighted (Villeneuve (1984)) quiescent pattern is shown in Figure 3, giving the directional power of the array. The relative amplitude weighting of each element for this pattern is shown in Figure 4.

Because of the wavelength dependency of (1), a small change $\delta\lambda$ in the wavelength is analogous to some change δu of u. Therefore, to impose a constraint on the signal power for all wavelengths over which the antenna may transmit or receive, it is necessary to apply the constraint over an interval in u proportional to the frequency bandwidth of the antenna.

3 The Algorithm

Essentially, the problem is to constrain the pattern (1) by altering the phase shifters ϕ_j to minimizing an error norm

$$\| P(u) - P_o(u) \|$$

where P_o is the quiescent pattern. The norm of approximation chosen is l_∞ (minimax) on a set of data points

$$u_k = \frac{2\pi}{\lambda} d(\sin\theta_k - \sin\theta'), \qquad \theta_k \in [-\frac{\pi}{2}, \frac{\pi}{2}], \qquad k = 1, \ldots, M.$$

[1]Boresight is the direction in which the array is steered.

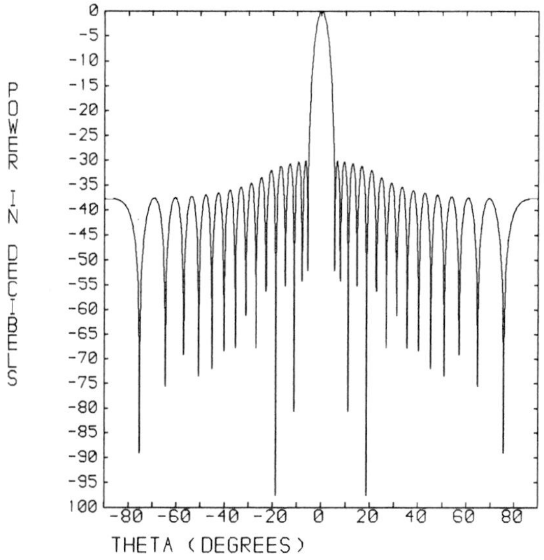

Figure 3: Quiescent Taylor weighted pattern.

Previous algorithms (for example Shore(1983), Mason *et al* (1989)) have, in general, applied a least squares measure of fit. The use of the l_∞ norm, although expensive in CPU time, establishes a maximum global error at the data points chosen. (The use of the l_2 norm, in contrast, would not ensure the latter.)

Define

$$P(u) - P_o(u) = R(u) + iI(u),$$

where

$$R(u) = \sum_{j=1}^{N} x_j[\cos(d_j u + \phi_j) - \cos(d_j u)],$$

$$I(u) = \sum_{j=1}^{N} x_j[\sin(d_j u + \phi_j) - \sin(d_j u)].$$

Following Barrodale *et al* (1978) we minimize:

$$E_M = \max_{k=1,\dots,M} \{| R(u_k) |, | I(u_k) |\}, \qquad \text{for } \phi_j \in \Re. \tag{2}$$

The exact l_∞ error is

$$E_{M,\infty} = \max_{k=1,\dots,M} | R(u_k) + iI(u_k) |, \tag{3}$$

and is bounded as follows:

$$E_M \leq E_{M,\infty} \leq \sqrt{2}E_M.$$

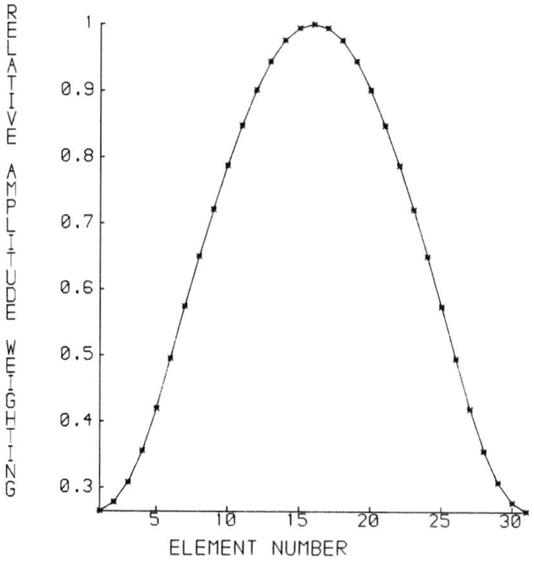

Figure 4: Current amplitudes of Taylor weighted pattern (Figure 3).

Streit(1985) uses the measure

$$\begin{aligned} E_{M,S} &= \max\{R(u_k)\cos\alpha_s + I(u_k)\sin\alpha_s\}, \\ & k = 1,\ldots,M \\ & s = 1,\ldots,S \end{aligned}$$

where

$$\alpha_s = (s-1)\frac{2\pi}{S}.$$

This is equivalent to (2) when $S = 4$ and although[2] it approaches $E_{M,\infty}$ as $S \to \infty$, it is computationally very expensive for even moderate values (≥ 8) of S.

The aim of the partial linearization of (3) is to aid convergence to a solution (see §5). To implement the minimization of (2), we introduce a variable Δ which bounds R and I in magnitude:

$$-\Delta \leq R(u_k) \leq \Delta,$$

$$-\Delta \leq I(u_k) \leq \Delta,$$

and then minimize Δ subject to constraints (see §4).

[2]In fact

$$E_{M,S} \quad \leq \quad E_{M,\infty} \quad \leq \quad E_{M,S}\sec\frac{\pi}{S}.$$

The physical loss (failure) of elements in an array can readily be counteracted by presetting the relevant weights to zero and applying phase changes to the remaining elements. Although this reduces the number of degrees of freedom available, it represents little pattern loss for a large array (e.g. 1000 elements).

4 Constraints

The constraints imposed are of two types:

1. Constraints on the pattern (over the required region)
2. Constraints on the interval within which each phase change must lie.

4.1 Constraints on the pattern

To constrain the pattern, P_u, over a designated range $[\theta_a, \theta_b]$ of θ we impose

$$\mid P(u_k) \mid \leq EPS \mid P(0) \mid \tag{4}$$

for ICON chosen points

$$u_k = \frac{2\pi}{\lambda} d(\sin\theta_k - \sin\theta'), \qquad \theta_k \in [\theta_a, \theta_b], \qquad k = M+1, \ldots, M+ICON.$$

EPS is given by

$$EPS = 10^{\frac{DBMAX}{20}},$$

where DBMAX is the required power constraint depth measured in decibels[3] and $\mid P(0) \mid^2$ is the boresight power of the array. The condition (4) is then replaced by the (stronger) linear conditions

$$-eps \leq R(u_k) \leq eps,$$

$$-eps \leq I(u_k) \leq eps,$$

where

$$eps = \frac{EPS}{\sqrt{2}}.$$

4.2 Constraints on the phases

The phase changes are restricted to a specified interval $[\phi_a, \phi_b]$ and for a global solution for $\phi_j \in \Re$, we require $\phi_b = \phi_a + 2\pi$. It would be expected that, given an interval length of 2π, the result obtained would be independent of ϕ_a if phases were reduced to their principal values.

[3]A decibel is a measure of power given by $10\log(\text{Power})$.

5 Results

The results for several test problems, solved by the partially linearized algorithm, have been compared with the true (nonlinear) solution of (3), i.e. the minimum of

$$E^2_{M,\infty} = \max_{k=1,\ldots,M} \{R(u_k)^2 + I(u_k)^2\}, \qquad \text{for } \phi_j \in \Re. \tag{5}$$

Both were solved using the sequential quadratic programming (SQP) routine E04VDF, from the NAG FORTRAN library (Mark 13).

Test	Constraint position ($\sin(\theta_k)$)	Constraint depth (DBMAX)	Number of SQP iterations
1	0.7(0.01)0.72	-100	20
2	0.7(0.01)0.74	-70	15
3	0.3(0.01)0.32	-100	12
4	0.1(0.01)0.12	-70	22

Table 1: Results for the partially linearized algorithm.

For each problem the number of elements of the array was 31 with boresight $\theta' = 0$ and inter-element spacing $d = \frac{\lambda}{2}$. The SQP algorithm was limited to a maximum of 500 iterations and the fitting points were equispaced over the interval $[-\frac{\pi}{2}, \theta_a] \cup [\theta_b, \frac{\pi}{2}]$. The quiescent pattern was chosen as a sinc pattern ($x_j = 0$) and the phase changes were unbounded and preset to zero on initiation of the algorithm. A summary of these results is given in Tables 1 and 2. The nonlinear algorithm was much slower and actually failed to converge within 500 iterations for tests 2 and 4.

Test	Constraint position ($\sin(\theta_k)$)	Constraint depth (DBMAX)	Number of SQP iterations
1	0.7(0.01)0.72	-100	176
2	0.7(0.01)0.74	-70	500
3	0.3(0.01)0.32	-100	277
4	0.1(0.01)0.12	-70	500

Table 2: Results for the nonlinear algorithm.

Returning to the partially linearized algorithm, the test problems were solved whilst restricting the interval of permitted phase change. Again the initial array phases were preset to zero. Now two quiescent patterns were considered: (a) a

Taylor weighted pattern (maximum sidelobe level 30DB and $\bar{n} = 6$) (see Figures 3 & 4) and (b) the sinc pattern. Defining the interval for ϕ_j to be $[0, 2\pi]$, we produced solutions with which many of the phase changes were zero. Table 3 lists the number of non-zero phase changes that occurred.

Test	Quiescent pattern	E04VDF iterations	Objective function (Δ)	Number of phase perturbations	Average phase pert. (rads)
1	Taylor	9	0.615	8	0.233
1	Sinc	9	2.355	10	0.244
2	Taylor	5	0.607	6	0.318
2	Sinc	6	2.468	6	0.431
3	Taylor	17	0.596	7	0.321
3	Sinc	23	2.312	6	0.491
4	Taylor	126	6.026	15	1.088
4	Sinc	152	6.385	25	0.789

Table 3: Restricting phase perturbations to $[0, 2\pi]$.

However, restricting the interval to be $[-\pi, \pi]$, we obtained skew symmetric results (i.e $\phi_j = -\phi_{N+1-j}$) for each of the test problems and non-zero phase changes occurred at each element (Table 4).

Test	Quiescent pattern	E04VDF iterations	Objective function (Δ)	Number of phase perturbations	Average phase pert. (rads)
1	Taylor	16	0.298	31	0.000
1	Sinc	20	1.146	31	0.000
2	Taylor	18	0.298	31	0.000
2	Sinc	15	1.316	31	0.000
3	Taylor	15	0.540	31	0.000
3	Sinc	12	2.208	31	0.000
4	Taylor	14	4.711	31	0.000
4	Sinc	22	5.909	31	0.000

Table 4: Restricting phase perturbations to $[-\pi, \pi]$.

In every case, the 'skew symmetric' solution was better (in the sense of a smaller objective function) than the solution resticting ϕ_j to the interval $[0, 2\pi]$. The results of test 2 (in Tables 3 & 4), using a Taylor weighted array, are shown in Figures 5 & 6.

Shore(1984) proved that for minimized weight perturbations in a phased-array, the optimum solution results in phases which are skew symmetric (with respect to

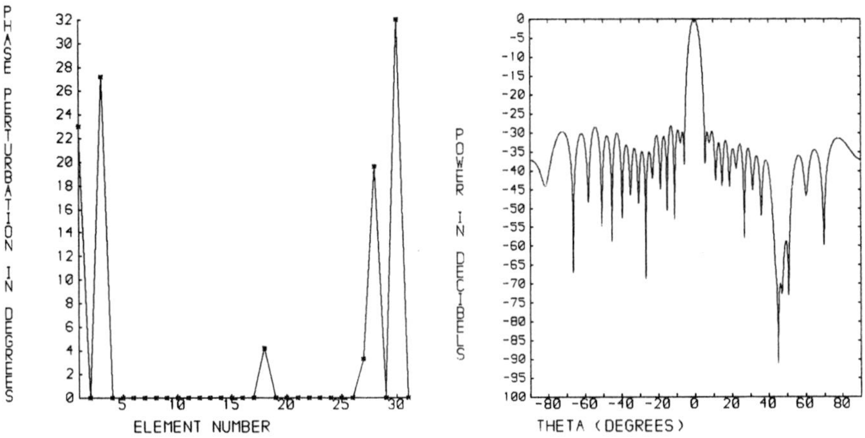

Figure 5: Test 2 - Phase interval $[0, 2\pi]$.

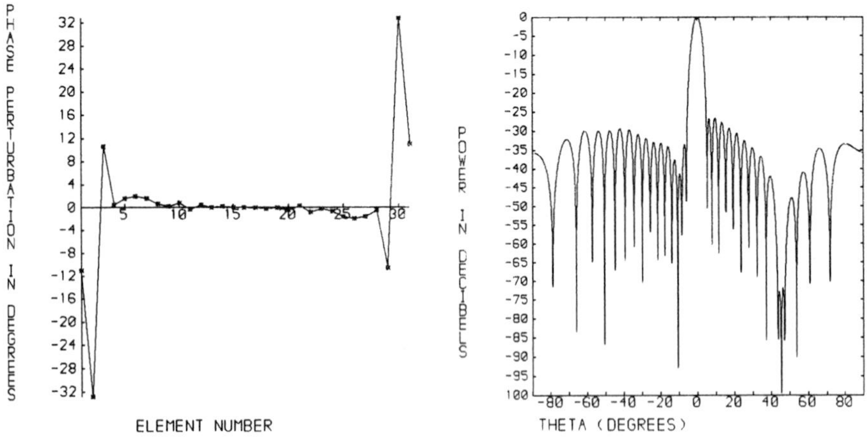

Figure 6: Test 2 - Phase interval $[-\pi, \pi]$.

a phase reference at the array centre). Numerical results lead us to conjecture that this is true for our 'minimax' algorithm with phase interval $[-\pi, \pi]$, although this has yet to be proved theoretically.

In conclusion, although convergence is not guaranteed, the algorithm appears to be robust for all test examples considered.

6 References

Elliot, R.S. (1981) Antenna Theory and Design, Prentice-Hall

Villeneuve, A.T. (1984) Taylor patterns for Discrete arrays. IEE Transactions on Antennas and Propagation, Vol. AP-32, No. 10 , pp. 1089-1093.

Barrodale, I., Delves, L.M. and Mason J.C. (1978) Linear Chebyshev approximation of complex-valued functions. Math. Comp., Vol. 32, pp. 853-863.

Mason, J.C. Trefethen, A.E. and Wilde, S.J. (1989) Constrained complex approximation algorithms in communication engineering. These proceedings.

Streit, R.L. (1985) An algorithm for the solution of systems of complex linear equations in the l_∞ norm with constraints on the unknowns. ACM Trans Math. Software 11, pp. 242-249.

Shore, R.A. (1983) Phase-only nulling as a nonlinear programming problem. RADC TR-83-37, AD-A130552.

Shore, R.A. (1984) A proof of the odd-symmetry of the phases for minimum weight perturbation phase only synthesis. IEE Transactions on Antennas and Propagation, Vol. AP-32, No. 5, pp. 528-530.

PART THREE
Catalogue of Algorithms

A catalogue of algorithms for approximation

E. Grosse

AT&T Bell Laboratories, Murray Hill, New Jersey, USA

Abstract This is an outline and a list of algorithms for numerical approximation, with references to the literature and pointers to available code. The database has been formatted here as an indented tree using $\TeX$; alternatively, it may be viewed in graphical form or may be traversed using a C program or Hypercard. Log in to `research.att.com` with userid `walk` by data network or telephone or search for keywords by `netlib`.

Key words: approximation algorithms, bibliography, database, library routines, subroutines, numerical consultant, netlib, walk

1. Introduction

There is a large gap between the classical polynomial, rational, and univariate spline fitting taught in textbooks and the variety of additional methods that have been developed for complicated real-world problems. Someday we may hope for a Knuth to write the definitive survey, but probably it is too soon for that. The following categorized list of algorithms can help bridge this gap in the meantime. This catalog is designed first to assist in numerical analysis consulting; second, to guide the acquisition of software for netlib[172]; and third, to help inventors of new methods search for prior art.

Notes in square brackets in the bibliography indicate where software may be found. Most commonly this will be a directory in netlib or one of the standard commercial libraries[19, 321, 404, 210]. For example, to get LSQR[414], send the one-line message `send lsqr from misc` by electronic mail to `research!netlib` or `netlib@research.att.com`. The name "toms" refers to the Collected Algorithms of the ACM, available via IMSL and netlib.

When known, references most closely related to final code are given. These will often not be the primary reference for the idea and hence many classic papers in approximation do not appear. Analytical papers are cited if they provide useful guidance on choosing between codes or if the topic seems ripe for computation although no algorithmic papers are known. Generally the comparative foundations do not exist for recommending a best method in each category. It would have been desirable to mention only methods for which convincing numerical evidence has been reported. Consistently adhering to such a policy would have made a very much shorter list but would necessarily have left out many good algorithms. Where NAG routines are apparently superseded by DASL routines, they have not been listed. Methods requiring an (expensive) software license have been de-emphasized, particularly when competitive public versions are available. Finally, it was not feasible to cite the papers in this proceedings; but they report a number of exciting developments and the reader who has not already done so is encouraged to look at the rest of the book!

2. Classification

In addition to the traditional characterization of practical approximation as fundamentally a choice of *form* and *norm*, this scheme adds a third category, *variable*, to reflect the important choices of sample designs and coordinate transforms. The primary classification is by form since it corresponds most directly to the literature. Some methods are not specific to a particular basis but are more readily characterized by, for example, the norm they optimize; these are listed under norm.

Another possible characterization would have been by the operation involved. Some of these are: evaluate, differentiate, integrate; interpolate, extrapolate, local fit; best or good fit in various norms; diagnostics, error estimation; conversions between different bases; i/o, graphics; zeros, maxima, other critical features.

The database is formatted below as an indented tree with cross-links. The bracketed numbers at the leaf nodes are lists of references. While this paper snapshot may be convenient for distribution, interactive software providing easier use is described in section 3.

```
                 form (basis of approximating space)
                     polynomial, rational
                         real polynomial
                             power
                                 evaluation [290, 199, 471]
                                 integrals [471]
                             Newton form of interpolating polynomial [237, 184, 200]
                             Chebyshev
                                 interpolation to function and derivative data [342]
                                 least squares [131]
                                 evaluation [231, 210, 19]
                                 derivatives [19]
                                 integrals [19]
                                 standard errors from least squares fit [19]
                                 endpoint constrained [343]
                             barycentric [300]
Bezier                       Bernstein-Bezier curves and patches [259, 459, 473, 471]
                             general orthogonal [207, 528, 210, 478, 147, 125, 505, 239]
polycon                      constrained [113, 376]
```

    - multivariate [109, 237, 91, 40, 179, 392]
  - real rational [509, 336]
    - Remez — Remez exchange [364, 384, 87, 99, 326]
      - Remez, first (single exchange) [176, 406]
      - Remez, second (multiple exchange) [134, 486, 103, 39, 104]
    - differential correction [210, 334, 103]
    - linear inequality method [348, 177]
    - interpolation [270, 271, 136, 138, 137, 57]
    - Padé [294, 375, 138, 510]
    - ratio of powers of polynomials [178]
  - complex polynomial, rational [492, 525]
    - polynomial [206, 434, 283, 200]
    - rational [436, 509]
  - zeros of polynomials [421, 362, 323]
- spline, finite element
  - discontinuities (jump, crease, edges)
    - known, fixed crease [68, 216, 480]
    - unknown crease, to be found [393, 480, 379, 94, 506, 54, 352, 351]
  - splines
    - univariate
      - piecewise linear
        - tabulate at arguments where values need few digits [234, 3]
        - segment — breakpoint placement [317, 418, 531, 416, 520]
      - pch: piecewise cubic Hermite [226]
        - u-der — derivative estimation [5, 187, 320]
      - pp: general piecewise polynomial [407]
        - almost smooth (jumps less than ε)
          - (no citations at hand, though surely some exist)
        - evaluation of general piecewise polynomial [70, 321]
        - one-pass least squares fitting [541]
      - B-splines [70]
        - fitting, interpolation
          - interpolation [70, 19, 321]
            - optinterp — "optimal" spline interpolation [70, 321]
            - many-knot interpolation [143]
          - end conditions for interpolation [358, 47]
          - local, variation diminishing [361, 143]
          - quasi-interpolation [70, 454, 108]
          - least squares fitting
            - fitting discrete data [70, 19, 321, 210]
            - fitting user function [210]
          - l_1 fitting to data points [133]
          - l_∞ fitting to a function [405, 62]
        - evaluation, derivatives, integral
          - spline evaluation [70, 19, 321, 210, 354]
          - B-spline basis evaluation [70, 210]

- spline derivatives [70, 19, 321, 210]
- B-spline basis derivatives [70, 210]
- spline integrals [19, 321, 210, 233]
- B-spline basis integrals [210]
- Fourier transform [238, 282, 395, 284, 154]
- `knotins` knot insertion and degree raising [122, 63, 345, 64, 141, 65, 360]

- zeros [167]
- subdivision, knot insertion →`knotins`
- degree raising [429, 123, 360]
- `B-diag` diagnostics, error estimation
    - standard errors from least squares fit [19]
    - function and spline [210]
    - two splines [210]
- rational B-splines [423, 355]

- Bernstein-Bezier curves →`Bezier`
- conversion between different bases [70, 223, 314, 225]
- specialized splines [455]
    - elastica [353, 366, 437, 254, 332, 90]
    - `nu-spline` ν, β, γ, Manning, Farin, W-F [396, 223, 306, 204, 225, 205]
    - taut splines, splines under tension [490, 116, 70, 441, 401, 439, 458]
    - `network` minimum norm network spline [399, 401, 402]
    - other [469, 17]
        - Catmull-Rom [97, 37]
        - discrete spline [180, 122]
        - EHB-spline (unusual smoothness constraints) [325]
        - `spec-spline` spectral spline [278]
        - conic spline [417, 427]

multivariate

- pp: general piecewise polynomial [12, 246]
    - Bernstein-Bezier patches →`Bezier`
- `tensor` tensor product splines [419, 71]
    - fitting, interpolation
        - interpolation [70, 19, 527]
        - least squares fitting [93, 293, 159, 280, 161, 19]
        - low rank [155]
    - evaluation, derivatives
        - evaluation [19, 321, 210]
        - derivatives [19, 321]
        - integrals [321]
- `plate` thin plate splines [89, 68, 213, 214, 349, 43]
- simplex splines [144, 145, 310, 267, 247, 249, 269]
- box splines; uniform grid [454, 65, 142, 324, 284]
    - evaluation [311]

interpolation and quasi-interpolation [108, 107, 139, 527]
triangular B-spline [218]
minimum norm network spline →`network`

`fem` finite elements [110, 51]
bivariate
triangle
quadratic, piecewise quadratic [425, 426, 483]
cubic, piecewise cubic [248, 268]
Hsieh-Clough-Tocher C^1 cubic macro-triangle [118, 56, 55, 369, 193, 536]
quintic, piecewise quintic
Argyris C^1 quintic triangle [55]
rational [381, 28, 370, 521, 14, 305]
other [398, 482]
rectangle [385, 41, 403]

`iso` isoparametric approximation ($\min_{P,Q} \|f \circ P - Q\|$) [263, 369, 219, 198, 111]

trivariate and general dimension
tetrahedron [8, 9, 536, 537, 449]

`blendmult` discretizations of blending on simplices [368, 32, 11, 274]

other specific approximating spaces
trigonometric [286, 210, 244]
FFT [126, 487, 500, 170, 497]
spectral methods [265, 266, 95]
special functions (approximation by, not approximation of)

`exp` sum of exponentials [286, 333, 450, 524, 488]
VARPRO [256, 337, 338, 255, 210]

`gaus` sum of gaussians [461]
sinc [92]

kernel smoothing [119]
moving least squares
distance weighted interpolation (Shepard) [479, 264, 28, 215, 357, 196, 33, 440]
(finite weight) least squares [380, 114, 344, 379, 115]
radial basis functions
Hardy's multiquadrics [289, 212, 203]
thin plate splines $r^2 \log r^2$ →`plate`
sum of gaussians →`gaus`
algorithms applicable to many radial bases [181, 183]
other [451]
recursive subdivision [341, 182, 387, 311]
knot insertion for splines →`knotins`

techniques for combining specific methods
extension of univariate methods to multivariate
tensor product and similar methods
rectangular grid data; tensor splines →`tensor`

slices parallel to coordinate axis [113, 19]
scattered data [298]
missing data : $\Delta u = 0$
curved mesh lines [132]
rational product [335, 302]
blending — blending, Coons patch, transfinite element
rectangular patch [98, 30, 28]
triangular patch [29, 28, 273, 397, 398, 14, 285, 202]
blendcur — network of curves [533]
slices parallel to coordinate plane [229, 67]
more than two variables →blendmult
boundary data as basis functions [503]
additive $f(x, y) \approx g(x) + h(y)$ [299, 103, 297, 413]
nomographic $\approx f(u(x) + b(y))$, ACE [103, 84]
projection pursuit, $\approx \sum f(\alpha \cdot x)$ [222, 502]
multistage methods
estimate nodal information for finite element
triang — triangulation [44]
Delaunay-Thiessen (dual to Voronoi polygons) [346, 6, 79, 523, 117, 32, 208]
searching general triangulation for given point [346, 471, 430]
estimate values, derivatives (for use with finite element)
univariate →u-der
least squares local quadratic [346]
weighted average of directional derivatives [6, 370, 7, 32]
variational method (minimize integral of derivatives) [10]
minimum norm network →network
finite element →fem
blending →blending
interpolate scattered data onto rectangular grid [468]
Boolean sum, other than blending [31, 28, 344]
smooth averaging of overlapping local patches [377, 330, 211, 512, 213, 146, 18]
iterated fitting of residual [222, 201, 453]
Ivakhnenko GMDH; adaptive learning networks [194]
isoparametric approximation ($\min_{P,Q} \|f \circ P - Q\|$) →iso
fillets [309, 522, 428, 446]
parametric, geometric/visual continuity
parametric smoothness desired [19]
curves
vparam — choice of parameterization variable [188, 371, 223, 315]
spline [167]
piecewise conic [382, 220]
surfaces [105]
irregular patches for otherwise rectangular grid [452]

- only geometric smoothness of locus of points needed
  - curves [72, 260]
    - ν, β, γ, Manning, Farin, W-F splines →`nu-spline`
  - surfaces [192, 324, 400, 481]
- abstract spaces
  - Haar system [150]
    - sum of exponentials →`exp`
    - Remez exchange →`Remez`
  - generating function methods (generalized Padé) [488]
  - `linalg` general linear basis functions
    - driver routine for user-supplied linear basis [19, 321]
    - linear equations [242, 257, 174, 70]
      - Linpack [173, 124]
      - solving Vandermonde systems [59, 235, 58, 498]
    - least squares [147, 295, 414, 251, 515, 288, 112, 27]
  - `opt` general non-linear basis functions [85, 157, 275, 158, 526]
  - implicit algebraic equations [252, 522, 428, 476, 415, 23]
  - differential equations [516]
  - functional approximation [388, 386]

norm (criteria of approximation, figure of merit, nonstandard data)
- classical L_p and analogous norms
  - l_2 [253]
  - total least squares, orthogonal distance regression [514, 296, 66]
  - l_∞, L_∞ [327]
    - Remez exchange →`Remez`
    - approximation by l_2 solution [435]
    - overdetermined linear system [34, 83, 492, 491]
    - overdetermined linear system [501]
    - Hankel norm ("CF" or "AAK" approximation) [509]
  - l_1 [1, 35, 38, 133]
  - piecewise smooth, including l_∞ and l_1 as specific cases [363]
  - Hausdorff $\max[\max_{y\in G}\min_{x\in F}||x-y||, \max_{y\in F}\min_{x\in G}||x-y||]$ [477]
  - statistically motivated metrics
    - maximum likelihood [156, 499, 412, 240]
    - AIC (Akaike's Information Criterion) [4, 442, 22]
    - Markov random field [241, 373]
  - bounding the relative error [465]
- penalties, smoothing, tolerances
  - choosing weights in l_2 and other norms [534, 502]
  - $l_2 + penalty$
    - L_2 norm of high derivative [303, 304, 70, 321]
    - l_2 and jumps in high derivative [162]
    - jumps in divided difference [530]
    - choice of smoothing parameter [424, 442, 432]

GCV generalized cross validation [243]

univariate [518, 411, 318, 535, 312, 340, 472]

bivariate [43, 539]

weighted norm of second derivative to approximate curvature [457]

min $\int g''^2$ subject to $|g(t_i) - z_i| \leq \varepsilon$ [513]

interpolation

at Chebyshev points [444, 300]

fair, pleasing [53, 383]

constraints

periodic [359, 379, 431, 167]

monotone, comonotone

quadratic [448, 470, 49]

cubic [227, 127, 319, 169, 517, 149]

$O(h^4)$ [185, 538, 48]

rational quadratic [152, 463]

bivariate [49, 130, 96]

monotone, other forms or general order [75, 130, 455, 439, 133, 433]

convex

quadratic [378, 448, 45]

cubic [70, 462]

with penalty term for smoothing [322, 16, 186]

convex, other forms [394, 160, 101, 474, 128, 439, 133, 197]

one-sided, non-negative [450, 46, 408, 463, 464, 410]

specialized constraints [410]

branch branching [485]

integer [279]

specific to polynomials →polycon

"essentially non-oscillatory" [292, 291]

ellipticity $\partial_{12}^2 \leq \partial_{11}\partial_{22}$ [135]

abstract methods [389]

semi-infinite programming [450, 245, 78]

data nonstandard data

cross sections [168, 250, 356]

given Fourier series coefficients, as in crystallography

rational in cos, sin [102]

spectral spline →spec-spline

derivative data

given gradients, get surface; "shape from shading" [313, 350, 129]

integral data [228]

given integrals over small subintervals [42, 456]

given flux, get velocities [287]

missing data [502]

multiple scales or resolutions [94, 54]

variable (coordinate transform, mesh generation)

"optimal" spline interpolation →`optinterp`
nonuniform sampling to ameliorate aliasing [391]
bivariate mesh generation [504, 328, 24]

diagnostics
recognizing noise [281, 409]
recognizing outliers and influential points [475, 365]
diagnostics specific to B-splines →`B-diag`
diagnostic graphics [367, 511, 15, 50]
contouring
irregular grid [372, 68, 451, 460, 148]
rectangular grid [380, 493, 489, 467, 494, 483, 543]
higher order than quadratic; general functions [191, 272]
contour surfaces for function of three variables [106, 195, 422]
color spectrum generation [277]
tools from other fields
linear algebra →`linalg`
optimization →`opt`
arithmetic [86]
random number generators [374, 209]
omitted topics
time series, digital filters, signal processing [170, 77, 532, 329]
finding intersections [81, 171, 316, 252, 100]
minimal surfaces [26]
quadrature [258, 151]
special functions [290, 121, 120, 231, 230]
extrapolation, acceleration of series [60, 88, 484]

3. Software for examining the catalog

The listing above is an adequate but hardly convenient way to look at the catalog. An interactive program for walking over the tree (actually, directed graph) is clearly preferable. Publicly available menu systems of this sort exist, but require numeric node labels and a specialized editor. Having mouse-based cut-and-paste, regular expression searching, undo, multiple windows, automatic backup against system crash, and other such features common to modern editors makes specialized editors unattractive. Instead I wrote my own in one evening, a cost quite insignificant compared to typing in the database, which itself is insignificant compared to reading all those papers! This program may be run by logging in as userid `walk` on `research.att.com`, reachable by phone or by various data networks. (Contact me for details.) No password is required.

Particularly when building a large graph with many cross links, it can be helpful to see a graphical representation of the walk database. There is an option to produce input for `dag`[236].

Another option produces a file which can be loaded into Hypercard. In order to fit onto the small Macintosh screen, there are a maximum of eight alternatives at any node. This might not be such a bad choice on psychological grounds as well. In one detail these interactive programs supply a bit more information than the typeset version; in place of the numeric reference number is a brief note that may indicate a particular routine or hint at important ideas of the algorithm not obvious from the title.

Keyword search is possible using Hypercard or a text editor, though the database wasn't really designed with this in mind. To do a search via `netlib`, send a message of the form `find Schumaker in approximation`. "Schumaker" may be replaced here by any other pattern to be matched against the BibTEX database using the Unix system command egrep.

A weakness of the current system for some users is the absence of a glossary. Faced with a difficult choice at some point in walking the tree, it might be useful to look at the precise definition of terms.

End users would prefer everything to be indexed under "problem" instead of "form" and "norm", but that will take a lot more thought. The present catalog is to such an expert system as the Merck manual is to the Mycin program. The following short list hints at some of the relevant questions such a consultant would ask the user.

- given data
  - arrangement of data
  - range of variables
- criteria and constraints
  - norm, weights
  - continuity class (C^k or G^k) of fitted surface with respect to
  - point of evaluation
  - data ordinates
  - data abscissae
- periodic
- shape fidelity (monotone, convex, "fair")
- local or global dependence on data
- other constraints

how approximation is to be used

- values, derivatives, integral, plotting, extrapolation

expected characteristics of data

- singularities or peaks

other considerations

- desired form; interpreting coefficients
- rotational, affine invariance in abscissae; linearity in ordinates
- speed, memory
- polynomial precision order, asymptotic convergence rate
- sensitivity to noise in data, rounding error
- failure modes
- available libraries, languages

Table 1
warmup attributes for Port.

attribute	values
form	rational; spline; Bspline
op	fit; eval; deriv; int; errest; mesh
dat	function; spline; accurate; noisy
shape	wiggles-allowed; monotone-in,monotone-out
dorder	values; derivatives; integral
spacing	uniform; piecewise-uniform; multiplicities

In some ways it would be more natural to classify the methods independently along several axes. But with so many methods, this would be awkward to display and traverse. A more classical expert system using production rules could probably handle this, but history suggests the result would be unreadable by others and hence not susceptible to peer review.

A partial solution is provided by the program `warmup`, which takes a list of routines with associated attributes and automatically builds a decision tree in the form of a `walk` database. The attributes used for describing approximation routines in the Port library are shown in Table 1. Using only these values, `warmup` reproduced practically the same `walk` tree that was created manually by Norm Schryer.

This strategy is motivated by my observation of typical consulting at Bell Labs. A physicist walks into my office and gives me about five sentences describing her problem before I interrupt with a few questions and offer a possible algorithm. The point is that she will naturally mention half a dozen attributes that drastically narrow the range of possibilities. If this is not the first time we have talked, I probably remember from previous sessions various attributes she prefers in a code. `Warmup` appears well suited to this mode, if we can ever get an English language front end that can translate from physics into numerical analysis jargon. Various attributes would be recognized as either required, prohibited, or irrelevant; restricting the database according to this user profile should yield a small decision tree.

4. Disclaimers

This is a catalog of a diverse collection, as distinguished from a table of contents of a high-quality mathematical software library. Certainly it would be pleasant if all routines adopted compatible calling styles, names, and data representations. But users see lots of interfaces anyway because they use collections of libraries. The enormous effort of making systematized interfaces means that libraries tend to fall behind the research front. Finally, even well-tested library routines may still have subtle numerical problems. Certainly the commercial libraries are valuable and well worth the rental. Library authors often do a much better job than the typical researcher when it comes to providing good example programs, carefully checking user input, and worrying about error messages. But there is also a place for research codes in exotic applications and to gain experience with different algorithms, in order to know what ought to go next into the libraries. Traditionally the time and expense of acquiring such research codes has been a major impediment, but netlib offers a solution.

Moreover, programming environments have improved so much that coding up a method straight from a well-written article is often easy. Still, it should be clearly understood that research codes must be used with the same professional caution you would want your doctor to exercise when trying a new procedure.

Another worthwhile project would be to collect and standardize test problems, as David Gay has done for linear programming in `netlib/lp/data` and Chris Fraley has done for unconstrained optimization in `netlib/uncon/data`. Gay found that it was best to compress the files and distribute C and Fortran programs for unpacking them into the standard MPS file format. Some places to start would be [5][424][380][346][6][227][212][216][15][224]. I would be willing to contribute an example coming from an important application in VLSI, monotone fitting of current as a function of the three independent terminal voltages for a MOSFET transistor [130].

Comments are welcome and will be incorporated in the online edition of the catalog. Send electronic mail to `ehg@research.att.com` or `uunet!research!ehg`. As well as pointing out missing papers, please note defects in the citations; in a list of this size, an inadequately classified paper is as good as lost. Since the list will always be incomplete, anyone seriously concerned with a particular topic should not restrict themselves to the references here but should follow the citations in the given papers and use Science Citation Index to check for more recent work.

Acknowledgments
I thank Peter Weinberger for discussions that led to *walk*, Anne Trefethen for suggesting this survey, and Al Aho, Bill Coughran, Dick Franke, Gene Golub, David Lee, Sam Marin, Victor Pereyra, Kishore Singhal, Nick Trefethen, and Grace Wahba for their suggestions. Surveys that I have found particularly helpful are [28][453][103][217][13][502]. Hypercard and Macintosh are trademarks of Apple Computer, Inc. Unix is a trademark of AT&T Bell Laboratories.

References

1. N. N. Abdelmalek. A Fortran subroutine for the L_1 solution of overdetermined systems of linear equations. *ACM Trans. on Mathematical Software*, 6:220–227, 1980. [toms/551].
2. S. Adjerid and J. E. Flaherty. Second-order finite element approximations and a posteriori error estimation for two-dimensional parabolic systems. *Numerische Mathematik*, 53:183–198, 1988.
3. R. C. Agarwal, J. W. Cooley, F. G. Gustavson, J. B. Shearer, G. Slishman, and B. Tuckerman. New scalar and vector elementary functions for the IBM System/370. *IBM J. Research and Development*, 30:126–144, 1986.
4. H. Akaike. Information theory and an extension of maximum likelihood principle. In B. N. Petrov and F. Csáki, eds., *2nd International Symposium on Information Theory*, pages 267–281. Akademia Kiadó, Budapest, 1973.
5. H. Akima. A new method of interpolation and smooth curve fitting based on local procedures. *J. ACM*, 17:589–602, 1970. [imsl/csakm].
6. H. Akima. A method of bivariate interpolation and smooth surface fitting for irregularly distributed data points. *ACM Trans. on Mathematical Software*, 4:148–159, 1978. [toms/526,

imsl/surf].
7. H. Akima. On estimating partial derivatives for bivariate interpolation of scattered data. *Rocky Mountain J. of Mathematics*, 14:41–52, 1984.
8. P. Alfeld. A discrete C^1 interpolant for tetrahedral data. *Rocky Mountain J. of Mathematics*, 14:5–16, 1984.
9. P. Alfeld. A trivariate Clough-Tocher scheme for tetrahedral data. *Computer Aided Geometric Design*, 1:169–181, 1984.
10. P. Alfeld. Derivative generation from multivariate scattered data by functional minimization. *Computer Aided Geometric Design*, 2:281–296, 1985.
11. P. Alfeld. Multivariate perpendicular interpolation. *SIAM J. on Numerical Analysis*, 22:95–106, 1985.
12. P. Alfeld. *The Multivariate Spline Newsletter*. Dept. Mathematics, Univ. Utah, Salt Lake City UT 84112, 1987–1988.
13. P. Alfeld. Scattered data interpolation in three or more variables. Technical report, Department of Mathematics, University of Utah, 1988.
14. P. Alfeld and R. E. Barnhill. A transfinite C^2 interpolant over triangles. *Rocky Mountain J. of Mathematics*, 14:17–39, 1984.
15. P. Alfeld and B. Harris. Microscope: a software system for multivariate analysis. MRC Technical Summary Report 2701, University of Wisconsin–Madison, 1984. [netlib/microscope/].
16. L.-E. Andersson and T. Elfving. An algorithm for constrained interpolation. *SIAM J. on Scientific and Statistical Computation*, 8:1012–1025, 1987.
17. P. M. Anselone and P. J. Laurent. A general method for the construction of interpolating or smoothing spline-functions. *Numerische Mathematik*, 12:66–82, 1968.
18. G. T. Anthony and M. G. Cox. The fitting of extremely large data sets by bivariate splines. In J. C. Mason and M. G. Cox, eds., *Algorithms for Approximation*, pages 5–20. Oxford Univ. Press, 1987.
19. G. T. Anthony and M. G. Cox. The National Physical Laboratory's data approximation subroutine library. In J. C. Mason and M. G. Cox, eds., *Algorithms for Approximation*, pages 669–687. Oxford Univ. Press, 1987. [DASL].
20. R. D. Armstrong and M. T. Kung. An algorithm to select the best subset for a least absolute value regression problem. In S. H. Zanakis and J. S. Rustagi, eds., *Optimization in Statistics*, pages 67–80. North-Holland, Amsterdam, 1982.
21. U. M. Ascher, R. M. M. Mattheij, and R. D. Russell. *Numerical Solution of Boundary Value Problems for Ordinary Differential Equations*. Prentice-Hall, Englewood Cliffs, New Jersey, 1988.
22. T. Atligan. Basis selection for density estimation and regression. Technical report, AT&T Bell Laboratories, 1988.
23. C. L. Bajaj, C. M. Hoffmann, R. E. Lynch, and J. E. H. Hopcroft. Tracing surface intersections. *Computer Aided Geometric Design*, 5:285–307, 1988.
24. B. Baker, E. Grosse, and C. Rafferty. Non-obtuse triangulation of polygons. *J. Discrete and Computational Geometry*, 3:147–168, 1988.
25. R. E. Bank, A. H. Sherman, and A. Weiser. Refinement algorithms and data structures for regular local mesh refinement. In R. Stepleman et al., eds., *Scientific Computing*, pages 3–17. North-Holland, Amsterdam, 1983.
26. J. Barbosa and M. Lucas. *Minimal Surfaces in R^3*. Springer-Verlag, New York, 1986.
27. J. L. Barlow and S. L. Handy. The direct solution of weighted and equality constrained least-squares problems. *SIAM J. on Scientific and Statistical Computation*, 9:704–716, 1988.
28. R. E. Barnhill. Representation and approximation of surfaces. In *Mathematical Software III*, pages 69–120. Academic Press, New York, 1977.
29. R. E. Barnhill, G. Birkhoff, and W. J. Gordon. Smooth interpolation in triangles. *J. of Approximation Theory*, 8:114–128, 1973.
30. R. E. Barnhill, J. H. Brown, and I. M. Klucewicz. A new twist in computer aided geometric

design. *Computer Graphics and Image Processing*, 8:78–91, 1978.

31. R. E. Barnhill and J. A. Gregory. Polynomial interpolation to boundary data on triangles. *Mathematics of Computation*, 29:726–735, 1975.
32. R. E. Barnhill and F. F. Little. Three- and four-dimensional surfaces. *Rocky Mountain J. of Mathematics*, 14:77–102, 1984.
33. R. E. Barnhill, B. R. Piper, and K. L. Rescorla. Interpolation to arbitrary data on a surface. In G. E. Farin, ed., *Geometric Modeling: Algorithms and New Trends*, pages 281–289. SIAM, Philadelphia, 1987.
34. I. Barrodale and C. Phillips. Solution of an overdetermined system of linear equations in the Chebyshev norm. *ACM Trans. on Mathematical Software*, 1:264–270, 1975. [toms/495].
35. I. Barrodale and F. D. K. Roberts. Solution of the constrained l_1 linear approximation problem. *ACM Trans. on Mathematical Software*, 6:231–235, 1980. [toms/552].
36. D. L. Barrow and P. W. Smith. Efficient L_2 approximation by splines. *Numerische Mathematik*, 33:101–114, 1979.
37. P. J. Barry and R. N. Goldman. A recursive evaluation algorithm for a class of Catmull-Rom splines. *Computer Graphics [SIGGRAPH]*, 22:199–204, 1988.
38. R. H. Bartels and A. R. Conn. Linearly constrained discrete l_1 problems. *ACM Trans. on Mathematical Software*, 6:594–608, 1980. [toms/563].
39. R. H. Bartels, A. R. Conn, and Y. Li. Primal methods are better than dual methods for solving overdetermined linear systems in the l_∞ sense? Research Report CS-87-12, University of Waterloo, 1987.
40. R. H. Bartels and J. J. Jezioranski. Least-squares fitting using orthogonal multinomials. *ACM Trans. on Mathematical Software*, 11:201–217, 1985. [toms/634].
41. G. Baszenski, H. Posdorf, and F. J. Delvos. Representation formulas for conforming bivariate interpolation. In *Approximation III*, pages 193–198. Academic Press, New York, 1980.
42. G. Baszenski and L. L. Schumaker. On a method for fitting an unknown function based on mean-value measurements. *SIAM J. on Numerical Analysis*, 24:725–736, 1987.
43. D. M. Bates, M. J. Lindstrom, G. Wahba, and B. Yandell. Gcvpack - routines for generalized cross validation. *Commun. Statist. Simul. Comput.*, 16:263–297, 1987. [netlib/gcv/gcvpack].
44. J. R. Baumgardner and P. O. Frederickson. Icosahedral discretization of the two-sphere. *SIAM J. on Numerical Analysis*, 22:1107–1115, 1985.
45. R. K. Beatson. Monotone and convex approximation by splines: Error estimates and a curve fitting algorithm. *SIAM J. on Numerical Analysis*, 19:1278–1285, 1982.
46. R. K. Beatson. Restricted range approximation by splines and variational inequalities. *SIAM J. on Numerical Analysis*, 19:372–380, 1982.
47. R. K. Beatson and E. Chacko. Which cubic spline should one use? Technical report, Math. Dept., Univ. of Canterbury, Christchurch, New Zealand, 1988.
48. R. K. Beatson and H. Wolkowicz. Post-processing piecewise cubics for monotonicity. *SIAM J. on Numerical Analysis*, page to appear, 1988.
49. R. K. Beatson and Z. Ziegler. Monotonicity preserving surface interpolation. *SIAM J. on Numerical Analysis*, 22:401–411, 1985. [netlib/misc/beatson].
50. R. A. Becker and W. S. Cleveland. Brushing scatterplots. *Technometrics*, 29:127–142, 1987. [dynam].
51. D. Bégis, F. Hecht, and M. Vidrascu. Presentation and evolution of the Club Modulef: a library of computer procedures for finite element analysis. In G. Birkhoff and A. Schoenstadt, eds., *Elliptic Problem Solvers II*, pages 23–33. Academic Press, New York, 1984.
52. M. Berger. Data structures for adaptive grid generation. *SIAM J. on Scientific and Statistical Computation*, 7:904–916, 1986.
53. S. A. Berger, W. C. Webster, R. A. Tapia, and D. A. Atkins. Mathematical ship lofting. *J. Ship Research*, 10:203–222, 1966.
54. F. Bergholm. Edge focusing. *IEEE Trans. on Pattern Analysis and Machine Intelligence*,

9:726–741, 1987.
55. M. Bernadou and M. M. Boisserie. *The Finite Element Method in Thin Shell Theory.* Birkhauser, Boston, 1982. [explicit basis].
56. M. Bernadou and K. Hassan. Basis functions for general Hsieh-Clough-Tocher triangles, complete or reduced. *International J. Numerical Methods in Engineering*, 17:784–789, 1981.
57. J.-P. Berrut. Rational functions for guaranteed and experimentally well-conditioned global interpolation. *Computers and Mathematics with Applications*, 15:1–16, 1988.
58. Å. Björck and T. Elfving. Algorithms for confluent Vandermonde systems. *Numerische Mathematik*, 21:130–137, 1973.
59. A. Bjorck and V. Pereyra. Solution of Vandermonde systems of equations. *Mathematics of Computation*, 24:893–904, 1970. [Algol listing].
60. P. E. Bjørstad, G. Dahlquist, and E. H. Grosse. Extrapolation of asymptotic expansions by a modified Aitken δ^2-formula. *BIT*, 21:56–65, 1981.
61. P. E. Bjørstad and E. H. Grosse. Conformal mapping of circular arc polygons. *SIAM J. on Scientific and Statistical Computation*, 8:19–32, 1987. [netlib/conformal/cap].
62. J. Blatter. An algorithm for best uniform approximation by splines with fixed knots. In C. K. Chui, L. L. Schumaker, and J. D. Ward, eds., *Approximation Theory V*, pages 263–266. Academic Press, New York, 1986.
63. W. Boehm. Inserting new knots into B-spline curves. *Computer Aided Design*, 12:199–201, 1980. [netlib/dierckx/fitpack (insert)].
64. W. Boehm and H. Prautzsch. The insertion algorithm. *Computer Aided Design*, 17:58–59, 1985. [simple code].
65. W. Boehm, H. Prautzsch, and P. Arner. On triangular splines. *Constructive Approximation*, 3:157–167, 1987.
66. P. T. Boggs, R. H. Byrd, and R. B. Schnabel. A stable and efficient algorithm for nonlinea orthogonal distance regression. *SIAM J. on Scientific and Statistical Computation*, 8:1052–1078, 1987.
67. J.-D. Boissonnat. Shape reconstruction from planar cross sections. *Computer Vision, Graphics, and Image Processing*, 44:1–29, 1988.
68. G. Bolondi, F. Rocca, and S. Zanoletti. Automatic contouring of faulted subsurfaces. *Geophysics*, 41:1377–1393, 1976.
69. C. de Boor. Good approximation by splines with variable knots, II. In G. A. Watson, ed., *Numerical Solution of Differential Equations*, pages 12–20. Springer-Verlag, New York, 1974. [netlib/pppack/newnot].
70. C. de Boor. *A Practical Guide to Splines.* Springer-Verlag, New York, 1978. [netlib/pppack/].
71. C. de Boor. Efficient computer manipulation of tensor products. *ACM Trans. on Mathematical Software*, 5:173–182, 1979.
72. C. de Boor, K. Höllig, and M. Sabin. High accuracy geometric Hermite interpolation. *Computer Aided Geometric Design*, 4:269–278, 1987.
73. C. de Boor and J. R. Rice. Least squares cubic spline approximation II - variable knots. CSD TR 21, Purdue University, Computer Sciences Department, 1968. [Fortran listing].
74. C. de Boor and J. R. Rice. An adaptive algorithm for multivariate approximation giving optimal convergence rates. *J. of Approximation Theory*, 25:337–359, 1979.
75. C. de Boor and B. Swartz. Piecewise monotone interpolation. *J. of Approximation Theory*, 21:411–416, 1977.
76. L. P. Bos. Bounding the Lebesgue function for Lagrange interpolation in a simplex. *J. of Approximation Theory*, 38:43–59, 1983.
77. N. K. Bose. *Digital Filters: Theory and Applications.* North-Holland, Amsterdam, 1985. [short Fortran listings].
78. K. W. Bosworth. Shape constrained curve and surface fitting. In G. E. Farin, ed., *Geometric Modeling: Algorithms and New Trends*, pages 247–263. SIAM, Philadelphia, 1987.
79. A. Bowyer. Computing Dirichlet tessellations. *Computer J.*, 24:162–166, 1981.

80. G. E. P. Box and N. R. Draper. *Empirical Model-Building and Response Surfaces*. John Wiley & Sons, New York, 1987.
81. J. W. Boyse and J. E. Gilchrist. GMSolid: interactive modeling for design and analysis of solids. *IEEE Computer Graphics and Applications*, March:27–40, 1982.
82. M. Brannigan. Criteria for adaptive approximation. In C. K. Chui, L. L. Schumaker, and J. D. Ward, eds., *Approximation Theory IV*, pages 381–386. Academic Press, New York, 1983.
83. M. Brannigan. Discrete Chebyshev approximation with linear constraints. *SIAM J. on Numerical Analysis*, 22:1–15, 1985.
84. L. Breiman and J. H. Friedman. Estimating optimal transformations for correlation and regression. *J. of the American Statistical Association*, 80:580–598, 1985. [ACE].
85. R. Brent. *Algorithms for Minimization Without Derivatives*. Prentice-Hall, Englewood Cliffs, New Jersey, 1973. [netlib/go/fmin,zeroin].
86. R. P. Brent. A Fortran multiple-precision arithmetic package. *ACM Trans. on Mathematical Software*, 4:57–81, 1978. [netlib/bmp/].
87. P. T. Breuer. A new method for real rational uniform approximation. In J. C. Mason and M. G. Cox, eds., *Algorithms for Approximation*, pages 265–283. Oxford Univ. Press, 1987.
88. G. Brezinski. A subroutine for the general interpolation and extrapolation problems. *ACM Trans. on Mathematical Software*, 8:290–301, 1982. [toms/585].
89. I. C. Briggs. Machine contouring using minimum curvature. *Geophysics*, 39:39–48, 1974.
90. A. M. Bruckstein and A. N. Netravali. On minimal energy trajectories. Technical report, AT&T Bell Laboratories, 1988.
91. J. R. Busch. Osculatory interpolation in $\mathbb{R}^n$. *SIAM J. on Numerical Analysis*, 22:107–113, 1985.
92. P. L. Butzer, W. Engels, S. Ries, and R. L. Stens. The Shannon sampling series and the reconstruction of signals in terms of linear, quadratic, and cubic splines. *SIAM J. of Applied Mathematics*, 46:299–, 1986.
93. E. S. Call and F. F. Judd. Surface fitting by separation. *J. of Approximation Theory*, 12:283–290, 1974.
94. J. Canny. A computational approach to edge detection. *IEEE Trans. on Pattern Analysis and Machine Intelligence*, 8:679–698, 1986.
95. C. Canuto, M. Y. Hussaini, A. Quarteroni, and T. A. Zang. *Spectral Methods in Fluid Dynamics*. Springer-Verlag, New York, 1987.
96. R. E. Carlson and F. N. Fritsch. BIMOND3: Monotone piecewise bicubic Hermite interpolation code. UCID-21143, Lawrence Livermore National Laboratory, 1987.
97. E. E. Catmull and R. J. Rom. A class of local interpolating splines. In R. E. Barnhill and R. F. Riesenfeld, eds., *Computer Aided Geometric Design*, pages 317–326. Academic Press, New York, 1974.
98. J. C. Cavendish. Local mesh refinement using rectangular blended finite elements. *J. Computational Physics*, 19:211–228, 1975.
99. B. Chalmers. The Remez exchange algorithm for approximation with linear restrictions. *Trans. of the American Mathematical Society*, 223:103–131, 1976.
100. V. Chandru and B. S. Kochar. Analytic techniques for geometric intersection problems. In G. E. Farin, ed., *Geometric Modeling: Algorithms and New Trends*, pages 305–318. SIAM, Philadelphia, 1987.
101. G.-z. Chang and P. J. Davis. The convexity of Bernstein polynomials over triangles. *J. of Approximation Theory*, 40:11–28, 1984.
102. R. J. Charron. Adapting rational approximants for Fourier series to data representation problems. *Computing*, 40:217–228, 1988.
103. E. W. Cheney. Algorithms for approximation. In C. de Boor, ed., *Approximation Theory*, volume 36 of *Proc. of Symposia in Applied Mathematics*, pages 67–80. American Mathematical Society, Providence, Rhode Island, 1986.
104. Y.-L. F. Chiang. A modified Remes algorithm. *SIAM J. on Scientific and Statistical Computa-*

tion, 9:1058–1072, 1988.

105. B. K. Choi, H. Y. Shin, Y. I. Yoon, and J. W. Lee. Triangulation of scattered data in 3D space. *Computer Aided Design*, 20:239–248, 1988.
106. H. N. Christiansen and M. B. Stephenson. MOVIE.BYU — a general purpose computer graphics display system. In L. C. Wellford, Jr., ed., *Applications of Computer Methods in Engineering, Vol. II*, pages 759–768. University of Southern California, 1977.
107. C. K. Chui, H. Diamond, and L. A. Raphael. Interpolation by multivariate splines. *Mathematics of Computation*, 51:203–218, 1988.
108. C. K. Chui and M. J. Lai. A multivariate analog of Marsden's identity and a quasi-interpolation scheme. *Constructive Approximation*, 3:111–122, 1987.
109. K. C. Chung and T. H. Yao. On lattices admitting unique lagrange interpolations. *SIAM J. on Numerical Analysis*, 14:735–743, 1977.
110. P. G. Ciarlet. *The Finite Element Method for Elliptic Problems*. North-Holland, Amsterdam, 1978.
111. E. Citipitioglu. Universal serendipity elements. *International J. Numerical Methods in Engineering*, 19:803–810, 1983.
112. D. I. Clark and M. R. Osborne. On linear restricted and interval least-squares problems. *IMA J. of Numerical Analysis*, 8:23–36, 1988.
113. C. W. Clenshaw and J. G. Hayes. Curve and surface fitting. *J. of the Institute of Mathematics and its Applications*, 1:164–183, 1965. [DASL/T1FCE, NAG/E02CAF].
114. W. S. Cleveland. Robust locally-weighted regression and smoothing scatterplots. *J. of the American Statistical Association*, 74:829–836, 1979. [netlib/go/lowess].
115. W. S. Cleveland, S. J. Devlin, and E. Grosse. Regression by local fitting: Methods, properties, and computational algorithms. *J. Econometrics*, 37:87–114, 1988. [netlib/a/loess].
116. A. K. Cline. Curve fitting in one and two dimensions using splines under tension. *Communications of the ACM*, 17:213–218, 1974. [fitpack (commercial, subset in netlib)].
117. A. K. Cline and R. L. Renka. A storage-efficient method for construction of a Theissen triangulation. *Rocky Mountain J. of Mathematics*, 14:119–139, 1984. [toms/624].
118. R. W. Clough and J. L. Tocher. Finite element stiffness matrices for analysis of plates in bending. In *Proc. of Conference on Matrix Methods in Structural Mechanics*. Air Force Institute of Technology, Wright-Patterson A.F.B., Ohio, 1965.
119. M. Clutton-Brock. Generalized Fejér and Lanczos kernels. *SIAM J. Mathematical Analysis*, 18:259–, 1987.
120. W. J. Cody. Software for special functions. *Rend. Sem. Mat. Univers. Politecn. Torino*, pages 92–116, 1987. [survey of Amoslib, Calgo, FUNPACK, IMSL, NAG, NUMAL, SFUN/fnlib, SLATEC, SPECFUN, other].
121. W. J. Cody and W. M. Waite. *Software Manual for the Elementary Functions*. Prentice-Hall, Englewood Cliffs, New Jersey, 1980. [netlib/elefunt/,specfun/].
122. E. Cohen, T. Lyche, and R. Riesenfeld. Discrete B-splines and subdivision techniques in computer-aided geometric design and computer graphics. *Computer Graphics and Image Processing*, 14:87–111, 1980.
123. E. Cohen, T. Lyche, and L. L. Schumaker. Algorithms for degree-raising of splines. *ACM Trans. on Graphics*, 4:171–181, 1985.
124. T. F. Coleman and C. Van Loan. *Handbook for Matrix Computations*. SIAM, Philadelphia, 1988.
125. S. D. Conte and C. de Boor. *Elementary Numerical Analysis: An Algorithmic Approach, 3rd edition*. McGraw-Hill, New York, 1980. [ortpol,p.263].
126. J. W. Cooley, P. A. Lewis, and P. D. Welch. The Fast Fourier Transform algorithm: Programming considerations in the calculation of since, cosine, and Laplace transforms. *J. of Sound and Vibration*, 12:315–337, 1970.
127. R. Correa F. and F. Utreras D. A variational approach to monotone interpolation. Technical report, Universidad de Chile, 1980.

128. P. Costantini. Co-monotone interpolating splines of arbitrary degree—a local approach. *SIAM J. on Scientific and Statistical Computation*, 8:1026–1034, 1987.
129. W. M. Coughran, Jr., W. Fichtner, and E. Grosse. Extracting transistor charges from device simulations by gradient fitting. *IEEE Trans. on Computer Aided Design*, to appear, 1988.
130. W. M. Coughran, Jr., E. Grosse, and D. J. Rose. Variation diminishing splines in simulation. *SIAM J. on Scientific and Statistical Computation*, 7:696–705, 1986. [netlib/port/vdss1, vdss2, vdss3].
131. M. G. Cox. Piecewise Chebyshev series. *Bulletin of the Institute of Mathematics and its Applications*, 22:396–411, 1986. [DASL/T1FE].
132. M. G. Cox. Data approximation by splines in one and two variables. In A. Iserles and M. J. D. Powell, eds., *The State of the Art in Numerical Analysis*, pages 111–138. Oxford Univ. Press, 1987.
133. M. G. Cox and H. M. Jones. Shape preserving spline approximation in the l_1-norm. In J. C. Mason and M. G. Cox, eds., *Algorithms for Approximation*, pages 115–129. Clarendon Press, Oxford, 1987. [DASL/planned].
134. A. Curtis and R. R. Osborne. The construction of minimax rational approximations to functions. *Computer J.*, 9:286–293, 1966. [harwell/pe05ad].
135. A. R. Curtis. Discretisation of the zonally-averaged transport equation for use in global atmosphereic pollution studies. AERE R 12524, Computer Science and Systems Division, Harwell Laboratory, 1987.
136. A. Cuyt. A recursive computation scheme for multivariate rational interpolants. *SIAM J. on Numerical Analysis*, 24:228–239, 1987.
137. A. Cuyt. A multivariate qd-like algorithm. *BIT*, 28:98–112, 1988.
138. A. Cuyt and L. Wuytack. *Nonlinear Methods in Numerical Analysis*. North-Holland, Amsterdam, 1987.
139. M. Dæhlen and T. Lyche. Bivariate interpolation with quadratic box splines. *Mathematics of Computation*, 51:219–230, 1988.
140. W. Dahmen. Approximation by smooth multivariate splines on non-uniform grids. In R. A. Devore and K. Scherer, eds., *Quantitative Approximation*, pages 99–114. Academic Press, New York, 1980.
141. W. Dahmen. Subdivision algorithms converge quadratically. *J. of Computational and Applied Mathematics*, 16:145–158, 1986.
142. W. Dahmen. Subdivision algorithms - recent results, some extensions and further developments. In J. C. Mason and M. G. Cox, eds., *Algorithms for Approximation*, pages 21–49. Oxford Univ. Press, 1987.
143. W. Dahmen, T. N. T. Goodman, and C. A. Micchelli. Compactly supported fundamental functions for spline interpolation. *Numerische Mathematik*, 52:639–664, 1988.
144. W. Dahmen and C. Micchelli. Numerical algorithms for least squares approximation by multivariate B-splines. In L. Collatz, G. Meinardus, and H. Werner, eds., *Numerical Methods of Approximation Theory*, volume 6. Birkhäuser Verlag, 1981.
145. W. A. Dahmen and C. A. Micchelli. On the linear independence of multivariate B-splines, 1. triangulations of simploids. *SIAM J. on Numerical Analysis*, 19:993–1012, 1982.
146. A. Daman. Extensions of smoothing spline methods using generalized cross validation. In C. K. Chui, L. L. Schumaker, and J. D. Ward, eds., *Approximation Theory V*, pages 311–314. Academic Press, New York, 1986.
147. W. C. Davidon. Fast least-squares algorithms. *American J. of Physics*, 45:260–262, 1977.
148. J. C. Davis. Contour mapping and SURFACE II. *Science*, 237:669–672, 1987. [commercial; see note 10 in article].
149. M. Davis and J. Dowden. Interpolation by a local taut cubic piecewise polynomial. *Computing*, 38:299–313, 1987. [Fortran listing].
150. P. J. Davis. *Interpolation and Approximation*. Dover Publications, New York, 1975.
151. P. J. Davis and P. Rabinowitz. *Methods of Numerical Integration: Second Edition*. Academic

Press, New York, 1984.

152. R. Delbourgo and J. A. Gregory. The determination of derivative parameters for a monotonic rational quadratic interpolant. *IMA J. of Numerical Analysis*, 5:397–406, 1985.

153. F. Delvos, H. Posdorf, and W. Schempp. Serendipity-type bivariate interpolation. In D. C. Handscomb, ed., *Multivariate Approximation*, pages 47–56. Academic Press, New York, 1978.

154. F.-J. Delvos. Bernoulli functions and periodic B-splines. *Computing*, 83:23–31, 1987.

155. S. Demko. Approximation by small rank tensor products of splines. In R. A. Devore and K. Scherer, eds., *Quantitative Approximation*, pages 115–120. Academic Press, New York, 1980.

156. A. P. Dempster, N. M. Laird, and D. B. Rubin. Maximum likelihood from incomplete data via the EM algorithm. *J. Royal Statistical Society, Series B*, 39:71–107, 1977.

157. J. E. Dennis, Jr., D. M. Gay, and R. E. Welch. An adaptive nonlinear least squares algorithm. *Trans. on Mathematical Software*, 7:348–368,369–383, 1981. [netlib/port/n2f,nsf,...].

158. J. E. Dennis, Jr. and T. Steihaug. On the successive projections approach to least-squares problems. *SIAM J. on Numerical Analysis*, 23:717–733, 1986.

159. P. Dierckx. An algorithm for least-squares fitting of cubic spline surfaces to functions on a rectilinear mesh over a rectangle. *J. of Computational and Applied Mathematics*, 3:113–129, 1977.

160. P. Dierckx. Algorithm 42: An algorithm for cubic spline fitting with convexity constraints. *Computing*, 24:349–371, 1980. [netlib/dierckx/fitpack(cocosp, concon), concon].

161. P. Dierckx. An algorithm for surface fitting with spline functions. *IMA J. of Numerical Analysis*, 1:267–283, 1981. [netlib/dierckx/surfac].

162. P. Dierckx. A fast algorithm for smoothing data on a rectangular grid while using spline functions. *SIAM J. on Numerical Analysis*, 19:1286–1304, 1982. [netlib/dierckx/smoopy, fitpack(curfit)].

163. P. Dierckx. Algorithms for smoothing data on the sphere with tensor product splines. *Computing*, 32:319–342, 1984. [netlib/dierckx/smosph].

164. P. Dierckx. An algorithm for fitting data on a circle using tensor product splines. *J. Computational and Applied Mathematics*, 15:161–173, 1986. [netlib/dierckx/smocir].

165. P. Dierckx. The spectral approximation of bicubic splines on the sphere. *SIAM J. on Scientific and Statistical Computation*, 7:611–623, 1986.

166. P. Dierckx. Fast algorithms for smoothing data over a disc or a sphere using tensor product splines. In J. C. Mason and M. G. Cox, eds., *Algorithms for Approximation*, pages 51–65. Oxford Univ. Press, 1987. [netlib/dierckx/sphery].

167. P. Dierckx. Fitpack user guide, part 1: Curve fitting routines. Report TW89, Katholieke Universiteit Leuven, 1987. [netlib/dierckx/fitpack].

168. P. Dierckx and P. Suetens. A fast algorithm for surface reconstruction from planar contours using tensor product splines. Report TW64, Katholieke Universiteit Leuven (Belgium), 1983. [netlib/dierckx/smocyl].

169. S. Dietze and J. W. Schmidt. Determination of shape preserving spline interpolants with minimal curvature via dual programs. Technical report, Technische Universität Dresden, 1985.

170. Digital Signal Processing Committee. *Programs for Digital Signal Processing*. IEEE Press, 1979.

171. T. Dokken. Finding intersections of B-spline represented geometries using recursive subdivision techniques. *Computer Aided Geometric Design*, 2:189–195, 1985.

172. J. J. Dongarra and E. Grosse. Distribution of mathematical software via electronic mail. *Communications of the ACM*, 30:403–407, 1987. [netlib/misc/netlib, netlib-paper].

173. J. J. Dongarra, C. B. Moler, J. R. Bunch, and G. W. Stewart. *LINPACK Users' Guide*. SIAM, Philadelphia, 1979.

174. I. S. Duff. Direct methods for solving sparse systems of linear equations. *SIAM J. on Scientific and Statistical Computation*, 5:605–619, 1984. [netlib/harwell/].

175. T. Duff. Splines in animation and modeling. In *State of the Art in Image Synthesis*. ACM SIGGRAPH, 1986.

176. C. B. Dunham. A Fortran program for discrete nonlinear Chebyshev approximation. *J. of Computational and Applied Mathematics*, 6:241–245, 1980. [Algorithm 017; Fortran listing].

177. C. B. Dunham. Stability of the linear inequality method for rational Chebyshev approximation. *J. of Computational and Applied Mathematics*, 11:139–143, 1984.

178. C. B. Dunham. Rationals with repeated poles. In J. C. Mason and M. G. Cox, eds., *Algorithms for Approximation*, pages 285–291. Oxford Univ. Press, 1987.

179. C. F. Dunkl. Orthogonal polynomials on the hexagon. *SIAM J. on Applied Mathematics*, 47:343–351, 1987. [lists expansions in monomials].

180. C. S. Duris. Fortran routines for discrete cubic spline interpolation and smoothing. *ACM Trans. on Mathematical Software*, 6:92–103, 1980. [toms/547].

181. N. Dyn and D. Levin. Bell-shaped basis functions for surface fitting. In Z. Ziegler, ed., *Approximation Theory and Applications*, pages 113–129. Academic Press, New York, 1981.

182. N. Dyn, D. Levin, and J. A. Gregory. A 4-point interpolatory subdivision scheme for curve design. *Computer Aided Geometric Design*, 4:257–268, 1987.

183. N. Dyn, D. Levin, and S. Rippa. Numerical procedures for surface fitting of scattered data by radial functions. *SIAM J. on Scientific and Statistical Computation*, 7:639–659, 1986.

184. O. Egecioglu, E. Gallopoulos, and C. K. Koc. Fast and practical parallel polynomial interpolation. report 646, Center for Supercomputing Research and Development, University of Illinois at Urbana-Champaign, 1987.

185. S. C. Eisenstat, K. R. Jackson, and J. W. Lewis. The order of monotone piecewise cubic interpolation. *SIAM J. on Numerical Analysis*, 22:1220–1237, 1985.

186. T. Elfving and L.-E. Andersson. An algorithm for computing constrained smoothing spline functions. *Numerische Mathematik*, 52:583–595, 1988.

187. R. M. R. Ellis and D. H. McLain. A new method of cubic curve fitting using local data. *ACM Trans. on Mathematical Software*, 3:175–178, 1977. [toms/514 algol listing].

188. M. P. Epstein. On the influence of parametrization in parametric interpolation. *SIAM J. on Numerical Analysis*, 13:261–268, 1976.

189. R. L. Eubank. Optimal grouping, spacing, stratification, and piecewise constant approximation. *SIAM Review*, 30:404–420, 1988.

190. B. M. Ewen-Smith. Algorithm for the production of contour maps from linearized data. *Nature*, 234:33–34, 1971.

191. K. J. Falconer. A general purpose algorithm for contouring over scattered data points. NAC6, National Physical Laboratory, 1971. [NAG/J06GFF].

192. G. Farin. Smooth interpolation to scattered 3d data. In R. E. Barnhill and W. Boehm, eds., *Surfaces in Computer Aided Geometric Design*. Oberwohlfach, North-Holland, Amsterdam, 1983.

193. G. Farin. A modified Clough-Tocher interpolant. *Computer Aided Geometric Design*, 2:19–27, 1985.

194. S. J. Farlow, ed. *Self-Organizing Methods in Modeling*. Marcel Dekker, New York, 1984. [SAS listing on pp.305–314].

195. E. J. Farrell. Visual interpretation of complex data. *IBM Systems J.*, 26:174–200, 1987.

196. R. Farwig. Multivariate interpolation of arbitrarily spaced data by moving least squares methods. *J. Computational and Applied Mathematics*, 16:79–93, 1986.

197. D. R. Ferguson, P. D. Frank, and A. K. Jones. Surface shape control using constrained optimization on the B-spline representation. *Computer Aided Geometric Design*, 5:87–103, 1988.

198. D. A. Field. Algorithms for determining invertible two- and three-dimensional quadratic isoparametric finite element transformations. *International J. Numerical Methods in Engineering*, 19:789–802, 1983.

199. C. T. Fike. *Computer Evaluation of Mathematical Functions*. Prentice-Hall, Englewood Cliffs, New Jersey, 1968.
200. B. Fischer and L. Reichel. Newton interpolation in Fejér and Chebhyshev points. *Mathematics of Computation*, in press, 1988.
201. T. A. Foley. Three-stage interpolation to scattered data. *Rocky Mountain J. of Mathematics*, 14:141–149, 1984.
202. T. A. Foley. A triangular surface patch with optimal error bounds. In C. K. Chui, L. L. Schumaker, and J. D. Ward, eds., *Approximation Theory V*, pages 343–346. Academic Press, New York, 1986.
203. T. A. Foley. Interpolation and approximation of 3-d and 4-d scattered data. *Comput. Math. Applic.*, 13:711–740, 1987.
204. T. A. Foley. Interpolation with interval and point tension controls using cubic weighted v-splines. *ACM Trans. on Mathematical Software*, 13:68–96, 1987.
205. T. A. Foley. A shape preserving interpolant with tension controls. *Computer Aided Geometric Design*, 5:105–118, 1988.
206. B. Fornberg. CPSC: complex power series coefficients. *ACM Trans. on Mathematical Software*, 7:542–547, 1981. [toms/579].
207. G. E. Forsythe. Generation and use of orthogonal polynomials for fitting data with a digital computer. *SIAM J. on Applied Mathematics*, 5:74–88, 1957. [imsl/opoly].
208. S. J. Fortune. A sweepline algorithm for voronoi diagrams. *Algorithmica*, 2:153–174, 1987. [netlib/voronoi/].
209. A. Fox. Implementation and relative efficiency of quasirandom sequence generators. *ACM Trans. on Mathematical Software*, 12:362–372, 1986. [toms/647].
210. P. A. Fox, ed. *The PORT Mathematical Subroutine Library*. AT&T Bell Laboratories, 1984. [netlib/port/].
211. R. Franke. Locally determined smooth interpolation at irregularly spaced points in several variables. *JIMA*, 19:471–482, 1977.
212. R. Franke. Scattered data interpolation: Tests of some methods. *Mathematics of Computation*, 38:181–200, 1982.
213. R. Franke. Smooth interpolation of scattered data by local thin plate splines. *Computers and Mathematics with Applications*, 8:273–281, 1982. [lotps].
214. R. Franke. Thin plate splines with tension. *Computer Aided Geometric Design*, 2:87–95, 1985.
215. R. Franke and G. Nielson. Smooth interpolation of large sets of scattered data. *International J. Numerical Methods in Engineering*, 15:1691–1704, 1980.
216. R. Franke and G. M. Nielson. Surface approximation with imposed conditions. In R. E. Barnhill and W. Boehm, eds., *Surfaces in Computer Aided Geometric Design*, pages 135–146. North-Holland, Amsterdam, 1983.
217. R. Franke and L. L. Schumaker. A bibliography of multivariate approximation. In C. K. Chui, L. L. Schumaker, and F. I. Utreras, eds., *Topics in Multivariate Approximation*, pages 275–335. Academic Press, New York, 1987.
218. P. O. Frederickson. Quasi-interpolation, extrapolation and approximation on the plane. In *Conference on Numerical Mathematics*, pages 159–167, 1971.
219. A. E. Frey, C. A. Hall, and T. A. Porsching. Some results on the global inversion of bilinear and quadratic isoparametric finite element transformations. *Mathematics of Computation*, 32:725–749, 1978.
220. W. H. Frey. A useful variant of McLaughlin's interpolant. GMR-5004, General Motors Research Laboratories, 1985.
221. J. H. Friedman. A tree-structured approach to nonparametric multiple regression. In T. Gasser and M. Rosenblatt, eds., *Smoothing Techniques for Curve Estimation*, pages 5–22. Springer-Verlag, New York, 1979. [CART].
222. J. H. Friedman, E. H. Grosse, and W. Stuetzle. Multidimensional additive spline approximation. *SIAM J. on Scientific and Statistical Computation*, 4:291–301, 1983. [MASA].

223. F. N. Fritsch. The Wilson-Fowler spline is a ν-spline. *Computer Aided Geometric Design*, 3:155–162, 1986.

224. F. N. Fritsch. Energy comparisons of Wilson-Fowler splines with other interpolating splines. In G. E. Farin, ed., *Geometric Modeling: Algorithms and New Trends*, pages 185–201. SIAM, Philadelphia, 1987.

225. F. N. Fritsch. Representations for parametric cubic splines. *Computer Aided Geometric Design*, page submitted, 1988.

226. F. N. Fritsch and J. Butland. A method for constructing local monotone piecewise cubic interpolants. *SIAM J. on Scientific and Statistical Computation*, 5:300–304, 1984. [pchip].

227. F. N. Fritsch and R. E. Carlson. Monotone piecewise cubic interpolation. *SIAM J. on Numerical Analysis*, 17:238–246, 1980.

228. M. Frontini, W. Gautschi, and G. V. Milovanović. Moment-preserving spline approximation on finite intervals. *Numerische Mathematik*, 50:503–518, 1987.

229. H. Fuchs, K. Z. M., and S. P. Uselton. Optimal surface reconstruction from planar contours. *Communications of the ACM*, 20:693–702, 1977.

230. L. W. Fullerton. A bibliography on the evaluation of mathematical functions. Comp. Sci. Tech. Rep. 86, AT&T Bell Laboratories, 1980.

231. L. W. Fullerton. Fnlib. User Manual, AT&T Bell Laboratories, 1981. [netlib/fn/].

232. G. M. Furnival and R. W. Wilson, Jr. Regressions by leaps and bounds. *Technometrics*, 16:499–511, 1974. [imsl/rbest].

233. P. W. Gaffney. The calculation of indefinite integrals of B-splines. *JIMA*, 17:37–41, 1976. [netlib/dierckx/fitpack.f (splint)].

234. S. Gal. Computing elementary functions: a new approach for achieving high accuracy and good performance. In W. L. Miranker and R. A. Toupin, eds., *Accurate Scientific Computations*, LNCS 235, pages 1–16. Springer-Verlag, New York, 1985.

235. G. Galimberti and V. Pereyra. Solving confluent Vandermonde systems of Hermite type. *Numerische Mathematik*, 18:44–60, 1971. [Algol listing].

236. E. R. Gansner, S. C. North, and K. P. Vo. Dag - a program that draws directed graphs. Technical Memorandum, AT&T Bell Laboratories, 1987.

237. M. Gasca and V. Ramirez. Interpolation systems in R^k. *J. of Approximation Theory*, 42:36–51, 1984.

238. W. Gautschi. Attenuation factors in practical Fourier analysis. *Numerische Mathematik*, 18:373–400, 1972.

239. W. Gautschi. On generating orthogonal polynomials. *SIAM J. on Scientific and Statistical Computation*, 3:289–317, 1982.

240. D. M. Gay and R. E. Welsch. Maximum likelihood and quasi-likelihood for nonlinear exponential family regression models. *J. of the American Statistical Association*, 1988.

241. S. Geman and D. Geman. Stochastic relaxation, Gibbs distributions, and the Bayesian restoration of images. *IEEE Trans. on Pattern Analysis and Machine Intelligence*, PAMI-6:721–741, 1984.

242. A. George and J. W. H. Liu. *Computer Solution of Large Sparse Positive Definite Systems.* Prentice-Hall, Englewood Cliffs, New Jersey, 1980. [netlib/sparspak/].

243. D. Girard. A fast 'Monte Carlo cross-validation' procedure for large least squares problems with noisy data. RR 687 -M-, Informatique et Mathématiques Appliquées de Grenoble, 1987.

244. G. Giunta and A. Murli. A package for computing trigonometric Fourier coefficients based on Lyness's algorithm. *ACM Trans. on Mathematical Software*, 13:97–107, 1987. [toms/649].

245. K. Glashoff and K. Roleff. A new method for Chebyshev approximation of complex-valued functions. *Mathematics of Computation*, 36:233–239, 1981.

246. R. Gmelig Meyling. Numerical experiments with cubic C^1-spline functions. In C. K. Chui, L. L. Schumaker, and J. D. Ward, eds., *Approximation Theory V*, pages 475–478. Academic Press, New York, 1986.

247. R. H. J. Gmelig Meyling. An algorithm for constructing configurations of knots for bivariate B-splines. *SIAM J. on Numerical Analysis*, 24:706–724, 1987.
248. R. H. J. Gmelig Meyling. Approximation by cubic C^1-splines on arbitrary triangulations. *Numerische Mathematik*, 51:65–85, 1987.
249. R. H. J. Gmelig Meyling. On algorithms and applications for bivariate B-splines. In J. C. Mason and M. G. Cox, eds., *Algorithms for Approximation*, pages 83–93. Oxford Univ. Press, 1987.
250. R. H. J. Gmelig Meyling and P. R. Pfluger. B-spline approximation of a closed surface. *IMA J. of Numerical Analysis*, 7:73–96, 1987.
251. D. Goldfarb and A. Idnani. A numerically stable dual method for solving strictly convex quadratic programs. *Mathematical Programming*, 27:1–33, 1983. [imsl/qprog (Powell)].
252. R. N. Goldman. The method of resolvents: A technique for the implicitization, inversion, and intersection of non-planar, parametric, rational cubic curves. *Computer Aided Geometric Design*, 2:237–255, 1985.
253. H. H. Goldstine. *A History of Numerical Analysis from the 16th through the 19th Century.* Springer-Verlag, New York, 1977. [Gauss].
254. M. Golomb and J. Jerome. Equilibria of the curvature functional and manifold of nonlinear interpolating spline curves. *SIAM J. on Mathematical Analysis*, 13:421–458, 1982.
255. G. H. Golub and R. J. LeVeque. Extensions and uses of the variable projection algorithm for solving nonlinear least squares problems. In *Proc. of the Army Numerical Analysis and Computers Conference, ARO Rep. 79-3*, pages 1–12, 1979. [netlib/misc/varp2].
256. G. H. Golub and V. Pereyra. Differentiation of psuedo-inverses and nonlinear least squares problems whose variables separate. *SIAM J. on Numerical Analysis*, 10:413–432, 1973. [netlib/misc/varpro].
257. G. H. Golub and C. F. Van Loan. *Matrix Computations.* Johns Hopkins University Press, Baltimore Maryland, 1983.
258. G. H. Golub and J. H. Welsch. Calculation of gaussian quadrature rules. *Mathematics of Computation*, 23:221–230, 1969. [netlib/go/gaussq].
259. H. H. Gonska and J. Meier. A bibliography on approximation of functions by Bernstein type operators (1955-1982). In C. K. Chui, L. L. Schumaker, and J. D. Ward, eds., *Approximation Theory IV*, pages 739–785. Academic Press, New York, 1983.
260. T. N. T. Goodman and K. Unsworth. Shape preserving interpolation by curvature continuous parametric curves. *Computer Aided Geometric Design*, 5:323–340, 1988.
261. J. H. Goodnight. A tutorial on the SWEEP operator. *American Statistician*, 33:149–158, 1979. [imsl/rstep].
262. W. J. Gordon. Distributive lattices and the approximation of multivariate functions. In I. J. Schoenberg, ed., *Proc. of the Symposium on Approximation with Special Emphasis on Splines*, pages 223–277. Univ. of Wisconsin, Academic Press, New York, 1969.
263. W. J. Gordon and C. A. Hall. Transfinite element methods: Blending-function interpolation over arbitrary curved element domains. *Numerische Mathematik*, 21:109–129, 1973.
264. W. J. Gordon and J. A. Wixom. Shepard's method of 'metric interpolation' to bivariate and multivariate interpolation. *Mathematics of Computation*, 32:253–264, 1978.
265. D. Gottlieb and S. A. Orszag. *Numerical Analysis of Spectral Methods: Theory and Applications.* SIAM, Philadelphia, 1977.
266. D. Gottlieb and E. Tadmor. Recovering pointwise values of discontinuous data within spectral accuracy. Report No. 85-3, ICASE, 1985.
267. T. A. Grandine. The computational cost of simplex spline functions. *SIAM J. on Numerical Analysis*, 24:887–890, 1987.
268. T. A. Grandine. An iterative method for computing multivariate C^1 piecewise polynomial interpolants. *Computer Aided Geometric Design*, 4:307–319, 1987.
269. T. A. Grandine. The stable evaluation of multivariate simplex splines. *Mathematics of Computation*, 50:197–205, 1988.

270. P. R. Graves-Morris. Efficient reliable rational interpolation. In M. G. de Bruin and H. van Rossum, eds., *Padé Approximation and its Applications Amsterdam 1980*, volume 888 of *Lecture Notes in Mathematics*, pages 28–63. Springer-Verlag, New York-Verlag, 1981.
271. P. R. Graves-Morris and T. R. Hopkins. Reliable rational interpolation. *Numerische Mathematik*, 36:111–128, 1981. [NAG E01RAF, E01RBF].
272. W. H. Gray and J. E. Akin. An improved method for contouring on isoparametric surfaces. *International J. Numerical Methods in Engineering*, 14:451–458, 1979.
273. J. A. Gregory. A blending function interpolant for triangles. In D. C. Handscomb, ed., *Multivariate Approximation*, pages 279–287. Academic Press, New York, 1978.
274. J. A. Gregory. Interpolation to boundary data on the simplex. *Computer Aided Geometric Design*, 2:43–52, 1985.
275. A. Griewank and P. L. Toint. Partitioned variable metric updates for large structured optimization problems. *Numerische Mathematik*, 39:119–137, 1982. [harwell/ve08ad, formerly pspmin].
276. W. D. Gropp. Local uniform mesh refinement with moving grids. *SIAM J. on Scientific and Statistical Computation*, 8:292–304, 1987.
277. E. Grosse. Colors for level plots. Numerical Analysis Manuscript 85-1, AT&T Bell Laboratories, 1985. [netlib/misc/rainbow.c].
278. E. Grosse. Spectral spline approximation. In C. K. Chui, L. L. Schumaker, and J. D. Ward, eds., *Approximation Theory V*, pages 363–366. Academic Press, New York, 1986. [crysalis, available from author].
279. E. Grosse and J. Hobby. Spline approximation with integer constraints on coefficients and knots. in preparation, AT&T Bell Laboratories, 1988.
280. E. H. Grosse. Tensor spline approximation. *Linear Algebra and Its Applications*, 34:29–41, 1980. [netlib/tensor/, imsl/bsls2].
281. V. Guerra and R. A. Tapia. A local procedure for error detectoion and data smoothing. MRC Technical Summary Report 1452, Mathematics Research Center, University of Wisconsin, Madison, 1974. [imsl/cssed].
282. M. H. Gutknecht. Two applications of periodic splines. In *Approximation Theory III*, pages 467–472. Academic Press, New York, 1980.
283. M. H. Gutknecht. An iterative method for solving linear equations based on minimun norm Pick-Nevanlinna interpolation. In C. K. Chui, L. L. Schumaker, and J. D. Ward, eds., *Approximation Theory V*, pages 371–374. Academic Press, New York, 1986.
284. M. H. Gutknecht. Attenuation factors in multivariate fourier analysis. *Numerische Mathematik*, 51:615–629, 1987.
285. H. Hagen. Geometric surface patches without twist constraints. *Computer Aided Geometric Design*, 3:179–184, 1986.
286. R. W. Hamming. *Numerical Methods for Scientists and Engineers, Second Edition.* McGraw-Hill, New York, 1973.
287. D. C. Handscomb. Recovery of fluid flow fields. In J. C. Mason and M. G. Cox, eds., *Algorithms for Approximation*, pages 531–540. Oxford Univ. Press, 1987.
288. R. J. Hanson. Linear least squares with bounds and linear constraints. *SIAM J. on Scientific and Statistical Computation*, 7:826–834, 1986.
289. R. L. Hardy. Multiquadric equations of topography and other irregular surfaces. *J. Geophys. Res.*, 76:1905–1919, 1971.
290. J. F. Hart, E. W. Cheney, C. L. Lawson, H. J. Maehly, C. K. Mesztenyi, J. R. Rice, H. G. Thacher, Jr., and C. Witzgall. *Computer Approximations.* John Wiley & Sons, New York, 1968.
291. A. Harten. Preliminary results on the extension of ENO schemes to two-dimensional problems. In C. Carasso, P.-A. Raviart, and D. Serre, eds., *Nonlinear Hyperbolic Problems*, LNM 1270, pages 23–40. Springer-Verlag, New York, 1986.
292. A. Harten, B. Engquist, S. Osher, and S. R. Chakravarthy. Uniformly high order accurate

essentially non-oscillatory schemes, III. *J. Computational Physics*, 71:231–303, 1987.

293. P. J. Hartley. Tensor product approximations to data defined on rectangular meshes in n-space. *Computer J.*, 19:348–352, 1975.

294. Harwell Laboratory, Computer Science and Systems Division. Harwell subroutine library. AERE - R 9185, 1987.

295. K. H. Haskell and R. J. Hanson. An algorithm for linear least squares problems with equality and nonnegativity constraints. *Mathematical Programming*, 21:98–118, 1981. [toms/587].

296. T. Hastie and W. Stuetzle. Principal curves. Technical Report, AT&T Bell Laboratories, 1987.

297. T. J. Hastie and R. J. Tibshirani. Generalized additive models. *Statistical Science*, 1:297–318, 1986. [netlib/a/gaim].

298. J. G. Hayes and J. Halliday. The least squares fitting of cubic spline surfaces to general data sets. *J. Institute of Mathematics and Its Applications*, 14:89–103, 1974.

299. W. J. Hemmerle. A comprehensive, matrix-free algorithm for analysis of variance. *ACM Trans. on Mathematical Software*, 8:383–401, 1982. [toms/591].

300. P. Henrici. *Essentials of Numerical Analysis with Pocket Calculator Demonstrations.* John Wiley & Sons, New York, 1982.

301. P. Henrici. *Applied and Computational Complex Analysis*, volume 3. John Wiley & Sons, New York, 1986.

302. J. N. Henry. Comparison of algorithms for multivariate rational approximation. *Mathematics of Computation*, 31:485–494, 1977.

303. J. G. Herriot and C. H. Reinsch. Procedures for natural spline interpolation. *Communications of the ACM*, 16:763–768, 1973. [toms/472].

304. J. G. Herriot and C. H. Reinsch. Procedures for quintic natural spline interpolation. *ACM Trans. on Mathematical Software*, 2:281–289, 1976. [toms/507].

305. G. Herron. A characterization of certain C^1 discrete triangular interpolants. *SIAM J. on Numerical Analysis*, 22:811–819, 1985.

306. J. D. Hobby. Smooth, easy to compute interpolating splines. *Discrete and Computational Geometry*, 1:123–140, 1986.

307. R. R. Hocking. Selection of the best subset of regression variables. In K. Enslein, A. Ralson, and H. S. Wilf, eds., *Statistical Methods for Digital Computers*, pages 39–57. John Wiley & Sons, New York, 1977.

308. P. Hoffman and K. C. Reddy. Numerical differentiation by high order interpolation. *SIAM J. on Scientific and Statistical Computation*, 8:979–987, 1987.

309. C. Hoffmann and J. Hopcroft. The potential method for blending surfaces and corners. In G. E. Farin, ed., *Geometric Modeling: Algorithms and New Trends*, pages 347–365. SIAM, Philadelphia, 1987.

310. K. Höllig. Multivariate splines. *SIAM J. on Numerical Analysis*, 19:1013–1031, 1982.

311. K. Höllig. Multivariate splines. In *Proc. of Symposia in Applied Mathematics*, pages 103–127. American Mathematical Society, Providence, Rhode Island, 1986.

312. F. R. de Hoog and M. F. Hutchinson. An efficient method for calculating smoothing splines using orthogonal transformations. *Numerische Mathematik*, submitted, 1987.

313. B. K. P. Horn and M. J. Brooks. The variational approach to shape from shading. *Computer Vision, Graphics, and Image Processing*, 33:174–208, 1986.

314. J. Hoschek. Approximate conversion of spline curves. *Computer Aided Geometric Design*, 4:59–66, 1987.

315. J. Hoschek. Intrinsic parameterization for approximation. *Computer Aided Geometric Design*, 5:27–31, 1988.

316. E. G. Houghton, R. F. Emnett, J. D. Factor, and C. L. Sabharwal. Implementation of a divide-and-conquer method for intersection of parametric surfaces. *Computer Aided Geometric Design*, 2:173–183, 1985.

317. D. J. Hudson. Fitting segmented curves whose join points have to be estimated. *J. of the American Statistical Association*, 61:1097–1129, 1966.

318. M. F. Hutchinson. Algorithm 642: A fast procedure for calculating minimum cross validation cubic smoothing splines. *ACM Trans. on Mathematical Software*, 12:150–153, 1986. [toms/642 (CUBGCV)].

319. J. M. Hyman. Accurate monotonicity preserving cubic interpolation. *SIAM J. on Scientific and Statistical Computation*, 4:645–654, 1983.

320. K. Ichida, T. Kiyono, and F. Yoshimoto. Curve fitting by a one-pass method with a piecewise cubic polynomial. *ACM Trans. on Mathematical Software*, 3:164–174, 1977.

321. IMSL. *MATH/LIBRARY User's Manual, Version 1.0.* 1987.

322. L. D. Irvine, S. P. Marin, and P. W. Smith. Constrained interpolation and smoothing. *Constructive Approximation*, 2:129–151, 1986. [imsl/cscon].

323. M. A. Jenkins. Algorithm 493: Zeros of a real polynomial. *ACM Trans. on Mathematical Software*, 1:178–189, 1975. [toms/493, imsl/zporc,zpocc].

324. T. Jensen. Assembling triangular and rectangular patches and multivariate splines. In G. E. Farin, ed., *Geometric Modeling: Algorithms and New Trends*, pages 203–220. SIAM, Philadelphia, 1987.

325. J. W. Jerome and L. L. Schumaker. Local support bases for a class of spline fucntions. *J. of Approximation Theory*, 16:16–27, 1976.

326. Z. Jing and A. T. Fam. An algorithm for computing Chebyshev approximations. *Mathematics of Computation*, 48:691–710, 1987.

327. Z. Jing and A. T. Fam. An algorithm for computing continuous Chebyshev approximations. *Mathematics of Computation*, 48:691–710, 1987.

328. B. Joe. Delaunay triangular meshes in convex polygons. *SIAM J. on Scientific and Statistical Computation*, 7:514–539, 1986.

329. R. H. Jones and P. V. Tryon. Continuous time series models for unequally spaced data applied to modeling atomic clocks. *SIAM J. on Scientific and Statistical Computation*, 8:71–81, 1987.

330. J. L. Junkins, G. W. Miller, and J. R. Jancaitis. A weighting function approach to modeling of irregular surfaces. *J. of Geophysical Research*, 78:1794–1803, 1973.

331. D. L. B. Jupp. Approximation to data by splines with free knots. *SIAM J. on Numerical Analysis*, 15:328–343, 1978.

332. M. Kallay. Plane curves of minimal energy. *ACM Trans. on Mathematical Software*, 12:219–222, 1986.

333. D. W. Kammler. L_1-approximation of completely monotonic functions by sums of exponentials. *SIAM J. on Numerical Analysis*, 16:30–45, 1979.

334. E. H. Kaufman, Jr., D. J. Leeming, and G. D. Taylor. A combined Remes–differential correction algorithm for rational approximation. *Mathematics of Computation*, 32:233–242, 1978. [diff-corr version in netlib/a/difcor].

335. E. H. Kaufman, Jr. and G. D. Taylor. Uniform rational approximation of functions of several variables. *International J. of Numerical Methods in Engineering*, 9:297–323, 1975.

336. E. H. Kaufman, Jr. and G. D. Taylor. Infinite-interval nonlinear approximations. *Constructive Approximation*, 4:211–221, 1988.

337. L. Kaufman. A variable projection method for solving separable nonlinear least squares problems. *BIT*, 15:49–57, 1975. [netlib/port/nsf,nsg].

338. L. Kaufman and V. Pereyra. A method for separable nonlinear leasw squares with separable nonlinear equality constraints. *SIAM J. on Numerical Analysis*, 15:12–20, 1978.

339. J. Kautsky and N. K. Nichols. Equidistributing meshes with constraints. *SIAM J. on Scientific and Statistical Computation*, 1:497–511, 1980.

340. R. Kohn and C. F. Ansley. A nerw algorithm for spline smoothing based on smoothing a stochastic process. *SIAM J. on Scientific and Statistical Computation*, 8:33–48, 1987.

341. P. A. Koparkar and S. P. Mudur. A new class of algorithms for the processing of parametric curves. *CAD*, 15:41–45, 1983. [Pascal listings].

342. F. T. Krogh. Efficient algorithms for polynomial interpolation and numerical differentiation.

Mathematics of Computation, 24:185–190, 1970. [NAG E01AEF, E02AKF, E02AHF, E02AJF].
343. M. A. Lachance. Chebyshev economization for parametric surfaces. *Computer Aided Geometric Design*, 5:195–208, 1988.
344. P. Lancaster and K. Šalkauskas. *Curve and Surface Fitting: An Introduction*. Academic Press, New York, 1986.
345. J. M. Lane and R. F. Riesenfeld. A theoretical development for the computer generation and display of piecewise polynomial surfaces. *IEEE Trans. on Pattern Analysis and Machine Intelligence*, PAMI-2:35–46, 1980.
346. C. L. Lawson. Software for C^1 surface interpolation. In J. R. Rice, ed., *Mathematical Software III*, pages 161–194. Academic Press, New York, 1977.
347. C. L. Lawson. C^1 surface interpolation for scattered data on a sphere. *Rocky Mountain J. of Mathematics*, 14:177–202, 1984.
348. C. M. Lee and F. D. K. Roberts. A comparison of algorithms for rational l_∞ approximation. *Mathematics of Computation*, 27:111–121, 1973.
349. D. Lee. Fast multiplication of a recursive block Toeplitz matrix by a vector and its application. *J. of Complexity*, 2:295–305, 1986.
350. D. Lee. Algorithms for shape from shading and occluding boundaries. Technical report, AT&T Bell Laboratories, 1988.
351. D. Lee. Coping with discontinuities in computer vision: their detection, classification, and measurement. In *Proc. 2nd International Conference on Computer Vision*, 1988.
352. D. Lee, T. Pavlidis, and K. Huang. Edge detection through residual analysis. In *Proc. of Computer Vision and Pattern Recognition*, pages 215–222. IEEE, 1988.
353. E. H. Lee and G. E. Forsythe. Variational study of nonlinear spline curves. *SIAM Review*, 15:120–133, 1973.
354. E. T. Y. Lee. Comments on some B-spline algorithms. *Computing*, 36:229–238, 1986.
355. E. T. Y. Lee. The rational Bézier representation for conics. In G. E. Farin, ed., *Geometric Modeling: Algorithms and New Trends*, pages 3–19. SIAM, Philadelphia, 1987.
356. D. Levin. Multidimensional reconstruction by set-valued approximations. *IMA J. of Numerical Analysis*, 6:173–184, 1986.
357. F. F. Little. Convex combination surfaces. In R. E. Barnhill and W. Boehm, eds., *Surfaces in Computer Aided Geometric Design*, pages 99–107. North-Holland, Amsterdam, 1983.
358. T. R. Lucas. Error bounds for interpolating cubic splines under various end conditions. *SIAM J. on Numerical Analysis*, 11:569–579, 1974.
359. T. R. Lucas. A posteriori improvements for interpolating periodic splines. *Mathematics of Computation*, 40:243–251, 1983.
360. T. Lyche and K. Mørken. A discrete approach to knot removal and degree reduction algorithms for splines. In J. C. Mason and M. G. Cox, eds., *Algorithms for Approximation*, pages 67–82. Oxford Univ. Press, 1987.
361. T. Lyche and L. Schumaker. Local spline approximation methods. *J. of Approximation Theory*, 15:294–325, 1975.
362. K. Madsen. A root-finding algorithm based on Newton's method. *BIT*, 13:71–75, 1973. [harwell/pa06].
363. K. Madsen. General algorithms for discrete non-linear parameter estimation. In J. C. Mason and M. G. Cox, eds., *Algorithms for Approximation*, pages 309–326. Oxford Univ. Press, 1987.
364. H. J. Maehly. Method for fitting rational approximations, parts II and III. *J. of the A. C. M.*, 10:257–277, 1963. [harwell/pe05].
365. H. H. Maindonald. *Statistical Computation*. John Wiley & Sons, New York, 1984. [imsl/rotin].
366. M. A. Malcolm. On the computation of nonlinear spline functions. *SIAM J. on Numerical Analysis*, 14:254–282, 1977. [code listing for uniformly spaced x_i].
367. C. L. Mallows. Some comments on C_p. *Technometrics*, 15:661–675, 1973.

368. L. Mansfield. Interpolation to boundary data in tetrahedra with applications to compatible finite elements. *J. Mathematical Analysis and Applications*, 56:137–164, 1976.
369. L. Mansfield. A Clough-Tocher type element useful for fourth order problems over nonpolygonal domains. *Mathematics of Computation*, 32:141:135–142, 1978.
370. L. Mansfield. Interpolation to scattered data in the plane by locally defined C^1 functions. In *Approximation Theory III*, pages 623–628. Academic Press, New York, 1980.
371. S. P. Marin. An approach to data parametrization in parametric cubic spline interpolation problems. *J. of Approximation Theory*, 41:64–86, 1983.
372. S. Marlow and M. J. D. Powell. A Fortran subroutine for plotting the part of a conic that is inside a given triangle. AERE R-8336, Atomic Energy Research Establishment, 1976. [harwell/ob14a].
373. J. Marroquin, S. Mitter, and T. Poggio. Probabilistic solution of ill-posed problems in computational vision. *J. of the American Statistical Association*, 82:76–89, 1987.
374. G. Marsaglia. A current view of random number generators. In L. Billard, ed., *Computer Science and Statistics: The Interface*, pages 3–10. North-Holland, Amsterdam, 1985.
375. J. C. Mason. Some applications and drawbacks of Padé approximants. In *Approximation Theory and Applications*. Academic Press, New York, 1981.
376. J. C. Mason and G. Opfer. An algorithm for complex polynomial approximation with nonlinear constraints. In C. K. Chui, L. L. Schumaker, and J. D. Ward, eds., *Approximation Theory V*, pages 471–474. Academic Press, New York, 1986.
377. A. D. Maude. Interpolation–mainly for graph plotters. *Computer J.*, 16:64–65, 1972.
378. D. F. McAllister and J. A. Roulier. Interpolation by convex quadratic splines. *Mathematics of Computation*, 32:1154–1162, 1978. [toms/574].
379. J. A. McDonald. Periodic smoothing of time series. *SIAM J. on Scientific and Statistical Computation*, 7:665–688, 1986.
380. D. H. McLain. Drawing contours from arbitrary data points. *Computer J.*, 17:318–324, 1974. [Algol listings; see also Sutcliffe].
381. D. H. McLain. Two dimensional interpolation from random data. *Computer J.*, 19:178–181, 1976. [Algol listing].
382. H. W. McLaughlin. Shape-preserving planar interpolation. *IEEE Computer Graphics and Applications*, 3:58–67, 1983.
383. H. Meier and H. Nowacki. Interpolating curves with gradual changes in curvature. *Computer Aided Geometric Design*, 4:297–305, 1987.
384. G. Meinardus. *Approximation of Functions: Theory and Numerical Methods*. Springer-Verlag, New York, 1967. [translated by L. L. Schumaker].
385. F. Melkes. Reduced piecewise bivariate Hermite interpolation. *Numerische Mathematik*, 19:326–340, 1972.
386. A. A. Melkman and C. A. Micchelli. Optimal estimation of linear operators in Hilbert spaces from inaccurate data. *SIAM J. on Numerical Analysis*, 16:87–105, 1979.
387. C. A. Micchelli and H. Prautzsch. Computing surfaces invariant under subdivision. *Computer Aided Geometric Design*, 4:321–328, 1987.
388. C. A. Micchelli and T. J. Rivlin. A survey of optimal recovery. In C. A. Micchelli and T. J. Rivlin, eds., *Optimal Estimation in Approximation Theory*, pages 1–53. Plenum Press, 1977.
389. C. A. Micchelli and F. I. Utreras. Smoothing and interpolation in a convex subset of a Hilbert space. *SIAM J. on Scientific and Statistical Computation*, 9:728–746, 1988.
390. C. A. Micchelli and G. Wahba. Design problems for optimal surface interpolation. In Z. Ziegler, ed., *Approximation Theory and Applications*, pages 329–348. Academic Press, New York, 1981.
391. D. P. Mitchell. Generating antialiased images at low sampling densities. *Computer Graphics (SIGGRAPH 87)*, 21:65–72, 1987.
392. G. Mühlbach. On multivariate interpolation by generalized polynomials on subsets of grids.

Computing, 40:201–215, 1988.

393. D. Mumford and J. Shah. Boundary detection by minimizing functionals, I. In *Proc. on Computer Vision and Pattern Recognition*. IEEE, 1985.

394. E. Neuman. Convex interpolating splines of arbitrary degree. In L. Collatz, G. Meinardus, and H. Werner, eds., *Numerical Methods of Approximation Theory*, volume 5, pages 211–222. Birkhäuser Verlag, 1980.

395. E. Neuman. Moments and Fourier transforms of B-splines. *J. of Computational and Applied Mathematics*, 7:51–, 1981.

396. G. M. Nielson. Some piecewise polynomial alternatives to splines under tension. In R. E. Barnhill and R. F. Riesenfeld, eds., *Computer Aided Geometric Design*, pages 209–235. Academic Press, New York, 1974.

397. G. M. Nielson. The side-vertex method for interpolation in triangles. *J. of Approximation Theory*, 25:318–336, 1979.

398. G. M. Nielson. Minimum norm interpolation in triangles. *SIAM J. on Numerical Analysis*, 17:44–62, 1980.

399. G. M. Nielson. A method for interpolating scattered data based upon a minimum norm network. *Mathematics of Computation*, 40:253–271, 1983.

400. G. M. Nielson. A transfinite, visually continuous, triangular interpolant. In G. E. Farin, ed., *Geometric Modeling: Algorithms and New Trends*, pages 235–246. SIAM, Philadelphia, 1987.

401. G. M. Nielson and R. Franke. A method for construction of surfaces under tension. *Rocky Mountain J. of Mathematics*, 14:203–221, 1984.

402. G. M. Nielson and R. Ramaraj. Interpolation over a sphere based upon a minimum norm network. *Computer Aided Geometric Design*, 4:41–57, 1987.

403. H. Nienhaus. A C^1-conforming finite element of the extended Melkes family. In C. K. Chui, L. L. Schumaker, and J. D. Ward, eds., *Approximation Theory V*, pages 503–506. Academic Press, New York, 1986.

404. Numerical Algorithms Group. *NAG Fortran Library Manual Mark 11*. 1984.

405. G. Nürnberger. Chebyshev approximation by splines with free knots and computation. In C. K. Chui, L. L. Schumaker, and J. D. Ward, eds., *Approximation Theory V*, pages 511–514. Academic Press, New York, 1986.

406. G. Nürnberger and M. Sommer. A Remez type algorithm for spline functions. *Numerische Mathematik*, 41:117–146, 1983.

407. G. Nürnberger, M. Sommer, and H. Strauss. An algorithm for segment approximation. *Numerische Mathematik*, 48:463–477, 1986.

408. D. P. O'Leary and B. W. Rust. Confidence intervals for inequality-constrained least squares problems, with applications to ill-posed problems. *SIAM J. on Scientific and Statistical Computation*, 7:473–489, 1986.

409. J. Oliver. An algorithm for numerical differentiation of a function of one real variable. *J. of Computational and Applied Mathematics*, 6:145–160, 1980. ["Algorithm 017" Algol listing].

410. G. Opfer and H. J. Oberle. The derivation of cubic splines with obstacles by methods of optimization and optimal control. *Numerische Mathematik*, 52:17–31, 1988.

411. F. O'Sullivan. Comments on Dr. Silverman's paper. *J. Royal Statistical Society B*, 47:39–40, 1985. [netlib/gcv/sbart].

412. F. O'Sullivan. Fast computation of fully automated log-density and log-hazard estimators. *SIAM J. on Scientific and Statistical Computation*, 9:363–379, 1988.

413. F. O'Sullivan, B. Yandell, and W. Raynor. Automatic smoothing of regression functions in generalized linear models. *J. of the American Statistical Association*, 81:96–103, 1986.

414. C. C. Paige and M. A. Saunders. LSQR: an algorithm for sparse linear equations and sparse least squares. *ACM Trans. on Mathematical Software*, 8:43–71, 1982. [netlib/misc/lsqr].

415. R. R. Patterson. Parametric cubics as algebraic curves. *Computer Aided Geometric Design*,

5:139–159, 1988.
416. T. Pavlidis. Polygonal approximations by Newton's method. *IEEE Trans. on Computers*, C-26:800–807, 1977.
417. T. Pavlidis. Curve fitting with conic splines. *ACM Trans. on Graphics*, 2:1–31, 1983.
418. T. Pavlidis and S. L. Horowitz. Segmentation of plane curves. *IEEE Trans. on Computers*, C23:860–870, 1974.
419. V. Pereyra and G. Scherer. Efficient computer manipulation of tensor products with applications to multidimensional approximation. *Mathematics of Computation*, 27:595–605, 1973.
420. V. Pereyra and G. Sewell. Mesh selection for discrete solution of boundary value problems in ode's. *Numerische Mathematik*, 23:261–268, 1975. [pasva3, imsl/bvpfd, nag/d02gaf].
421. G. Peters and J. H. Wilkinson. Practical problems arising in the solution of polynomial equations. *JIMA*, 8:16–35, 1971. [harwell/pa06].
422. C. S. Petersen, B. R. Piper, and A. J. Worsey. Adaptive contouring of a trivariate interpolant. In G. E. Farin, ed., *Geometric Modeling: Algorithms and New Trends*, pages 385–395. SIAM, Philadelphia, 1987.
423. L. Piegl and W. Tiller. Curve and surface constructions using rational B-splines. *Computer Aided Design*, 19:485–498, 1987.
424. M. J. D. Powell. Curve fitting by splines in one variable. In J. G. Hayes, ed., *Numerical Approximation to Functions and Data*, pages 65–83. Institute of Mathematics and Its Applications, The Athlone Press, University of London, 1970. [harwell/ts01].
425. M. J. D. Powell. Piecewise quadratic surface fitting for contour plotting. In D. J. Evans, ed., *Software for Numerical Mathematics*. Academic Press, New York, 1974. [harwell].
426. M. J. D. Powell and M. A. Sabin. Piecewise quadratic approximations on triangles. *ACM Trans. on Mathematical Software*, 3:316–325, 1977.
427. V. Pratt. Techniques for conic splines. In *SIGGRAPH Proc.* ACM, 1985.
428. V. Pratt. Direct least-squares fitting of algebraic surfaces. *Computer Graphics (SIGGRAPH 87)*, 21:145–152, 1987.
429. H. Prautzsch. Degree elevation of B-spline curves. *Computer Aided Geometric Design*, 1:193–198, 1984.
430. F. P. Preparata and R. Tamassia. Fully dynamic techniques for point location and transitive closure in planar sturctures. In *29th Annual Symposium on Foundations of Computer Science*, pages 558–567. IEEE, 1988.
431. K. v. Radziewski. On periodic Hermite spline interpolation. In C. K. Chui, L. L. Schumaker, and J. D. Ward, eds., *Approximation Theory V*, pages 531–534. Academic Press, New York, 1986.
432. D. L. Ragozin. The discrete *k*-functional and spline smoothing of noisy data. *SIAM J. on Numerical Analysis*, 22:1241–1254, 1985.
433. J. Ramsey. Monotone splines in action. *Statistical Science*, page to appear, 1988.
434. L. Reichel. An asymptotically orthonormal polynomial family. *BIT*, 24:647–655, 1984.
435. L. Reichel. On polynomial approximation in the uniform norm by the discrete least squares method. *BIT*, 26:349–368, 1986.
436. L. Reichel. Some computational aspects of a method for rational approximation. *SIAM J. on Scientific and Statistical Computation*, 7:1041–1057, 1986.
437. K.-D. Reinsch. *Numerische Berechnung von Biegelinien in der Ebene*. PhD thesis, Institut für Mathematik, Technishen Universität München, 1981.
438. R. J. Renka. Interpolation of data on the surface of a sphere. *ACM Trans. on Mathematical Software*, 10:417–436, 1984. [toms/623].
439. R. J. Renka. Interpolatory tension splines with automatic selection of tension factors. *SIAM J. on Scientific and Statistical Computation*, 8:393–415, 1987.
440. R. J. Renka. Multivariate interpolation of large sets of scattered data. *ACM Trans. on Mathematical Software*, 14:139–148, 1988. [toms/660,661].

441. P. Rentrop. An algorithm for the computation of the exponential spline. *Numerische Mathematik*, 35:81–93, 1980.
442. J. Rice. Methods for bandwidth choice in nonparametric kernel regression. In J. E. Gentle, ed., *Computer Science and Statistics: 15th Interface*, pages 186–190. North-Holland, Amsterdam, 1983.
443. J. R. Rice. Adapt, adaptive smooth curve fitting. *ACM Trans. on Mathematical Software*, 4:82–94, 1978. [toms/525].
444. T. J. Rivlin. *An Introduction to the Approximation of Functions*. Dover Publications, New York, 1969.
445. K. S. Roberts, G. Bishop, and S. K. Ganapathy. Smooth interpolation of rotational motions. Technical Report, AT&T Bell Laboratories, 1987.
446. A. P. Rockwood and J. C. Owen. Blending surfaces in solid modeling. In G. E. Farin, ed., *Geometric Modeling: Algorithms and New Trends*, pages 367–383. SIAM, Philadelphia, 1987.
447. J. B. Rosenberg. Geographical data structures compared: A study of data structures supporting region queries. *IEEE Trans. on Computer Aided Design*, CAD-4:53–67, 1985. [C listing].
448. J. A. Roulier. Constrained interpolation. *SIAM J. on Scientific and Statistical Computation*, 1:333–344, 1980.
449. V. Ruas. A quadratic finite element method for solving biharmonic problems in $\mathbf{R}^n$. *Numerische Mathematik*, 52:33–43, 1988.
450. A. Ruhe. Fitting empirical data by positive sums of exponentials. *SIAM J. on Scientific and Statistical Computation*, 1:481–498, 1980.
451. M. A. Sabin. Contouring: A review of methods for scattered data. In K. W. Brodlie, ed., *Mathematical Methods in Computer Graphics and Design*. Academic Press, New York, 1980.
452. M. A. Sabin. Non-rectangular surface patches suitable for inclusion in a b-spline surface. In P. J. W. ten Hagen, ed., *Eurographics '83*, pages 57–69. Eurographics Association, , North-Holland, Amsterdam, 1983.
453. M. A. Sabin. Contouring - the state of the art. In R. A. Earnshaw, ed., *Fundamental Algorithms for Computer Graphics*, pages 411–482. NATO ASI Series F17, Springer-Verlag, New York, 1985.
454. P. Sablonnière. Bernstein-Bézier methods for the construction of bivariate spline approximants. *Computer Aided Geometric Design*, 2:29–36, 1985.
455. M. Sakai and M. C. L. d. Silanes. A simple rational spline and its application to monotonic interpolation to monotonic data. *Numerische Mathematik*, 50:171–182, 1986.
456. M. Sakai and R. A. Usmani. A shape preserving area true approximation of histogram by rational splines. *BIT*, 28:329–339, 1988.
457. K. Salkauskas. C^1 splines for interpolation of rapidly varying data. *Rocky Mountain J. of Mathematics*, 14:239–250, 1984.
458. N. S. Sapidis and P. D. Kaklis. An algorithm for constructing convexity and monotonicity-preserving splines in tension. *Computer Aided Geometric Design*, 5:87–103, 1988.
459. R. F. Sarraga. G^1 interpolation of generally unrestricted cubic Bézier curves. *Computer Aided Geometric Design*, 4:23–39, 1987.
460. I. P. Schagen. Automatic contouring from scattered data points. *Computer J*, 25:7–11, 1982.
461. I. P. Schagen. Internal modelling of objective functions for global optimization. *J. of Optimization Theory and Applications*, 11:345–353, 1986. [interpolation by sum of equal-width gaussians; derivatives of penalized surface].
462. J. W. Schmidt. An unconstrained dual program for computing convex C^1-spline approximants. *Computing*, 39:133–140, 1987. [explicit formulas for Newton's method].
463. J. W. Schmidt and W. Heß. Positive interpolation with rational quadratic splines. *Computing*, 38:261–267, 1987.
464. J. W. Schmidt and W. Heß. Positivity of cubic polynomials on intervals and positive spline

interpolation. *BIT*, 28:340–352, 1988.

465. J. L. Schonfelder and M. Razaz. Error control with polynomial approximations. *IMA J. of Numerical Analysis*, 1:105–114, 1980.
466. N. L. Schryer. SSAF – smooth spline approximations to functions. Numerical Analysis Memorandum 86-8, AT&T Bell Laboratories, 1986. [SSAF].
467. H. Schultheis and R. Schultheis. Algorithm 35: An algorithm for non-smoothing contour representations of two-dimensional arrays. *Computing*, 19:381–387, 1978. [Fortran listing].
468. L. L. Schumaker. Fitting surfaces to scattered data. In G. G. Lorentz, C. K. Chui, and L. L. Schumaker, eds., *Approximation Theory II*, pages 203–268. Academic Press, New York, 1976.
469. L. L. Schumaker. On hyperbolic splines. *J. of Approximation Theory*, 38:144–166, 1983.
470. L. L. Schumaker. On shape preserving quadratic spline interpolation. *SIAM J. on Numerical Analysis*, 20:854–864, 1983.
471. L. L. Schumaker. Numerical aspects of spaces of piecewise polynomials on triangulations. In J. C. Mason and M. G. Cox, eds., *Algorithms for Approximation*, pages 373–406. Oxford Univ. Press, 1987.
472. L. L. Schumaker and F. Utreras. Asymptotic properties of complete smoothing splines and applications. *SIAM J. on Scientific and Statistical Computation*, 9:24–38, 1988.
473. L. L. Schumaker and W. Volk. Efficient algorithms for evaluating multivariate polynomials. *Computer Aided Geometric Design*, 3:149–154, 1986.
474. D. S. Scott. The complexity of interpolating given data in three space with a convex function of two variables. *J. of Approximation Theory*, 42:52–36, 1984.
475. G. A. F. Seber. *Linear Regression Analysis*. John Wiley & Sons, New York, 1977.
476. T. W. Sederberg. Algebraic geometry for surface and solid modeling. In G. E. Farin, ed., *Geometric Modeling: Algorithms and New Trends*, pages 29–42. SIAM, Philadelphia, 1987.
477. B. Sendov. Approximation relative to Hausdorff distance. In *Approximation Theory*. Academic Press, New York, 1970.
478. L. F. Shampine. Discrete least squares polynomial fits. *Communications of the ACM*, 18:179–180, 1975. [imsl/rcurv].
479. D. Shepard. A two-dimensional interpolation function for irregularly-spaced data. In *Proc. of ACM National Conference*, pages 517–524, 1968.
480. J.-J. H. Shiau. *Smoothing Spline Estimation of Functions with Discontinuities*. PhD thesis, University of Wisconsin, Madison, 1985.
481. L. A. Shirman and C. H. Séquin. Local surface interpolation with Bezier patches. *Computer Aided Geometric Design*, 4:279–295, 1987.
482. R. Sibson. A brief description of natural neighbor interpolation. In V. Barnett, ed., *Interpreting Multivariate Data*, pages 21–36. John Wiley & Sons, New York, 1981.
483. R. Sibson and G. D. Thomson. A seamed quadratic element for contouring. *Computer J*, 24:378–382, 1981.
484. A. Sidi, W. F. Ford, and D. A. Smith. Acceleration of convergence of vector sequences. *SIAM J. on Numerical Analysis*, 23:178–196, 1986.
485. B. W. Silverman and J. T. Wood. The nonparametric estimation of branching curves. *J. of the American Statistical Association*, 82:551–558, 1987.
486. J. C. Simpson. Fortran translation of algorithm 409, discrete Chebyshev curve fit. *ACM Trans. on Mathematical Software*, 2:95–97, 1976. [toms/501].
487. R. C. Singleton. An algorithm for computing the mixed radix fast fourier transform. *IEEE Trans. on Audio and Electroacoustics*, AU-17:93–103, 1969. [netlib/go/fft,realtr].
488. R. D. Small. The generating function method of nonlinear approximation. *SIAM J. on Numerical Analysis*, 25:235–244, 1988.
489. W. V. Snyder. Contour plotting. *ACM Trans. on Mathematical Software*, 4:290–294, 1978. [toms/531].
490. H. Späth. *Spline-Algorithmen zur Konstruktion glatter Kurven und Flächen*. R. Oldenbourg

Verlag, München, 1973.

491. R. L. Streit. An algorithm for the solution of systems of complex linear equations in the L_∞ norm with constraints on the unknowns. *ACM Trans. on Mathematical Software*, 11:242–249, 1985. [toms/635].

492. R. L. Streit and A. H. Nuttall. A general Chebyshev complex function approximation procedure and an application to beamforming. *J. of the Acoustical Society of America*, 72:181–190, 1982.

493. D. C. Sutcliffe. A remark on a contouring algorithm. *Computer J*, 19:3–333–335, 1976. [Algol bug fix to McLain].

494. D. C. Sutcliffe. Contouring over rectangular and skewed rectangular grids – an introduction. In K. W. Brodlie, ed., *Mathematical Methods in Computer Graphics and Design*, pages 39–62. Academic Press, New York, 1980.

495. P. N. Swarztrauber. On the spectral approximation of discrete scalar and vector functions on the sphere. *SIAM J. on Numerical Analysis*, 16:934–949, 1979.

496. P. N. Swarztrauber. The approximation of vector functions and their derivatives on the sphere. *SIAM J. on Numerical Analysis*, 18:191–210, 1981.

497. P. N. Swarztrauber. Vectorizing the ffts. In G. Rodrigue, ed., *Parallel Computations*, pages 51–83. Academic Press, New York, 1982. [netlib/fftpack].

498. W. P. Tang and G. H. Golub. The block decomposition of a Vandermonde matrix and its applications. *BIT*, 21:505–517, 1981.

499. R. A. Tapia and J. R. Thompson. *Nonparametric Probability Density Estimation*. Johns Hopkins University Press, Baltimore Maryland, 1978.

500. L. F. Ten Eyck. Crystallographic fast fourier transforms. *Acta Crystallographica*, A29:183–191, 1973. [widely distributed in crystallographic community].

501. J. P. Thiran and S. Thiry. Strict Chebyshev approximation for general systems of linear equations. *Numerische Mathematik*, 51:701–725, 1987.

502. R. A. Thisted. *Elements of Statistical Computing: Numerical Computation*. Chapman and Hall, New York, 1988.

503. D. H. Thomas. A natural tensor product interpolation formula and the pseudoinverse of a matrix. *Linear Algebra and Its Applications*, 13:239–250, 1976.

504. J. F. Thompson, Z. U. A. Warsi, and C. W. Mastin. *Numerical Grid Generation*. North-Holland, Amsterdam, 1985.

505. B. Y. Ting and Y. L. Luke. Conversion of polynomials between different polynomial bases. *IMA J. of Numerical Analysis*, 1:229–234, 1981.

506. V. Torre and T. A. Poggio. On edge detection. *IEEE Trans. on Pattern Analysis and Machine Intelligence*, 8:147–163, 1986.

507. C. R. Traas. Smooth approximation of data on the sphere with splines. *Computing*, 38:177–184, 1987.

508. L. N. Trefethen. Numerical computation of the Schwarz-Christoffel transformation. *SIAM J. on Scientific and Statistical Computation*, 1:82–102, 1980. [netlib/conformal/scpack, scdoc, sclib].

509. L. N. Trefethen. Matlab programs for CF approximation. In C. K. Chui, L. L. Schumaker, and J. D. Ward, eds., *Approximation Theory V*, pages 599–602. Academic Press, New York, 1986.

510. L. N. Trefethen and M. H. Gutknecht. Padé, stable Padé, and Chebyshev-Padé approximation. In J. C. Mason and M. G. Cox, eds., *Algorithms for Approximation*, pages 228–264. Oxford Univ. Press, 1987.

511. J. W. Tukey. *Exploratory Data Analysis*. Addison-Wesley Publishing, Reading Massachusetts, 1977.

512. F. Utreras. Cross-validation techniques for smoothing spline functions in one and two dimensions. In Gasser and Rosenblatt, eds., *Smoothing Techniques for Curve Estimation*, pages 196–232. Heidelberg, 1979.

513. F. I. Utreras. On computing robust splines and applications. *SIAM J. on Scientific and Statistical Computation*, 2:153–163, 1981.

514. S. Van Huffel. *Analysis of the Total Least Squares Problem and its Use in Parameter Estimation*. PhD thesis, Katholieke Universiteit Leuven, 1987. [netlib/vanhuffel/].

515. C. Van Loan. On the method of weighting for equality-constrained least-squares problems. *SIAM J. on Numerical Analysis*, 22:851–864, 1985.

516. J. M. Varah. A spline least squares method for numerical parameter estimation in differential equations. *SIAM J. on Scientific and Statistical Computation*, 3:28–46, 1982.

517. M. L. Varas. A modified dual algorithm for the computation of the monotone cubic spline. In C. K. Chui, L. L. Schumaker, and J. D. Ward, eds., *Approximation Theory V*, pages 607–610. Academic Press, New York, 1986.

518. G. Wahba. Smoothing noisy data by spline functions. *Numerische Mathematik*, 24:383–393, 1975. [imsl/csscv].

519. G. Wahba. Surface fitting with scattered noisy data on euclidean d-space and on the sphere. *Rocky Mountain J. of Mathematics*, 14:281–299, 1984.

520. K. Wall and P. Danielsson. A fast sequential method for polygonal approximation of digitized curves. *Computer Vision, Graphics, and Image Processing*, 28:220–227, 1984.

521. C. Y. Wang. C^1 rational interpolation over an arbitrary triangle. *CAD*, 15:33–36, 1983.

522. J. D. Warren. On algebraic surfaces meeting with geometric continuity. TR 86-770, Dept. Computer Science, Cornell University, 1986.

523. D. F. Watson. Computing the n-dimensional Delaunay tesselation with application to Voronoi polytopes. *Computer J*, 24:167–172, 1981.

524. G. A. Watson. Data fitting by positive sums of exponentials. In J. C. Mason and M. G. Cox, eds., *Algorithms for Approximation*, pages 337–356. Oxford Univ. Press, 1987.

525. G. A. Watson. Numerical methods for chebyshev approximation of complex-valued functions. Numerical Analysis Report NA/116, University of Dundee, 1988.

526. L. T. Watson, S. C. Billups, and A. P. Morgan. HOMPACK: a suite of codes for globally convergent homotopy algorithms. *ACM Trans. on Mathematical Software*, 13:281–310, 1987. [toms/652, netlib/hompack/].

527. A. Weiser and S. E. Zarantonello. A note on piecewise linear and multilinear table interpolation in many dimensions. *Mathematics of Computation*, 50:189–196, 1988.

528. M. Weisfeld. Orthogonal polynomials in several variables. *Numerische Mathematik*, 1:38–40, 1959.

529. A. B. White, Jr. On selection of equidistributing meshes for two-point boundary-value problems. *SIAM J. on Numerical Analysis*, 16:472–502, 1979.

530. E. T. Whittaker. On a new method of graduation. *Edinburgh Mathematical Society Proc.*, 41:63–75, 1923.

531. D. G. Wilson. Piecewise linear approximations to tabulated data. *ACM Trans. on Mathematical Software*, 2:388–391, 1976. [toms/510].

532. G. T. Wilson. Problems in time-series analysis. In J. L. Mohamed and J. E. Wlash, eds., *Numerical Algorithms*, pages 293–313. Clarendon Press, Oxford, 1986.

533. J. A. Wixom. Interpolation to networks of curves in E^3. *SIAM J. on Numerical Analysis*, 15:1178–1193, 1978.

534. R. Wolke and H. Schwetlick. Iteratively reweighted least squares: Algorithms, convergence analysis, and numerical comparisons. *SIAM J. on Scientific and Statistical Computation*, 9:907–921, 1988.

535. H. Woltring. A FORTRAN package for generalized, cross-validatory spline smoothing and differentiation. *Advances in Engineering Software*, 8:104–113, 1986. [netlib/gcv/gcvspl].

536. A. J. Worsey and G. Farin. An n-dimensional Clough-Tocher interpolant. *Constructive Approximation*, 3:99–110, 1987.

537. A. J. Worsey and B. Piper. A trivariate Powell-Sabin interpolant. *Computer Aided Geometric Design*, 5:177–186, 1988.

538. Z. Yan. Piecewise cubic curve fitting algorithm. *Mathematics of Computation*, 49:203–213, 1987.

539. B. S. Yandell. Block diagonal smoothing splines. TR 812, Dept. Statistics, University of Wisconsin, 1987.

540. M. A. Yerry and M. S. Shephard. A modified quadtree approach to finite element mesh generation. *IEEE Computer Graphics and Applications*, 3:39–46, 1983.

541. F. Yoshimoto. Least squares approximation by one-pass methods with piecewise polynomials. In J. C. Mason and M. G. Cox, eds., *Algorithms for Approximation*, pages 213–224. Oxford Univ. Press, 1987.

542. H. Yserentant. Hierarchical bases of finite-element spaces in the discretization of nonsymmetric elliptic boundary value problems. *Computing*, 35:39–49, 1985.

543. M. J. Zyda. A decomposable algorithm for contour surface display generation. *ACM Trans. on Mathematical Software*, 7:129–148, 1988.